CHEMICAL METHODS
IN BACTERIAL SYSTEMATICS

A complete list of titles in the Society for Applied Bacteriology appears at the end of this volume.

THE SOCIETY FOR APPLIED BACTERIOLOGY
TECHNICAL SERIES NO. 20

CHEMICAL METHODS IN BACTERIAL SYSTEMATICS

Edited by

MICHAEL GOODFELLOW

Department of Microbiology,
University of Newcastle upon Tyne,
Newcastle upon Tyne, UK

AND

DAVID E. MINNIKIN

Department of Organic Chemistry,
University of Newcastle upon Tyne,
Newcastle upon Tyne, UK

1985

ACADEMIC PRESS
(Harcourt Brace Jovanovich, Publishers)

London · Orlando · San Diego · New York
Toronto · Montreal · Sydney · Tokyo

ACADEMIC PRESS INC. (LONDON) LTD.
24–28 Oval Road
LONDON NW1 7DX

United States Edition published by
ACADEMIC PRESS, INC.
Orlando, Florida 32887

British Library Cataloguing in Publication Data

Chemical methods in bacterial systematics.–
(The Society for Applied Bacteriology technical series)
1. Microbiology–Classification
2. Biological chemistry
I. Goodfellow, M. II. Minnikin, D. E.
III. Series
576'.012 QR12

Library of Congress Cataloging in Publication Data
Main entry under title:

Chemical methods in bacterial systematics.

Presented at a poster session during the summer conference of the Society of Applied Bacteriology, at the University of Heriot-Watt, July 1983.
Includes bibliographical references and index.
1. Bacteriology--Classification--Congresses.
2. Chemotaxonomy--Congresses. I. Goodfellow, M.
II. Minnikin, D. E. III. Society of Applied Bacteriology.
QR81.C47 1984 589.9'0012 84-11032
ISBN 0–12–289675–0 (alk. paper)

PRINTED IN THE UNITED STATES OF AMERICA

85 86 87 88 9 8 7 6 5 4 3 2 1

Contributors

R. C. W. Berkeley, *Department of Microbiology, The Medical School, University Walk, University of Bristol, Bristol BS8 1TD, UK*

I. J. Bousfield, *National Collection of Industrial & Marine Bacteria Ltd., Torry Research Station, Aberdeen AB9 8DG, UK*

K. Bryn, *National Institute of Public Health, Geitmyrsveien 75, 0462 Oslo 4, Norway*

M. A. Carver, *Department of Biochemistry, University of Leicester, University Road, Leicester LE1 7RH, UK*

M. D. Collins, *Department of Microbiology, National Institute for Research in Dairying, Shinfield, Reading RG2 9AT, UK*

T. R. Dando, *National Collection of Industrial & Marine Bacteria Ltd., Torry Research Station, Aberdeen AB9 8DG, UK*

V. E. Dent, *Sub-Department Microbiology, Department of Botany, The Queen's University of Belfast, Belfast BT7 1NN, Northern Ireland*

G. Dobson, *Departments of Microbiology and Organic Chemistry, University of Newcastle upon Tyne, Newcastle upon Tyne NE1 7RU, UK*

D. B. Drucker, *Department of Bacteriology and Virology, University of Manchester, Manchester M13 9PT, UK*

Valerie J. Fowler, *Institute of Botany and Microbiology, The Technical University, Arcisstrasse 16, D-8000 Munich 2, Federal Republic of Germany*

D. A. Geddes, *University of Glasgow Dental Hospital and School, Glasgow G2 3JZ, UK*

M. Goodfellow, *Department of Microbiology, University of Newcastle upon Tyne, Newcastle upon Tyne NE1 7RU, UK*

W. D. Grant, *Department of Microbiology, University of Leicester, Leicester LE1 7RH, UK*

C. S. Gutteridge, *Cadbury Schweppes PLC Group Research, Lord Zuckerman Research Centre, University of Reading, Whiteknights, Reading RG6 2LA, UK*

J. M. Hardie, *Oral Microbiology, London Hospital Medical College, London E1 2AD, UK*

J. E. Harris, *AFRC Food Research Institute, Colney Lane, Norwich NR4 7UA, UK*

P. J. H. Jackman, *Department of Bacteriology, Institute of Dermatology, Homerton Grove, London E9 6BX, UK*

E. Jantzen, *National Institute of Public Health, Geitmyrsveien 75, 0462 Oslo 4, Norway*

C. W. JONES, *Department of Biochemistry, University of Leicester, University Road, Leicester LE1 7RH, UK*

R. M. KEDDIE, *Department of Microbiology, University of Reading, Reading RG1 5AQ, UK*

R. M. KROPPENSTEDT, *German Collection of Microorganisms, Gesellschaft für Biotechnologische, Forschung mbH, Braunschweig Schnittspahnstrasse 9, D-6100 Darmstadt, Federal Republic of Germany*

W. LUDWIG, *Institute of Botany and Microbiology, The Technical University, Arcisstrasse 16, D-8000 Munich 2, Federal Republic of Germany*

T. W. MACFARLANE, *University of Glasgow Dental Hospital and School, Glasgow G2 3JZ, UK*

M. MAGNUSSON, *Tuberculin Department, Statens Seruminstitut, Amager Boulevard 80, DK-2300 Copenhagen, Denmark*

D. E. MINNIKIN, *Department of Organic Chemistry, University of Newcastle upon Tyne, Newcastle upon Tyne NE1 7RU, UK*

S. M. MINNIKIN, *Department of Organic Chemistry, University of Newcastle upon Tyne NE1 7RU, UK*

M. MORDARSKI, *Institute of Immunology and Experimental Therapy, Polish Academy of Sciences, Czerska 12, 53-114 Wrocław, Poland*

R. A. NASH, *Oral Microbiology, London Hospital Medical College, London E1 2AD, UK*

J. R. NORRIS, *Cadbury Schweppes PLC Group Research, Lord Zuckerman Research Centre, University of Reading, Whiteknights, Reading RG6 2LA, UK*

A. G. O'DONNELL, *Departments of Microbiology and Organic Chemistry, University of Newcastle upon Tyne, Newcastle upon Tyne NE1 7RU, UK*

R. J. OWEN, *National Collection of Type Cultures, Central Public Health Laboratory, London NW9 5HT, UK*

J. H. PARLETT, *Departments of Microbiology and Organic Chemistry, University of Newcastle upon Tyne, Newcastle upon Tyne NE1 7RU, UK*

D. PITCHER, *National Collection of Type Cultures, Central Public Health Laboratory, London NW9 5HT, UK*

M. RIDELL, *Department of Medical Microbiology, University of Gothenburg, S-413 46 Gothenburg, Sweden*

H. N. M. ROSS, *Department of Microbiology, University of Leicester, Leicester LE1 7RH, UK*

K. P. SCHAAL, *Institute of Hygiene, University of Cologne, Goldenfelstrasse 21, D-5000 Cologne 41, Federal Republic of Germany*

K. H. SCHLEIFER, *Institute of Botany and Microbiology, The Technical University, Arcisstrasse 21, D-8000 Munich 2, Federal Republic of Germany*

P. H. SEIDL, *Institute of Botany and Microbiology, The Technical University, Arcisstrasse 21, D-8000 Munich 2, Federal Republic of Germany*

H. N. SHAH, *Oral Microbiology, London Hospital Medical College, London E1 2AD, UK*

S. SHAW, *Watney Mann and Truman Brewers Ltd., London SW14 8JD, UK*
LESLEY A. SHUTE, *Department of Microbiology, The Medical School, University Walk, University of Bristol, Bristol BS8 1TD, UK*
E. STACKEBRANDT, *Institute of Botany and Microbiology, The Technical University, Arcisstrasse 16, D-8000 Munich 2, Federal Republic of Germany*
A. TAVAKKOL,[1] *Department of Bacteriology and Virology, University of Manchester, Manchester M13 9PT, UK*
D. A. WEETMAN, *University of Glasgow Dental Hospital and School, Glasgow G2 3JZ, UK*
R. A. D. WILLIAMS, *Department of Biochemistry, The London Hospital Medical College, London E1 2AD, UK*

[1]Present address: Basic Medical Sciences, School of Medicine, Memorial University of Newfoundland, St. John's, Newfoundland, Canada A1B 3V6.

Preface

A poster session on the topic "The Use of Chemotaxonomic Methods for Bacteria" was held during the Summer Conference of the Society of Applied Bacteriology at the University of Heriot-Watt in July 1983. The invited posters, which were given by leading specialists in the field, were designed to display up-to-date chemotaxonomic procedures currently used in bacterial systematics.

Chemical methods not only are well established in bacterial systematics but are also rapidly changing our views on how bacteria should be classified and identified. Chemotaxonomy, which owes much to the availability of new and highly sensitive analytical techniques, can be used to establish relationships between bacteria at all levels in the taxonomic hierarchy. However, many of the new approaches are designed for the identification of bacteria and some are quantitative, employing computers and a variety of statistical techniques. This book contains detailed procedures on many of the chemical methods currently being used for the classification and identification of bacteria. As such, it will be of value generally to microbiologists and is an essential reference for those actively engaged in diagnostic bacteriology.

We are grateful to those who contributed posters and subsequently converted them into manuscripts. We also wish to thank Dr. F. G. Priest of the Department of Brewing and Biological Sciences, Heriot-Watt University, for assistance with the staging of the poster session.

January 1984

M. Goodfellow
D. E. Minnikin

Contents

Introduction to Chemosystematics

M. GOODFELLOW AND D. E. MINNIKIN

Departments of Microbiology and Organic Chemistry, University of Newcastle upon Tyne, Newcastle upon Tyne, UK

Introduction

Systematics is one of the oldest branches of the biological sciences. It is that part of biology that is concerned with the scientific study of the diversity of organisms and the relationships between them. It includes classification, the arrangement of organisms into groups or taxa, and taxonomy. The latter is the science of classification including its principles, procedures, and purposes. To facilitate communication, names are given to the groups defined in classification. Once groups or taxa have been defined and named, unknown organisms can be related to or identified with them. This latter process is known as identification. Classifications can be either phenetic or phylogenetic. In the former, organisms are classified on the basis of shared properties and in the latter, according to their degree of common ancestry.

Until recently, bacterial systematics was built upon a host of morphological and behavioural properties of microorganisms. Strains were assigned to groups based on colony and micromorphology, staining properties, the presence and absence of spores, the ability to produce acid from sugars, to grow on a variety of compounds as sole sources of carbon, and to grow in the presence of inhibitory compounds. This approach was not without its successes but overreliance on small numbers of subjectively chosen characters yielded narrow data bases, which in turn caused serious problems in the classification of many taxa, notably the coryneform bacteria and actinomycetes. Indeed, it was only with the introduction and application of chemical procedures that actinomycetes were detached from the fungi and seen to be bona fide prokaryotes.

Chemosystematics, chemotaxonomy, or chemical taxonomy, is the study of the chemical variation in living organisms and the use of chemical charac-

CHEMICAL METHODS IN
BACTERIAL SYSTEMATICS

0-12-289675-0

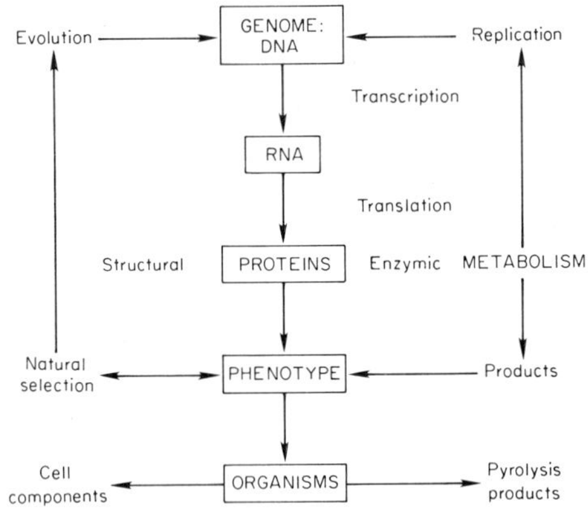

FIG. 1. General interrelationships of macromolecules and other chemical components of value in bacterial systematics.

ters in classification and identification. The subject is not a new one and in a sense most bacterial taxonomies are based on chemical properties. Thus morphological traits are influenced by the chemical composition of wall polymers, pigmentation is usually a reflection of pigment production, and relationships derived from serological and nucleic acid analyses are also expressions of chemical similarity and dissimilarity. In a narrower sense, chemosystematics is concerned with the discontinuous distribution of specific chemicals such as amino acids, sugars, lipids and proteins that can provide grist to the taxonomic mill. The interrelationships between both macromolecules and other chemical components of value in bacterial systematics are shown in Fig. 1.

Initially, many of the chemical characters used in bacterial classification were merely by-products of biochemical and chemical work. It was only with the introduction of new and sensitive analytical techniques that chemosystematics became an integral part of bacterial systematics. The breakthrough in studying the distribution of chemical compounds came with the modern development of chromatography in the immediate postwar period. Similarly, at about the same time, the rapid screening of protein extracts became possible given the introduction of starch-gel and polyacrylamide-gel electrophoresis and the establishment of gel diffusion techniques allowed the development of immunological procedures for the comparison of proteins from different bacteria. A host of sensitive analytical techniques are now available for collecting chemical data from whole organisms, the genome and other cell fractions (Fig. 1) and examples of those currently being developed and employed are described in the subsequent chapters of this book. Chemical methods vary in speed and ease of application but collectively they yield valuable data for constructing classifications, phylogenies and identification schemes. At present, some of the more involved methods can only readily be applied to a few representative strains. The latter should always be chosen with care, for all too frequently the significance of otherwise excellent work is marred by the use of misclassified or unrepresentative bacteria. One successful strategy has been to evaluate numerical phenetic taxonomies by analysing representatives of defined clusters, using chemical techniques (Goodfellow & Wayne 1982).

Chemical information can be used to establish relationships at all levels in the taxonomic hierarchy and it seems likely that chemical properties will become an essential ingredient in the minimal descriptions of many genera and species. It is, however, already clear that the discriminatory power of chemical criteria may vary between taxa. Thus, for example, menaquinone composition of aerobic, endospore-forming bacteria and polar lipid patterns of staphylococci are quite uniform, but menaquinones and polar lipids are of value in the subgeneric classification of *Bacillus* and *Staphylococcus*, respectively (Minnikin & Goodfellow 1981; Nahaie *et al.* 1984). It is also possible that certain chemical differences may have more taxonomic significance than others. Thus the replacement of lysine in peptidoglycans by the chemically very similar amino acid ornithine is probably less significant than substitution by a chemically very different component such as diaminopimelic acid.

Chemosystematics at the Macromolecular Level

Nucleic acids

Estimates of the mean overall base composition of DNA and of the extent of nucleotide base pairing found in homology studies are two of the approaches widely used in bacterial classification (Table 1). The richest source of information in the DNA macromolecule lies in the nucleotide sequences of complete bacterial genomes but sequencing the latter is not a practical proposition at present. Thus, comparison of nucleotide sequences by DNA–DNA pairing is the only direct measure of the percentage of complementary bases of two different genomes. Bacteria can also be distinguished on the basis of genome size, that is, the amount of chromosomal DNA they contain expressed in daltons or in numbers of nucleotide pairs (Owen & Hill 1979).

It is well established that the ratio of adenine (A) plus thiamine (T) to guanine (G) plus cytosine (C), or the mol % guanine plus cytosine [$\% \text{ GC} = 100(G + C)/(A + T + G + C)$], of prokaryote DNA varies between different taxa within the range 23–78% G plus C. A variety of well-tested methods are available for the extraction and analysis of DNA (see Chapter 4), and

TABLE 1

Chemotaxonomic information from bacterial nucleic acids and proteins

Macromolecule	Technique	Value at
DNA	Mean base composition (% GC) DNA–DNA resassociation Genome size	Generic level
	Electrophoresis of fragments generated by restriction endonucleases Plasmid profiles Phage host range studies genetic analysis *in vivo*	Species and subspecific levels
	DNA–RNA reassociation	Generic and suprageneric levels
RNA	Sequence catalogues Secondary structures	
Proteins	Specific proteins Amino acid sequences Immunological analysis	Generic and specific levels
	Groups of protein Electrophoretic patterns Enzyme patterns and activities	Generic and specific levels

several broad generalisations can be made in light of the studies to date. The essential point is that base ratio determinations must be interpreted within the context of data derived from independent taxonomic studies as similarity in % G plus C of DNA samples does not necessarily imply biological relatedness because unrelated strains can have identical GC ratios. The significant point is that dissimilarity in % GC of DNA from two strains establishes that they are different. As a general rule, bacteria with DNA differing by more than 5% G plus C should not be classified in the same species and those showing differences of over 10% should not be assigned to the same genus.

Given the specificity of base-pair formation, the base sequence of one DNA strand determines the sequence of the complementary strand. Complementary strands can be separated by heating or by alkali treatment but under suitable conditions they can reassociate to give the original double-stranded macromolecule. DNA homology studies make use of this renaturation property so that DNA from genetically closely related strains will almost completely reanneal, that is, seek complementary base sequences, whereas sparse duplex formation is taken as an indication of low homology and hence distant genetic relatedness. Current methods recommended for DNA-pairing studies are also described in Chapter 4. In practice, DNA-reassociation experiments have been most effective in establishing and evaluating taxonomies at the generic level in the case of overclassified taxa and at the rank of species in most other genera. They are also of value in detecting new centres of variation (see Chapter 17).

DNA homology studies can only be used to chart relationships within relatively narrow ranges of variation. It has been repeatedly shown that homology values fall to quite low levels, below 30%, for species that are only moderately different phenotypically. In marked contrast, rRNA–DNA reassociation studies yield data for establishing affinities between bacteria at both generic and suprageneric levels. Similarities in nucleotide sequences of rRNA preparations provide a means of determining relationships among representatives of diverse taxa as base sequences of rRNA cistrons are more highly conserved than the bulk of the genes forming the bacterial genome. Several techniques are available for determining ribosomal RNA similarities (see Chapter 3). A further development in the analysis of nucleic acids for taxonomic purposes is based on the work of Woese, Stackebrandt, and their colleagues, who have constructed bacterial phylogenies on the basis of similarities found between the nucleotide sequences of 16S rRNA (Stackebrandt & Woese 1981). The procedures currently recommended for rRNA sequencing studies are the subject of Chapter 2.

Plasmids are widely distributed amongst bacteria but surprisingly few attempts have been made to exploit the discontinuous distribution of extrachromosomal DNA for taxonomic purposes. Plasmid profiles have been shown to be of value in the identification of lactic streptococci (Davies *et al.* 1981) and in epidemiological studies of *Staphylococcus epidermidis* (Parisi & Hecht 1980) and some Gram-negative bacilli (Schaberg *et al.* 1981). Indeed, Parasi and Hecht believed that typing of plasmid profiles might prove to be more successful in epidemiological studies than phage typing or biotyping.

In vivo genetic analyses can also be used to generate taxonomic data (Bøvre 1980). Gene transfer can be achieved by transduction, transformation and conjugation and, as a general rule, gene transfer and successful recombination are at a much higher frequency between members of closely related taxa than between representatives of groups that are only peripherally related. Techniques for monitoring gene transfer have not been widely used in bacterial systematics as negative data have little, if any, meaning because the inability to exchange genes may be due to factors other than the lack of genetic homology (Sakaguichi 1980). Taxonomic deductions can also be drawn from the comparison of chromosomal maps and from phage host range studies. The emphasis in phage host range studies, as opposed to phage typing, is to generate taxonomic information in order to highlight similarities between strains. Batteries of polyvalent phages have been used to detect lytic patterns at the generic level (Prauser 1984).

Proteins

DNA is the source of all taxonomic information, but the translation of the code into sequences of amino acids yields a secondary macromolecule that can be exploited for taxonomic purposes. Proteins are usually easier to work with than nucleic acids because they can be more readily purified, have relatively short chains, and can be compared using a variety of techniques most of which are simple and relatively inexpensive. At this secondary, or epigenetic, level of investigation comparative analyses of specific homologous proteins provide an accurate means of establishing relationships between bacteria. More complex epigenetic analyses involve the comparison of groups of proteins by gel electrophoresis and of regulatory mechanisms governing the synthesis of enzymes operative in a given pathway.

Changes in the amino-acid sequence of specific proteins can reflect evolutionary divergence between organisms. These changes can be measured when specific proteins such as cytochrome *c* are isolated from representative strains and their sequences determined and compared, usually using numerical techniques. The most extensive protein sequencing studies on bacteria have been conducted by Ambler and his colleagues (Ambler 1985). The impact of such studies on bacterial systematics has not been great but is likely to become of increasing importance in the future. Changes in amino-acid sequences of specific proteins can also be measured using serological techniques (Wayne & Diaz 1979). To date, reasonably good correspondence has been found between the extent of immunological resemblance and the degree of sequence homology of specific proteins, and between the latter and 16S rRNA cataloguing data. Further, the comparison of catalytically isofunctional enzymes from different organisms in terms of their allosteric properties provides an additional way of evaluating their homology.

Protein electrophoresis is a powerful, relatively simple, and cost-effective method that is widely used in bacterial systematics. The method has the advantage that electrophoretic mobilities of large numbers of proteins, encoded by a sizeable fraction of the bacterial genome, can readily be compared in well-defined standardised conditions. Proteins of genetically closely related bacteria give similar or almost identical electrophoretograms. The polyacrylamide gel electrophoresis technique adopted by the National Collection of Type Cultures as the foundation of a system for the comparative identification of medically important bacteria is described in Chapter 6. Protein gel electrophoresis has also been used to detect new centres of variation within established genera (see Chapter 17). It is also possible to stain gels to reveal the presence of specific enzyme patterns of taxonomic value (Williams & Shah 1980).

Chemotaxonomic Information from the Wall Envelope

All organic molecules are an expression of information coded in the nucleic acids, and all such molecules can be exploited in bacterial systematics provided they are discontinuously distributed. The nature of the layer binding the bacterial cell has held a special place in the development of systematics since the discovery of the Gram stain 100 years ago. The scientific basis to the stain has still to be explained in detail but it is evident that the Gram reaction is correlated with significant differences in the structure and chemical composition of the prokaryote wall. Little was known of the chemical composition of the outer layers of bacteria until Salton (1952) introduced methods for the isolation and purification of walls. Since that time the walls of many bacteria have been analysed and several chemotaxonomic procedures are now available to detect the various wall envelope components (Table 2). Data from such studies allow prokaryotes to be divided into Gram-positive and Gram-negative bacteria, mollicutes, which lack a cell wall, and archaebacteria which have a fairly rigid wall that consists of a sulphated heteroglycan or a heteropolysaccharide (Rogers *et al.* 1980).

The fundamental polymer common to the walls of Gram-positive and Gram-negative bacteria is peptidoglycan. This consists of heteropolymers

TABLE 2

Chemotaxonomic information from bacterial cell envelopes

Site	Component
Plasma membrane	Isoprenoid quinones
	Lipid soluble pigments
	Lipoteichoic acids and analogues
	Polar lipids: Fatty acids; Isoprenoid ethers; Other long-chain components
	Proteins
Walls	Peptidoglycan and analogues
	Polysaccharides—sugar patterns
	Teichoic acids and analogues
Outer membrane Gram-negative bacteria	Lipopolysaccharides: K antigens; O antigens
	Polar lipids
Outer membrane Gram-positive bacteria	Bound lipids—mycolic acids
	Free lipids: Glycolipids; Sulphoglycolipids; Waxes

composed of an amino-sugar backbone cross-linked through tetrapeptide side chains. The amino-sugars form glycan strands composed of alternating residues of *N*-acetylglucosamine and *N*-acetylmuramic acid. The glycan moiety is relatively uniform, though atypical structures have been reported. Thus the substitution of *N*-acetylmuramic acid by *N*-glycolylmuramic acid in some mycolic acid-containing actinomycetes is of potential value in classification and identification (Goodfellow & Cross 1984). The tetrapeptide side chains are linked to the carboxyl groups of the muramic acid. The most common sequence in the peptide subunit is L-alanine, D-glutamic acid, an L-diamino acid with D-alanine in position 4. Variation in the composition of the tetrapeptide, especially that at position 3 where a diamino acid usually occurs, is extensively exploited in the classification of Gram-positive bacteria. Indeed, the diamino acid present in the peptidoglycan of actinomycetes, including coryneform bacteria, is the major taxonomic character provided by qualitative wall analysis and there are good grounds for believing that organisms containing different diamino acids should not usually be classified in the same genus.

In both Gram-positive and Gram-negative bacteria, adjacent peptide side chains are usually cross-linked by peptide binding between the terminal D-alanine at position 4 on one unit and the free amino group on another unit. In many Gram-positive bacteria, the peptide subunits are cross-linked through interpeptide bridges composed of 1–6 amino-acid units. These cross-linkages usually occur between amino acids at positions 3 and 4, though linkages between the free carboxyl group of D-glutamic acid and D-alanine are not uncommon. Most variations of the peptide moiety occur in the interpeptide bridge and in the method of cross-linkage. These variations, taken with the discontinuous distribution of sugars and teichoic acids in the peptidoglycan matrix, make the Gram-positive wall a rich reservoir for chemical markers. In marked contrast, the peptidoglycan of Gram-negative bacteria is not only remarkably uniform but forms a relatively small fraction of the total wall envelope.

Simple rapid qualitative methods of wall analysis can be used to detect diagnostic diamino acids and sugars in Gram-positive bacteria. Such procedures can readily be applied in examinations of quite large numbers of organisms and those recommended for the analysis of coryneform bacteria and clinically significant actinomycetes are given in Chapters 11 and 18, respectively. Thus chromatographic analyses of acid hydrolysates of whole organisms are recommended for the detection of marker chemicals such as diaminopimelic acid and arabinose, which are located mainly or entirely in the wall, though acid hydrolysates of purified walls are necessary should other diamino acids or sugars be present. Elucidation of the primary structure of the peptidoglycan provides more taxonomically useful information than that obtained using simple qualitative procedures, but the more detailed analyses require the use of more specialised techniques (see Chapter 10) that cannot readily be applied to large numbers of strains. However, the analysis of wall amino acids by gas chromatography provides a fast and convenient way of obtaining quantitative data on wall amino acid composition without recourse to an amino acid analyser (see Chapter 7).

The wall envelopes of Gram-positive bacteria also contain teichoic acids, which are water-soluble, phosphate-containing polymers. They usually consist of polyols (glycerol, mannitol, ribitol), which are linked through phosphodiester bridges and substituted by amino-sugars, sugars, or D-alanine. Two groups of teichoic acids are recognised, the cell wall teichoic acids which are covalently linked to the peptidoglycan through muramic acid, and the lipoteichoic acids, which are covalently linked to lipids and are associated with the plasma membrane. Wall teichoic acids show more structural diversity than lipoteichoic acids and are consequently of greater value both as chemical markers and as specific surface antigens. Chemical analyses of both teichoic acids and peptidoglycan have been invaluable in staphylococcal systematics (Schumacher-Perdreau *et al.* 1983). Immunological studies, exploiting these variations in wall composition, have also been found useful in the classification of staphylococci (Schleifer & Seidl 1977).

The discontinuous distribution of various lipid classes in the cell envelopes of bacteria provides another rich source of chemical data. The walls of Gram-negative bacteria are especially rich in lipoproteins, lipopolysaccharides, and polar and nonpolar lipids that are located in the plasma and outer membranes. In contrast, the wall envelope lipids of Gram-positive bacteria are concentrated in the plasma membrane, though some actinomycetes, notably mycobacteria, have a well-defined outer membrane consisting of bound and free lipids (Table 2). The essential components of plasma membranes are amphipathic polar lipids, which consist of a balanced proportion of hydrophilic headgroups linked to two apolar, hydrophobic long-chain fatty acids. The amphipathic lipids combine to form a bilayer with a hydrophobic interior and thereby provide a matrix for the location of specific functional proteins and nonpolar lipids such as isoprenoid quinones and lipid-soluble pigments. Both the polar and nonpolar lipids are free lipids, that is, they are readily extractable simply by soaking in appropriate organic solvents.

The extraction of polar lipids from bacteria and their analysis by TLC often yield patterns of diagnostic value (Minnikin *et al.* 1984). The most common polar lipids are phospholipids; other polar lipid types include glycolipids and certain amino-acid amides. Some polar lipids such as phosphatidylethanolamine, phosphatidylglycerol and diphosphatidylglycerol are widely distributed but compounds with a more restricted occurrence, and hence of greater diagnostic value, include the phosphatidylinositol mannosides of actinomycetes and related bacteria (Minnikin & O'Donnell 1984). Several procedures are offered for the analysis of the polar lipids of eubacteria (see Chapter 7) and archaebacteria (see Chapter 14). Archaebacterial lipids can be distinguished from those of eubacteria by the presence of the characteristic isoprenoid ethers and the absence of fatty-acid glycerol ester lipids. It is also possible to distinguish halophilic, methanogenic and thermophilic archaebacteria from one another by the structure of the core diether and tetraether lipids (Chapter 14). It is also interesting that the groups based on these polar lipid patterns correspond to those derived from ribosomal RNA–DNA homology studies.

Isoprenoid quinones are also characteristic components of the plasma membranes of aerobic bacteria and play an important role in electron transport. The two most widely distributed types of isoprenoid quinones are the menaquinones or vitamin K_2 and the ubiquinones or coenzyme Q. Both of these isoprenoid quinone types form large classes of molecules in which the length of the polyprenyl side chain varies between 1 and 14 isoprene units (Collins & Jones 1981). In addition to variations in the length of the side chain, various degrees of saturation or hydrogenation of the C_3 multiprenyl side chain occur in menaquinones. The structural variations shown by both menaquinones and ubiquinones, together with a number of more unusual structural modifications, are used for taxonomic purposes (Chapters 9 and 13). Isoprenoid quinones are readily extracted from bacterial cells using standard lipid solvents (Chapter 7) and the components of natural mixtures of menaquinones and ubiquinones can readily be determined by using a variety of chromatographic procedures (see Chapter 13). Until recently, most analyses of isoprenoid quinones generated only qualitative or semiquantitative information but it is now possible to obtain quantitative profiles using high performance liquid chromatography. Lipid-soluble pigments also occur widely in bacterial plasma membranes (Rogers *et al.* 1980), but further systematic studies are needed if the full chemotaxonomic potential of these compounds is to be realised.

Most bacterial cytochromes are also associated with the cytoplasmic membrane. The cytochromes are involved in a wide variety of redox processes such as photosynthetic electron transfer and chemolithotrophic, heterotrophic and anaerobic respiration. Bacterial respiratory chains contain a number of different cytochromes and the variations in cytochrome patterns between different bacteria can be exploited by applying chemotaxonomic techniques (see Chapter 19). Similarly, the structural variations found in the iron-binding compounds of mycobacteria, the mycobactins, can be exploited in the subgeneric classification of *Mycobacterium* (Ratledge 1982).

The outer membranes of mycobacteria and related organisms contain substantial amounts of long-chain fatty acids, the mycolic acids, which are considered to be esterified to an arabinogalactan linked to the peptidoglycan. Mycolic acids, high-molecular-weight 3-hydroxy acids with a long alkyl branch in position 2, are restricted to members of the genera *Caseobacter*, *Corynebacterium*, *Mycobacterium*, *Nocardia* and *Rhodococcus*. These characteristic complex lipids vary widely in structure and a variety of techniques have been developed to distinguish between the various types (Minnikin & Goodfellow 1980). Mycobacterial mycolic acids are the most complex. They have from 60 to 90 carbon atoms and encompass mycolates with carboxy, epoxy, keto, and methoxy groups, whereas those from the remaining taxa have 1–5 double bonds as the only functional groups in addition to the 3-hydroxy unit. The discontinuous distribution of the various mycolic acid types is of value not only at the generic level (see Chapter 18) but also in the classification and identification of mycobacterial species. The distribution of the various types of mycobacterial mycolic acids can readily be assessed by TLC of whole-organism methanolysates (see Chapter 12).

The outer membrane formed by the mycolic acids possibly provides a matrix for other complex lipids, especially in the case of mycobacteria (Minnikin 1982; Minnikin & O'Donnell 1984). These complex lipids include glycopeptidolipids, sulphoglycolipids, glycosylated phenolphthiocerol esters and waxes based on phthiocerols, all of which have chemotaxonomic value (Chapter 12). The more polar glycopeptidolipids are especially important inasmuch as they are serologically active and type specific and correspond to antigens used in a serological procedure for the identification of certain mycobacteria.

All Gram-negative bacteria have a complex outer membrane composed of phospholipids, certain proteins and the characteristic lipopolysaccharide O antigens, which consist of three covalently linked segments: the oligosaccharide side chain, the core and lipid A (Westphal *et al.* 1983). The side chain carries the determinant of the O-antigenic specificity and is the most important of the three cell-surface antigens used in the serological classification of the *Enterobacteriaceae*. The variety, substitution and linkage of the 3–6 sugars forming the side chains are responsible for the antigenic and chemical specificity of the moiety. Less chemical variation is found in lipid A and in the core fraction of the lipopolysaccharide, which is composed of a short, acidic heterooligosaccharide. Lipid A consists of a β-1,6-linked D-glucosamine disaccharide carrying phosphate residues in positions 1 and 4 with long-chain fatty acids linked to both the amino and hydroxy groups of glucosamine. The patterns of fatty acids, which form about 15–20% of the total cellular fatty acids of Gram-positive bacteria, vary between taxa. Given these properties it is not surprising that the structure, chemical composition and serological specificity of lipopolysaccharides are extensively used in the classification and identification of many Gram-negative taxa. Methods recommended for the determination of the fatty acid and sugar composition of lipopolysaccharides are given in Chapter 8. Capsular polysaccharide K antigens are also anchored to the outer membrane by a lipid unit and offer another rich source of chemotaxonomic data (Jann & Jann 1983).

The discontinuous distribution of the many different types of fatty acids found in eubacteria provide a further rich seam of chemotaxonomic data (Drucker 1981). This is particularly so in the case of the nonhydroxylated fatty acids that have up to around 20 carbon atoms. These compounds form the lipid component of the structural plasma membrane and Gram-negative outer membrane amphipathic polar lipids and are an integral part of the lipid A fraction of lipopolysaccharides. Most eubacteria can be assigned to one of two broad groups on the basis of their nonhydroxylated fatty acid components. One group encompasses strains with major amounts of straight-chain and monounsaturated fatty acids, in some cases supplemented with tuberculostearic and cyclopropane acids, and the second group comprises organisms with major amounts of iso- and anteiso-acids, though small amounts of straight-chain fatty acids are also usually present. Fatty acids are routinely analysed by gas chromatography of their methyl esters, and the latter are formed during acid or alkaline methanolysis (Lambert & Moss 1983). Methods for the analysis of whole-cell fatty acid composition can be found in Chapters 8 and 9 and procedures for the detection of the fatty acid composition of lipopolysaccharides in Chapter 8. Numerical analyses of total fatty acid profiles are being increasingly used as an aid in the identification of bacteria (Bousfield *et al.* 1983).

Information from Whole Organisms and Volatile End Products of Metabolism

Whole organisms

Several of the chemical methods applied in bacterial systematics employ a number of statistical methods to analyse the data produced. This is particularly true of pyrolysis mass spectrometry and pyrolysis gas chromatography. Pyrolysis used in conjunction with these two techniques provides one of the few ways of analysing bacteria directly. This approach is clearly attractive given the increasing emphasis placed on the need to characterise micro-organisms quickly. Pyrolysis gas chromatography is discussed in detailed reviews (Gutteridge & Norris 1979; O'Donnell & Norris 1981).

When bacteria grown under standard conditions are thermally degraded in an inert atmosphere, and the products analysed by mass spectrometry, the resulting pyrogram provides a chemical profile of the organism in the form of a number of peaks representing the different products of pyrolysis. Most bacteria tend to produce essentially the same peaks but the quantitative data, represented as peak heights, vary significantly and reproducibly between members of different taxa. The quantitative comparison of data in pyrograms between strains provides a way of establishing inter- and intra-group relationships between them. Pyrolysis mass spectrometry allows rapid identification because the pyrograms of unknown strains can readily be compared with spectra held in a data base. Details of the principles and operational procedures of pyrolysis mass spectrometry are given in Chapter 5.

Volatile end products of metabolism

The fermentation end products of aerobic eubacteria are usually insufficiently diverse to be of value in classification but anaerobic bacteria generally use organic compounds, rather than oxygen, as terminal electron acceptors, producing acids, alcohols and ketones that are of considerable taxonomic value. The fatty acids formed as end products of protein and carbohydrate metabolism are especially important in the classification of anaerobes inasmuch as the types of acids produced are reproducible under standard conditions. The detection, using a variety of procedures and techniques, of acid end products of metabolism of anaerobic Gram-negative bacteria is considered in Chapter 16. Similarly, the detection of volatile end products of Gram-positive bacteria, including new methods for the analysis of amines, is the subject of Chapter 15.

Summary

The application of chemical methods is making significant improvements to classification and identification, particularly in those bacterial groups where the more established methods based on form and function have failed to provide workable systems. Chemical methods also provide the only way of constructing bacterial phylogenies and made a substantial contribution to the recognition of the archaebacteria. A variety of well-tested chemical methods are now available, each one of which has its strengths and weaknesses depending, in part, on the purpose in hand. The fact that different chemical characters have different discriminatory abilities makes it attractive to maximise the amount of chemical information derived from the same small sample of biomass. Indeed, the continued wide-scale use of chemical markers in bacterial systematics depends on the development of simple but comprehensive extraction and analytical procedures and the application of appropriate statistical techniques to handle objectively the large amounts of data generated.

References

AMBLER, R. P. 1985 Protein sequencing and taxonomy. In *Current Topics in Computer Assisted Systematics* ed. Goodfellow, M., Jones, D. & Priest, F. G. London: Academic Press. (In press.)

BOUSFIELD, I. J., SMITH, G. L., DANDO, T. R. & HOBBS, G. 1983 Numerical analysis of total fatty acid profiles in the identification of coryneform, nocardioform and some other bacteria. *Journal of General Microbiology* **129**, 375–394.

BØVRE, K. 1980 Progress in classification and identification of *Neisseriaceae* based on genetic affinity. In *Microbiological Classification and Identification* ed. Goodfellow, M. & Board, R. G., pp. 55–72. London: Academic Press.

COLLINS, M. D. & JONES, D. 1981 Distribution of isoprenoid quinone structural types in bacteria and their taxonomic implications. *Microbiological Reviews* **45**, 316–354.

DAVIES, F. L., UNDERWOOD, H. M. & GASSON, M. J. 1981 The value of plasmid profiles for strain identification in lactic streptococci and the relationship between *Streptococcus lactis* 712, ML3 and C2. *Journal of Applied Bacteriology* **51**, 325–337.

DRUCKER, D. B. 1981 *Microbiological Applications of Gas Chromatography*. Cambridge: University Press.

GOODFELLOW, M. & CROSS, T. 1984 Classification. In *The Biology of the Actinomycetes* ed. Goodfellow, M., Mordarski, M. & Williams, S. T., pp. 7–164. London: Academic Press.

GOODFELLOW, M. & WAYNE, L. G. 1982 Taxonomy and nomenclature. In *The Biology of the Mycobacteria* ed. Ratledge, C. & Stanford, J. L., Vol. 1, pp. 471–521. London: Academic Press.

GUTTERIDGE, C. S. & NORRIS, J. R. 1979 The application of pyrolysis techniques to the identification of micro-organisms. *Journal of Applied Bacteriology* **47**, 5–43.

JANN, K. & JANN, B. 1983 The K antigens of *Escherichia coli*. *Progress in Allergy* **33**, 53–79.

LAMBERT, M. A. & MOSS, C. W. 1983 Comparison of the effects of acid and base hydrolyses on hydroxy and cyclopropane fatty acids in bacteria. *Journal of Clinical Microbiology* **18**, 1370–1377.

MINNIKIN, D. E. 1982 Lipids: complex lipids, their chemistry, biosynthesis and roles. In *The Biology of the Mycobacteria* ed. Ratledge, C. & Stanford, J. L., Vol. 1, pp. 95–184. London: Academic Press.

MINNIKIN, D. E. & GOODFELLOW, M. 1980 Lipid composition in the classification and identification of acid-fast bacteria. In *Microbiological Classification and Identification* ed. Goodfellow, M. & Board, R. G., pp. 189–256. London: Academic Press.

MINNIKIN, D. E. & GOODFELLOW, M. 1981 Lipids in the classification of *Bacillus* and related taxa. In *The Aerobic Endospore-Forming Bacteria* ed. Berkeley, R. C. W. & Goodfellow, M., pp. 59–90. London: Academic Press.

MINNIKIN, D. E. & O'DONNELL, A. G. 1984 Actinomycete envelope lipid and peptidoglycan composition. In *The Biology of the Actinomycetes* ed. Goodfellow, M., Mordarski, M. & Williams, S. T., pp. 337–388. London: Academic Press.

MINNIKIN, D. E., O'DONNELL, A. G., GOODFELLOW, M., ALDERSON, G., ATHALYE, M., SCHAAL, A. & PARLETT, J. H. 1984 An integrated procedure for the extraction of bacterial isoprenoid quinones and polar lipids. *Journal of Microbiological Methods* **2**, 233–241.

NAHAIE, M., GOODFELLOW, M. & MINNIKIN, D. E. 1984 Polar lipid and isoprenoid quinone composition in the classification of *Staphylococcus*. *Journal of General Microbiology* **130**, 2427–2437.

O'DONNELL, A. G. & NORRIS, J. R. 1981 Pyrolysis gas-liquid chromatographic studies. In *The Aerobic Endospore-Forming Bacteria: Classification and Identification* ed. Berkeley, R. C. W. & Goodfellow, M., pp. 141–179. London: Academic Press.

OWEN, R. J. & HILL, L. R. 1979 The estimation of base compositions, base pairing and genome sizes of bacterial deoxyribonucleic acids. In *Identification Methods for Microbiologists*, ed. Skinner, F. A. & Lovelock, D. W., 2nd ed., pp. 277–296. London: Academic Press.

PARISI, J. T. & HECHT, D. W. 1980 Plasmid profiles in epidemiological studies of infections by *Staphylococcus epidermidis*. *Journal of Infectious Diseases* **141**, 637–643.

PRAUSER, H. 1984 Phage host ranges in the classification and identification of Gram-positive branched and related bacteria. In *Actinomycete Biology* ed. Bojalil, L. and Ortiz-Ortiz, L. New York: Academic Press. (In press.)

RATLEDGE, C. 1982 Nutrition, growth and metabolism. In *The Biology of the Mycobacteria* ed. Ratledge, C. and Stanford, J. L., Vol. 1, pp. 185–271. London: Academic Press.

ROGERS, H. J., PERKINS, H. R. & WARD, J. B. 1980 *Microbial Cell Walls and Membranes*. London: Chapman and Hall.

SAKAGUICHI, K. 1980 Species barriers to the maintenance and expression of foreign DNA. In *Molecular Breeding and Genetics of Applied Microorganisms* ed. Sakaguichi, K. & Okanishi, M., pp. 1–7. London: Academic Press.

SALTON, M. R. J. 1952 Studies on the bacterial cell wall: Preliminary investigation of the chemical constitution of the cell wall of *Streptococcus faecalis*. *Biochimica et Biophysica Acta* **8**, 510–519.

SCHABERG, D. R., TOMPKINS, L. S. & FALKOW, S. 1981 Use of agarose gel electrophoresis of plasmid deoxyribonucleic acid to fingerprint Gram-negative bacilli. *Journal of Clinical Microbiology* **13**, 1105–1108.

SCHLIEFER, K. H. & SEIDL, H. P. 1977 Structure and immunological aspects of peptidoglycans. In *Microbiology 1977* ed. Schlessinger, D., pp. 339–343. Washington D.C.: American Society for Microbiology.

SCHUMACHER-PERDREAU, F., ROTERING, H. & PULVERER, G. 1983 Cell wall analysis and taxonomy of staphylococci. *Zentralblatt für Bakteriologie, Mikrobiologie und Hygiene, Series A* **256**, 25–36.

STACKEBRANDT, E. & WOESE, C. R. 1981 The evolution of prokaryotes. In *Molecular and Cellular Aspects of Microbial Evolution* ed. Carlile, M. J., Collins, J. F. & Moseley, B. E. B., pp. 1–31. Cambridge: University Press.

WAYNE, L. G. & DIAZ, G. A. 1979 Reciprocal immunological distances of catalase derived from *Mycobacterium avium*, *M. tuberculosis* and closely related species. *International Journal of Systematic Bacteriology* **29**, 19–24.

WESTPHAL, O. W., JANN, K. & HIMMELSPACH, K. 1983. Chemistry and immunochemistry of bacterial lipopolysaccharides as cell wall antigens and endotoxins. *Progress in Allergy* **33**, 9–39.

WILLIAMS, R. A. D. & SHAH, H. N. 1980 Enzyme patterns in bacterial classification and identification. In *Microbiological Classification and Identification* ed. Goodfellow, M. & Board, R. G., pp. 299–318. London: Academic Press.

Ribosomal Ribonucleic Acid Cataloguing in Bacterial Systematics: The Phylogeny of *Actinomadura*

VALERIE J. FOWLER, W. LUDWIG AND E. STACKEBRANDT

Institute of Botany and Microbiology, The Technical University, Munich, Federal Republic of Germany

In the early 1970s, Carl Woese and his collaborators developed a method the application of which is changing our way of thinking about the phylogeny of prokaryotes. The method, comparative analysis of RNase T_1 resistant oligonucleotides of 16S ribosomal RNA, is also called oligonucleotide cataloguing (Uchida *et al.* 1974; Fox *et al.* 1977). The 16S rRNA molecule has turned out to be the most suitable for phylogenetic studies, since it can be used to determine both close and very distant relationships (Woese & Fox 1977; Balch *et al.* 1979; Fox *et al.* 1980; Stackebrandt & Woese 1981a). The molecule is ubiquitously distributed, it is highly conserved in sequence, it is genetically stable, and the sequencing of fragments obtained by enzymatic digestion of 16S rRNA no longer presents any problem.

Disadvantages of the original method of sequencing oligonucleotides (Sanger *et al.* 1965; Uchida *et al.* 1974) included the *in vivo* labelling of organisms with relatively high dosages of ^{32}P, which meant that symbiotic, parasitic and slow-growing microorganisms and the 18S rRNA of eukaryotes could not be included in this kind of study, and the rather complicated sequence determination of fragments containing more than 12 bases. However, the introduction of *in vitro* methods for labelling nucleic acids (Jay *et al.* 1974; Silberklang *et al.* 1977, 1979) in principle allows all organisms to be investigated, and as a consequence of the different labelling technique, sequence determinations have become much easier.

Phylogenetic studies initiated some years earlier in this laboratory were based on the improved labelling technique (Stackebrandt *et al.* 1981, 1982). The sequencing data obtained can be compared directly with those reported from the laboratories of Carl Woese (Urbana, Illinois, USA) and Ford Doolit-

CHEMICAL METHODS IN BACTERIAL SYSTEMATICS

0-12-289675-0

tle (Halifax, Nova Scotia, Canada), where the classical sequencing method was employed. Using all of these data, a phylogenetic tree of the prokaryotes, consisting of about 400 species, can be constructed. In this communication some information about the background of this kind of phylogenetic study is given, the *in vitro* labelling and sequencing methods are presented in detail, and the usefulness of the method is demonstrated by reference to the phylogeny of the genus *Actinomadura*.

16S rRNA Cataloguing

The 16S rRNA of *Escherichia coli* has 1642 nucleotides (Brosius *et al.* 1978); this is probably the approximate size of all prokaryotic 16S rRNAs. RNase T_1 cleaves 16S rRNA into 550 to 600 fragments that range from a single guanosine up to oligonucleotides with 20 bases. The number of fragments of a certain length decreases with increasing fragment length. At the same time, the possible permutation of sequences increases according to the expression 3^n, where n is the number of nucleotides preceding the 3′ terminal guanosine. It follows that shorter oligonucleotides (pentamers or smaller) of a given sequence can easily occur as more than one copy, but for larger oligonucleotides the probability of more than one copy occurring is extremely small. Thus, for example, in *E. coli*, 52 tetramer fragments are found but the number of possible combinations is only 27 (3^3). Several oligonucleotides must therefore be present as more than one copy. Alternatively, in the case of hexamers, 22 fragments are found, and the number of theoretically possible combinations is 243 (3^5). Subsequent analysis of the primary structure of 16S rRNA of *E. coli* (Brosius *et al.* 1978) has confirmed the observation that oligonucleotides with six or more bases occur only once each. It follows that the same large oligonucleotides present in the 16S rRNA catalogues of two different organisms probably occupy the same position in the respective primary structures (Woese *et al.* 1976). In addition, the short oligonucleotides present in the catalogues of all organisms are nonspecific and consequently have no phylogenetic significance. It is for this reason that only hexamers and larger oligonucleotides are used for the calculation of similarity coefficients (S_{AB} values, see below). This means that approximately 65–80 oligonucleotides totalling 500–650 bases, i.e., 30–40% of the 16S rRNA macromolecule, are examined. It can be shown that these fragments are distributed evenly over the entire 16S rRNA primary structure.

The calculation of S_{AB} values on the basis of the large fragments tends to produce an underestimation of the actual nucleotide homology. The comparison of 16S rRNAs of *E. coli* and *Zea mays* chloroplasts revealed 72%

homology (Schwarz & Kössel 1980), but the corresponding S_{AB} value was only 0.28. Clearly, phylogenetic differences are emphasised in such studies.

The Method

Isolation and purification of 16S rRNA

Wet cells (2–3 g) are resuspended in up to 5 ml of buffer (0.04 *M* Tris, 0.02 *M* Na acetate, 0.001 *M* Na_2EDTA, 0.2% SDS, pH 7.2) and opened by passing through a French pressure cell at 140 MPa. The lysate is subjected to two or three phenol extractions (shaken and then centrifuged at 20,000 *g* for 20 min, the phenol phase and the protein interface being rejected). The nucleic acids are then precipitated by three volumes of ethanol for at least two h at −20°C.

Separation of the nucleic acid species is achieved by one-dimensional polyacrylamide slab gel electrophoresis (Bishop *et al.* 1967). Polyacrylamide (2.8% PA, 0.14% *N,N'*-methylene bisacrylamide in 0.04 *M* Tris, 0.02 *M* Na acetate, 0.001 *M* Na_2EDTA, pH 7.2) is degassed and polymerized by the addition of 0.8 μl/ml *N,N,N',N'*-tetramethylethylenediamine and 1.6 μl/ml ammonium peroxodisulfate (50%). Gels (e.g. 20 × 20 × 0.4 cm) should be allowed to polymerize for at least 1 h. Preelectrophoresis is carried out for 1 h at a current of 70 mA (constant voltage); the electrophoresis buffer is 0.04 *M* Tris, 0.02 *M* Na acetate, 0.001 *M* Na_2EDTA, 0.2% SDS, pH 7.2.

The nucleic acids are redissolved in up to 2 ml of electrophoresis buffer, mixed with 10 μl of bromophenol blue saturated in distilled water, made up to 10% glycerol, and then applied to the top of the gel. Electrophoresis is carried out for 8–16 h at a current of 60–70 mA (constant voltage) toward the anode. For visualisation of the nucleic acids, the gel is laid on a background fluorescent at 254 nm and illuminated by UV (254 nm). The gel material containing 16S rRNA is cut out and embedded in a cylindrical (Ø 1 cm) agarose (0.5% in 0.1 *M* NH_4 formate) gel for electrophoretic elution. The 16S rRNA is collected in 2–3 ml of elution buffer (0.1 *M* NH_4 formate), which is separated from the reservoir by a dialysis membrane. Electrophoresis is toward the anode at a current of 10 mA per tube for 5–6 h. Next, the voltage is reversed for 3–5 min in order to release the RNA from the dialysis membrane. The solution containing 16S rRNA is subjected to two phenol extractions to remove SDS and traces of contaminants, and phenol is removed by extraction with four volumes of ether. Finally, the 16S rRNA is precipitated by the addition of three volumes of cold ethanol. The RNA is redissolved in sterile water, it is checked for purity, and the yield is determined spectrophotometrically at 260 nm.

RNase T_1 digestion and in vitro *labelling of 16S rRNA*

16S rRNA (40 μg) is redissolved in 80 μl of sterile distilled water, mixed with 5 μl of 1 *M* Tris, pH 8.0, and 2.5 μl of RNase solution (Calbiochem, 1000 U/ml in 0.05 *M* Tris, pH 8.0), and incubated at 37°C for three h. The oligonucleotides generated are denatured by heating to 100°C for 2 min followed by chilling in ice. For 5′ ^{32}P labelling of the oligonucleotides, a 5-μl aliquot of the digest is mixed with 4 μl of buffer (0.25 *M* Tris, pH 8.0, 0.25 m*M* ATP, 0.75 *M* mercaptoethanol, 0.42 *M* spermidine, 0.05 *M* $MgCl_2$), 10 μl of γ-^{32}P-ATP (NEN, 1 mCi, 1000–3000 Ci/mmol), and 1 μl of phosphatase-free T4 polynucleotide kinase (NEN, 8000 U/ml, 6000 U/mg protein) (Cameron *et al.* 1979). The mixture is incubated at 37°C for 30 min, and the reaction is terminated by heating to 100°C for 2 min. The mixture is then frozen and dried over P_2O_5 in a vacuum desiccator.

Primary analysis of oligonucleotides: the "fingerprint"

Separation of labelled RNase T_1-generated oligonucleotides is done in two dimensions by a combination of high-voltage electrophoresis in the first dimension (Sanger *et al.* 1965) and gradient thin-layer chromatography in the second dimension (Domdey & Gross 1979). Oligonucleotides are separated in the first dimension according to size and charge, and in the second mainly by size (for further details, see Brownlee 1972; Silberklang *et al.* 1977, 1979; and Domdey *et al.* 1978).

For separation in the first dimension, a cellulose acetate strip (Schleicher & Schuell) (3 × 100 cm) is wetted with electrophoresis buffer (5% acetic acid, 0.05% pyridine, 7 *M* urea, pH 3.5), and an area 5–7 cm from the end of the strip is blotted on both sides with absorbent paper to remove surplus buffer. This region of the strip is supported on two glass rods 2 cm apart. The sample (1 μl), dissolved in 2.5 μl of sterile water, is then applied in a line 1 cm long across the width of the strip, 6 cm from the end, using a fine glass capillary. At the same time, on either side of the radioactive sample, spots of a mixture of markers, 1% xylene cyanol, 2% fuchsin and 1% methyl orange in 10% ethanol, are applied. After the sample has been absorbed onto the strip, excess buffer is removed.

High voltage electrophoresis is carried out in a Savant apparatus at 5000 V until the xylene cyanol marker has migrated 25 cm. The methyl orange

---

FIG. 1. Fingerprints of RNase T_1-resistant oligonucleotides of 16S rRNAs from 3 *Actinomadura* strains. Note the similarity between the fingerprints of *Actinomadura pusilla* DSM 43357 and *Actinomadura roseoviolacea* DSM 43144 in contrast to that of *Actinomadura madurae* DSM 43067. The numbers refer to identical oligonucleotides (≥ nonamers) in the catalogues of the strains investigated (Table 1). The borderline between hexamers and pentamers is indicated.

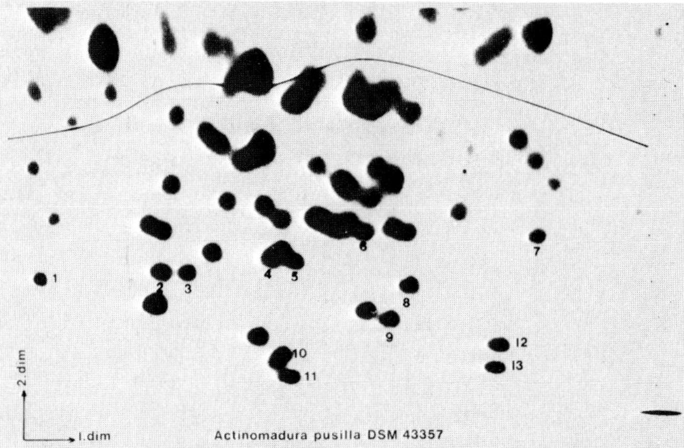

Actinomadura pusilla DSM 43357

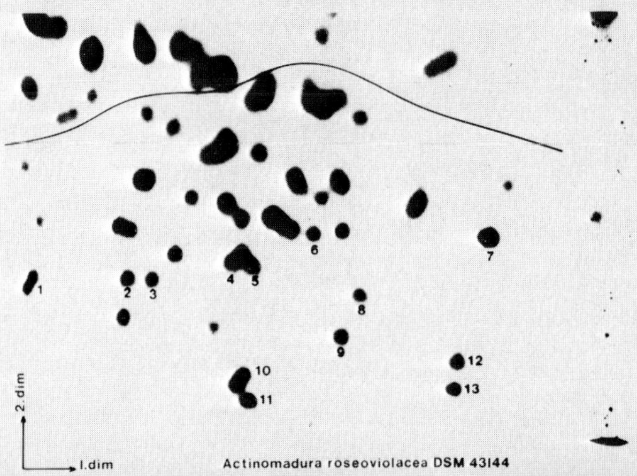

Actinomadura roseoviolacea DSM 43144

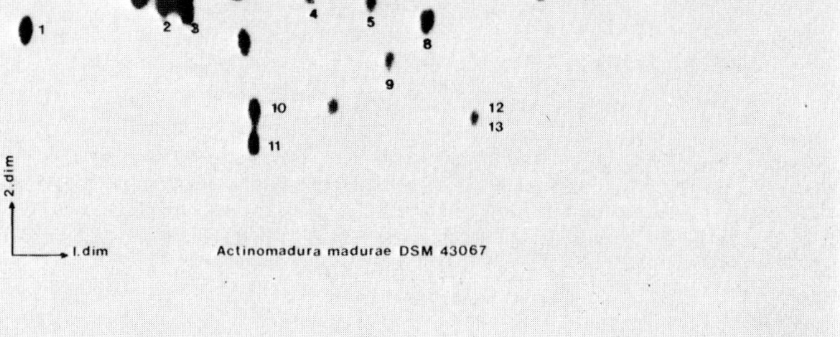

Actinomadura madurae DSM 43067

marker migrates approximately three times as quickly, and the oligonucleotides lie between the blue and yellow markers.

For separation in the second dimension, the oligonucleotides are transferred from the strip to the longer side of a 20 × 40 cm DEAE-cellulose thin-layer glass plate (Macheray & Nagel, Düren, FRG). The strip is laid lengthwise on a glass rod (Ø 0.4 cm) and strips of Whatman 3 MM filterpaper (5 × 45 cm), soaked in distilled water, are laid on either side to overlap the strip by 1–2 mm in the region between the blue and yellow markers. The thin-layer plate is placed onto the strip so that the line of contact between the strip and plate runs 2 cm from the long edge of the plate (Stackebrandt *et al.* 1982). The oligonucleotides are transferred to the plate by capillary action and immediately bound to the DEAE-cellulose. After transfer, which takes about 20 min, this part of the DEAE-cellulose thin-layer plate is washed with 0.25 litre of ethanol to remove the high voltage buffer salts, and allowed to dry.

Oligonucleotides are separated in the second dimension by NH_4 formate-gradient thin-layer chromatography carried out at 70°C. After prechromatography in distilled water for approximately 20 min, i.e. until the water front has reached the middle of the plate, the latter is transferred to a solution of 0.3 *M* NH_4 formate in 9 *M* urea, and 1 m*M* Na_2EDTA. An equal volume of a solution of 0.5 *M* NH_4 formate in 9 *M* urea and 1 m*M* Na_2EDTA is added dropwise over a period of 3 h, resulting in a concentration gradient of 0.3–0.4 *M* NH_4 formate. To ensure good mixing of the solvent, the thin-layer plate is supported on notched plastic rods and the solution is stirred in the chromatography tank.

The front, because of the water prechromatography, runs ahead of the smallest oligonucleotides and is absorbed by a filter paper wick (three sheets 10 × 40 cm) fastened onto the top of the chromatography plate. Chromatography should be terminated when the methyl orange marker has migrated about 16 cm. After removing the wick and drying the plate, the positions of the oligonucleotides are visualised by autoradiography for 10–20 min. Examples of 16S rRNA fingerprints are shown in Fig. 1.

Isolation and controlled hydrolysis of the oligonucleotides

The individual spots containing labelled oligonucleotides are scraped from the DEAE-cellulose (Domdey *et al.* 1978; Stackebrandt *et al.* 1982). The cellulose is sucked by means of a waterpump into the open end of drawn-out, cotton-plugged glass tubes (0.4 × 3–4 cm). The oligonucleotides are then eluted from the DEAE-cellulose "mini" columns by 3 × 80 µl of 1 *M* NaCl being centrifuged through them into Eppendorf tubes (Beckman TJ 6, 750 g for 2 min). Next, the RNA fragments are precipitated by three volumes of ethanol at −20°C for at least 3 h. After centrifugation at 1000 **g** for 20 min and rejection of the supernatant, the precipitate is dried in a vacuum desiccator over P_2O_5. For controlled alkaline hydrolysis (Wengler *et al.* 1979), the oligonucleotides are heated in 10 μl of 0.05 *M* NaOH, pH 9.3, to 100°C for 20 min, then chilled in ice. The samples are dried *in vacuo* after the addition of 1.2 μl of 2.5 *N* HCl to cleave any remaining cyclic phosphate groups.

Secondary analysis

The alkali-generated fragments must be separated in two dimensions to determine the oligonucleotide sequences. The technique involved is known as the "mobility shift" method (Jay *et al.* 1974; Silberklang *et al.* 1977, 1979; Gross *et al.* 1978). Sequences can be deduced from the two-dimensional patterns of the fragments, which are, once again, separated according to their size and charge.

Each alkaline hydrolysate is dissolved in 2.4 μl of sterile distilled water, and a 1 μl aliquot is applied to a cellulose acetate strip for high voltage electrophoretic separation in the first dimension. The procedures used are the same as those described for fingerprinting, but in this case three different samples can be applied to a strip (55 cm) 6, 18 and 30 cm from the end. The xylene cyanol marker is allowed to migrate 6 cm. All three fragments separated on a strip are transferred to a DEAE-cellulose thin-layer plate (20 × 40 cm) as described for the fingerprint.

In the second dimension, the plates are developed in an NH_4 formate gradient. Up to 25 plates, with a wick fastened to the top, are placed in a suitable tank. The plates are supported on rods to allow the solvent to be mixed. Water not completely used in the prechromatography is sucked off when the water front has migrated about 7 cm. Next, 1 litre of NH_4 formate solution (0.28 *M* NH_4 formate in 9 *M* urea, 1 m*M* Na_2EDTA, 1 *M* boric acid, pH 4.3, adjusted with formic acid) is added, and a linear NH_4 formate gradient (2–0.5 *M*, total 200 ml) is added dropwise over 3 h at 70°C, up to a final concentration of 0.44 *M* NH_4 formate. This procedure establishes an approximately asymptotic increase in the NH_4 formate concentration in the tank. By this time the xylene cyanol dye has almost reached the upper edge of the plates and the water front has been absorbed by the wicks. The latter are then removed, the plates are dried, and the positions of the 5′ labelled fragments are visualised by autoradiography (12–24 h). Sequences can be deduced from the relative positions of the spots (see Fig. 2).

In oligonucleotides with more than 11 bases, the larger fragments are usually insufficiently resolved. In such cases, a sample of the oligonucleotide is separated in two dimensions as described above, but the NH_4 formate

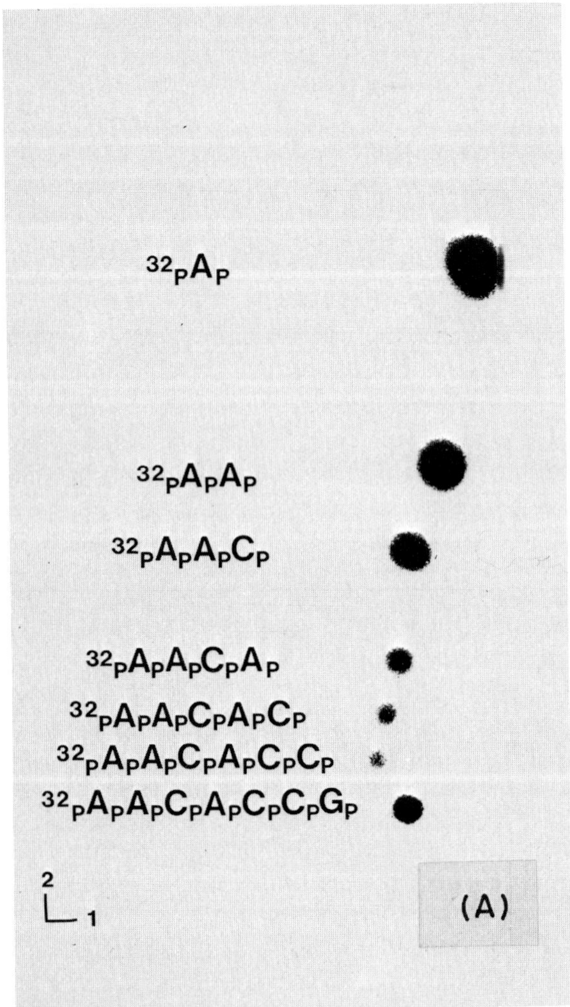

FIG. 2. Autoradiogram of the sequence analysis of two oligonucleotides from controlled alkaline hydrolysis. A. Oligonucleotide ^{32}PAACACCG, showing the nucleotide composition of the individual fragments. Note the influence of fragment size and composition (pyrimidines versus purines) on the separation behaviour. B. Oligonucleotide ^{32}PAAUUAUUG, showing the influence of uridine residues on the separation behaviour of the fragments.

gradient is replaced by a "homomix" (10:20:30:50:75 m*M* KOH = 0.5:1:1:1:1). Homomix is prepared by hydrolysis of a 3% solution of yeast RNA in 7 *M* urea by KOH (10, 20, 30, 50 or 75 m*M* KOH) at 65°C for 20 h. The pH is then set to 4.7 with glacial acetic acid (Jay *et al.* 1974; Silberklang *et al.* 1977). Homomix produces a better separation of long fragments than does the

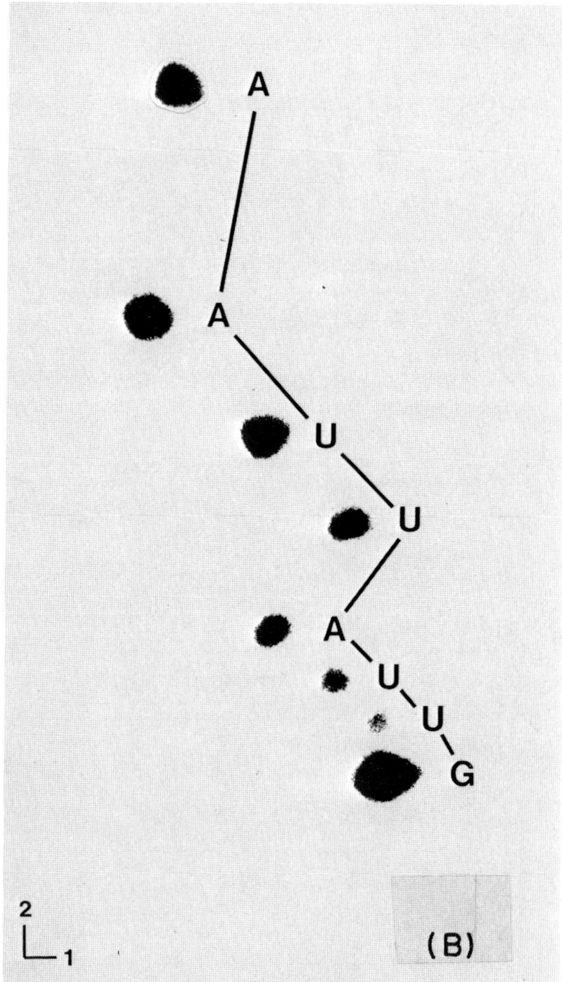

FIG. 2. (*Continued*)

gradient. Fragments up to heptanucleotides are allowed to migrate into the wick, which results in a sufficiently good resolution of the 3′ terminal fragments (Fig. 3).

Tertiary analysis

In those cases where the sequences cannot be determined unambiguously from the secondary analysis, it is necessary to perform a tertiary analysis by one or a combination of procedures as follows:

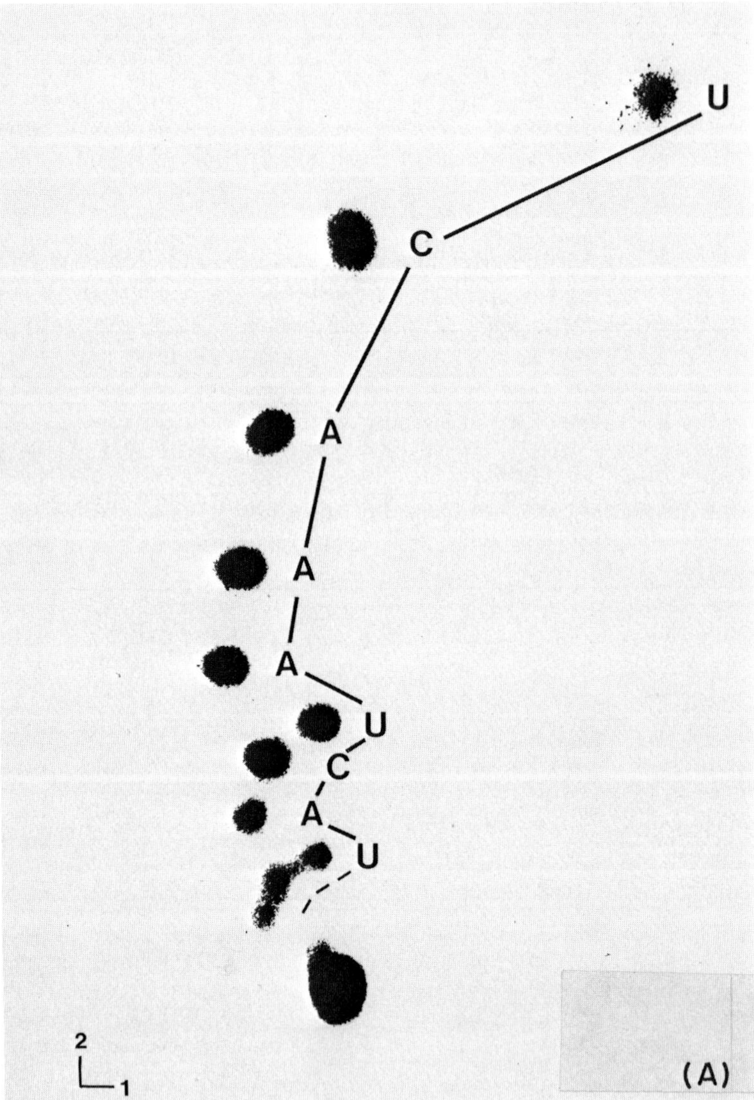

FIG. 3. Autoradiogram of the sequence analysis of the oligonucleotide ^{32}PUCAAAUCAUCAUG (3A). The 3′ terminal fragments can be determined unambiguously only after a second, prolonged, separation in the second dimension (3B).

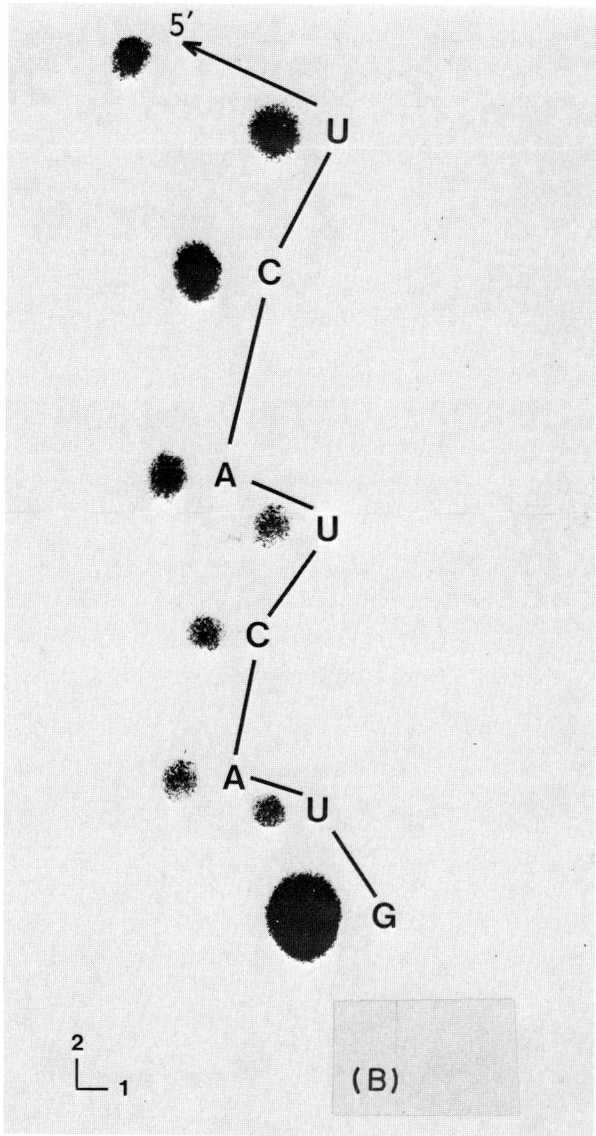

FIG. 3. (*Continued*)

(1) For the determination of the 5′ terminus, the complete oligonucleotide or fragments thereof are scraped from the thin-layer plate, and isolated as described. The precipitated and dried 5′ labelled material is completely digested by 4 μl of nuclease P1 (25 μg/ml in 0.05 *M* NH_4 acetate, pH 4.5) at 37°C for several h. The 5′ labelled nucleotides are identified by

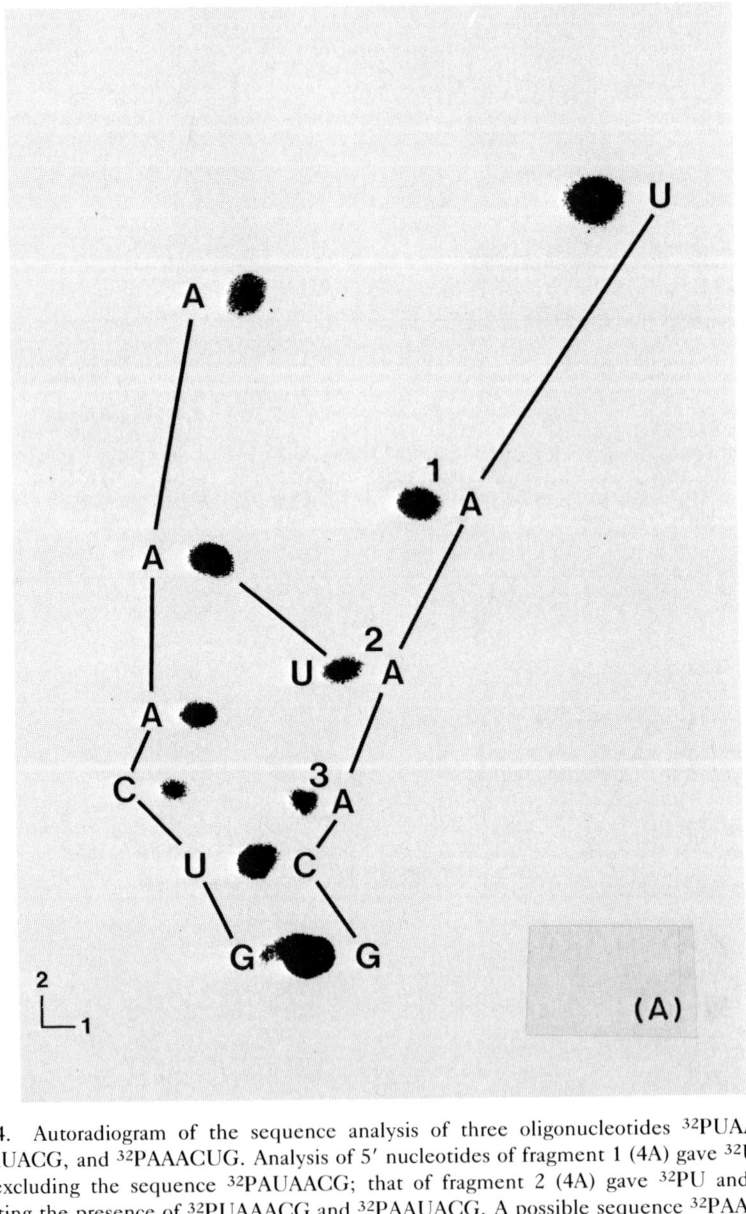

FIG. 4. Autoradiogram of the sequence analysis of three oligonucleotides ^{32}PUAAACG, ^{32}PAAUACG, and ^{32}PAAACUG. Analysis of 5′ nucleotides of fragment 1 (4A) gave ^{32}U only, thus excluding the sequence ^{32}PAUAACG; that of fragment 2 (4A) gave ^{32}PU and ^{32}PA, indicating the presence of ^{32}PUAAACG and ^{32}PAAUACG. A possible sequence ^{32}PAAAUCG could be excluded by alkaline hydrolysis of fragment 3 (4A), followed by two-dimensional analysis of fragment 3 (4B). It can be seen that the sequence ^{32}PAAAU is missing.

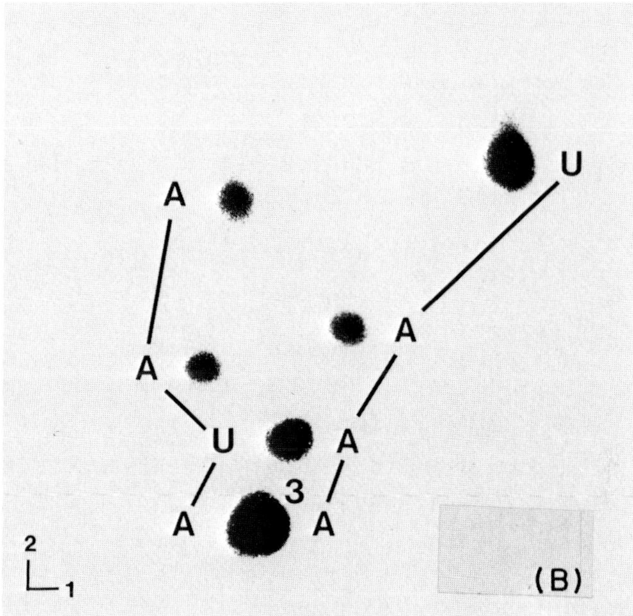

FIG. 4. (*Continued*)

thin-layer chromatography in one dimension on 20 × 20-cm cellulose plates in tertiary butanol–H_2O–HCl (14:3:3) (Domdey *et al.* 1978). In this solvent system the motility is: uridine > guanosine > cytidine > adenosine. The determination of the 5′ terminal nucleotides is necessary when the identity of the 5′ terminus is not obvious from the secondary analysis (it is sometimes difficult to distinguish between 5′ terminal adenosine and cytidine) or in cases where two or more different oligonucleotides with a different 5′ terminal oligonucleotide occupy the same spot in the fingerprint.

(2) It is sometimes impossible, however, to determine the sequences of oligonucleotides sharing the same spot in the fingerprint. This is so when coinciding or overlapping oligonucleotides share common intermediates of alkaline hydrolysis. In such cases, fragments are isolated from the plate as described above and subjected to a second controlled alkaline hydrolysis. The resulting tertiary fragments are again separated in two dimensions using the methods described for the secondary analysis. Because of the smaller size of the fragments, four samples can be applied to each cellulose acetate strip, 6, 15, 24 and 33 cm from one end, and the xylene cyanol marker is allowed to migrate 4.5 cm. The gradient thin-layer chromatography is as described for the secondary analysis. An analysis of a set of hexamers is shown in Fig. 4.

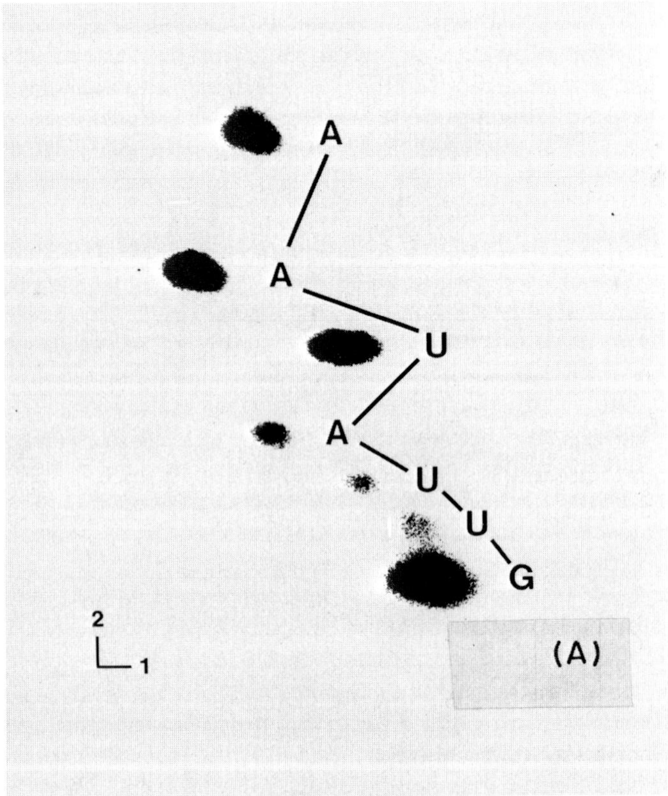

FIG. 5. Autoradiogram of the sequence analysis of the oligonucleotide ^{32}PAAUAUUG (5A). This sequence was verified by one-dimensional chromatography of the oligonucleotide, proving the absence of cytidine residues within the sequence (5B).

(3) With uridine-rich oligonucleotides, difficulties are sometimes experienced in discriminating between adenosines and cytidines in the middle or at the end of uridine-rich stretches. Such sequences can be elucidated by one-dimensional thin-layer chromatography of the alkaline hydrolysate, since pyrimidines can easily be distinguished from purines in this way. The presence of a uridine residue in the sequence is seen from the two-dimensional separation pattern, and guanosine occurs at the 3′ terminus of an oligonucleotide only. Adenosine and cytidine can thus be readily distinguished. The hydrolysate of an oligonucleotide is dissolved in 100 μl of sterile distilled water, and the RNA is precipitated with three volumes of ethanol for 16 h at −20°C. The dried fragments are redissolved in 2.5 μl of distilled water, applied to DEAE-cellulose thin-

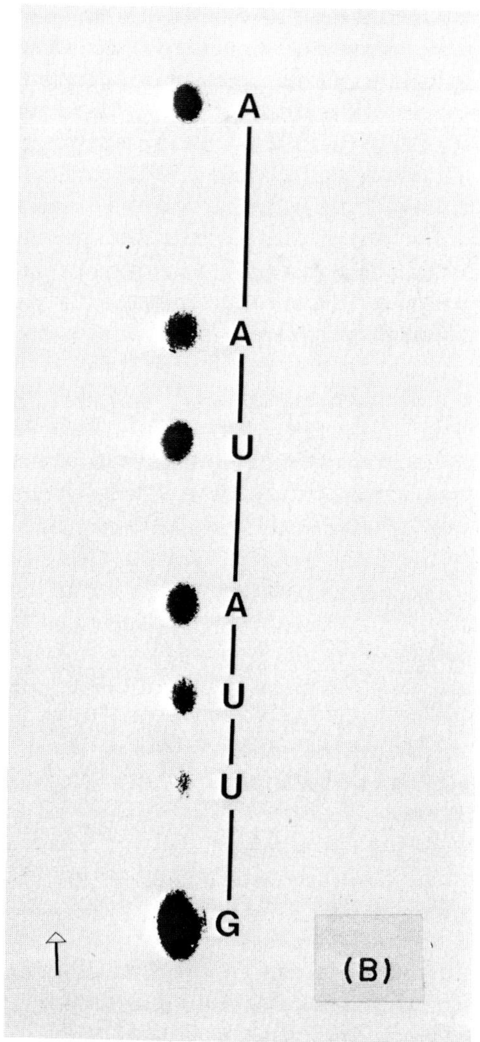

FIG. 5. (*Continued*)

layer glass plates (20 × 20 cm for fragments up to nonanucleotides; 20 × 30 cm for longer fragments) and separated in one dimension by homochromatography (equal volumes of 20, 30, 50 and 75 m*M* KOH homomixes) at 70°C until the xylene cyanol marker has almost reached the upper edge of the plate. An example is shown in Fig. 5.

TABLE 1

Oligonucleotide catalogues of some Actinomadura *and* Nocardiopsis *strains**

Sequence	Strains
Hexamers	
CCCACG	1,2
CACAAG	1–6
ACACAG	1,2,6
CCCCUG	1–6
UCCACG	1,3–5
CCUACG	1
CUAACG	1–6
UAACCG	1,2
ACACUG	1,5
ACAUCG	2
AUACCG	1–3,5,6
UAAACG	1–6
AAUACG	3–6
AAACUG	1–4
CCUUCG	1–6
CUCCUG	5,6
UUCCCG	1–6
UCUCAG	1–6
UCCAUG	5,6
UCAUCG	2
UCACUG	3,4
UACCUG	1,3,4
ACUCUG	1–6
AUCCUG	1,2,5,6
AUUCCG	1
UAAUCG	1–6
UAACUG	3,4
AUUACG	5,6
ACAUUG	1
AACUUG	2
UCUCUG	5,6
UUUCCG	5,6
Heptamers	
AACACCG	1–6
CAACCUG	1,2,4–6
CAACUCG	1–6
UAACACG	1–6
U̇AACAAG†	1–6
CCCUUCG	1–6
CUUCCAG	5,6
CUCCAUG	4
CACUUCG	1,2
ACUCAUG	1–6
UUAAACG	1
CAUUAAG	1–6
UACAAUG	5,6
UAAUACG	1–6
AUAUCAG	1,3–6
AAACUUG	5,6
UUCCCUG	5,6
UUCCAUG	1,2
AAUAUUG	1,3,4
UUCUCUG	1,5,6
CUUUUUG	1
UUUUUCG	1,5,6
Octamers	
ACACACCG	1
CCACACUG	3–6
ACCCCAUG	3–6
CAAACAUG	1–6
AAAACCUG	1
CCCCUUCG	1
CACUCUCG	1
CACCUUCG	4
CCUAUCAG	1–6
CUUAACCG	1,2
UCAACUCG	1–6
AUACCCUG	1–6
AAUUCCCG	1,2
CUACAAUG	1–6
CUCCUUAG	3,4
UCAUCAUG	5,6
CCUUAAUG	1
UCUAACUG	3,4
CUUAACUG	3,4
AAUUCCUG	3–6
UAAUUCAG	3
CCUUUAUG	5,6
UCAUUUAG	5,6
AAUUAUUG	1–6
UCCUUUUG	2
Nonamers	
CCCAACCCG	4
(1) CAACACCCG	1–6
CAACCCUCG	1,2
(2) CCUACCAAG	1–6

(*continued*)

TABLE 1 (*Continued*)

Sequence	Strains
CAACCCUUG	3–6
(4) ACUCCUACG	1–4
(5) CUAACUACG	1–6
CCCUCUUCG	3
(6) CCCCUUAUG	1–6
UCUUCCACG	3
CCUUACCUG	3–6
CCUUUAAUG	4
(7) CUUAAUUCG	1–6
UCCUUUUUG	1
Decamers	
(3) UCACACCACG	1–6
ACCACUCCAG	3
ACAUCACCCG	3–6
UCCUUCCACG	2
CUUAACUCCG	5,6
ACUUUCCACG	5,6
(8) UCUAAUACCG	1–6
UUCUUACACG	1
Undecamers	
AAAACCAUCAG	2
CCCAACCCUUG	1
ACCCUUCCACG	4
CUUAACCCUAG	3
(9) CUUAACACAUG	1–6
Dodecamers	
CACAACCACUUG	3
(10) AACCUUACCAAG	1–6
CCUAACCCCUUG	5,6
AAUCCCUUAAAG	1,2
AAUCCCUAAAAG	3–6
Tridecamers	
(11) CUAAAACUCAAAG	1–6
Tetradecamers	
(13) UAAACCUCUUUCAG	1–4
(12) AUCACCUCCUUUCU$_{OH}$	1–6

*Organisms are designated by numbers following the sequences: 1, *Actinomadura madurae* DSM 43067; 2, *Actinomadura verrucosospora* DSM 43358; 3, *Actinomadura pusilla* DSM 43357; 4, *Actinomadura roseoviolacea* DSM 43144; 5, *Nocardiopsis dassonvillei* DSM 43111; 6, *Nocardiopsis dassonvillei* DSM 43235. The numbers in parentheses before the sequences refer to those in Fig. 1 and indicate the position of the oligonucleotides in the 16S rRNA fingerprints.
†Modified base, $\overset{*}{U}$.

Construction of phylogenetic trees

The oligonucleotide sequences (≥ hexamers) are compiled in an oligonucleotide catalogue that is specific for each organism investigated. Catalogues of representative *Actinomadura* species and two *Nocardiopsis dassonvillei* strains are shown in Table 1. The similarity coefficient, S_{AB}, is calculated for each binary comparison. The S_{AB} value is defined as follows: $S_{AB} = 2N_{AB}/N_A + N_B$, where N_A and N_B are the total numbers of bases of oligonucleotides, hexamers and above, in catalogues of strain A and of strain B, respectively, and N_{AB} is the total number of bases in all of the oligonucleotides (≥ hexamer) identical in the catalogues of strains A and B (Fox *et al.* 1977). Values for S_{AB} range from 1.0 for identical RNAs to around 0.03 for randomly related sequences; the higher the S_{AB} value, the greater the degree of relatedness between the two organisms compared.

The S_{AB} values for the six *Actinomadura* and *Nocardiopsis* strains and for various actinomycete marker strains are shown in Table 2. The S_{AB} values were analysed using standard clustering algorithms to produce a dendrogram

TABLE 2

Number of coincident nucleotides and the S_{AB} *values of* Actinomadura, Nocardiopsis, *and various reference actinomycete strains**

Organisms	Organisms												
	1	2	3	4	5	6	7	8	9	10	11	12	13
Actinomadura madurae DSM 43067	X	389	343	333	340	338	316	275	293	302	236	275	276
Actinomadura verrucosospora DSM 43358	0.81	X	314	310	314	305	315	251	258	265	221	227	258
Thermomonospora curvata DSM 43183	0.69	0.64	X	304	298	296	291	256	263	272	224	237	257
Actinomadura pusilla DSM 43357	0.65	0.66	0.63	X	427	361	356	310	317	332	247	253	287
Actinomadura roseoviolacea DSM 43144	0.68	0.66	0.62	0.89	X	362	350	310	311	326	247	254	281
Nocardiopsis dassonvillei DSM 4311 (=43111)	0.64	0.62	0.59	0.71	0.69	X	338	295	310	305	241	263	282
Streptosporangium roseum DSM 43021	0.62	0.65	0.58	0.73	0.71	0.66	X	336	429	334	265	262	288
Streptomyces griseus strain 1947	0.54	0.51	0.51	0.62	0.62	0.57	0.64	X	331	363	259	254	291
Streptoverticillium baldaccii DSM 40845	0.56	0.52	0.52	0.63	0.62	0.59	0.63	0.82	X	397	255	265	298
Chainia antibiotica DSM 43022	0.57	0.54	0.54	0.66	0.65	0.59	0.64	0.70	0.75	X	255	259	321
Actinoplanes philippinensis DSM 43019	0.46	0.45	0.44	0.49	0.49	0.46	0.54	0.50	0.48	0.49	X	367	274
Micromonospora chalcea NRRL-B2344	0.54	0.47	0.48	0.51	0.51	0.52	0.51	0.49	0.51	0.51	0.75	X	266
Dermatophilus congolensis DSM 43037	0.53	0.51	0.49	0.55	0.55	0.52	0.54	0.54	0.55	0.60	0.51	0.50	X

*Data in the upper triangle are the total number of nucleotides present in the coincident oligomers; those in the lower triangle are S_{AB} values. The oligonucleotide catalogues of the reference strains have been published (Stackebrandt & Woese 1981b; Stackebrandt *et al.* 1983).

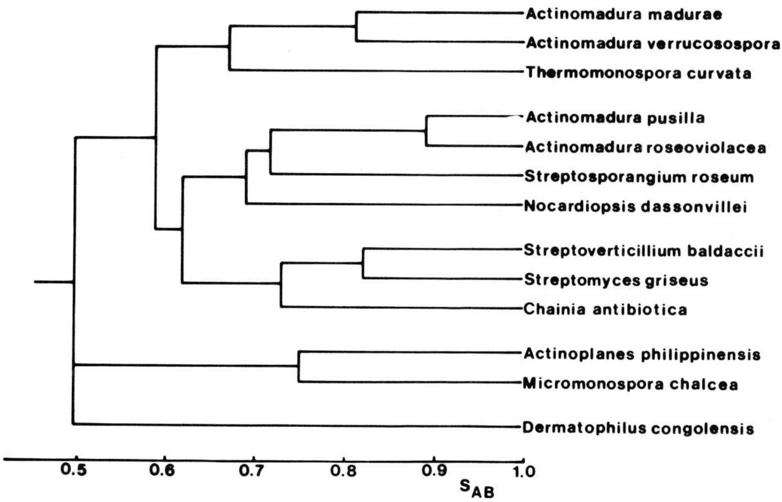

FIG. 6. Dendrogram, based on S_{AB} values and average linkage clustering, showing the heterogeneity of the genus *Actinomadura*.

that shows the phylogenetic relationships of the organisms in question (see Fig. 6).

Phylogeny of the Genus *Actinomadura*

The value of oligonucleotide cataloguing in the determination of relationships is demonstrated using the genus *Actinomadura* as an example. The oligonucleotide catalogues of the strains investigated (Table 1), the corresponding S_{AB} values (Table 2), and the dendrogram derived therefrom (Fig. 6) have been considered above.

Actinomadurae are actinomycetes that produce a branched, usually stable, substrate mycelium, which does not carry spores but bears aerial hyphae, some of which may be differentiated into short chains of arthrospores; the peptidoglycan contains *meso*-diaminopimelic acid, and whole-cell hydrolysates contain madurose (Williams & Wellington 1981). *Actinomadura* is separated from other actinomycete genera mainly by morphological features, cell-wall chemotype and phospholipid composition (Lechevalier & Lechevalier 1981).

Numerical phenetic surveys (e.g. Goodfellow & Pirouz 1982) gave the first indication that the genus was heterogeneous. The numerical investigations were soon supported by polar lipid and menaquinone data and the results of DNA–DNA and DNA–rRNA hybridization studies (Fischer *et al.*

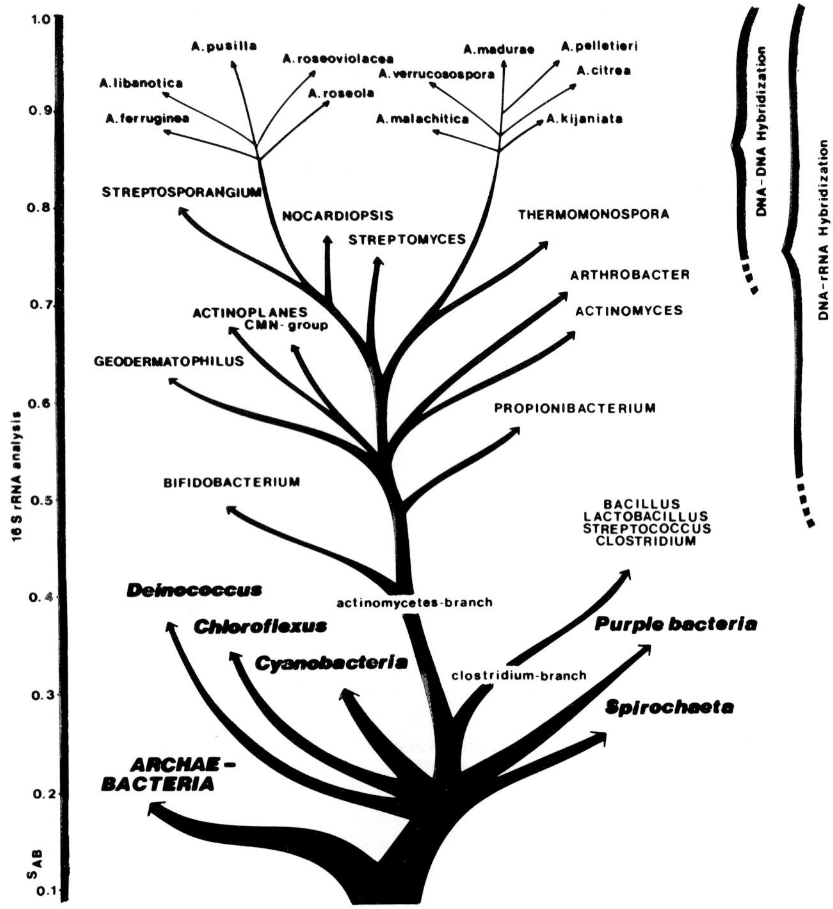

FIG. 7. Schematic family tree of prokaryotes showing those regions that can meaningfully be investigated by 16S rRNA analyses, DNA–rRNA cistron similarities and DNA–DNA hybridization.

1983) showing that *Actinomadura* species fell into two distinct aggregate clusters. The molecular genetic data, however, did not indicate whether the two aggregate clusters merited generic status. Ribosomal rRNA cataloguing studies not only underlined the heterogeneity of *Actinomadura* but clearly showed that *Actinomadura madurae*, the type species of the genus, and related species were more closely related to *Thermomonospora curvata* (Figs. 6 and 7) than to *Actinomadura pusilla* and allied taxa. Indeed, the latter were moderately related to *Streptosporangium roseum* and *Nocardiopsis dassonvillei* (the two strains of *N. dassonvillei*, whose catalogues are presented in Table 1, are highly related, showing an S_{AB} value of 0.96) and more distantly to the *Streptomyces* group (Figs. 6 and 7). The chemical, hybridization and sequencing studies indicate that the genus *Actinomadura* should be restricted to *A. madurae* and related species. The second aggregate cluster, which is also worthy of generic status, has still to be described.

The phylogenetic relationships of the major groups of actinomycetes and other Gram-positive bacteria are also shown in Fig. 7 (Stackebrandt & Woese 1981b; Stackebrandt *et al.* 1983).

Methodological and Phylogenetic Implications

The comparative analysis of large numbers of RNase T_1-resistant oligonucleotides of 16S rRNA is time consuming, laborious and expensive, but the method is invaluable for evolutionary and phylogenetic studies.

(1) The oligonucleotide cataloguing method presented has the following advantages: *in vitro* labelling of oligonucleotides allow sequence determination by the mobility shift method (Jay *et al.* 1974; Silberklang *et al.* 1977, 1979; Gross *et al.* 1978); partial hydrolysis of oligonucleotides by alkali yields a more uniform cleavage pattern than enzymatic partial digestion; difficulties in ordering secondary and tertiary fragments within large oligonucleotides experienced in the classical method for sequencing RNA fragments (Sanger *et al.* 1965) are circumvented since sequences can now be determined directly. Further, apart from the high-voltage electrophoresis apparatus, little instrumentation is required.

(2) Because of the ubiquitous distribution of ribosomal RNAs, any organism can in principle be investigated, and the *in vitro* labelling methods presented allow investigation of slow-growing, parasitic, and symbiotic living organisms. Indeed, studies are not restricted to prokaryotes, since the technique can also be used for the investigation of eukaryotic 18S rRNA (Stöcklein 1983).

(3) Phylogenetic conclusions derived from the analysis of rRNAs are more reliable than those based on characters of unknown genetic stability.

(4) Oligonucleotide catalogues are cumulative; they can be stored and retrieved whenever necessary. Though expensive, the method is certainly more economical in the long run than hybridization techniques. With the latter, laboratories working with the same group of organisms at different times must grow the strains, isolate and label the nucleic acids, and perform the hybridization studies independently; the same reference strains must be included repeatedly, and redundant data are accumulated.

However, because only one or a small number of strains from a whole group of organisms can be investigated by oligonucleotide cataloguing, the selection of strains must be done carefully to avoid phylogenetic conclusions based on data from atypical strains. A second source of error may arise from differences in the mutation rates (often described as different "clocks") of 16S rRNA. The clock of the majority of eubacteria runs more or less isochronically; however, the similarity coefficients and the corresponding branching patterns found when comparing organisms with higher (or lower) mutation rates with organisms whose clocks run isochronically do not accurately reflect true phylogenetic relationships (Stackebrandt & Woese 1981a; Woese *et al.* 1980). The existance of organisms with a "fast clock" within a group of isochronic organisms can be detected (see Stackebrandt & Woese 1981a).

The analysis of single conserved molecules such as cytochrome *c*, ferredoxin, 5S rRNA or 16S rRNA provides a useful first approximation of phylogenetic relationships. A more precise and convincing picture, however, is achieved by a combination of data obtained from two or more conserved molecules.

In principle it is possible to develop a taxonomy that is solely based on 16S rRNA. This has been demonstrated with methanogenic bacteria (Balch *et al.* 1979), a group of archaebacteria, which were themselves found to be a unique group on the basis of 16S rRNA analysis (Woese & Fox 1977). Studies on cell wall structure, lipid composition, intermediary metabolism, enzyme patterns, and 5S rRNA have substantiated the hierarchy of relationships of the methanogenic bacteria (Balch *et al.* 1979; Woese 1981).

Since the ultimate goal of taxonomy is to establish a phylogeny, work on constructing natural relationships must continue. Such studies should include the development of techniques suitable for elucidating different levels of relationships. So far, only three methods are available that can readily be applied to this task, namely DNA–DNA hybridization, DNA–rRNA cistron similarity, and 16S rRNA cataloguing (Fig. 7).

This work was supported by a grant from the Deutsche Forschungsgemeinschaft.

References

BALCH, W. E., FOX, G. E., MAGRUM, L. J., WOESE, C. R. & WOLFE, R. S. 1979 Methanogens: Reevaluation of a unique biological group. *Microbiological Reviews* **43**, 260–296.

BISHOP, D. H. L., CLAYBROOK, J. R. & SPIEGELMAN, S. 1967 Electrophoretic separation of viral nucleic acids on polyacrylamide gels. *Journal of Molecular Biology* **26**, 373–387.

BONEN, L., DOOLITTLE, W. F. & FOX, G. E. 1979 Cyanobacterial evolution: results of 16S ribosomal ribonucleic acid sequence analysis. *Canadian Journal of Biochemistry* **57**, 879–888.

BROSIUS, J., PALMER, M. L., KENNEDY, P. J. & NOLLER, H. F. 1978 Complete nucleotide sequence of a 16S rRNA gene from *Escherichia coli*. *Proceedings of the National Academy of Sciences of the United States of America* **75**, 4801–4805.

BROWNLEE, G. G. 1972 In *Determination of Sequences in RNA* ed. Work, T. S., pp. 18–21. New York: American Elsevier Publishing.

CAMERON, V., SOLTIS, D., & UHLENBECK, O. C. 1979 Polynucleotide kinase from a *T4* mutant which lacks the 3′-phosphatase activity. *Nucleic Acids Research* **5**, 825–834.

DOMDEY, H. & GROSS, H. J. 1979 Gradient thin-layer chromatography an alternative to homochromatography of oligonucleotides on DEAE-cellulose. *Analytical Biochemistry* **98**, 346–352.

DOMDEY, H., JANK, P., SÄNGER, H. L. & GROSS, H. J. 1978 Studies on the primary and secondary structure of potato spindle tuber viroid, produced at digestion with RNase A and RNase T_1 and modification with bisulfite. *Nucleic Acids Research* **5**, 1221–1236.

FISCHER, A., KROPPENSTEDT, R. M. & STACKEBRANDT, E. 1983 Molecular-genetic and chemotaxonomic studies on *Actinomadura*. *Journal of General Microbiology* **129**, 3433–3446.

FOX, G. E., PECHMAN, K. R. & WOESE, C. R. 1977 Comparative cataloguing of 16S ribosomal ribonucleic acid: molecular approach to procaryotic systematics. *International Journal of Systematic Bacteriology* **27**, 44–57.

FOX, G. E., STACKEBRANDT, E., HESPELL, R. B., GIBSON, J., MANILOFF, J., DYER, T., WOLFE, R. S., BALCH, W., TANNER, R., MAGRUM, L. J., ZABLEN, L. B., BLAKEMORE, R., GUPTA, R., LUEHRSEN, K. R., BONEN, L., LEWIS, B. J., CHEN, K. N. & WOESE, C. R. 1980 The phylogeny of prokaryotes. *Science* **209**, 457–463.

GOODFELLOW, M. & PIROUZ, T. 1982 Numerical classification of sporoactinomycetes containing *meso*-diaminopimelic acid in the cell wall. *Journal of General Microbiology* **128**, 503–527.

GROSS, H. J., DOMDEY, H., LOSSOW, C., JANK, P., RABA, M., ALBERTY, H. & SÄNGER, H. L. 1978 Nucleotide sequence and secondary structure of potato spindle tuber viroid. *Nature, London* **277**, 203–208.

JAY, E., BAMBARA, R., PADMANAGHAN, R. & WU, R. 1974 DNA sequence analysis: a general, simple, and rapid method for sequencing large oligodeoxyribonucleotide fragments by mapping. *Nucleic Acids Research* **1**, 331–353.

LECHEVALIER, H. A. & LECHEVALIER, M. P. 1981 Introduction to the Order *Actinomycetales*. In *The Prokaryotes* eds. Starr, M. P., Stolp, H., Trüper, H. G., Balows, A. & Schlegel, H. G., pp. 1915–1922. Berlin-Heidelberg-New York: Springer.

SANGER, F., BROWNLEE, G. G. & BARRELL, B. G. 1965 A two-dimensional fractionation procedure for radioactive nucleotides. *Journal of Molecular Biology* **13**, 373–398.

SCHWARZ, Z. & KÖSSEL, H. 1980 The primary structure of 16S rDNA from *Zea mays* chloroplast is homologous to *Escherichia coli* 16S rRNA. *Nature, London* **283**, 739–742.

SILBERKLANG, M., GILLUM, A. M. & RAJBHANDARY, U. L. 1977 Studies on the sequence of the 3′-terminal region of Turnip-Yellow-Mosaic-Virus RNA. *Nucleic Acids Research* **4**, 4091–4108.

SILBERKLANG, M., GILLUM, A. M. & RAJBHANDARY, U. L. 1979 Use of *in vitro* ^{32}P labelling in the sequence analysis of nonradioactive tRNAs. In *Methods in Enzymology* ed. Moldave, K., Vol. 59, Part G, pp. 58–109. New York: Academic Press.

STACKEBRANDT, E. & WOESE, C. R. 1981a The evolution of prokaryotes. In *Molecular and Cellular Aspects of Microbial Evolution* eds. Carlile, M. J., Collins, J. F. & Moseley, B. E. B., pp. 1–31. Cambridge: University Press.

STACKEBRANDT, E. & WOESE, C. R. 1981b Towards a phylogeny of actinomycetes and related organisms. *Current Microbiology* **5**, 131–136.

STACKEBRANDT, E., LUDWIG, W., SCHLEIFER, K. H. & GROSS, H. J. 1981 Rapid cataloguing of ribonuclease T_1 resistant oligonucleotides from ribosomal RNAs for phylogenetic studies. *Journal of Molecular Evolution* **17**, 227–236.

STACKEBRANDT, E., SEEWALDT, E., LUDWIG, W., SCHLEIFER, K. H. & HUSER, B. A. 1982 The phylogenetic position of *Methanothrix soehngenii*. Elucidated by a modified technique of sequencing oligonucleotides from 16S rRNA. *Zentralblatt für Bakteriologie und Hygiene. I. Abteilung, Originale* **C3**, 90–100.

STACKEBRANDT, E., LUDWIG, W., SEEWALDT, E. & SCHLEIFER, K. H. 1983 Phylogeny of sporeforming members of the Order *Actinomycetales*. *International Journal of Systematic Bacteriology* **33**, 173–180.

STÖCKLEIN, L. 1983 Phylogenetische Untersuchungen an Eukaryonten mit Hilfe der vergleichenden Analyse der 18S rRNA. Ph.D. Thesis, Technical University Munich, Munich.

UCHIDA, T., BONEN, L., SCHAUP, H. W., LEWIS, B. J., ZABLEN, L. & WOESE, C. R. 1974 The use of ribonuclease U_2 in RNA sequence determination: some corrections in the catalog of oligomers produced by RNase T_1 digestion of *Escherichia coli* 16S ribosomal RNA. *Journal of Molecular Evolution* **1**, 173–184.

WENGLER, G., WENGLER, G. & GROSS, H. J. 1979 Replication form of Semliki-Forest-Virus RNA contains unpaired guanosin. *Nature, London* **282**, 754–756.

WILLIAMS, S. T. & WELLINGTON, E. M. H. 1981 The genera *Actinomadura, Actinopolyspora, Excellospora, Microbispora, Micropolyspora, Microtetraspora, Nocardiopsis Saccharopolyspora,* and *Pseudonocardia*. In *The Prokaryotes* eds. Starr, M. P., Stolp, H., Trüper, H. G., Balows, A. & Schlegel, H. G., pp. 21O3–2117. Berlin-Heidelberg-New York: Springer.

WOESE, C. R. 1981 Archaebacteria. *Scientific American* **244**, 94–106.

WOESE, C. R. & FOX, G. E. 1977 Phylogenetic structure of the prokaryotic domain: the primary kingdoms. *Proceedings of the National Academy of Sciences of the United States of America* **74**, 5088–5090.

WOESE, C. R., SOGIN, M., STAHL, D., LEWIS, B. J. & BONEN, L. 1976 A comparison of the 16S ribosomal RNAs from mesophilic and thermophilic bacilli: some modifications in the Sanger method for RNA sequencing. *Journal of Molecular Evolution* **7**, 197–213.

WOESE, C. R., MANILOFF, J. & ZABLEN, L. B. 1980 Phylogenetic analysis of the mycoplasmas. *Proceedings of the National Academy of Sciences of the United States of America* **77**, 494–498.

Detection of Ribosomal Nucleic Acid Homologies

M. MORDARSKI

Institute of Immunology and Experimental Therapy, Polish Academy of Sciences, Wrocław, Poland

The affinities between microorganisms can be measured at several levels by *in vitro* studies of their nucleic acids. Estimates of DNA overall nucleotide similarity, expressed as the percentage of homology between reference and donor DNA and the similarity of ribosomal transcript units, e.g. the DNA region comprising genes coding for ribosomal RNA (rDNA), have proved to be especially popular and valuable.

In 1961 Hall & Spiegelman reported the development of a method for the direct analysis of genes coding for RNA. The first experiments were performed on hybrids formed between phage DNA and RNA synthesised in *Escherichia coli* infected with the phage. In subsequent experiments Yankofsky & Spiegelman (1962a) developed procedures that permitted the identification and isolation of DNA sequences complementary to known RNA molecules; the sequences formed 0.01% of the total genome of *E. coli*.

A few years later Gillespie & Spiegelman (1965) proposed a method based on the immobilization of DNA on nitrocellulose membranes (Nygaard & Hall 1964) and the formation of DNA–rRNA hybrids directly on such membranes. The membrane filter method for DNA–rRNA hybridization was modified and improved by De Ley and his associates (De Ley & Tijtgat 1970; De Ley & De Smedt 1975; De Smedt & De Ley 1977). The method, because of its relative simplicity, was also adapted and used in a number of other laboratories (Legault-Demare *et al.* 1967; Mordarski *et al.* 1976).

General Characteristics of DNA Cistrons Coding for rRNA

Several workers have shown the highly conservative character of the genes coding for rRNA compared with those forming the bulk of the bacterial

CHEMICAL METHODS IN
BACTERIAL SYSTEMATICS

0-12-289675-0

genome. Comparative analysis of DNA–rRNA hybrid similarities, therefore, affords a way of determining the nature and degree of phylogenic relationships between taxa at and above the generic level (Doi & Igarashi 1965; Woese *et al.* 1975; De Smedt & De Ley 1977; Glötz *et al.* 1981).

Most prokaryotes contain multiple genes specifying the rRNA components (Kennell 1968). In all prokaryotes examined so far, the fraction of the genome that is complementary to rRNA (rDNA) is remarkably constant, comprising about 0.3–0.4% of the total DNA (Nomura 1970, 1972), i.e. about 0.6–0.8 of the potential genetic information. In *E. coli*, the amount of DNA hybridizing with rRNA corresponds to about six copies of each of the rRNA genes. In contrast, in *Mycoplasma*, which has a genome about one-fourth the size of that of *E. coli*, the amount of rDNA corresponds to only one gene for each of the rRNA components (Ryan & Morowitz 1969). The number of genes coding for 16S, 23S and 5S rRNA is the same, which means that all three components are produced in equimolar concentrations (Pace 1973).

The genes coding for rRNA are organized into transcription units. The primary *E. coli* transcript sediments at 30S and has a molecular mass of about 2.0×10^6 daltons. It contains the entire sequence of 23S, 16S and 5S RNA and several other low-molecular-weight species that represent tRNA. The gene order within the transcription unit has been shown to be promotor, 16S, tRNA, 23S and 5S rRNA (Lund *et al.* 1976).

Ribosomes of Prokaryotes

Ribosomes from most prokaryotes, including the cyanobacteria, are very similar but are distinctly different from those of eukaryotic organisms (Taylor & Storck 1964). All prokaryotic ribosomes consist of characteristic 70S particles, which, under proper conditions, dissociate into 50S and 30S subunits. The 50S component is composed of 75% RNA and 25% proteins and the 30S component of 50% RNA and 40% protein (Wittmann 1976).

Ribonucleic acids from the ribosomes of prokaryotes are similar in dimension but differ considerably in primary structure. The ribosomal proteins from diverse prokaryotes are similar in number i.e. over 50, but differ in antigenic properties (Pace 1973). From the evolutionary point of view the most important point is that the functional homology of the prokaryotic ribosome has been retained so that hybrid ribosomal subunits active in protein synthesis can be reconstructed from proteins and rRNA isolated from microorganisms with little DNA in common (Nomura *et al.* 1968). Such a reconstruction of the basic function of ribosomes cannot be achieved with rRNA from eukaryotic organisms (Traub & Nomura 1968; Wittmann 1976).

Bacterial ribosomes usually contain three rRNA components classified by sedimentation velocity: 23S, 5S and 16S. The first two are derived from the 50S ribosomal subunits, the latter from the 30S subunit (Kurland 1972). The molecular weights of 23S, 16S and 5S rRNAs are 1.1, 0.55 and 0.04 × 10^6 daltons, respectively. These weights correspond to chain lengths of about 3300, 1650 and 120 nucleotides, respectively. A few photosynthetic prokaryotes have 23S rRNA only transiently (Marrs & Kaplan 1970).

Nucleotide Composition of Bacterial rRNA

The conserved character of rRNA cistrons from different microorganisms is reflected in the similarity of their G + C content, especially when compared with the extremely variable base composition of the total bacterial genome (Table 1) (Hill 1966; Johnson & Horowitz 1971). Indeed, the average G + C content of the total rRNA from diverse prokaryotes is remarkably constant around 53%. The G + C content of the 23S and 16S components is nearly the same, but some differences have been detected (Woese 1961; Stanley & Bock 1965). In the various bacteria studied, about 1% of the total nucleotides are methylated (Dubin & Günlap 1967; Fellner & Sanger 1968).

Ribosomal RNA from diverse prokaryotes may show similar base composition, but this does not necessarily mean that they have a similar primary structure. Indeed, the nucleotide sequences of rRNA molecules from diverse microorganisms may differ distinctly, the differences can readily be

TABLE 1

*Base composition of total rRNA and total DNA from various bacteria**

	rRNA G + C	DNA G + C (mol %)
Mycoplasma hominis	46.4	27–29
Proteus vulgaris	53.1	38
Bacillus megaterium	53.1	38
Streptococcus pneumoniae	50.1	39
Vibrio marinus	51.1	40
Bacillus subtilis	53.3	43
Escherichia coli	53.7	52
Bacillus stearothermophilus	59.2	52
Spirillum intersonii	57.0	55
Aerobacter aerogenes	53.4	56
Pseudomonas aeruginosa	53.3	65
Micrococcus lysodeikticus	53.9	72

*Taken from Pace (1973).

detected in DNA–rRNA hybridization studies. Physical measurements indicate that 60–70% of the bases in both 16S and 23S rRNA are involved in hydrogen bonding (Cox 1970; Hartman & Thomas 1970). Adjacent sequences of 10–30 nucleotides tend to be complementary and in solution may form clusters of hairpin loops (Pace 1973).

The complete sequences of 16S and 23S rRNA from a number of diverse strains was made recently and a direct comparison of the two forms of rRNA was made. The results, especially those concerning the secondary structural domains of the ribosomal RNA, strongly underlined the remarkable conservation of ribosomal nucleic acids during evolution (Glötz *et al.* 1981; Zwieb *et al.* 1981).

Methods of DNA–RNA Hybridization

Preparation of cell mass

Base composition and size of bacterial genomes are independent of both the medium and growth conditions provided mutagens are not involved. It is, therefore, possible to select the growth conditions and media most suitable for the cultivation of particular microorganisms. For cell mass production, it is important to obtain quick and extensive growth free from substances difficult to separate, such as polysaccharides. In order to obtain labelled DNA or rRNA, different media are often used to facilitate high levels of incorporation of the radioactive precursors (De Ley 1971; Owen & Hill 1978). Strains are usually cultivated in 100–200 ml of liquid media in 1-litre flasks or on solid media in Roux bottles. Solid media are preferred by some investigators given the problem of detecting contaminants in liquid media (De Ley 1971).

The amount of cell mass obtained depends on the organism. Usually it is possible to obtain 1–5 g of wet packed cells from 1 litre of liquid medium. Yields of pure DNA vary but about 1 mg can be expected from 1-g wet-cell mass. It is advisable to arrest cell growth before the end of the exponential growth phase since older cells are resistant to disruption by both mechanical and enzymatic methods. Incubation times vary from a few hours to several days or longer depending on the organisms under study (De Ley 1971).

Preparation of reference labelled organisms

For labelling, strains grown at optimal temperature in reciprocal shakers are controlled turbidometrically, and at the beginning of the exponential growth phase the radioactive precursor is added. The strains are allowed to grow almost to the end of log phase when they are harvested by centrifugation and washed 2–3 times in the usual way. Washed cells can be lyophilized and preserved until use (De Ley & Tijtgat 1970). Rigorous safety precautions should be taken when working with radioactive materials: in the disposal of waste, in washing glassware and the like. It is advisable to sterilize radioactive precursors by filtration through membrane filters before adding them to media. Carbon-14 and tritium in the forms of [^{14}C]thymidine and methyl[^{3}H]thymidine are commonly used to label DNA. Thymidine incorporation may be inhibited because of thymidine-degrading enzymes; in such cases [^{32}P]orthophosphate or [^{14}C]uracil should be used. The former is relatively cheap, gives high specific activities, but its short half-life necessitates the use of additional controls and requires work to be carried out quickly.

Radioactively labelled rRNA is essential for all DNA–rRNA hybridization experiments. Given the small quantities of RNA bound in DNA–rRNA-pairing experiments, it is important to obtain rRNA preparations with high specific activities. Consequently, [^{32}P]orthophosphate or [^{14}C]uracil are commonly used. In some cases ^{14}C-labelled adenine or guanine are added to the medium (De Ley 1971; Midgley 1971). For a satisfactory level of specific activity of DNA or rRNA, 40–50 μCi/ml or carrier-free [^{32}P]orthophosphate is added to phosphate-low media or 1–2 μCi/ml of [^{14}C]uracil, thymidine or adenine to minimal media. [^{3}H]thymidine is used at a higher concentration, e.g. 5 μCi/ml of medium, and is added three times during the early exponential growth phase (Owen & Hill 1978).

Isolation of DNA

The purity of the DNA and rRNA used in reassociation experiments is critical since contaminating proteins, polysaccharides or metal ions can markedly alter reassociation results and reproducibility. In general, the isolation of nucleic acids suitable for analysis depends on the use of satisfactory procedures for the disintegration of bacterial cells and for freeing the DNA or rRNA from contaminating substances. Some bacteria can be lysed by lysozyme or by combinations of lysozyme, pronase and sodium dodecyl sulphate (SDS) (Owen & Hill 1978). It may be helpful to use high concentrations of lysozyme, 5 mg/g wet-cell mass, for prolonged incubation periods. It is important, however, that the enzymes used for cell lysis or nucleic acid purification be free of deoxyribonuclease activity when DNA is prepared and of ribonucleases when rRNA is being isolated.

When bacteria are resistant to the action of lysozyme and detergent, nucleic acids can be isolated after two or three cycles of freezing and thawing. With very resistant cells a French-Press, Hughens-Press or X-Press can be used. Other possibilities such as ultrasonic disintegration or mechanical vibrators are not recommended since DNA and rRNA molecules are sheared by such treatment and may then be unsuitable for hybridization (Owen & Hill 1978).

Several relatively simple methods have been developed for the isolation and purification of bacterial DNA. The method described by Marmur (1961) is popular and has been modified in a number of laboratories. In this method, proteins are dissociated from nucleic acids by treating the cell extract with 1 *M* sodium perchlorate ($NaClO_4$). Most of the proteins are denatured and are removed by shaking the viscous extract with chloroform and isoamyl alcohol. Contamination with RNA is effectively eliminated by digesting preparations with ribonuclease and precipitating the polimerized DNA with ethanol; the ribonucleotides are left in solution. The partially purified DNA is dissolved in acetate buffer. Similarly, isopropanol, in the presence of acetate ions, selectively precipitates DNA strands, leaving RNA in solution. Chelating agents such as ethylenediaminetetraacetate (EDTA) and citrate are added to the buffers during the isolation procedure to minimize endogenous deoxyribonuclease activity (Bradley & Mordarski 1976).

De Ley & De Smedt (1975) modified Marmur's method in several ways. Thus DNA was treated 3–6 times to remove protein and after RNase treatment, 40 μg of pronase per ml of liquid was added and incubated overnight at 37°C. The final step of DNA purification was equilibrium density centrifugation in a CsCl gradient.

In our laboratory, pure, nondegraded DNA (Mordarski *et al.* 1976) is obtained using a modification of the method of Saito & Miura (1963). In this method (Table 2), wet cells (200 mg) are suspended in 1 ml of buffer A (0.03 *M* Tris–HCl and 0.1 *M* EDTA, pH 8.0) containing 1–5 mg of lysozyme (Sigma or Serva) per g of wet mass and incubated for 18 h at 37°C. Strains resistant to lysozyme are disintegrated by passing through an X-Press once or, when necessary, twice. After disintegration or lysis, SDS (2%, w/v) is added, the suspension is heated for 30 min at 60°C and then rapidly cooled when an equal volume of freshly distilled phenol saturated with Tris–SDS buffer (0.01 *M* Tris, 1% w/v SDS and 0.1 *M* NaCl, pH 9.0) is added. The resulting mixture is then shaken in a gyratory incubator for 20 min and the emulsion formed separated in a cooled centrifuge for 10 min at 10,000 *g*. The aqueous phase is removed and treated twice with an equal volume of phenol saturated with the Tris–SDS buffer. Nucleic acids are precipitated from the aqueous fraction by the addition of 2 volumes of cooled ethanol (−10°C). The precipitate is spooled onto a glass rod, dissolved in 0.1 × SSC (SSC: 0.15 *M* NaCl, 0.015 *M* trisodium citrate) and the salt concentration is increased to 1 × SSC by the addition of 10 × SSC. The solution is then incubated at 37°C for 30 min with 50 μg of pancreatic ribonuclease (Sigma) per ml. The ribonuclease is preheated at 100°C, pH 5.0, for 10 min to destroy DNase activity.

TABLE 2

Isolation of DNA

Packed cells + buffer: 0.03 *M* Tris–HCl, pH 8.0, + 0.1 *M* EDTA
↓
Disintegration in X-Press or treatment with lysozyme 18 h at 37°C
↓
2% w/v SDS, 30 min at 60°C
↓
Deproteination with phenol saturated with Tris–HCl buffer, pH 9.0, + 1% SDS, repeated two to three times
↓
Precipitation with cold ethanol at −10°C
↓
Pancreatic ribonuclease treatment, 1 × SSC, pH 7.0
↓
Phenol extraction three to four times
↓
Precipitation with cold ethanol, repeated three times
↓
Precipitation with isopropanol, 0.54 vol, from acetate buffer solution, redissolved and repeated as necessary
↓
Ethanol wash

After treatment with RNase, an equal volume of phenol saturated with water is added, the mixture is shaken for 10 min at 0–5°C, the resulting emulsion is centrifuged, and the aqueous phase is separated. Phenol extraction is repeated several times, usually three, until the interphase layer disappears. The partially purified DNA is precipitated with ethanol (−10°C), redissolved in 0.1 × SSC, and a 0.1 volume of 3.0 *M* sodium acetate–0.001 *M* EDTA solution, pH 7.0, is added. The DNA is then precipitated by slowly adding a 0.54 volume of isopropanol, wound on a glass rod and washed in ethanol solutions of increasing concentration (70, 80 and 90%). After drying, the DNA is redissolved in 0.1 × SSC, and the treatment with isopropanol is repeated until a preparation having E_{260}/E_{280} of 1.9 is obtained and E_{230}/E_{255} is greater than 2.0. The purified DNA is stored at 5°C in 0.1 SSC containing a few drops of chloroform.

^{14}C-labelled DNA is prepared in the same way.

Isolation of labelled rRNA

Strains are grown in a suitable liquid medium supplemented with a radioactive precursor in the form of [^{32}P]orthophosphate, ^{3}H- or ^{14}C-labelled uracil, adenine or guanine. At the end of the exponential growth phase, cells are

TABLE 3

Isolation of rRNA

Packed cells + inhibitors of RNase (bentonite or PVS) + DNase in 0.01 *M* Tris; 0.01 *M* $MgCl_2$; 0.01 *M* KCl, pH 7.2

↓

Disintegration in X-Press

↓

Deproteination with phenol + 1% SDS, pH 7.0, repeated three times

↓

Precipitation with cold ethanol at −10°C overnight

↓

DNase and pronase treatment

↓

Deproteination with phenol three times

↓

Precipitation with ethanol repeated two to three times

harvested and washed three times in buffer (0.01 *M* Tris–HCl, 0.01 *M* $MgCl_2$, 0.01 *M* KCl, pH 7.2).

In most laboratories, rRNA is isolated and purified (Table 3) according to the procedure of Moore & McCarthy (1967). Cells are suspended in the same buffer (1:3, w/v) as above with the addition of 10–20 μg/ml of DNase free of ribonuclease. Commercial DNase can be purified according to Polatnick & Bachrach (1961). Suspended cells are then disrupted using either an X-Press or a French-Press and the biomass is treated with RNase inhibitors, Bentonite (1 mg/ml) and poly(vinyl sulphate) (PVS) (10 μg/ml). The resultant mixture is treated with SDS at 1% w/v and shaken twice at room temperature for 5 min with 1 volume of phenol saturated with water. The emulsion is then centrifuged at 10,000 rev/min for 10 min, the upper (aqueous) phase is removed and RNA is precipitated by 2 volumes of cold ethanol (95%, w/v). The RNA is collected by centrifugation after allowing the solution to stand overnight at −10°C. The crude RNA preparation is redissolved in 0.1 SSC solution, treated with DNase (10 μg/ml), incubated at 37°C for 15 min, treated with pronase (50 μg/ml) and incubated for a further 15 min. The mixture is then treated once again with aqueous phenol, centrifuged and the RNA is precipitated with 95% ethanol. The deproteinization with phenol is repeated three times, the RNA is again precipitated with ethanol (2 or 3 times), and at the end the purified RNA is resuspended in SSC buffer.

The purified RNA is fractionated by sucrose-gradient centrifugation (McConkey 1967). RNA (about 1 mg) is placed on the surface of a sucrose linear gradient (15–30% w/v sucrose in 0.1 *M* sodium acetate buffer with 0.05 *M* NaCl, pH 5.0–5.1) and centrifuged in a preparative ultracentrifuge at

TABLE 4

Separation of ribosomal RNA

Sucrose gradient
centrifugation in 15–30% sucrose linear gradient, 21 ml, in sodium acetate buffer, pH 5.0, 1 mg RNA, 109,000 *g*, 18 h, 4°C
↓
Collection of fractions
↓
Determination of radioactivity to locate 23S and 16S RNA
↓
Dialysis against 2 × SSC

109,000 *g* for 20 h at 4°C. Fractions (0.5 ml) are collected and the radioactivity of each one is measured (Table 4). Fractions containing the 23S and 16S rRNA are separately dialysed three times against 4 liters of 2 × SSC buffer at 4°C and then stored at −12°C until required (Mordarski *et al.* 1980). Dialysis of the RNA solutions is carried out using Visking dialysis tubing that has been treated with boiling 1 m*M* EDTA and washed thoroughly with distilled water.

Fixing of denatured, high-molecular-weight DNA to membrane filters

The original filter method of Gillespie & Spiegelman (1965), as modified by De Ley & De Smedt (1975), is applied in many laboratories (Mordarski *et al.* 1980, 1981a,b; Kilpper *et al.* 1980; Fiedler *et al.* 1981; Kilpper-Bälz & Schleifer 1981). This method has several advantages. Thus the denatured DNA is bound to membrane filters that can be directly placed in scintillation vials after samples have been hybridized and washed. Nitrocellulose membrane filters can absorb up to nearly 100 μg of single-stranded DNA per 1 cm^2 and the filters plus DNA, dried first at room temperature, then under vacuum at 80°C for 1 h, are stable when stored in a vacuum at 4°C for several months. In addition, there is little nonspecific absorption of RNA to the filters alone, and hybridized DNA may be quantitatively removed from filters at low ionic strength (Midgley 1971).

Denaturation and immobilization of DNA

Nitrocellulose membrane filters (24, 50 or 100 mm diameter, Schleifer and Schuell, Millipore or Sartorius) are presoaked for at least 1 h in 6 × SSC, transferred to filter holders and washed with 6 × SSC. Solutions containing 150 μg DNA/ml are denatured for 10 min at 106°C, chilled quickly on ice, made up to 6 × SSC (with 10 × SSC) and passed through membrane filters under gravity. Immobilization of DNA should be performed at 3°C, otherwise the efficiency of DNA retention is lowered by at least a factor of 10 (Midgley 1971). The loaded filters are washed with 100–500 ml of 6 × SSC, allowed to dry at room temperature for a few hours and then dried for 4 h at 80°C under vacuum.

The approximate amount of fixed DNA is calculated from the difference in absorbance of the DNA solution before and after filtration, assuming that A_{260} of 1.0 is equal to a concentration of 50 μg of DNA/ml. For hybridization, small discs are punched out of the main filters. The diameter is selected so that each "small" disc contains approximately 50 μg of high-molecular-weight, single-stranded DNA (De Ley & De Smedt 1975).

Chemical determination of DNA on filters

It is necessary to determine the exact amount of DNA fixed on nitrocellulose membranes since the results of DNA–rRNA reassociation are expressed as a percentage of bound rRNA to immobilized DNA. Determinations must be carried out both before and after reassociation as the quantity of single-stranded DNA on the membranes may change owing to "leaching" during incubation. The amount of lost DNA depends on the source and degree of purity of the nucleic acid and on the incubation temperature (De Ley & De Smedt 1975). At higher temperatures, a considerable amount of DNA is released from membranes, especially if a CsCl preparative ultracentrifugation step is routinely used in DNA purification.

For chemical determinations, the DNA remaining on filters is released by treatment with 1 ml of 1.6% perchloric acid ($HClO_4$) at 70°C for 30 min (Meys & Schilperoort 1971). After cooling, 2 ml of diphenylamine reagent is added (Burton 1956). The filter discs are then removed and the mixture is incubated at 30°C for 17 h, i.e. overnight. The optical density of the DNA is measured at 600 nm. The method of Chen *et al.* (1956) can also be used for the chemical determination of DNA on membrane filters. According to Midgley (1971), this very sensitive method (less than 0.5 μg of phosphate can be assayed) overcomes the possible interference of materials eluted from the nitrocellulose filters in the diphenylamine test.

Samples to be determined (0.5 ml) are ashed by the addition of 1 drop of concentrated sulphuric acid and 2 drops of 1 *N* $HClO_4$. After heating, white fumes appear, tubes containing samples are then cooled, and 1 ml of distilled water is added. The sample is then heated in boiling water for 10 min and the contents of the tubes are then made up to 4 ml with water.

Immediately before use, the following reagents are prepared: ascorbinic acid 10% w/v (A); 3 *M* H_2SO_4 (B) and 2.5% w/v ammonium molybdate (C).

The three reagents are then mixed: 1 volume A + 1 volume B + 1 volume C + 2 volumes of distilled water. Four ml of the mixture is then added to 4 ml of the sample, and tubes are stoppered, shaken and incubated at 100°C for 2 h. After cooling, the absorbancy is read at 820 nm (Midgley 1971).

DNA–rRNA hybridization on nitrocellulose membrane filters

In the standard procedure (Table 5), filter discs loaded with 50 μg of high-molecular-weight, single-stranded DNA, punched from the "main" filter, are incubated in 1 ml of 2 × SSC in 20% formamide (w/v) and 10 μg of ^{14}C-labelled rRNA in small airtight, screw-capped vials. The vials are immersed in a water bath at 45 or 50°C and rotated slowly at 50 rpm for 16 h.

The optimal renaturation temperature (T_{OR}) depends on the base composition of the reacting nucleic acid molecules and on the ionic composition of the reassociation mixture (Marmur & Doty 1961; Gillis *et al.* 1970). The value of T_{OR} is approximately 25°C lower than the melting temperature ($T_m - 25°C$). For less stringent conditions $T_m - 35°C$ is used.

The melting temperature (T_m) and optimum renaturation temperature (T_{OR}) can be calculated according to the following equations:

$$T_m = 0.41 \times \%(G + C) + 78.7 + 13.3 \log cNa^+ \text{ (Gillis } et\ al. \text{ 1970)}$$

and

$$T_{OR} = 1.24 \times T_{m,x} - 43.8 \ (T_{OR,x} \text{ and } T_{m,x} \text{ for } x\text{SSC})$$

TABLE 5

DNA–rRNA hybridization on nitrocellulose membrane filters

Heat denaturation of unlabeled DNA, 100°C, 10 min, 0.1 SSC

↓

Adjustment to 6 × SSC

↓

Immobilization of single-stranded DNA on presoaked 6 × SSC nitrocellulose membranes, wash and dry

↓

Incubation of filters with fixed DNA in the medium containing labelled rRNA in 2 × SSC–20% formamide, 18 h at 50°C

↓

Pancreatic ribonuclease treatment of discs with bound DNA–rRNA hybrids, wash and dry

↓

Radioactivity measurement in scintillation counter

↓

Filters removed from scintillation fluid, washed and dried

↓

Chemical determination of the DNA content on the discs

therefore, for 2 × SSC, T_{OR} = 0.51 × %(G + C) + 47.0 (Gillis & De Ley 1975).

De Ley & De Smedt (1975) proposed that 2 × SSC solution containing 20% of formamide should be used for DNA–rRNA reassociation if the moieties being studied had G + C values around 53%. In formamide, duplex T_m values decrease almost linearly by 0.7°C per 1% formamide, which means that the T_{OR} in 2 × SSC plus 20% formamide is low enough, about 55°C, to eliminate the leaching problem and to prevent the rRNA molecules from thermal disintegration. Sometimes temperatures around 50°C are applied in order to obtain additional nonspecific duplexing of taxonomic interest between the DNA and rRNA (De Ley & De Smedt 1975).

When hybridization is completed, filter discs are removed, incubated for 5 min at 50°C in 2 × SSC + 20% formamide to remove nonspecifically bound rRNA, and rinsed twice in 50 ml 2 × SSC solution at room temperature. After washing, the filters are treated with 250 μg of RNase (crystalline, DNase-free, bovine pancreas, Sigma) in 1 × SSC at 37°C for 1 h.

Determination of thermal stability of DNA–rRNA hybrids

After RNase treatment, each filter disc is removed and rinsed twice in 2 × SSC at room temperature and then dried. The thermal stability of duplexes formed between the filter-bound DNA and reference rRNA is determined from the $T_{m(e)}$, i.e. the temperature at which half of the paired rRNA becomes dissociated and eluted from the test DNA bound to the filter (Table 6). Thus the dried filter discs are incubated separately for 15 min in each of a

TABLE 6

Determination of the thermal stability of DNA–rRNA duplexes [$T_{m(e)}$]*

Nitrocellulose membranes with fixed DNA–rRNA duplexes, after ribonuclease treatment and washing

↓

Heat in 1.5 SSC–20% formamide in water baths at 60–95°C; aliquots withdrawn at 5°C intervals

↓

Radioactivity determination of the eluted fluids in a scintillation counter

↓

After the last elution, at 95°C, membranes are washed and the remaining radioactivity measured

↓

Drawing of thermal denaturation curves

*$T_{m(e)}$ is the temperature at which half of the paired rRNA becomes dissociated and can be eluted from the test DNA bound to the filter.

series of small vials containing 2 ml of 1.5 × SSC and 20% formamide, in a series of water baths of increasing temperature (5°C steps) from the hybridization temperature up to 95°C. The radioactivity of each vial is measured after the addition of 2 ml of water and 10 ml of scintillation solution (500 ml of Triton-X 100, 500 mg of dimethyl-POPOP and 5 g of PPO in 1 litre of toluene). The radioactivity on the filter discs is determined in the same scintillation fluid. The time of counting depends on the specific radioactivity of the rRNA used.

The total amount of bound radiolabelled rRNA, expressed as μg rRNA per 100 μg DNA retained on the filters, is called the "percentage of rRNA binding." The DNA remaining after hybridization on the filter is released using the chemical method as described before, and determined colorimetrically.

Precautions to be taken with the membrane–filter method

The method is relatively simple in principle but gives reproducible and exact results only when performed carefully. Thus comparatively low hybridization temperatures, obtained by the addition of formamide, are needed to diminish leaching of DNA from the filters. It is also important to check the purity of the formamide since unknown impurities decrease the thermal stability of duplexes, giving distinctly lower $T_{m(e)}$ values. Storage of rRNA solutions requires special care because glassware is easily contaminated by traces of RNase. It is, therefore, necessary to sterilize the glassware used to store RNA solution.

New nitrocellulose membranes usually bind less DNA and cause reduced hybridization efficiency compared with stored membranes. Indeed, nonspecific absorption of RNA may be initially as high as 8% of the RNA input. The efficiency of duplexing may be improved and the nonspecific RNA absorption may be reduced by soaking membranes at 3°C in SSC solution for not less than 24 h (Midgley 1971).

Basic proteins, such as RNase, lysozyme and methylated albumin, commonly used for the isolation and purification of nucleic acids, can occur as contaminants of both DNA and RNA. The presence of these compounds can generate difficulties by virtue of their ability to absorb RNA and to reduce the efficiency with which the resulting complex is retained on the membrane filter. Passage through an MAK column (methylated albumin Kieselguhr) is extremely effective since RNase remains on the column. Further, incubation of DNA with pronase destroys residual RNase activity. When pronase is used, it is important to remove any nucleases present by self-digestion. Only DNA preparations that have been assayed for RNase activity should be employed in DNA–rRNA reassociation experiments. Both DNase and

RNase must be assayed to be certain that each is free of significant contamination by the other.

Comparison of DNA–rRNA hybrids

Ribosomal RNA similarity is expressed by two parameters: $T_{m(e)}$ and the percent of rRNA binding. The extent of the latter depends not only on the actual rRNA homology but also on the size of the genome, the replication state of the genome, and the number of rRNA cistrons per genome. The percentage of binding represents a certain amount of rRNA bound to the rDNA, is typical for the taxon under study, but it should not be called rRNA homology (De Smedt & De Ley 1977).

The correlation between data obtained in DNA–rRNA hybridization experiments is best revealed by plotting, for each labelled reference rRNA, the $T_{m(e)}$ values against the percent of binding (De Smedt & De Ley 1977). The resulting graphs are called "rRNA similarity maps" (Fig. 1). Each strain has a fixed position, within the limits of reproducibility [$T_{m(e)}$ ± 0.5°C; percent of rRNA binding ± 0.015%] on the maps. After investigation of

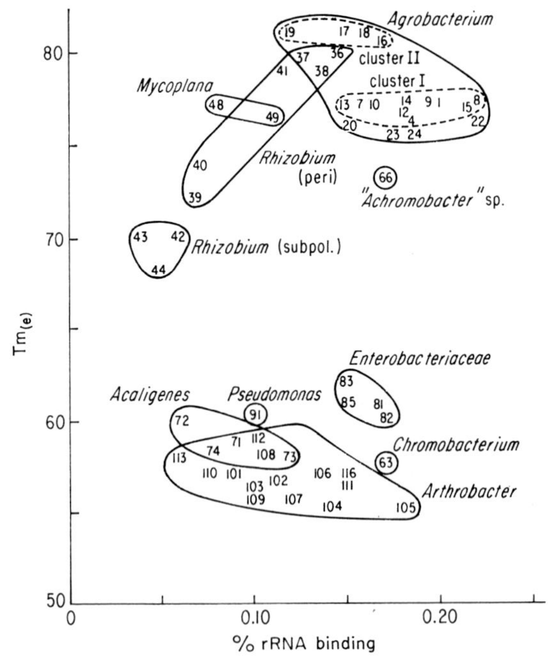

FIG. 1. Similarity map of hybrids between the 16S ^{14}C-labelled rRNA fraction of *Agrobacterium rhizogenes* and DNA from a variety of bacteria. Abbreviations: subpol., subpolar flagellation; peri., peritrichous flagellation. (From De Smedt & De Ley 1977.)

hundreds of different strains, De Ley & De Smedt (1977) stated that cultures belonging to the same species occupied very similar or identical positions on similarity maps (Fig. 1). Similarly, species belonging to the same genus are recovered close together. The shape and size of the area occupied by a taxon is a measure of its homogeneity. A few good representative strains suffice to place a genus on a map. The percentage of rRNA bound helps to distinguish genera defined at the same $T_{m(e)}$ level. The $T_{m(e)}$ similarity between RNA cistrons from different bacteria is a measure of their relationship at the inter- and suprageneric level and can be used both for classification and identification (Gillis & De Ley 1980; Kersters & De Ley 1980). In one study (Stackebrandt *et al.* 1981), the relationships between representative actinomycetes were illustrated in dendrograms derived from average linkage clustering of the $T_{m(e)}$ values of the DNA–rRNA hybrids. Principal-components analysis has also been used to analyse the results of DNA reassociation data (Grimont & Popoff 1980).

The competition–hybridization method

In the competition–hybridization method, sheared, denatured, immobilized DNA is allowed to hybridize with labelled rRNA in the presence of excess unlabelled homologous or heterologous rRNA. If the progressive addition of unlabelled RNA to the fixed amounts of labelled RNA reduces the hybridization level of the labelled RNA the fractions are present in both labelled and unlabelled RNA with a considerable degree of nucleotide sequence homology. If labelled RNA is completely eliminated from the hybrids by a large excess of the unlabelled competitor, then the RNA species in the labelled fraction must all be present in the unlabelled competitor. Alternatively, if excess unlabelled RNA does not influence the hybridization level of the labelled RNA, then this fraction cannot be present in the unlabelled RNA (De Ley 1971; Midgley 1971; Yankofsky & Spiegelman 1962b).

Studies of 16S and 23S nucleotide sequence similarity illustrate the competition–hybridization method (Fig. 2). Dennis & Nordan (1976) developed an assay to distinguish specifically between both main fractions of rRNA. The method involved hybridization of radioactive rRNA to specific DNA isolated from λ ilv 5 transducing phage. The latter carries a complete copy of the *E. coli* B rRNA transcription unit (Wu & Davidson 1975; Collins *et al.* 1976). The length of the double-stranded phage DNA is about 4.6×10^4 base pairs and the combined length of mature 16S and 23S rRNA is about 4.5×10^3 bases. This means that about 5% of the DNA (rRNA) is complementary to the 16S and 23S sequences in this model.

For hybridization, purified phage DNA is denatured and immobilized on nitrocellulose filters. Hybridization is performed in 2 ml of 2 × SSC contain-

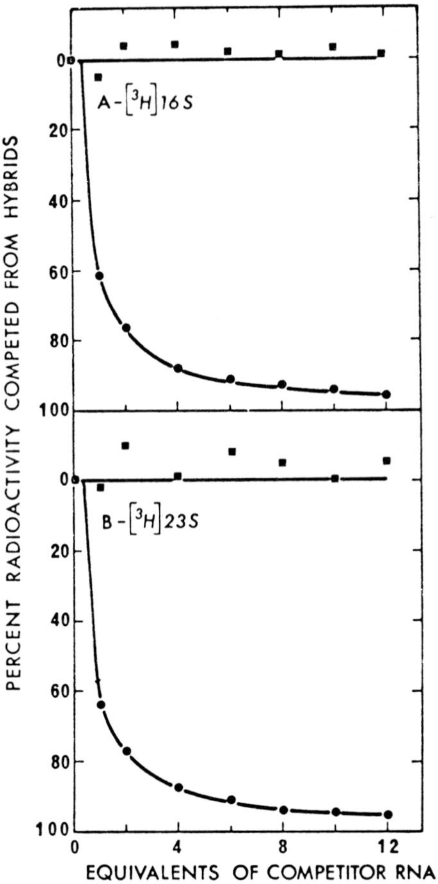

FIG. 2. Competition–hybridization with 16S and 23S rRNA. (A) Hybridization of ^{3}H-labelled 16S rRNA to λ ilv5 DNA in the presence of homologous 16S (●) and heterologous 23S (■) competitor RNA. (B) Hybridization of ^{3}H-labelled 23S rRNA to λ ilv5 DNA in the presence of homologous 23S (●) and heterologous 16S (■) competitor RNA. Hybridizations were done in the presence of 0–12 μg-equivalent of competitor RNA. A microgram-equivalent of 16S rRNA is 0.37 μg and that of 23S rRNA is 0.63 μg; these values represent equimolar amounts of the two rRNA species. (From Dennis & Nordan 1976.)

ing ^{3}H-labelled rRNA for 20 h at 67°C. The filters are then washed, treated with ribonuclease, washed again, dried and counted in a scintillation counter. The hybridization efficiency of this system is extremely high: in conditions of increasing amounts of labelled rRNA and an excess of phage DNA, 80–85% of the homologous 16S and 23S RNA enter into the rRNA–DNA hybrids. Both 16S and 23S RNA are equally efficient in forming rRNA–DNA hybrids, and at saturation of the DNA equimolar amounts of 16S and 23S rRNA are hybridized.

In competition–hybridization experiments, a small but constant amount of labelled 16S or 23S rRNA is incubated with phage DNA (10 μg) in the presence of a 0–20-fold excess of the homologous or heterologous 16S or 23S nonradioactive competitor (see Fig. 2). Competition with a 20-fold excess of the homologous competitor leads to nearly 100% competition. On the other hand, competition is not observed if heterologous rRNA is used as the competitor even in a 20-fold excess. These results, supported by 23S and 16S rRNA "fingerprint" analysis (Fellner & Ebel 1970), definitely show that 16S and 23S rRNA do not cross hybridize and that the sequence homology between 16S and 23S rRNA is negligible.

Competition assays, based on modifications of the same principle, have been used to unravel details of the nucleotide sequence similarity of the DNA region coding for rRNAs in distantly related microorganisms (Doi & Igarashi 1966; De Ley 1971; Midgley 1971; Kourilsky *et al.* 1974; Oostra *et al.* 1976).

Nuclease S_1–DE 81 method

New possibilities for the exact assessment of nucleotide sequence similarity were provided by the introduction of single-strand specific nucleases, usually S_1 endonuclease from *Aspergillus oryzae,* for the exhaustive digestion of unhybridized DNA or RNA followed by the binding of hybrids to diethylaminoethylcellulose (DEAE-cellulose; Crosa *et al.* 1973; Maxwell *et al.* 1978). Several nucleases other than S_1 perform the hydrolysis of single-stranded DNA, but in practice, all have some disadvantages (Sutton 1971).

Endonuclease S_1, which specifically splits the phosphodiester bonds of single-stranded DNA and RNA, attacks short unhybridized regions, different "loops" and "tails" of DNA and RNA, leaving double-stranded nuclease-resistant duplexes (Ando 1966). For DNA–rRNA hybrids, the most suitable method involves binding the hybrids to DEAE-cellulose (DE 81 paper discs, Whatman). Other methods commonly and successfully used in DNA–DNA reassociation analysis, such as precipitation with trichloroacetic acid (TCA) or chromatography on hydroxyapatite (HA) can have serious disadvantages if applied to DNA–rRNA assays. Precipitation with TCA followed by the collection of the precipitate on cellulose membranes or glass fibre filters leads to the retention of a significant fraction of the digestion products, usually in the range of 2–10% of the input radioactivity on the filters (Salzberg *et al.* 1977; Maxwell *et al.* 1978). High contamination of this nature cannot be tolerated in experiments involving only very small regions of the total genome. Chromatography on hydroxyapatite may lead to errors, especially in the case of short duplexes that are not efficiently bound to the hydroxyapatite (Maxwell *et al.* 1978). Further, many controls are required to establish the conditions for each batch of HA in each laboratory, given the variations that exist between batches of HA (Britten *et al.* 1974).

TABLE 7

DNA–rRNA hybridization nuclease S_1–DE 81 method

Reassociation of denatured, sheared DNA with rRNA in aqueous reassociation mixture or in 70–80% formamide, according to "free solution" methods

↓

Exhaustive dialysis against 0.001 *M* NaCl or dilution with S_1 buffer

↓

Nuclease S_1 treatment or pancreatic RNase, followed by nuclease S_1

↓

Fixing of hybrids on DE 81 filter discs, wash with 5% Na_2HPO_4 and distilled water, dry

↓

Deposition of discs in scintillation vials and counting

In contrast, relatively short duplexes bind efficiently to DEAE-cellulose filters. In the experiments of Maxwell *et al.* (1978) fractions of double-stranded *E. coli* DNA (cloned operator of the lac operon, only 36 base pairs in length) were quantitatively bound (96–99%) to DEAE-cellulose filter paper. Since the nonspecific background was low, the DEAE-cellulose filter method seems to be very suitable for reassociation experiments in which only a small fraction of the total input tracer is hybridized (Maxwell *et al.* 1978). Such small duplexes can be expected in comparative studies of the overall nucleotide sequence similarity of the rDNA region in distantly related microorganisms.

In comparative DNA–rRNA reassociation studies, it is important to know the sensitivity of the hybrids formed to the action of S_1 nuclease, especially the specificity of the latter toward hybridized rRNA. Investigations performed with ^{32}P–^{33}P dual-labeled DNA and ribosomal RNA, isolated from *Bacillus cereus,* showed that after nuclease S_1 treatment the hybrids still contained about 14% more RNA than DNA. When only ribonuclease was used for duplex digestion, the remaining duplexes contained 22% more DNA than RNA. Only after treatment with both pancreatic ribonuclease and S_1 endonuclease is it possible to obtain hybrids in which the molar DNA/RNA ratio is precisely 1.0 (Wittelsberger & Hansen 1977). It is clear from these findings that accurate measurements of hybridized RNA require ribonuclease treatment in addition to S_1 nuclease digestion. An abbreviated nuclease S_1–DE 81 procedure used for DNA–rRNA hybridization studies is shown in Table 7.

An additional advantage of the nuclease S_1–DE 81 method is the possibility of eluting the hybrid molecules from the filters in the native form; over 70% of the radioactivity of RNA–DNA hybrid molecules is eluted at 1.5 *M* KCl at 60°C. Further, it is easy to perform the $T_{m(e)}$ determination of the hybrids obtained.

DNA–rRNA hybrid formation

The first step in the nuclease S_1–DE 81 method applied to DNA–rRNA hybridization is the reassociation reaction performed in "free solution" according to one of several methods (Nygaard & Hall 1963; Yankofsky & Spiegelman 1963; McConkey & Dubin 1966; Brenner 1973) or in the modifications of Cassey & Davidson (1977) and Vogelstein & Gillespie (1977).

It has been shown that for any given RNA sequence, complementary DNA must be present in the reaction mixture in a vast excess to achieve reasonably complete hybridization of RNA. Optimum rates of formation of DNA–rRNA hybrids in aqueous solution are achieved at nearly 70°C but at this temperature some degradation of RNA may occur. It was for this reason that hybridization in a medium containing 70–80% formamide (FA) was proposed. In the modified medium, the reassociation temperature is low enough to permit extended incubation without appreciable rRNA instability. In addition, medium containing 70–80% FA practically stops DNA self-annealing; the latter was one of the main obstacles to the introduction of the free solution hybridization to comparative DNA–rRNA analyses.

The hybridization mixture proposed by Vogelstein & Gillespie (1977) consists of 7 parts of DNA in 100% FA and 3 parts of rRNA in 1.2 *M* phosphate buffer at pH 6.8. Hybridization is performed in sealed ampoules that are heated first at 100°C for 5 min, then incubated at the hybridization temperature (40 to 50°C) for several h, usually 16. The rate of DNA–RNA hybridization is only two times slower in FA–phosphate medium at 46°C than in aqueous phosphate at 60°C. Degradation of RNA does not occur even after several days incubation at 46°C in FA–phosphate. DNA is highly soluble in formamide (over 100 mg/ml) and all problems concerning high concentration of DNA in case of dilution (if needed) can be easily solved.

Digestion of DNA–rRNA duplexes

Phosphate buffer, SDS, and chelating agents influence the enzymatic activity of nuclease S_1 and can cause erratic results. It is, therefore, important to dialyze all DNA and rRNA preparations exhaustively prior to their assay with endonuclease S_1. Dialysis can be achieved against 0.001 SSC at pH 7.0.

The activity of S_1 nuclease is optimal at pH 4.5 and is 50 times lower at pH 6.4 (Crosa *et al.* 1973). The enzyme is relatively insensitive to the ionic strength of the mixture. In order to obtain optimal conditions for S_1 activity, a small sample of the hybridization mixture (5–10 μl) should be diluted 20–100 times with S_1 nuclease buffer (0.25 *M* NaCl, 1 m*M* Zn acetate, 0.03 *M* Na acetate, pH 4.5). The dilution should be high enough to bring the phosphate concentration to less than 4 m*M* in cases where the reassociation mixture contains phosphate buffer (Maxwell *et al.* 1978).

The action of nuclease S_1 depends on the DNA and rRNA concentration. At very low levels of sheared, denatured DNA (0.01–0, 0.1 μg/ml) only 75% of the nucleic acid is hydrolyzed (Crosa *et al.* 1973). However, if "carrier" DNA is added (calf thymus DNA, sheared and denatured), the enzyme exhibits increased activity and at carrier concentrations of 20 μg/ml, 90–92% of the labelled DNA is digested. The "core" resistant to nuclease S_1 action is about 8–10% and cannot be reduced (Crosa *et al.* 1973). The degree of homoduplex formation is minimally affected by added carrier DNA. Heteroduplex formation is affected and results should be corrected for the S_1 resistant core and normalized to the homoduplex volume.

According to Wittelsberger & Hansen (1977) the reassociation sample, prior to nuclease S_1 treatment, should be treated with pancreatic ribonuclease (50 μg/ml, 2 × SSC, 1 h at 25°C), then dialysed against nuclease S_1 buffer and digested with S_1 enzyme.

Application of samples to DE 81 filter discs

Small aliquots (100 μl) of hybridization mixtures, after digestion with RNase and nuclease S_1, are distributed on DE 81 cellulose filters (24 mm diameter). The filters are then placed in beakers and washed three times with 5% Na_2HPO_4. After two additional washings with distilled water, the filters are dried for 1 h at 80°C. Cooled filters are placed in vials with scintillation fluid and counted (Popoff & Coynault 1980).

The extent of the homologous reaction is calculated by determining the ratio between the average counts remaining on filters loaded with nuclease-treated and nuclease-untreated samples. This value is corrected for the S_1 resistant core and the corrected value is taken as 100%. All other reassociation values, expressed as relative binding ratios, are corrected in the same way and normalized to the homologous reaction (Popoff & Coynault 1980).

Determination of $T_{m(e)}$

The thermal stability of DNA–rRNA duplexes is determined in a way analogous to that applied in the nitrocellulose membrane filter method (see Table 8). Thus hybrids formed during incubation, after RNase and S_1 nuclease treatment, are heated in a water bath in 5°C steps from 60 to 95°C. At 5-min intervals, aliquots (0.3 ml) are withdrawn, diluted with S_1 buffer and the amount of nucleic acids hydrolyzed with nuclease S_1 is determined. The conditions of the enzymatic digestion are the same as described earlier (p. 59) and include the obligatory addition of calf thymus DNA (20 μg/ml) to the reaction mixture as a carrier. The addition of the carrier is necessary because of the very low concentration of the labelled nucleic acid in the reaction mixture (Crosa *et al.* 1973).

TABLE 8

Determination of the $T_{m(e)}$ value in the nuclease S_1–DE 81 method*

DNA–rRNA hybrids after RNase and S_1 nuclease digestion are heated in 1 × SSC solution in a water bath heated in 5°C steps from 60 to 95°C
↓
Aliquots are withdrawn at each of the 5°C steps, diluted with S_1 buffer and "carrier" DNA (denatured, sheared calf thymus DNA, 20 μg/ml) added
↓
Treatment with nuclease S_1 and deposition on DE 81 filter discs
↓
Wash with 5% Na_2HPO_4 and distilled water, dry
↓
Measurement of radioactivity in scintillation counter

*$T_{m(e)}$ is the temperature at which 50% of the DNA–rRNA hybrids become hydrolized by S_1 nuclease under standard conditions.

The thermal stability of the DNA–rRNA hybrids, characterized by the $T_{m(e)}$ value, in the nuclease S_1–DE 81 method is assessed by determining the temperature at which 50% of the reassociated nucleic acids are hydrolyzed by the S_1 nuclease. The results obtained with nuclease S_1 are in full agreement with those from the hydroxyapatite method when applied to DNA–DNA duplexes (Crosa *et al.* 1973). According to Popoff & Coynault (1980) the S_1–DE 81 method is efficient, fast and reproducible in DNA–DNA reassociation experiments. About 50 hybridization tubes can be routinely assayed, in duplicate, by one person during the course of a working day. The method, therefore, allows the determination of the polynucleotide sequence similarities of a large number of strains simultaneously.

Conclusions

The comparison of overall nucleotide sequence similarities has provided valuable data for microbial systematics. Thus, the results of DNA–DNA-pairing experiments have been widely used to circumscribe species and have helped to resolve taxonomic dilemmas in several major bacterial taxa (Brenner 1973; Brenner *et al.* 1978; Palleroni *et al.* 1973; Bradley 1975, 1980; Stackebrandt *et al.* 1981). Similarly the comparative analysis of the evolutionary more conserved ribonucleic acid cistrons (rDNA) is providing an exciting insight into the phylogenic relationships amongst prokaryotes (Woese *et al.* 1976). These modern approaches to systematics combined with the application of numerical phenetic techniques are providing the basis of a natural classification of prokaryotes. However, a note of caution is needed. Molecular methods are a little complicated and one of many technical errors can lead to highly erroneous results. Thus great care is needed at all stages in work of this nature from the choice of bacterial strains used to the final evaluation and interpretation of the results.

References

ANDO, T. 1966 A nuclease specific for heat-denaturated DNA isolated from a product of *Aspergillus oryzae*. *Biochimica et Biophysica Acta* **114**, 158–168.

BRADLEY, S. G. 1975 Significance of nucleic acid hybridization to systematics of actinomycetes. *Advances in Applied Microbiology* **19**, 59–70.

BRADLEY, S. G. 1980 DNA reassociation and base composition. In *Microbiological Classification and Identification* eds. Goodfellow, M. & Board, R. G., pp. 11–26. London: Academic Press.

BRADLEY, S. G. & MORDARSKI, M. 1976 Association of polydeoxyribonucleotides of deoxyribonucleic acids from nocardioform bacteria. In *Biology of the Nocardiae* eds. Goodfellow, M., Brownell, G. H. & Serrano, J. A., pp. 310–336. London: Academic Press.

BRENNER, D. J. 1973 Deoxyribonucleic acid reassociation in the taxonomy of enteric bacteria. *International Journal of Systematic Bacteriology* **23**, 298–307.

BRENNER, D. J., FARMER, J. J., FANNING, G. R., STEIGERWALT, A. G., KLYKKEN, P., WATHEN, H. G., HICKMAN, F. W. & EWING, W. H. 1978 Deoxyribonucleic acid relatedness of *Proteus* and *Providencia* species. *International Journal of Systematic Bacteriology* **28**, 269–282.

BRITTEN, R. J., GRAHAM, D. E. & NEUFELD, B. R. 1974 Analysis of repeating DNA sequences by reassociation. In *Methods in Enzymology* eds. Colowick, S. P. & Kaplan, N. O., Vol. 29, Part E, London: Academic Press.

BURTON, K. 1956 A study of the conditions and mechanisms of the diphenylamine reaction for the colorimetric estimation of deoxyribonucleic acid. *Biochemical Journal* **62**, 315–323.

CASSEY, J. & DAVIDSON, N. 1977 Rates of formation and thermal stabilities of RNA-DNA and DNA:DNA duplexes at high concentrations of formamide. *Nucleic Acids Research* **4**, 1539–1552.

CHEN, P. S., JR., TORIBARA, T. Y. & WARNER, H. 1956 Microdetermination of phosphorus. *Analytical Chemistry* **28**, 1756–1758.

COLLINS, J., FILL, N., JORGENSEN, P. & FRIESEN, J. 1976 Gene cloning of *Escherichia coli* chromosomal genes important in regulation of ribosomal RNA synthesis. *Alfred Benzon Symposium* **IX.**

COX, R. A. 1970 A spectrophotometric study of the secondary structure of ribonucleic acid isolated from the smaller and larger ribosomal subparticles of rabbit reticulocytes. *Biochemical Journal* **117**, 101–118.

CROSA, J. H., BRENNER, D. J. & FALKOW, S. 1973 Use of single-strand specific nuclease for analysis of bacterial and plasmid deoxyribonucleic acid homo- and heteroduplexes. *Journal of Bacteriology* **115**, 904–911.

DE LEY, J. 1971 Hybridization of DNA. In *Methods in Microbiology* eds. Norris, J. R. & Ribbons, D. W., Vol. 5A, pp. 311–329. London: Academic Press.

DE LEY, J. & DE SMEDT, J. 1975 Improvements to the membrane filter method for DNA:rRNA hybridization. *Antonie van Leeuwenhoek* **41**, 287–307.

DE LEY, J. & TIJTGAT, R. 1970 Evaluation of membrane filter methods for DNA-DNA hybridization. *Antonie van Leeuwenhoek* **36**, 461–474.

DENNIS, P. P. & NORDAN, D. H. 1976 Characterization of the hybridization between purified 16S and 23S ribosomal ribonucleic acid and ribosomal deoxyribonucleic acid from *Escherichia coli*. *Journal of Bacteriology* **128**, 28–34.

De Smedt, J. & De Ley, J. 1977 Intra- and intergeneric similarities of *Agrobacterium* ribosomal nucleic acid cistrons. *International Journal of Systematic Bacteriology* **27**, 222–240.
Doi, R. H. & Igarashi, T. 1965 Conservation of ribosomal and messenger ribonucleic acid cistrons in *Bacillus* species. *Journal of Bacteriology* **90**, 384–390.
Doi, R. H. & Igarashi, R. T. 1966 Heterogeneity of the conserved ribosomal ribonucleic acid sequences of *Bacillus subtilis*. *Journal of Bacteriology* **92**, 88–96.
Dubin, D. T. & Günlap, A. 1967 Minor nucleotide composition of ribosomal precursor, and ribosomal ribonucleic acid in *Escherichia coli*. *Biochimica et Biophysica Acta* **134**, 106–123.
Fellner, P. & Ebel, J. P. 1970 Observations on the primary structure of the 23S ribosomal RNA from *E. coli*. *Nature, London* **325**, 1131–1132.
Fellner, P. & Sangar, F. 1968 Sequence analysis of specific areas of the 16S and 23S ribosomal RNAs. *Nature, London* **219**, 236–238.
Fiedler, F., Schäffler, M. J. & Stackebrandt, E. 1981 Biochemical and nucleic acid hybridization studies on *Brevibacterium linens* and related strains. *Archives of Microbiology* **129**, 85–93.
Gillespie, D. & Spiegelman, S. 1965 A quantitative assay for DNA-RNA hybrids with DNA immobilized on a membrane. *Journal of Molecular Biology* **12**, 829–842.
Gillis, M. & De Ley, J. 1975 Determination of the molecular complexity of double-stranded phage genome DNA from initial renaturation rates. The effect of DNA base composition. *Journal of Molecular Biology* **98**, 447–464.
Gillis, M. & De Ley, J. 1980 Intra- and intergeneric similarities of ribosomal ribonucleic acid cistrons of *Acetobacter* and *Gluconobacter*. *International Journal of Systematic Bacteriology* **30**, 7–27.
Gillis, M., De Ley, J. & De Cleene, M. 1970 The determination of molecular weight of bacterial genome DNA from renaturation rates. *European Journal of Biochemistry* **12**, 143–153.
Glötz, C., Zwieb, C., Brimacombe, R., Edwards, K. & Koessel, H. 1981 Secondary structure of the large subunit ribosomal RNA from *Escherichia coli*, *Zea mays* chloroplast and human and mouse mitochondrial ribosomes. *Nucleic Acids Research* **9**, 3287–3306.
Grimont, P. A. D. & Popoff, M. Y. 1980 Use of principal component analysis in interpretation of deoxyribonucleic acid relatedness. *Current Microbiology* **4**, 337–342.
Hall, B. D. & Spiegelman, S. 1961 Sequence complementarity of T2-DNA and T2 specific RNA. *Proceedings of the National Academy of Sciences of the United States of America* **46**, 461–471.
Hartman, K. A. & Thomas, G. J., Jr. 1970 Secondary structure of ribosomal RNA. *Science* **170**, 740–741.
Hill, L. R. 1966 An index to deoxyribonucleic acid base composition of bacterial species. *Journal General Microbiology* **44**, 419–437.
Johnson, J. D. & Horowitz, J. 1971 Characterization of ribosomes and RNAs from *Mycoplasma hominis*. *Biochimica et Biophysica Acta* **247**, 262–279.
Kennell, D. E. 1968 Titration of the gene size on DNA by DNA:rRNA hybridization. II. The *Escherichia coli* chromosome. *Journal of Molecular Biology* **34**, 85–103.
Kersters, K. & De Ley, J. 1980 Classification and identification of bacteria by electrophoresis of their proteins. In *Microbiological Classification and Identification* eds. Goodfellow, M. & Board, R. G., pp. 273–297. London: Academic Press.
Kilpper, R., Buhl, U. & Schleifer, K. H. 1980 Nucleic acid homology studies between *Peptococcus saccharolyticus* and various anaerobic and facultative anaerobic Gram-positive cocci. *FEMS Microbiology Letters* **8**, 205–210.
Kilpper-Bälz, R. & Schleifer, K. H. 1981 DNA:rRNA hybridization studies among staphylococci and some other Gram-positive bacteria. *FEMS Microbiology Letters* **10**, 357–362.
Kourilsky, P., Mercereau, O., Gros, D. & Tremblay, G. 1974 Hybridization on filter with competitor DNA in the liquid phase in a standard and a micro assay. *Biochemie* **56**, 1215–1221.

KURLAND, C. G. 1972 Structure and function of the bacterial ribosome. *Annual Review of Biochemistry* **41**, 377–408.

LEGAULT-DEMARE, J., DESSEAUX, B., HEYMAN, T., SEROR, S. & RESS, G. P. 1967 Studies on hybrid molecules of nucleic acids. I. DNA-DNA hybrids on nitrocellulose filters. *Biochemical and Biophysical Research Communications* **28**, 550–557.

LUND, E., DAHLBERG, I., LINDAHL, L., JASKUNAS, R. S., DENNIS, P. & NOMURA, M. 1976 Transfer RNA genes between 16S and 23S rRNA genes in rRNA transcription units of *E. coli*. *The Cell* **7**, 165–177.

MCCONKEY, E. H. 1967 The fractionation of RNAs by sucrose gradient centrifugation. *Methods in Enzymology* ed. Grossman, L., Vol. 12, pp. 620–634. London: Academic Press.

MCCONKEY, E. H. & DUBIN, D. T. 1966 RNA-DNA hybrid formation with methyl-deficient and mature ribosomal RNA from *Escherichia coli*. *Journal of Molecular Biology* **15**, 102–110.

MARMUR, J. 1961 A procedure for the isolation of deoxyribonucleic acid from microorganisms. *Journal of Molecular Biology* **3**, 208–218.

MARMUR, J. & DOTY, P. 1961 Thermal renaturation of deoxyribonucleic acids. *Journal of Molecular Biology* **3**, 585–594.

MARRS, B. & KAPLAN, S. 1970 23S precursor ribosomal RNA of *Rhodopseudomonas spheroides*. *Journal of Molecular Biology* **49**, 297–317.

MAXWELL, I. H., VAN NESS, I. & HAHN, H. 1978 Assay of DNA:RNA hybrids by S_1 nuclease digestion and absorption to DEAE cellulose filters. *Nucleic Acids Research* **5**, 2033–2038.

MEYS, W. H. & SCHILPEROORT, R. A. 1971 Determination of the amount of DNA on nitrocellulose membrane filters. *FEBS Letters* **12**, 166–168.

MIDGLEY, J. E. M. 1971 Hybridization of microbial RNA and DNA. In *Methods in Microbiology* eds. Norris, J. R. and Ribbons, D. W., Vol. 5A, pp. 331–360. London: Academic Press.

MOORE, R. & MCCARTHY, B. 1967 Comparative study of ribosomal ribonucleic acid cistrons in enterobacteria and mycobacteria. *Journal of Bacteriology* **94**, 1066–1074.

MORDARSKI, M., SZYBA, K., PULVERER, G. & GOODFELLOW, M. 1976 Deoxyribonucleic acid reassociation in the classification of the "*rhodochrous*" complex and allied taxa. *Journal of General Microbiology* **94**, 235–245.

MORDARSKI, M., GOODFELLOW, M., TKACZ, A., PULVERER, G. & SCHAAL, K. P. 1980 Ribosomal ribonucleic acid similarities in the classification of *Rhodococcus* and related taxa. *Journal of General Microbiology* **118**, 313–319.

MORDARSKI, M., TKACZ, A., GOODFELLOW, M., PULVERER, G. & SCHUMACHER-PERDREAU, F. 1981a Ribosomal ribonucleic acid similarities in the classification of *Staphylococcus*. *FEMS Microbiology Letters* **11**, 159–163.

MORDARSKI, M., TKACZ, A., GOODFELLOW, M., SCHAAL, K. P. & PULVERER, G. 1981b Ribosomal ribonucleic acid similarities in the classification of actinomycetes. *Zentralblatt für Bakteriologie, Mikrobiologie und Hygiene. I. Abteilung, Supplement* **11**, 79–95.

NOMURA, M. 1970 Bacterial ribosome. *Bacteriological Reviews* **34**, 228–277.

NOMURA, M. 1972 Assembly of bacterial ribosomes. *Federation Proceedings, Federation of American Societies for Experimental Biology* **31**, 18–20.

NOMURA, M., TRAUB, P. & BECHMANN, H. 1968 Hybrid 30S ribosomal particles reconstituted from components of different bacterial origins. *Nature, London* **219**, 793–799.

NYGAARD, A. P. & HALL, B. D. 1963 A method for detection of RNA:DNA complexes. *Biochemical and Biophysical Research Communications* **12**, 98–104.

NYGAARD, A. P. & HALL, B. D. 1964 Formation and properties of RNA:DNA complexes. *Journal of Molecular Biology* **9**, 125–142.

OOSTRA, B. A., ZENTINGE, B., VAN GOOR, A. L., VAN OOYEN, A. J. & GRUBER, M. 1976 A rapid and sensitive determination method of bacterial rRNA by means of hybridization-competition. *Analytical Biochemistry* **74**, 496–502.

OWEN, N. J. & HILL, L. R. 1978 The estimation of base compositions, base pairing and genome sizes of bacterial deoxyribonucleic acids. In *Identification Methods for Microbiologists* eds. Skinner, F. A. & Lovelock, D. W., pp. 277–296. London: Academic Press.

PACE, N. R. 1973 Structure and synthesis of the ribosomal ribonucleic acid of prokaryotes. *Bacteriological Reviews* **37**, 562–603.

PALLERONI, N. J., KUNISAWA, R., CONTOPOULOU, R. & DOUDOROFF, M. 1973 Nucleic acid homologies in the genus *Pseudomonas*. *International Journal of Systematic Bacteriology* **23**, 333–339.

POLATNICK, J. & BACHRACH, H. L. 1961 Ribonuclease contamination of crystalline deoxyribonuclease, trypsin and of partially purified foot-and-mouth disease virus preparations. *Analytical Biochemistry* **2**, 161–168.

POPOFF, M. & COYNAULT, C. 1980 Use of DEAE cellulose filters in the S_1 nuclease method for bacterial deoxyribonucleic acid hybridization. *Annales Microbiologiae, Institut Pasteur, Paris* **131A**, 151–155.

RYAN, J. L. & MOROWITZ, H. J. 1969 Partial purification of native rRNA and rRNA cistrons from *Mycoplasma* species (KID). *Proceedings of the National Academy of Sciences of the United States of America* **63**, 12828–12829.

SAITO, H. & MIURA, K. I. 1963 Preparation of transforming deoxyribonucleic acid by phenol treatment. *Biochimica et Biophysica Acta* **72**, 619–629.

SALZBERG, S., LEVI, Z., ABOUD, M. & GOLDBERGER, A. 1977 Isolation and characterization of DNA-DNA and DNA-RNA hybrid molecules formed in solution. *Biochemistry* **16**, 25–29.

STACKEBRANDT, E., WUNNER-FÜSSL, B., FOWLER, V. J. & SCHLEIFER, H. K. 1981 Deoxyribonucleic acid homologies and ribosomal nucleic acid similarities among sporeforming members of the order *Actinomycetales*. *International Journal of Systematic Bacteriology* **31**, 420–431.

STANLEY, W. M., JR. & BOCK, R. M. 1965 Isolation and physical properties of the ribosomal ribonucleic acid of *Escherichia coli*. *Biochemistry* **4**, 1302–1311.

SUTTON, W. D. 1971 A crude nuclease preparation suitable for use in DNA reassociation experiments. *Biochimica et Biophysica Acta* **240**, 522–531.

TAYLOR, M. M. & STORCK, R. 1964 Uniqueness of bacterial ribosomes. *Proceedings of the National Academy of Sciences of the United States of America* **52**, 958–965.

TRAUB, P. & NOMURA, M. 1968 Structure and function of *Escherichia coli* ribosomes. V. Reconstruction of functionally active 30S ribosomal particles from RNA and proteins. *Proceedings of the National Academy of Sciences of the United States of America* **59**, 777–84.

VOGELSTEIN, B. & GILLESPIE, D. 1977 RNA-DNA hybridization in solution without reannealing. *Biochemical and Biophysical Research Communications* **75**, 1127–1132.

WITTELSBERGER, S. C. & HANSEN, J. R. 1977 The specificity of S_1 nuclease toward RNA-DNA hybrids as studied using isotopes of phosphorus-32 and phosphorus-33. *Nucleic Acids Research* **4**, 1829–1835.

WITTMANN, H. 1976 Structure, function and evolution of ribosomes. *European Journal of Biochemistry* **61**, 1–13.

WOESE, C. R. 1961 Composition of various ribonucleic acid fractions from microorganisms of different deoxyribonucleic acid composition. *Nature, London* **189**, 920–921.

WOESE, C. R., FOX, G. E., ZABLEN, L., UCHIDA, T., BONEN, L., PECHMAN, K., LEWIS, B. J. & STAHL, D. 1975 Conservation of primary structure in 16S ribosomal rRNA. *Nature, London* **254**, 83–86.

WOESE, C. R., SOGIN, M., STAHL, D., LEWIS, B. J. & BONEN, L. 1976 A comparison of the 16S ribosomal RNAs from mesophilic and thermophilic bacilli: some modifications in the Sanger methods for RNA sequencing. *Journal of Molecular Evolution* **7**, 197–213.

WU, M. & DAVIDSON, N. 1975 Use of gene 32 protein staining of single strand polynucleotides for gene mapping by electron microscopy: application to the $_{\phi}80d_3$ ilvsu 7 system. *Proceedings of the National Academy of Sciences of the United States of America* **72**, 4506–4510.

YANKOFSKY, S. A. & SPIEGELMAN, S. 1962a The identification of the ribosomal RNA cistron by sequence complementarity. I. Specificity of complex formation. *Proceedings of the National Academy of Sciences of the United States of America* **48**, 1069–1078.

YANKOFSKY, S. A. & SPIEGELMAN, S. 1962b The identification of the ribosomal RNA cistron by sequence complementarity. II. Saturation of and competitive interaction at the RNA cistron. *Proceedings of the National Academy of Sciences of the United States of America* **48**, 1466–1472.

YANKOFSKY, S. A. & SPIEGELMAN, S. 1963 Distinct cistrons for the two ribosomal RNA components. *Proceedings of the National Academy of Sciences of the United States of America* **49**, 538–544.

ZWIEB, C., GLÖTZ, C. & BRIMACOMBE, R. 1981 Secondary structure comparisons between small subunit ribosomal RNA molecules from six different species. *Nucleic Acid Research* **9**, 3621–3640.

Current Methods for Estimating DNA Base Composition and Levels of DNA–DNA Hybridization

R. J. OWEN AND D. PITCHER

National Collection of Type Cultures,
Central Public Health Laboratory, London, UK

Since the early 1960s, direct analysis of the chromosomal deoxyribonucleic acid (DNA) complement of the microbial cell has proved a valuable approach to classification. The two features of DNA most widely used for this purpose are the overall base composition (mol % G + C content) and comparative base sequence similarities obtained by *in vitro* DNA–DNA molecular hybridization. In theory, the DNA approach should give a greater understanding of the genetic relationships of a bacterium because it deals directly with the constituent molecules of the genes rather than with the products of their expression, as is the case with most other taxonomic methods. Conventional bacteriological tests measure at best only 10% of the genetic capability of a bacterium (Harvey & Pickett 1980), so the study of DNA is especially appropriate in the case of bacterial groups for which there are few tests to distinguish species. For a more detailed discussion of the uses of genome analysis in bacterial classification the reader is referred to articles by Mandel (1969), Jones & Sneath (1970), Bradley & Enquist (1974), Moore (1974), Bradley (1980) and Sneath (1983). The impact of DNA methods in clinical microbiology is discussed by Brenner (1983).

DNA Extraction and Purification

Cultivation of bacteria

In most cases the culture requirements are selected to give efficient rates of growth and high yields of cells. These criteria depend on the microorganism

CHEMICAL METHODS IN
BACTERIAL SYSTEMATICS

0-12-289675-0

and are outside the scope of this contribution but as a guide, suitable media and culture conditions can be obtained from standard bacteriology texts such as *Bergey's Manual of Determinative Bacteriology* (Buchanan & Gibbons 1974). Also some culture collection catalogues (for example, the *American Type Culture Collection Catalogue of Strains* 1983) are useful sources of media recipes for individual strains. Instances that necessitate a modification of standard media are the addition of supplements to induce sensitivity to lysozyme and defined media to enhance the uptake and incorporation of labelled nucleotide precursors. The yield of bacterial cells can vary considerably from species to species but as a general guide, quantities of the order of 2–4 g wet weight of packed cells are sufficient for DNA extraction.

Cell disruption

Harvested bacterial cells are washed with buffer to remove traces of the culture medium. Saline–EDTA (150 m*M* NaCl, 100 m*M* EDTA, pH 8.0) is suitable for all Gram-negative bacteria. However, in saline–EDTA Gram-positive bacteria may form aggregates of cells that are difficult to disperse and are resistant to subsequent lysozyme or SDS treatment. The problem can usually be avoided by harvesting and washing cells in 20 m*M* Tris–HCl buffer, pH 8.2, which provides a uniform suspension of cells.

The most difficult part of the preparation of DNA from bacteria is breaking the cells to release the intracellular components. The stability of bacterial cell walls to mechanical and chemical action varies considerably even between strains within the same genus. In general, physical methods such as sonication, grinding with glass powder, and pressure cell disruption should be avoided because DNA is very sensitive to shearing forces. Such methods may be unavoidable in some cases; for example, in disrupting streptococci. For these and other physically resistant microorganisms, vortexing with glass beads is often the preferred method (Ranhand 1974).

Gram-negative bacteria are more labile than Gram-positive bacteria and are lysed by treatment with 2% (w/v) sodium dodecyl sulphate (SDS) at 60°C. Gram-positive bacteria are first treated with lysozyme (0.1 mg/ml) at 37°C in 1 *M* sucrose for 30 min, then with SDS at 60°C for 10 min. Difficulties may be encountered with genera such as *Corynebacterium*, *Lactobacillus*, *Mycobacterium* and *Streptococcus*, which have resistant cell walls. Thus modifications to the above method are necessary to render them more susceptible to lysis. The addition of glycine, lysine or threonine to the culture medium impairs cell-wall biosynthesis and results in a weakened cell wall (Yamada & Komagata 1970; Chassy & Guiffreda 1980). The disadvantage of this method is that the particular amino acids in relatively high concentrations are inhibitory to growth and a reduced yield of bacterial cells may result. Pretreatment with antibiotics such as penicillin G or methicillin, which inhibit cell-wall biosynthesis, also render cells more susceptible to lysis by the action of lysozyme (Silvestri & Hill 1965; Komatsu 1979). The sensitivity to lysozyme of *Mycobacterium* species, which have a high content of cell-wall lipids and polysaccharides, may be improved by washing with 2-propanol (Baess 1979). Chassy & Guiffreda (1980) used poly(ethylene glycol) (PEG) to enhance lysis of particularly resistant bacteria. However, the removal of PEG from the lysate is necessary because it may interfere in subsequent analyses.

All cellular debris is removed after lysis by high-speed centrifugation, and the resulting viscous, clear lysate containing a mixture of macromolecules is treated with pronase, which is useful at this stage to break down residual proteins and to inhibit DNase activity. The activity of the pronase is enhanced by the presence of SDS in the lysate (Butterworth 1976).

DNA purification

Most purification protocols for bacterial DNA are based on the method of Marmur (1961) in which chloroform is the deproteinizing agent, or on the methods of Kirby (1964) and Saito & Miura (1963) in which phenol is employed. A general technique for the purification of DNA is given in Table 1. The volumes given should be sufficient for approximately 2–4 g wet weight of cells but can be adjusted proportionally as required. For *in vivo* labelled radioactive DNA (see below) the volumes may require scaling down by a factor of 10 or more.

Other procedures may be necessary in special cases. For example, in *Mycobacterium* a large amount of polysaccharide is often closely associated with DNA and this can be dissociated by selective precipitation of the DNA with cetyltrimethylammonium bromide (Hill *et al.* 1972). The technique has also been used by Meyer & Schleifer (1975) for purifying staphylococcal and micrococcal DNA. Gel filtration methods are an alternative and reliable way of preparing pure DNA. Sepharose 4B (Pharmacia), with an exclusion limit of 5×10^6 daltons, separates DNA from RNA and proteins and if the columns have sufficiently large bed volumes, fairly crude lysates can be used (Zadražil *et al.* 1973). However, it is preferable to carry out a preliminary purification before loading because Sepharose 4B becomes heavily contaminated when crude preparations are run repeatedly. Polysaccharide may be eluted with the DNA but is eliminated by repeated precipitation with 2-propanol (Gibson & Ogden 1979).

Storage of DNA

Samples of DNA are stored for several months dry in sealed tubes at 4°C or in 0.1 × SSC or 20 m*M* NaCl as a concentrated solution at 4 or −20°C

TABLE 1

*Protocol for DNA purification**

(a) Wash the bacterial cells three times with 10 ml of 20 m*M* Tris buffer, pH 8.2; resuspend the pellet in 2 ml of Tris.
(b) Add 4 ml of 1 *M* sucrose and 10 mg of lysozyme; mix well and incubate at 37°C for 1 h or until the solution is clear and viscous.
(c) Add 1 ml of TE buffer (50 m*M* Tris, 50 m*M* EDTA, pH 8.0), 0.1 ml of pronase B (Sigma, 2 mg/ml in 0.15 *M* NaCl, self-digested by incubation at 80°C/2 min and 37°C/30 min).
(d) Add 1 ml of 20% (w/v) SDS (BDH specially purified) in 10 m*M* Tris, 10 m*M* EDTA. Incubate at 37°C for 2 h. If lysis (evident from increased viscosity) is not apparent, heat to 55°C for 5 min, cool, and repeat the process several times.
(e) Add 3 ml of sodium perchlorate [66.5% (w/v) in 10 m*M* Tris, 10 m*M* EDTA] and 10 ml of chloroform–2-pentanol [24:1 (v/v)]. Shake gently for 5–6 min until a stable emulsion forms.
(f) Centrifuge at 28,000 *g* for 20 min.
(g) Remove the upper phase and overlay with 2 volumes of ice-cold 95% (v/v) ethanol.
(h) Collect the DNA by gently stirring with a glass rod. The fibres of DNA will adhere to the rod.
(i) Wash the DNA with 2 changes of ice-cold 95% (v/v) ethanol and dry in a cool air draught.
(j) Dissolve the DNA overnight in 4.6 ml of 0.1 × SSC (SSC = 150 m*M* NaCl, 15 m*M* trisodium citrate).
(k) Add 0.1 ml of RNase (2 mg/ml; BDH bovine pancreatic ribonuclease I in 150 m*M* NaCl treated at 80°C/2 min to inactivate any DNase). Incubate at 37°C for 1 h.
(l) Add 0.2 ml of 10 × SSC and 0.1 ml of pronase. Incubate for a further 2 h.
(m) Add 3 ml of phenol reagent (phenol 100 g, SDS 1 g, TE buffer 1 ml, 8-hydroxyquinoline 0.1 g) and 3 ml of chloroform. Mix gently for 5 min and centrifuge the emulsion at 5000 *g* for 10 min.
(n) Remove the upper layer and wash it with an equal volume of chloroform–2-pentanol (24:1 v/v).
(o) Centrifuge at 5000 *g* for 10 min.
(p) It is not necessary to deproteinize any further if gel filtration is to be carried out at this stage. Otherwise repeat stages (h) to (o) until no white protein deposit is visible at the interface.
(q) Finally, overlay the DNA solution with 2 volumes of 2-ethoxyethanol and spool the DNA by stirring with a glass rod. Wash three times with ethanol and air dry the sample.

*Based on the methods of Marmur (1961), Hansen & Olsen (1978), Gibson & Ogden (1979) and Saito & Miura (1963).

(Crombach 1973). Storage over chloroform at 4°C may prevent microbial contamination but care is needed to ensure that the chloroform is removed if the DNA is to be used in the nick translation (see below) or other enzymatic reactions.

Concentrating DNA

DNA is concentrated by ultrafiltration under nitrogen pressure (0.7 atm) in a stirred cell containing an ultrafilter disc (Amicon or Millipore) with a molecular cutoff point of 1×10^5 daltons. Volumes can be reduced to as little as 0.5 ml in a few min by this method without a change in solvent concentration. An alternative but slower method of concentrating DNA is to dialyse against 60% (w/v) Carbowax 20M [poly(ethylene glycol), 20,000 daltons]. Dialysis is carried out in 10×1.5 cm lengths of glass tubing with a dialysis (visking) membrane stretched over one end and secured with a tight O-ring. Dialysis tubing is prepared according to Maniatis *et al.* (1982). The dialysing surface is placed just below the meniscus of 100 ml of 60% (w/v) Carbowax in $0.1 \times$ SSC or 20 m*M* NaCl in a beaker on a magnetic stirrer at 4°C. The DNA solution is layered above the membrane. If the volume of the DNA sample is small, which is usually the case with radioactive DNA, it can be precipitated with ice-cold ethanol after concentration (Maniatis *et al.* 1982).

The estimation of DNA purity

Before thermal denaturation or hybridization, it is essential to check the purity of the DNA sample. A rapid method is to determine the 230:260 and 280:260 nm absorbance ratios, for which values of 0.45 and 0.515, respectively, indicate a pure DNA (Marmur 1961). The chemical estimation of DNA concentration is described by Maniatis *et al.* (1982), and methods for the estimation of the concentrations of contaminating RNA, protein and polysaccharide are reviewed by Butterworth (1976).

Preparation of radioactively labelled DNA

To make radioactive DNA for hybridization studies, a precursor of the DNA containing one or more isotope atoms is incorporated into the DNA. The aim is to obtain a high specific activity probe with physicochemical properties similar to those of the native unlabelled DNA.

In vivo *labelling*

Various methods exist for the *in vivo* preparation of radioactive DNA. The method of choice is largely dictated by the media requirements and the biochemical and physiological activity of the particular organism. Generally, it is best to use a chemically defined medium containing all the factors necessary for growth as well as a radioactively labelled deoxyribonucleotide precursor. Phosphorus-32, tritium or carbon-14 and occasionally iodine-131 are used to label DNA. When ^3H or ^{14}C are used, aliquots of the radioactive nucleotide such as [^3H]methylthymidine are added to the growing culture during the log phase, and the cells are harvested when the growth is maximal. It is advisable to carry out a trial run to determine the optimum conditions and the amount of labelled nucleotide to produce DNA of suitable specific activity. Experimental details are given by De Ley (1971) and Owen & Hill (1979).

In vitro *labelling*

Purified DNA from any source can be labelled *in vitro* to high specific activity by the nick translation reaction (Rigby *et al.* 1977). In the presence of free nucleotides, one of which is labelled with ^3H or ^{32}P, a single-strand break is introduced into the DNA by the action of DNase I leaving a 3′-hydroxyl terminus. *Escherichia coli* DNA polymerase I in the reaction mixture then catalyses the addition of a nucleotide to the terminus and at the same time a new nick is created on the 5′ side of the nucleotide, leaving another 3′-hydroxyl terminus. As the reaction proceeds, the nick is translated along the DNA strand. The reaction can be stopped after a suitable period by freezing and the DNA is separated from free nucleotides by ethanol precipitation or Sephedex G50 (Pharmacia) filtration as described by Maniatis *et al.* (1982). Commercially prepared kits with all the components (except possibly the label) and instructions for nick translation are available from Amersham International Ltd. (Amersham, UK), Bethesda Research Laboratories (Cambridge, UK), and New England Nuclear (Southampton) UK.

Base Composition Estimation

Certain physical properties of a DNA molecule can be correlated with its base composition (mol % G + C content). The most important are the buoyant density in a caesium chloride (CsCl) gradient and the transition temperature of double-stranded to single-stranded DNA (thermal denaturation temperature). These two techniques offer reliable and accurate ways of estimating G + C content, provided care is taken to ensure reproducibility of the test conditions. De Ley (1970) and Mandel *et al.* (1970) confirmed excellent linear correlations between chemical estimations of base compositions and both thermal denaturation temperature (T_m) and buoyant density in CsCl. No other techniques have gained as wide a general acceptance. However, chemical determinations of base composition by quantitative paper chromatography are essential to provide the basis for the empirical methods above (De Ley & Schell 1963). Differences in the spectra of pure DNA samples can also be used to estimate base composition and, provided standards of known G + C content are used, the method gives rapid and accurate estimations (Ulitzer 1972). Although spectrophotometric methods are not widely used, they are a convenient alternative when equipment for T_m or ρ determinations is not available.

A comprehensive list of G + C contents of bacteria is given by Holländer & Pohl (1980).

Thermal denaturation method

The thermal denaturation method is based on thermal disruption of the hydrogen bonding between the complementary strands of native DNA, thus allowing the separation of the strands. The single-stranded DNA has approximately 40% more absorbance at 260 nm than the double-strand DNA. As the native DNA is heated, the proportion of single-stranded DNA in the solution increases with a concomitant rise in absorbance at 260 nm (hyperchromism). The stability of the hydrogen bonding between guanine and cytosine is greater than between adenine and thymine. Thus the greater the amount of G + C pairing in a DNA molecule, the greater the energy required to dissociate the strands. Eventually, when only single-stranded DNA is present, the absorbance becomes constant. The thermal denaturation value (T_m) is the temperature at 50% hyperchromism, and it is linearly related to G + C content for DNA with a base composition between 30 and 70%.

The method for determining T_m described by Marmur & Doty (1962) is widely used. The DNA solution is adjusted to a concentration of approximately 25 μg/ml ($A_{260} = 0.50$) in a 1-cm silica cell with a tightly fitting stopper and thermistor probe, its tip placed just beneath the fluid surface. The cell is placed in a thermostatically controlled cell holder in a spectrophotometer linked to a pen recorder. The temperature is raised to 25°C and the absorbance of the sample at 260 nm is recorded against a blank containing solvent only. The temperature is increased slowly (0.5°C/min) and the absorbance is monitored continuously. The temperature measurement at 50% hyperchromicity gives the T_m value.

Estimation of base composition

Since G + C content and T_m are linearly related, it is a relatively simple calculation to obtain the base composition of an unknown DNA. The calculations are influenced by the fact that T_m is also logarithmically related to sodium ion concentration (Schildkraut & Lifson 1965), so it is essential to use standard buffers for thermal denaturation experiments. SSC is the most commonly used buffer and the relationship between T_m and G + C content in this buffer is well established (De Ley *et al.* 1970) and given by the equation:

$$(G + C)\% = 2.44T_m - 169.00$$

DNA samples with G + C contents of 51 mol % G + C or over give melting temperatures in SSC exceeding 90°C, so for high G + C content DNA it is preferable to use more dilute buffers such as 0.33 or 0.1 × SSC. Equations for these buffers, derived from the general relationships between T_m, G + C, and [Na^+] (Schildkraut & Lifson 1965; Owen *et al.* 1969; Frank-Kamenetskii 1971) are given by:

$$0.33 \times \text{SSC: } (G + C)\% = 2.24T_m - 135.14$$
$$0.1 \times \text{SSC: } (G + C)\% = 2.08T_m - 106.40$$

TABLE 2

Standardized equations for estimating DNA base composition

Buffer	Equation	Reference T_m* (°C)
SSC	$G + C = 51.0^{\dagger} + 2.44(T_{mx} - T_{mr})$	90.5
0.33 × SSC	$G + C = 51.0 + 2.24(T_{mx} - T_{mr})$	83.4
0.1 × SSC	$G + C = 51.0 + 2.08(T_{mx} - T_{mr})$	74.6

*The *E. coli* NCTC 9001 reference DNA (T_{mr}) data are from Owen *et al.* (1969). Values should be checked to ensure they correspond to within ± 1°C. T_{mx} is the denaturation temperature of the unknown DNA.
†A mean value of 51.0 ± 0.7 (n = 4) for chemical estimates of *E. coli* DNA.

As a check on reproducibility, the G + C content of DNA from a standard strain should be estimated frequently. DNA of *Escherichia coli* NCTC 9001 (ATCC 11775) should be used for this purpose because it gives reproducible, sigmoid-shaped curves. Its T_m in SSC is 90.5°C and its buoyant density is 1.710 (De Ley 1970). To allow interlaboratory comparison of results, data on this strain should be used in the equations for the various concentrations of SSC given in Table 2.

Buoyant density method

When DNA molecules are centrifuged at high speeds in a CsCl solution, they will eventually band at an equilibrium position in the linear CsCl density gradient that forms. The band is positioned where the density of the DNA molecules is equal to the density of the CsCl in that part of the gradient. The density of the DNA at the midpoint of this band is called its buoyant density and is linearly related to the G + C content. This method was fully described by Schildkraut *et al.* (1962) and has been little modified since then.

Measurement of density

Solid CsCl is dissolved in a suitable buffer such as 50 m*M* Tris, pH 7.3, to a density of ~1.700 g cm^{-3}; this is achieved by dissolving 6.3 g of CsCl in a volume of buffer adjusted to a final weight of 10 g. To 1 ml of this solution, 2–3 μg of the DNA to be investigated and a similar amount of marker DNA are added, and the density is readjusted to 1.700 g cm^{-3}. This is simply checked by using a refractometer; a density of 1.700 is equivalent to a refractive index of 1.3994. The solution is transferred to a clean, dry 10-mm single-sector cell, placed in an analytical rotor, and accelerated to 45,000 rev/min in an analytical ultracentrifuge. Equilibrium in the gradient normally takes 20 h at 25°C.

Estimation of G + C content

The buoyant density of the unknown DNA is calculated using the difference between its position and that of the marker DNA. A marker DNA should be selected so that its band does not coincide with that of the test DNA. If the expected G + C content is between 30 and 55 mol %, a reference DNA with a high G + C content is needed; for example, DNA of *Micrococcus luteus* NCTC 2665 (ATCC 4698) with a density of 1.731g cm^{-3} (72% G + C). If the expected G + C content is likely to be higher than 55 mol %, then the marker DNA should have a lower density: for example, DNA of *E. coli* NCTC 9001 (ATCC 11775) with density of 1.710 g cm^{-3} (51% G + C). The base composition of the unknown is then calculated from the equation relating buoyant density to % G + C derived by Schildkraut *et al.* (1962), checked by De Ley (1970) and expressed as:

$$\rho = 1.660 + 0.098(G + C)$$

Details of the calculation of buoyant density are usually available from the centrifuge manufacturer or from a standard text on centrifugation (Bowen 1970).

Comments on method

The method has an accuracy and reproducibility comparable to that of the T_m method (De Ley 1970). The main disadvantage is the cost of purchasing and maintaining an analytical ultracentrifuge. However, if access to a machine is possible, the method offers several attractive advantages:

(1) Only small amounts of DNA are required (1–2 μg), and the DNA does not necessarily have to be highly purified. Pure DNA is preferable because the presence of contaminating macromolecules makes it difficult to estimate DNA concentrations and causes an uneven background on the traces.
(2) Up to five DNA samples can be run simultaneously.
(3) Extra-chromosomal DNA that differs in density from the main band is readily observable either as a distinct separate band or as a "shoulder" to the main band. Its G + C content and amount relative to the main band can be calculated.
(4) The height and width of the DNA band is a sensitive measure of the intermolecular heterogeneity of the DNA. Broad bands would suggest that the DNA was degraded and unsuitable for G + C estimation. It should be noted that abnormal density values are obtained if the DNA contains unusual bases, although these are thought to be rare in bacterial DNA.

DNA–DNA Molecular Hybridization

The measurement of the extent to which single-strand DNA molecules from two different microorganisms hybridize to form double-stranded DNA molecules *in vitro* is used widely in taxonomy. It is not possible to compare the nucleotide sequences of complete bacterial genomes because the largest sequence so far reported, that of bacteriophage λ (48,502 base pairs), is smaller by at least one order of magnitude than a typical bacterial genome. Thus comparison of nucleotide sequences by molecular hybridization is at present the only direct means of estimating the percentage of complementary bases present in two different genomes.

The agar gel method of Bolton & McCarthy (1962) was the first attempt at a general method of quantitatively measuring interstrain DNA–DNA hybridization. The four main methods of DNA–DNA hybridization currently favoured by molecular taxonomists fall broadly into two groups: methods in which DNA is fixed to a macroporous support and free-solution methods. There is no single method that is universally favoured since each has inherent advantages and disadvantages. Temperature, ionic concentration, DNA concentration, fragment size and incubation period have an effect on the rate and level of DNA–DNA hybridization (Marmur & Doty 1961; Britten &

TABLE 3

Temperatures for DNA–DNA hybridization

DNA G + C content (mol %)	Hybridization temperature (°C)*,†		
	T_{OR}	T_s	T_{ns}
30	62.3	72.3	52.3
35	64.9	74.9	54.9
40	67.4	77.4	57.4
45	70.0	80.0	60.0
50	72.5	82.5	62.5
55	75.1	85.1	65.1
60	77.6	87.6	67.6
65	80.2	90.2	70.2
70	82.7	92.7	72.7
75	85.3	95.3	75.3

*Hybridization temperature in 2 × SSC (390 m*M* – [Na^+]) buffer is calculated from T_{OR} = [0.51 × (G + C)%] + 47.0 (Gillis *et al.* 1970).

†Abbreviations used: T_{OR}, optimum temperature; T_s, stringent reassociation temperature; T_{ns}, nonstringent reassociation temperature.

TABLE 4

DNA–DNA hybridization methods and selected examples of bacterial genera investigated

Genus		Solvent*	Reference
Membrane–filter method			
Arthrobacter		3 × SSC/25%FM	Stackebrandt & Fiedler (1979)
Corynebacterium		3 × SSC/30%FM	Döpfer *et al.* (1982)
Actinoplanes		2 × SSC/35%FM	Stackebrandt *et al.* (1981)
Gardnerella		2 × SSC/50%FM	Greenwood & Pickett (1980)
Bacteroides		2 × SSC/30%DMSO	Coykendall *et al.* (1980)
Nocardia		2 × SSC/30%DMSO	Mordarski *et al.* (1977)
Spiroplasma		2 × SSC	Rahimian & Gumph (1980)
Bacillus		2 × SSC	O'Donnell *et al.* (1980)
Hydroxyapatite batch procedure			
Staphylococcus		280 m*M* PB	Kloos & Wolfshohl (1979)
Enterobacter		280 m*M* PB	Farmer *et al.* (1980)
Legionella		280 m*M* PB	Brenner *et al.* (1978)
Flavobacterium		280 m*M* PB	Owen & Snell (1976)
Bordetella		280 m*M* PB	Kloos *et al.* (1981)
Rhizobium		280 m*M* PB	Jarvis *et al.* (1980)
Escherichia		280 m*M* PB	Brenner *et al.* (1982)
S1 endonuclease assay			
Campylobacter		420 m*M* NaCl	Belland & Trust (1982)
		420 m*M* NaCl	Leaper & Owen (1982)
Pseudomonas		420 m*M* NaCl	Owen & Jackman (1982)
Bacteroides		420 m*M* NaCl	Holdeman & Johnson (1982)
Cedecea		420 m*M* NaCl	Grimont *et al.* (1978)
Aeromonas		420 m*M* NaCl	MacInnes *et al.* (1979)
Flavobacterium		420 m*M* NaCl	Levine *et al.* (1980)
Peptococcus		420 m*M* NaCl	Crosa *et al.* (1979)
Serratia		420 m*M* NaCl	Grimont *et al.* (1982)
Micrococcus		420 m*M* NaCl	Brooks *et al.* (1980)
Spectrophotometric method			
Flavobacterium *Cytophaga*	RR-AP	2 × SSC	Callies & Mannheim (1980)
Pseudomonas	RR-AP	510 m*M* NaCl	Owen *et al.* (1978)
Mycobacterium	RR-COT	3 × SSC/25%FM	Baess & Bentzon (1978)
	RR-COT	6 × SSC/25%FM	Imaeda *et al.* (1982)
Nocardia	RR-COT	6 × SSC/25%FM	Bradley *et al.* (1978)
Alcaligenes	RR-COT RR-AP	0.5 *M* PB/25%FM	Auling *et al.* (1980)

*Abbreviations used: SSC, standard saline citrate; FM, formamide; DMSO, dimethyl sulphoxide; PB, equimolar phosphate buffer; RR-COT, renaturation rate, COT method; RR-AP, renaturation rate, apparent rate method.

Kohne 1966; Wetmur 1976). Most of these variables are fixed by the particular hybridization method used except for the temperature, which is dependent on the base composition of the DNA and on the degree of stringency required. Stringent conditions (T_s) favour base pairing between well-matched nucleotide sequences, whereas less stringent (T_{ns}) conditions permit the reassociation of partially matched base sequences (Johnson & Ordal 1968; De Ley *et al.* 1973). Table 3 lists the optimum, stringent and nonstringent temperatures for hybridizing DNA with G + C contents between 30 and 75 mol %. Maximum rates of reassociation occur approximately 25°C below the denaturation temperature (T_m) for a given buffer. The stringent and nonstringent temperature conditions are not precisely defined but may be calculated by assuming $T_s = T_m - 15°C$, and $T_{ns} = T_m - 35°C$. Under T_s conditions, strains belonging to the same species should hybridize to over 70%. Selected examples of recent applications of DNA–DNA hybridization methods in bacterial classification are listed in Table 4. The list is not exhaustive but readers are referred to these papers for detailed information on how the methods, in particular the solvent conditions, have been modified to investigate different groups of bacteria.

Membrane filter method

First introduced by Gillespie & Spiegelman (1965) to measure hybridization between DNA and RNA, nitrocellulose membrane filters are used extensively to immobilize single-stranded, high-molecular-weight DNA. The single-stranded DNA is retained on the membrane after filtration and is "fixed" by baking at 80°C in a vacuum oven for 2 h. The filters are then ready for hybridization with labelled RNA but not with DNA because single-stranded DNA fragments bind nonspecifically to the filters as well as hybridizing specifically to complementary "fixed" DNA sequences. The problem of nonspecific binding of DNA was eliminated by a modification introduced by Denhardt (1966), who found that treatment of DNA-containing membranes with a solution containing 0.02% each of ficoll, poly(vinyl pyrrolidone) and bovine serum albumin blocked nonspecific binding sites on the nitrocellulose. The polymer–albumin mixture is referred to as Denhardt's reagent. De Ley & Tijtgat (1970) evaluated three modifications of the nitrocellulose membrane filter technique for DNA–DNA hybridization and they concluded that only the dimethyl sulphoxide (DMSO) filter method of Legault-Démare *et al.* (1967) could be recommended for the entire temperature and G + C range. It was found that 30% (v/v) DMSO must be used or much of the DNA was spontaneously released from the filters during incubation at high temperatures, and the heterologous DNA counts were unreliable. In 2 × SSC plus 50% DMSO the melting temperature of DNA is 19°C lower than in 2 × SSC alone.

The effective hybridization temperature can also be reduced by including formamide in the mixture (Bonner *et al.* 1967; McConaughy *et al.* 1969). For each 1% (v/v) formamide present, the temperature is reduced by 0.9°C. The use of formamide has proved extremely successful and is favoured in many qualitative and quantitative systems that combine the Gillespie and Spiegelman membrane filter "fixed" DNA method with preincubation in Denhardt's reagent. There are many variations to the technique in the literature but the basic stages are outlined below in Table 5.

Other recent developments in hybridization technology have provided alternatives to the conventional nitrocellulose membranes as macroporous supports for DNA, notably charge-modified nylon membranes (Pall Biodyne; Bio Rad Zeta-Probe) and a modified form of cellulose paper in which the cellulose is substituted by aminobenzyloxymethyl (ABM) groups (Schleicher & Schuell Transa-bind). The groups are activated before use by diazotizing to form diazobenzyloxymethylcellulose (DBM-cellulose), which covalently couples single-stranded nucleic acids. Grunstein & Hogness (1975) described an important adaptation of the nitrocellulose membrane filter technique, called colony hybridization, to detect specific plasmid DNA sequences in colonies formed on the filters. The technique has been successfully used to screen for homologous sequences in recombinant DNA research and has interesting possibilities for bacterial identification. Re-

TABLE 5

Procedure for DNA–DNA filter hybridization

(a) Soak the filter membranes (e.g. Millipore, HAWP) in 6 × SSC for 3–4 h.
(b) Denature DNA in 0.1 × SSC by heating for 10 min in a boiling water bath, chill quickly on ice and make up to 6 × SSC.
(c) Pass solutions containing 100 μg of denatured DNA slowly through the filters, wash with 6 × SSC and dry overnight at room temperature, then dry at 80°C for 2 h in a vacuum oven.
(d) Sonicate reference and radioactive DNA at 20–25 kHz for 2 min to fragment, and denature by heating as in (b) above.
(e) DNA-loaded filters are placed in 0.2 ml of 50% (v/v) formamide plus 2 × SSC, pH 7.0, and 1 μg of the appropriate reference labelled DNA is added.
(f) Mixtures are incubated at 40°C (stringent conditions for a 50% G + C DNA) for 20 h. Tubes may be overlayed with mineral oil to prevent evaporation.
(g) Transfer filter to 50 ml of 2 × SSC plus 50% (v/v) formamide and wash with stirring for 30 min at room temperature.
(h) Repeat with six to eight washes in small volumes of 2 × SSC.
(i) Dry filters and count radioactivity.
(j) The percent relative hybridization is expressed as the percentage of labelled DNA reassociated with heterologous DNA, compared with that reassociated with the homologous DNA (100%). Blank filters are included in the above procedure to check for nonspecific binding.

cently, Hodgson & Roberts (1983) used the technique to identify rhizobia in mixed bacterial populations and to identify specific *Rhizobium* strains in the presence of other rhizobia.

Hydroxyapatite batch procedure

The general principle of free-solution hybridization is that the unlabelled DNA is in high concentrations, whereas the labelled DNA is in significantly smaller amounts. A high ratio of unlabelled to labelled DNA of at least 2000:1 is essential to ensure that self-reassociation of the label is kept to a minimum. The approach has two advantages. First, it avoids having to "fix" DNA to some kind of support with the inherent problem of leaching at high temperatures. Second, hybridization rates are one order of magnitude faster than comparable reactions done on macroporous supports (Bünemann 1982), presumably because the unlabelled DNA is more accessible to the probe when in free solution. Hydroxyapatite (HAP), a form of calcium phosphate, is a compound to which nucleic acids bind electrostatically in phosphate buffers and, depending on the phosphate molarity, either double- or single-stranded DNA can be eluted or bound. Because of this property, HAP is used widely in nucleic acid research for chromatographic separation. Brenner *et al.* (1969) described a simple development to allow batch processing of hybridization mixtures. After free-solution hybridization is completed, the mixture is added directly to a suspension of HAP in 140 m*M* sodium phosphate and at that concentration double- but not single-stranded DNA molecules are bound. The unbound DNA fragments are removed by thorough washing with 140 m*M* sodium phosphate and then the hybridized DNA molecules are released from the HAP by increasing the phosphate concentration to 400 m*M*. The fractions containing the single- and double-stranded DNA molecules are collected and the amount of radioactivity in each is determined. The method is described in Table 6. In the method it is essential that hybridization values are corrected for zero-time binding (DNA that binds irreversibly to the HAP) and that temperatures are maintained above 55°C, otherwise inaccuracies due to nonspecific binding of DNA to the HAP will occur. A simple modification was described by Lachance (1980) in which a microcolumn containing a small quantity of HAP under controlled temperature was used to separate hybrid DNA.

The hydroxyapatite batch procedure has been used extensively by Brenner and co-workers in DNA–DNA hybridization studies of enteric bacteria, and examples of the genera examined by them and other workers are listed in Table 4. Only phosphate buffer systems are used with HAP. High incubation temperatures are possible, although well-sealed containers must be used to prevent evaporation of the hybridization mixtures. It is often the practice

TABLE 6

DNA–DNA hybridization by the hydroxyapatite batch procedure

(a) Shear the DNA by sonication as described in Table 5.
(b) Mix 150–200 μg of unlabelled, sheared DNA with 0.05 μg of sheared, labelled DNA. Denature DNA mixture at 95°C for 10 min and adjust the buffer concentration to 280 m*M* with 500 m*M* sodium phosphate buffer.
(c) Incubate the mixture at the required temperature for 20 h.
(d) Adjust mixture to 140 m*M* sodium phosphate, and add to 10 ml of HAP (Bio Rad DNA grade Bio-Gel HTP). Mix well and equilibrate at 65°C for 10 min.
(e) Increase volume to 20 ml with 140 m*M* sodium phosphate, mix, and centrifuge briefly at 60°C. Collect and pool supernatants.
(f) Repeat with two further washes of 140 m*M* phosphate at 65°C to remove all single-strand DNA. Pool the three supernatants.
(g) Wash three times with 20-ml portions of 0.4 *M* sodium phosphate at 65°C to elute double-stranded DNA from the HAP. Supernatants are collected by centrifugation at 60°C and pooled.
(h) Add yeast RNA (200 μg) as carrier to each pooled supernatant and precipitate DNA with cold 5% (v/v) trichloracetic acid (TCA).
(i) Collect precipitates by suction filtration on Whatman GF/B discs, wash, dry thoroughly and determine radioactivity.
(j) The percentage hybridization ($\%H$) is calculated from the equation $H = 100X/(X + Y)$, where X is the number of counts recovered from the 400-m*M* washes and Y is the number of counts recovered from the 140-m*M* washes. The heterologous hybridization values are expressed relative to the homologous value for the labelled probe DNA.

to do hybridizations at 55 or 60°C (nonstringent conditions) and to repeat selected mixtures at 70 or 80°C (stringent conditions).

Endonuclease assay technique

An alternative approach to the analysis of DNA–DNA hybrids formed in homogeneous systems is to assay the amount of double-stranded DNA by the endonuclease technique (Sutton 1971; Crosa *et al.* 1973). The principle of the method is to digest the single-strand DNA remaining in the hybridization mixture with a single-strand specific endonuclease. The double-stranded DNA is then separated from the endonuclease digestion products either by trichloracetic acid precipitation (S1–TCA method) or by collecting on Whatman DE81 filters (S1–DE81 method). The latter method, described by Popoff & Coynault (1980), is based on the fact that free nucleotides, but not DNA, are eluted from DEAE-cellulose with a phosphate buffer (Barth & Grinter 1975; Salzberg *et al.* 1977). It has the advantage of being more reproducible than the TCA method. The procedure used in our laboratory is outlined in Table 7.

At least four determinations should be made for each DNA pair hybri-

TABLE 7

Protocol for DNA–DNA hybridization by the S1 endonuclease assay

(a) 1 μg of labelled DNA is mixed with 150 μg of unlabelled DNA in 0.3-ml-vol Reactivials (Pierce & Warriner, Chester, UK), then heat denature at 90°C for 10 min, and cool in an ice bath.
(b) Adjust sodium concentration to 420 m*M* in a final volume of 200 μl by adding 10 μl of 4.2 *M* NaCl. Incubate at the optimum temperature for 18 h.
(c) Add 200 μl of enzyme assay buffer (0.62 m*M* sodium acetate buffer, pH 4.8, and 2.05 m*M* zinc sulphate), 20 mg/ml of DNA in 10 m*M* Tris–HCl buffer, pH 7.5), and 50 units of S1 endonuclease (Sigma). Mix thoroughly.
(d) Remove a zero-time sample (50 μl) and incubate the mixture for 20 min at 60°C.
(e) Stop reaction by chilling in an ice bath, take 70 μl of sample and transfer to dry 2.4-cm DE81 filter disc in a scintillation vial. Wash each disc twice with 5-ml aliquots of 5% (w/v) Na_2HPO_4 and twice with 5-ml portions of distilled water.
(f) Dry discs thoroughly at 60°C and determine radioactivity in a scintillation counter.

dized, and each set of hybridizations should include a negative control to check for self-reassociation of the labelled DNA, which should not exceed 5%. The method yields an estimate of the total amount of labelled duplex DNA, in contrast to the membrane filter and HAP methods, which may include noncomplementary regions with labelled single-strand loops and tails.

The percentage DNA–DNA hybridization is calculated as follows. If X is the amount of radioactivity (cpm) in the mixture prior to S1 digestion, and Y is the mean cpm of two separate aliquots from the hybridization mixture after S1 treatment, then the apparent absolute hybridization (%H) is given by $H = [100\ Y/X]$. A corrected value for each of the heterologous mixtures is then expressed relative to the corrected absolute hybridization value of the homologous reference DNA. Both modifications (S1–TCA, S1–DE81) of the S1 endonuclease assay are used and examples of the genera examined are given in Table 4. Neither formamide nor DMSO is used in this method, presumably because these compounds affect the activity of the endonuclease.

Spectrophotometric method

In the spectrophotometric method, which is useful when radioactive DNA is not available, the amount of hybridization is determined spectrophotometrically. The principle is that the rate of reassociation of two single-strand DNA samples with many noncomplementary base sequences will be slower than the rate of the homologous mixture or a heterologous mixture of two complementary DNAs. As DNA absorbance decreases during reassociation, the

TABLE 8

Protocol for DNA–DNA hybridization from renaturation rates

(a) Prepare two sheared, DNA samples (A and B) at an accurately known concentration between 60 and 90 μg ml^{-1} in 1.8 ml of 0.1 × SSC buffer and a third sample (C) containing an equimolar mixture of DNA samples A and B at the same total DNA concentrations as A or B.
(b) Filter through a nitrocellulose membrane (0.2 μm pore size).
(c) Denature the DNA at 95°C for 10 min in stoppered glass tubes.
(d) Cool rapidly on ice and adjust the sodium ion concentration to 510 m*M* NaCl with 0.2 ml of 5 *M* NaCl. Mix well.
(e) Heat to 60°C and quickly transfer equal volumes of A, B and C with a heated pipette into preheated, semimicro silica cells (1-cm light path) in the spectrophotometer cell compartment. The cells must then be tightly stoppered to avoid evaporation.
(f) The reassociation of the DNA samples at the optimum temperature is monitored simultaneously in all three samples by the decrease in absorbance at 260 or 280 nm.

changes can be monitored at 260 nm. Both DNAs are sheared into lengths with molecular weights of about 2–5 × 10^5 and incubated at the optimum temperature for reassociation. The procedures used in taxonomic studies to measure rates of DNA reassociation are based either on the methods of De Ley *et al.* (1970) or Seidler & Mandel (1971). A comprehensive review of the theoretical aspects of hybridization kinetics of nucleic acids is given by Wetmur (1976). A convenient procedure for measuring the reassociation of DNA fragments is given in Table 8. The reassociation results are then handled in one of two ways for the estimation of percentage hybridization.

Initial renaturation rate method

The method described by De Ley *et al.* (1970) records hybridization for a period of 40 min, from which an initial renaturation rate ($\acute{V}$) is calculated. The rate is expressed as decrease in absorbance at 260 nm/min which is the slope of the reassociation curve obtained by regression analysis. The percentage hybridization (%*H*) is then calculated from the formula:

$$\%H = 100[4\acute{V}\mathrm{M} - (\acute{V}\mathrm{A} + \acute{V}\mathrm{B})]/2\sqrt{\acute{V}\mathrm{A} \times \acute{V}\mathrm{B}}$$

where $\acute{V}$A, $\acute{V}$B and $\acute{V}$M are the initial renaturation rates of samples A, B and the mixture M, respectively. Modified forms of this equation can be used to express the hybridizations relative to A or relative to B if the genomes have significantly different sizes (De Ley *et al.* 1970). Some examples of genera studied by this method are given in Table 4. The method was recently evaluated by Huss *et al.* (1983), who confirmed it was excellent for hybridizations above 30%.

Cot ½ method

An alternative method of calculation is based on the use of the Cot ½ value, which is defined as the concentration of DNA, in moles of nucleotide per litre, multiplied by the time in seconds required for 50% reassociation (Britten & Kohne 1966). If two bacterial strains have identical DNAs, the Cot ½ of the mixture of their DNAs is the same as the Cot ½ of each single DNA. If two bacterial DNAs have no mutual sequences, the Cot ½ of the mixture will be the sum of their Cot ½ values. The percent hybridization (%*H*) can be calculated from the expression:

$$\%H = 200\,(\text{Cot}\,\tfrac{1}{2}\,\text{A} + \text{Cot}\,\tfrac{1}{2}\,\text{B} - \text{Cot}\,\tfrac{1}{2}\,\text{M})\,/\,(\text{Cot}\,\tfrac{1}{2}\,\text{A} + \text{Cot}\,\tfrac{1}{2}\,\text{B})$$

The equation was developed by Bradley (1972) and is similar to that of Seidler & Mandel (1971). The Cot ½ values are calculated from the standard second-order rate plots of each sample (Wetmur 1976) or from Cot plots (Britten & Kohne 1968).

Comparisons of known sequences

It is not yet possible to sequence complete bacterial genomes but sequences are known for specific regions of some species, notably *E. coli*. At present, the systematic application of such data is limited with the exception of studies on ribosomal RNA, which has proved extremely useful in the investigation of evolutionary relationships amongst bacterial genera and families (Fox *et al.* 1980). The approach is facilitated by the availability of various computer programs to store and analyse nucleic acid sequence data (Söll & Roberts 1982). One feature of the programs relevant to taxonomy is the facility by which sequences are aligned and levels of homology calculated. Information on sequence divergence in terms of base substitution, deletion and insertion at individual sites can be obtained from computer-generated dot matrices, which are pair-wise comparisons of sequences in a graphical form (Novotny 1982).

Theoretical estimation

In 1969, De Ley presented a mathematical method for calculating the upper limit of DNA homology between any two bacteria, which depended on the mean base composition, the molecular weight and the compositional nucleotide distribution of chromosomal DNA. The method gives the degree of overlapping in G + C composition between two bacterial DNAs. If the compositional distribution curves of two bacterial DNAs do not overlap, it can be predicted that they will not display sequence homology. The relation between maximum possible percent DNA homology and G + C difference is linear for differences between 0 and 11% G + C but deviates for larger G + C differences. If it is assumed that the population of DNA molecules or fragments from each bacterium are normally distributed with respect to their mean base composition, it is possible to estimate theoretical maximum hybridization by assuming that a 1% G + C content difference in the mean value is equivalent to a decrease in common base sequence of 9%; a similar figure was previously obtained by Sueoka (1961). Data from our literature survey of several thousand cross hybridizations show that: (1) complete or nearly complete levels of hybridization occur only between strains that have DNA with almost identical mean G + C contents; (2) strains with differences in mean G + C contents of 9% or more do not hybridize; and (3) observed intermediate hybridization levels rarely exceed the theoretical maximum values and may even be significantly lower.

Analysis of DNA Hybrids

The accuracy of base-pair formation within DNA–DNA hybrids is checked by thermal denaturation analysis. This analysis is carried out either on hybrids formed on filters or in free solution (HAP and S1 endonuclease methods) but not by the spectrophotometric method because there is no means of separating the homologous from the heterologous hybrids (Crombach 1974). Hybrid analysis involves stepwise thermal denaturation and the construction of a denaturation profile. The hybrid thermal stability is expressed as the denaturation temperature [$T_{m(e)}$], which is the temperature at which 50% of the hybrid denatures. The thermal stability of a heterologous hybrid is compared with that of the homologous hybrid and their difference [$\Delta T_{m(e)}$] is given by the relationship

$$\Delta T_{m(e)} = \text{homologous } T_{m(e)} - \text{heterologous } T_{m(e)}$$

To determine the thermal stability, hybrids are denatured by temperature increments of 3–5°C. The conditions and methods depend on the hybridization technique, and exact methods are given by Brenner *et al.* (1969) (HAP technique), De Ley *et al.* (1973) (MF method), and Crosa *et al.* (1973) (S1 endonuclease method). In the membrane filter technique, for example, the filters are removed after hybridization and washed in 2 × SSC plus 50% DMSO. They are then incubated for 15 min in each of a series of small flasks containing 2 ml of 2 × SSC plus 50% DMSO in a number of water baths of increasing temperatures in 3°C steps from the hybridization temperature up to 90°C. The amount of radioactivity removed at each temperature is determined.

The relationship between percent DNA hybridization at about T_m − 25°C and hybrid stability (De Ley *et al.* 1973; Johnson 1973; Grimont *et al.* 1980) is linear between 35 and 100% relatedness but nonlinear at values lower than 35%. As would be expected, the accuracy of base pairing within hybrids decreases as the overall level of similarity between two organisms decreases. The thermal stability can also be used as a means of estimating base sequence divergence within related sequences because a change in 1% of the base pairs decreases $T_{m(e)}$ by about 1.6°C.

Reliability and Accuracy

A key factor in the assessment of DNA results is the level of agreement between different methods. This is less of a problem with G + C determination, which can be accurately checked by use of a standard reference DNA, than with hybridization because of the considerable choice in methods, conditions and reference strains available. Even so, comparisons between the DNA–DNA hybridization methods described above show surprisingly good agreement of within at least 5% in most cases (De Ley *et al.* 1970; Seidler & Mandel 1971; Gibbins & Gregory 1972; Bradley 1973; Baess & Bentson 1978; Garvie 1978; Coykendall & Munzenmaier 1978). In a detailed study, Grimont *et al.* (1980) demonstrated that the S1–DE81 method was more reproducible than the HA method. Hybridization by both methods agreed well at the stringent temperature but deviated under nonstringent conditions: for example, at 50% hybridization with the S1–DE81 method, the corresponding HA method estimate was 70%. The above workers also observed that $T_{m(e)}$ determination was subject to less variation. Similarly, a nonlinear relationship existed between hybridization data obtained with S1–TCA and HA procedures at both optimal and stringent conditions. In a recent analysis of the spectrophotometric method of DNA hybridization, Huss *et al.* (1983) found excellent agreement with the membrane filter technique. The main disadvantage of the optical method is a high background of 25–30%, so low levels of relatedness cannot be determined with accuracy.

Concluding Comments

There are a variety of well-tested methods now available for estimating base composition and DNA–DNA hybridization, but even so their usefulness is to some extent limited by their inherent laboriousness. The resultant analysis and interpretation of results can prove difficult. Incomplete data sets based on a few selected reference strains and problems associated with comparing data from different methods mean that hybridization results cannot be analysed by the sophisticated statistical techniques now commonly used to handle other forms of taxonomic information. Analysis of incomplete matrices is discussed by Sneath (1983). There is scope to develop methods for screening larger numbers of strains; this would assist in selecting suitable reference strains for more detailed hybridization analysis. Simpler, speedier methods would also find application in routine identification if specific DNA probes were available for hybridization.

Once DNA data have been analysed, the observations must be reconciled with the results of conventional phenotypic testing and perhaps with data from other chemotaxonomic approaches. These aspects are outside the scope of this contribution but it is relevant to consider the extent that DNA characteristics correlate with levels in the taxonomic hierarchy. Experience has shown that DNA–DNA hybridization data are primarily of importance in investigating relationships within and between species, although the statistical significance of species based on such data has not been rigorously tested. Some basic guidelines for interpreting DNA characteristics are given in Table 9. In general, it appears impossible to be categorical about the congruence between DNA–DNA hybridization and levels in the taxonomic

TABLE 9

Guidelines for interpreting DNA data in taxonomy

(a) Differences in mean G + C content of 10% or more indicate two bacteria are unrelated and that their DNAs will not hybridize.

(b) If two bacteria have identical G + C contents, DNA sequence similarities of 70% or more, and low base sequence divergence within the hybrids, they are closely related and belong to the same species.

(c) If two bacteria have 25% or less DNA sequence similarity and significant amounts of base sequence divergence within hybrids, they almost certainly do not belong in the same species, even though they may have similar G + C contents. Differences in G + C content of more than 2% would indicate a separate species.

(d) Closely related bacteria give high levels of DNA hybridization irrespective of stringent or nonstringent conditions. Likewise, unrelated bacteria give low hybridization under both conditions.

(e) Observed hybridization levels should not exceed the theoretical maximum estimates obtained from mean G + C content differences. Each G + C difference of 1% corresponds to a 9% decrease in maximum hybridization.

(f) If two bacteria have 5% or less DNA sequence similarity, they probably belong in different genera even though they may have similar G + C contents. Differences in G + C content of more than 5% would also suggest a different genus.

(g) The classification of two strains with DNA similarities between 30 and 65% is questionable. They may be viewed as DNA groups within a species or as closely related species depending on (1) the extent they are related under stringent and nonstringent conditions, (2) base sequence divergence in hybrids, (3) differences in G + C contents, and (4) obvious correlations with other taxonomic characters.

hierarchy, simply because the conventional criteria for defining species vary considerably from group to group. Investigations into the agreement between DNA similarities and similarities based on numerical phenetic analyses and other kinds of chemotaxonomic data are clearly needed to allow future progress in the interpretation of DNA data.

References

AULING, G., DETTBRENNER, M., MAARZAHL, M., NOKHAL, T. & REH, M. 1980 Deoxyribonucleic acid relationships among hydrogen-oxidizing strains of the genera *Pseudomonas, Alcaligenes* and *Paracoccus. International Journal of Systematic Bacteriology* **30**, 123–128.

BAESS, I. 1979 Deoxyribonucleic acid relatedness amongst species of slowly-growing mycobacteria. *Acta Pathologica et Microbiologica Scandinavica, Section B* **87**, 221–226.

BAESS, I. & BENTZON, M. W. 1978 Deoxyribonucleic acid hybridization between different species of mycobacteria. *Acta Pathologica et Microbiologica Scandinavica* **86**, 71–76.

BARTH, P. T. & GRINTER, N. J. 1975 Assay of deoxyribonucleic acid homology using a single-strand-specific nuclease at 75°C. *Journal of Bacteriology* **121**, 434–441.

BELLAND, R. J. & TRUST, T. J. 1982 Deoxyribonucleic acid sequence relatedness between thermophilic members of the genus *Campylobacter. Journal of General Microbiology* **128**, 2515–2522.

BOLTON, E. T. & MCCARTHY, B. J. 1962 A general method for the isolation of RNA complementary to DNA. *Proceedings of the National Academy of Sciences of the United States of America* **48**, 1390–1397.

BONNER, J., KUNG, G. & BEKLOR, I. 1967 A method for the hybridization of nucleic acid molecules at low temperature. *Biochemistry* **6**, 3650–3653.

BOWEN, T. J. 1970 *An Introduction to Ultracentrifugation*. London: John Wiley & Sons.

BRADLEY, S. G. 1972 Reassociation of deoxyribonucleic acid from selected mycobacteria with that from *Mycobacterium bovis* and *Mycobacterium farcinica. Annual Review of Respiratory Diseases* **106**, 122–124.

BRADLEY, S. G. 1973 Relationships among mycobacteria and nocardiae based upon deoxyribonucleic acid reassociation. *Journal of Bacteriology* **113**, 645–651.

BRADLEY, S. G. 1980 DNA reassociation and base composition. In *Microbiological Classification and Identification* eds. Goodfellow, M. & Board, R. G. pp. 11–26. London: Academic Press.

BRADLEY, S. G. & ENQUIST, L. W. 1974 Systematic implications of DNA analysis. In *Molecular Microbiology* ed. Enquist, L. W., pp. 47–100. New York: John Wiley & Sons.

BRADLEY, S. G., ENQUIST, L. W. & SCRIBNER, H. E., III 1978 Heterogeneity among deoxyribonucleotide sequences of Actinomycetales. In *Genetics of the Actinomycetales*, eds. Freerksen, E., Tarnok, I. & Thumim, J. H., pp. 207–224. New York: Gustav Fischer.

BRENNER, D. J. 1983 Impact of modern taxonomy on clinical microbiology. *American Society for Microbiology News* **49**, 58–63.

BRENNER, D. J., FANNING, G. R., RAKE, A. V. & JOHNSON, K. E. 1969 Batch procedure for thermal elution of DNA from hydroxyapatite. *Analytical Biochemistry* **28**, 447–459.

BRENNER, D. J., STEIGERWALT, A. G., WEAVER, R. E., MCDADE, J. E., FEELEY, J. C. & MANDEL, M. 1978 Classification of the Legionnaires' Disease bacterium: an interim report. *Current Microbiology* **1**, 71–75.

BRENNER, D. J., DAVIS, B. R., STEIGERWALT, A. G., RIDDLE, C. R., MCWHORTER, A. C., ALLEN, S. D., FARMER, J. J., III, SAITCH, Y. & FANNING, R. 1982 Atypical biogroups of *Escherichia coli* found in clinical specimens and description of *Escherichia hermannii* sp. nov. *Journal of Clinical Microbiology* **15**, 703–713.

BRITTEN, R. J. & KOHNE, D. E. 1966 Nucleotide sequence repetition in DNA. *Carnegie Institute, Washington, Yearbook* **65**, 78–106.

BRITTEN, R. J. & KOHNE, D. E. 1968 Repeated sequences in DNA. *Science* **161**, 529–540.

BROOKES, B. W., MURRAY, R. G. E., JOHNSON, J. L., STACKEBRANDT, E., WOESE, C. R. & FOX, G. E. 1980 Red-pigmented micrococci: a basis for taxonomy. *International Journal of Systematic Bacteriology* **30**, 627–646.

BUCHANAN, R. J. & GIBBONS, N. E., eds. 1974 *Bergey's Manual of Determinative Bacteriology*, 8th edn. Baltimore: Williams & Wilkins.

BÜNEMANN, H. 1982 Immobilization of denatured DNA to macroporous supports: II. steric and kinetic parameters of heterogeneous hybridization reactions. *Nucleic Acids Research* **10**, 7181–7196.

BUTTERWORTH, P. H. W. 1976 Isolation of DNA from eucaryotic cells. In *Subnuclear Components, Preparation and Fractionation* ed. Birnie, G. D., pp. 295–323. London: Butterworths.

CALLIES, E. & MANNHEIM, W. 1980 Deoxyribonucleic acid relatedness of some menaquinone-producing *Flavobacterium* and *Cytophaga* strains. *Antonie van Leeuwenhoek* **46**, 41–49.

CHASSY, B. M. & GUIFFREDA, A. 1980 Method for the lysis of Gram-positive, asporogenous bacteria with lysozyme. *Applied and Environmental Microbiology* **39**, 153–158.

COYKENDALL, A. L. & MUNZENMAIER, A. J. 1978 Deoxyribonucleic acid base sequence studies on glucan-producing and glucan-negative strains of *Streptococcus mitior*. *International Journal of Systematic Bacteriology* **28**, 511–515.

COYKENDALL, A. L., KACZMARCK, F. S., & STOTS, J. 1980 Genetic heterogeneity in *Bacteroides asaccharolyticus* (Holdeman and Moore 1970) Finegold and Barnes 1977 (Approved Lists, 1980) and proposal of *Bacteroides gingivalis* sp. nov. and *Bacteroides macacae* (Stots and Genco) comb. nov. *International Journal of Systematic Bacteriology* **30**, 559–564.

CROMBACH, W. H. J. 1973 Deep freezing of bacterial DNA for thermal denaturation and hybridization experiments. *Antonie van Leeuwenhoek* **39**, 249–255.

CROMBACH, W. H. J. 1974 Thermal stability of homologous and heterologous bacterial DNA duplexes. *Antonie van Leeuwenhoek* **40**, 133–144.

CROSA, J. H., BRENNER, D. J. & FALKOW, S. 1973 Use of a single-strand specific nuclease for analysis of bacterial and plasmid deoxyribonucleic acid homo- and hetero-duplexes. *Journal of Bacteriology* **115**, 904–911.

CROSA, J. H., WILLIAMS, B. L., JORGEMAN, J. J. & EVANS, C. A. 1979 Comparative study of deoxyribonucleic acid homology and physiological characteristics of strains of *Peptococcus saccharolyticus*. *International Journal of Systematic Bacteriology* **29**, 328–332.

DE LEY, J. 1969 Compositional nucleotide distribution and the theoretical prediction of homology in bacterial DNA. *Journal of Theoretical Biology* **22**, 89–116.

DE LEY, J. 1970 Re-examination of the association between melting point, buoyant density, and chemical base composition of deoxyribonucleic acid. *Journal of Bacteriology* **101**, 738–754.

DE LEY, J. 1971 Hybridization of DNA. In *Methods in Microbiology* eds. Norris, J. R. & Ribbons, D. W., Vol. 5A, pp. 311–329. London & New York: Academic Press.

DE LEY, J. & SCHELL, J. 1963 Deoxyribonucleic acid composition of acetic acid bacteria. *Journal of General Microbiology* **33**, 243–253.

DE LEY, J. & TIJTGAT, R. 1970 Evaluation of membrane filter methods for DNA-DNA hybridization. *Antonie van Leeuwenhoek* **41**, 287–307.

DE LEY, J., CATTOIS, H. & REYNAERTS, A. 1970 The quantitative measurements of DNA hybridization from renaturation rates. *European Journal of Biochemistry* **12**, 133–142.

DE LEY, J., TIJTGAT, R., DE SMEDT, J. & MICHIELS, M. 1973 Thermal stability of DNA-DNA hybrids within the genus *Agrobacterium*. *Journal of General Microbiology* **78**, 241–252.

DENHARDT, D. T. 1966 A membrane-filter technique for the detection of complementary DNA. *Biochemical and Biophysical Research Communications* **23**, 641–646.

DÖPFER, H. G., STACKEBRANDT, E. & FIEDLER, R. 1982 Nucleic acid hybridization studies on *Microbacterium, Curtobacterium, Agromyces* and related taxa. *Journal of General Microbiology* **128,** 1697–1708.

FARMER, J. J., III, ASTBURY, M. A., HICKMAN, F. W., BRENNER, D. J. & THE *ENTEROBACTERIACEAE* STUDY GROUP 1980 *Enterobacter sakazakii:* a new species of '*Enterobacteriaceae*' isolated from clinical specimens. *International Journal of Systematic Bacteriology* **30,** 569–584.

FOX, G. E., STACKEBRANDT, E., HESPELL, R. B., GIBSON, J., MANILOFF, J., DYER, T. A., WOLFE, R. S., BALCH, W. E., TANNER, R. S., MAGRUM, L. J., ZABLEN, L. B., BLAKEMORE, R., GUPTA, R., BONEN, L., LEWIS, B. J., STAHL, D. A., LUEHRSEN, K. R., CHEN, K. N., & WOESE, C. R. 1980 The phylogeny of prokaryotes. *Science* **209,** 457–463.

FRANK-KAMENETSKII, M. D. 1971 Simplification of the empirical relationship between melting temperature of DNA, its GC content and concentration of sodium ions in solution. *Biopolymers* **10,** 2623–2624.

GARVIE, E. I. 1978 *Streptococcus raffinolactis* Orla-Jensen and Hansen, a group N. *Streptococcus* found in raw milk. *International Journal of Systematic Bacteriology* **28,** 190–193.

GIBBINS, A. M. & GREGORY, K. F. 1972 Relatedness among *Rhizobium* and *Agrobacterium* species determined by three methods of nucleic acid hybridization. *Journal of Bacteriology* **111,** 129–141.

GIBSON, D. M. & OGDEN, I. D. 1979 A rapid method for purifying bacterial deoxyribonucleic acid. *Journal of Applied Bacteriology* **46,** 421–423.

GILLESPIE, D. & SPIEGELMAN, S. 1965 A quantitative assay for DNA-DNA hybrids with DNA immobilized on a membrane. *Journal of Molecular Biology* **12,** 829–842.

GILLIS, M., DE LEY, J. & DE CLEENE, M. 1970 The determination of molecular weight of bacterial genome DNA from renaturation rates. *European Journal of Biochemistry* **12,** 143–153.

GREENWOOD, J. R. & PICKETT, M. J. 1980 Transfer of *Haemophilus vaginalis* Gardner and Dukes to a new genus *Gardnerella: G. vaginalis* (Gardner and Dukes) comb. nov. *Journal of Systematic Bacteriology* **30,** 170–178.

GRIMONT, P. A. D., GRIMONT, F. & STARR, M. P. 1978 *Serratia proteamaculans* (Paine and Stansfield) comb. nov. a senior subjective synonym of *Serratia liquefaciens* (Grimes and Hennerty) Bascomb *et al. International Journal of Systematic Bacteriology* **28,** 503–510.

GRIMONT, P. A. D., POPOFF, M. Y., GRIMONT, F., COYNAULT, C. & LEMELIN, M. 1980 Reproducibility and correlation study of three deoxyribonucleic acid hybridization procedures. *Current Microbiology* **4,** 325–330.

GRIMONT, P. A. D., IRINO, K. & GRIMONT, F. 1982 The *Serratia liquefaciens—S. proteomaculans—S. grimesii* complex DNA relatedness. *Current Microbiology* **7,** 63–68.

GRUNSTEIN, M. & HOGNESS, D. S. 1975 Colony hybridization: a method for the isolation of cloned DNAs that contain a specific gene. *Proceedings of the National Academy of Sciences of the United States of America* **72,** 3961–3965.

HANSEN, J. B. & OLSEN, R. H. 1978 Isolation of large bacterial plasmids and characterization of the P2 incompatibility group plasmids pMGI. *Journal of Bacteriology* **135,** 227–238.

HARVEY, S. & PICKETT, M. J. 1980 Comparison of Adansonian analysis and deoxyribonucleic acid hybridization results in the taxonomy of *Yersinia enterocolitica. International Journal of Systematic Bacteriology* **30,** 86–102.

HILL, E. B., WAYNE, L. G. & GROSS, W. M. 1972 Purification of mycobacterial deoxyribonucleic acid. *Journal of Bacteriology* **112,** 1033–1039.

HODGSON, A. L. M. & ROBERTS, W. P. 1983 DNA colony hybridization to identify *Rhizobium* strains. *Journal of General Microbiology* **129,** 207–212.

HOLDEMAN, L. V. & JOHNSON, J. L. 1982 Description of *Bacteroides loescheii* sp. nov. and emendation of the descriptions of *Bacteroides melaninogenicus* (Oliver and Wherry) Roy and Kelly 1939 and *Bacteroides denticola* Shah and Collins 1981. *International Journal of Systematic Bacteriology* **32,** 399–409.

HOLLÄNDER, R. & POHL, S. 1980 Deoxyribonucleic acid base composition of bacteria. *Zentralblatt für Bakteriologie, Mikrobiologie und Hygiene I. Abteilung Originale. A* **246**, 236–275.

HUSS, V. A. R., FESTL, H. & SCHLEIFER, K. H. 1983 Studies on the spectrophotometric determination of DNA hybridization from renaturation rates. *Systematic and Applied Microbiology* **4**, 184–192.

IMAEDA, T., BARKSDALE, L. & KIRCHHEIMER, W. F. 1982 Deoxyribonucleic acid of *Mycobacterium lepraemurium:* its genome size, base ratios and homology with those of other mycobacteria. *International Journal of Systematic Bacteriology* **32**, 456–458.

JARVIS, B. D. W., DICK, A. G. & GREENWOOD, R. M. 1980 Deoxyribonucleic acid homology among strains of *Rhizobium trifolii* and related species. *International Journal of Systematic Bacteriology* **30**, 42–52.

JOHNSON, J. L. 1973 Use of nucleic-acid homologies in the taxonomy of anaerobic bacteria. *International Journal of Systematic Bacteriology* **23**, 308–315.

JOHNSON, J. L. & ORDAL, E. J. 1968 Deoxyribonucleic acid homology in bacterial taxonomy: effect of incubation temperature on reaction specificity. *Journal of Bacteriology* **95**, 893–900.

JONES, D. & SNEATH, P. H. A. 1970 Genetic transfer and bacterial taxonomy. *Bacteriological Reviews* **34**, 40–81.

KIRBY, K. S. 1964 Isolation and fractionation of nucleic acids. *Progress in Nucleic Acid Research* **3**, 1–31.

KLOOS, W. E. & WOLFSHOHL, J. F. 1979 Evidence for deoxyribonucleotide sequence divergence between staphylococci living on human and other primate skin. *Current Microbiology* **3**, 167–172.

KLOOS, W. E., MOHAPATRA, N., DOBROGOSZ, W. J., EZZELL, J. W. & MANCLARK, C. R. 1981 Deoxyribonucleotide sequence relationships among *Bordetella* species. *International Journal of Systematic Bacteriology* **31**, 173–176.

KOMATSU, Y. 1979 Complete lysis of glutamic acid-producing bacteria by the use of antibiotics which inhibit the biosynthesis of cell walls. *Journal of General Microbiology* **113**, 407–408.

LACHANCE, M. A. 1980 Simple method for the determination of deoxyribonucleic acid relatedness by thermal elution in hydroxyapatite microcolumns. *International Journal of Systematic Bacteriology* **30**, 433–436.

LEAPER, S. & OWEN, R. J. 1982 Differentiation between *Campylobacter jejuni* and allied thermophilic campylobacters by hybridization of deoxyribonucleic acids. *FEMS Microbiology Letters* **15**, 203–208.

LEGAULT-DÉMARE, J., DESSEAUX, B., HEYMAN, T., SEROR, S., & RESS, G. F. 1967 Studies on hybrid molecules of nucleic acids. I. DNA-DNA hybrids in nitrocellulose filters. *Biochemical and Biophysical Research Communications* **28**, 550–557.

LEVINE, M. G., PICKETT, M. J. & MANDEL, M. 1980 Taxonomy of nonfermentative bacilli: the IIk-2 group. *Current Microbiology* **4**, 41–44.

MCCONAUGHTY, B. L., LAIRD, C. B. & MCCARTHY, B. J. 1969 Nucleic acid reassociation in formamide. *Biochemistry* **8**, 3289–3295.

MACINNES, J. I., TRUST, T. J. & CROSA, J. C. 1979 Deoxyribonucleic acid relationships among members of the genus *Aeromonas*. *Canadian Journal of Microbiology* **25**, 579–586.

MANDEL, M. 1969 New approaches to bacterial taxonomy: perspective and prospects. *Annual Review of Microbiology* **23**, 239–274.

MANDEL, M., IGAMBI, L., BERGENDAHL, J., DODSON, M. L. & SCHELTGEN, E. 1970 Correlation of melting temperature and cesium chloride buoyant density of bacterial deoxyribonucleic acid. *Journal of Bacteriology* **101**, 333–338.

MANIATIS, T., FRITSCH, E. F. & SAMBROOK, J. 1982 *Molecular Cloning: A Laboratory Manual.* New York: Cold Spring Harbour Publications.

MARMUR, J. 1961 A procedure for the isolation of deoxyribonucleic acid from microorganisms. *Journal of Molecular Biology* **3**, 208–218.

MARMUR, J. & DOTY, P. 1961 Thermal renaturation of deoxyribonucleic acids. *Journal of Molecular Biology* **3**, 585–594.

MARMUR, J. & DOTY, P. 1962 Determination of the base composition of deoxyribonucleic acid from its thermal denaturation temperature. *Journal of Molecular Biology* **5**, 109–118.

MEYER, S. A. & SCHLEIFER, K. H. 1975 Rapid procedure for the approximate determination of the deoxyribonucleic acid base composition of micrococci, staphylococci and other bacteria. *International Journal of Systematic Bacteriology* **25**, 383–385.

MOORE, R. L. 1974 Nucleic acid reassociation as a guide to genetic relatedness among bacteria. *Current Topics in Microbiology and Immunology* **64**, 105–128.

MORDARSKI, M., GOODFELLOW, M., SZYBA, K., PULVERER, G. & TKACZ, A. 1977 Classification of the '*rhodochrous*' complex and allied taxa based upon deoxyribonucleic acid reassociation. *International Journal of Systematic Bacteriology* **27**, 31–37.

NOVOTNY, J. 1982 Matrix program to analyse primary structure homology. In *The Applications of Computers to Research on Nucleic Acids* eds. Söll, D. & Roberts, R. J., pp. 127–131. Oxford: IRL Press.

O'DONNELL, A. G., NORRIS, J. R., BERKELEY, R. C. W., CLAUS, D., KANEKO, T., LOGAN, N. A. & NOZAKI, R. 1980 Characterization of *Bacillus subtilis, Bacillus pumilus, Bacillus licheniformis,* and *Bacillus amyloliquefaciens* by pyrolysis gas-liquid chromatography, deoxyribonucleic acid-deoxyribonucleic acid hybridization, biochemical tests, and API systems. *International Journal of Systematic Bacteriology* **30**, 448–459.

OWEN, R. J. & HILL, L. R. 1979 The estimation of base compositions, base pairing and genome sizes of bacterial deoxyribonucleic acids. In *Identification Methods for Microbiologists* eds. Skinner, F. A. & Lovelock, D. W., pp. 277–296. London: Academic Press.

OWEN, R. J. & JACKMAN, P. J. H. 1982 The similarities between *Pseudomonas paucimobilis* and allied bacteria derived from analysis of deoxyribonucleic acids and electrophoretic patterns. *Journal of General Microbiology* **128**, 2945–2954.

OWEN, R. J. & SNELL, J. S. S. 1976 Deoxyribonucleic acid reassociation in the classification of flavobacteria. *Journal of General Microbiology* **93**, 89–102.

OWEN, R. J., HILL, L. R. & LAPAGE, S. P. 1969 Determination of DNA base compositions from melting profiles in dilute buffers. *Biopolymers* **7**, 503–516.

OWEN, R. J., LEGROS, R. M. & LAPAGE, S. P. 1978 Base composition, size and sequence similarities of genome deoxyribonucleic acids from clinical isolates of *Pseudomonas putrefaciens*. *Journal of General Microbiology* **104**, 127–138.

POPOFF, M. & COYNAULT, C. 1980 Use of DEAE-cellulose filters in the S1 nuclease method for bacterial deoxyribonucleic acid hybridization. *Annales Microbiologie* **131A**, 151–155.

RAHIMIAN, H. & GUMPF, D. J. 1980 Deoxyribonucleic acid relationship between *Spiroplasma citri* and the corn stunt spiroplasma. *International Journal of Systematic Bacteriology* **30**, 605–608.

RANHAND, J. M. 1974 Simple, inexpensive procedure for the disruption of bacteria. *Applied Microbiology* **28**, 66–69.

RIGBY, P. W. J., DIECKMANN, M., RHODES, C. & BERG, P. 1977 Labelling deoxyribonucleic acid to high specific activity *in vitro* by nick translation with DNA polymerase I. *Journal of Molecular Biology* **113**, 237–251.

SAITO, H. & MIURA, K. 1963 Preparation of transforming deoxyribonucleic acid by phenol treatment. *Biochimica et Biophysica Acta* **72**, 619–629.

SALTZBERG, S., LEVI, Z., ABOUD, M. & GOLDBERGER, A. 1977 Isolation and characterization of DNA-DNA and DNA-RNA hybrid molecules formed in solution. *Biochemistry* **16**, 25–29.

SCHILDKRAUT, C. L. & LIFSON, S. 1965 Dependance of the melting temperature of DNA on salt concentration. *Biopolymers* **3**, 195–208.

SCHILDKRAUT, C. L., MARMUR, J. & DOTY, P. 1962 Determination of the base composition of deoxyribonucleic acid from its buoyant density in CsCl. *Journal of Molecular Biology* **4**, 430–443.

SEIDLER, R. J. & MANDEL, M. 1971 Quantitative aspects of deoxyribonucleic acid renaturation: base composition, state of chromosomal replication, and polynucleotide homologies. *Journal of Bacteriology* **106**, 608–614.

SILVESTRI, L. G. & HILL, L. R. 1965 Agreement between deoxyribonucleic acid base composition and taxometric classification of Gram-positive cocci. *Journal of Bacteriology* **90**, 136–140.

SNEATH, P. H. A. 1983 Distortions of taxonomic structure from incomplete data on a restricted set of reference strains. *Journal of General Microbiology* **129**, 1045–1073.

SÖLL, D. & ROBERTS, R. J. eds. 1982 *The Application of Computers to Research on Nucleic Acids*. Oxford: IRL Press.

STACKEBRANDT, E. & FIEDLER, F. 1979 DNA-DNA homology studies among strains of *Arthrobacter* and *Brevibacterium*. *Archives of Microbiology* **120**, 289–295.

STACKEBRANDT, E., WUNNER-FUSSL, B., FOWLER, V. J. & SCHLEIFER, K. H. 1981 Deoxyribonucleic acid homologies among spore-forming members of the order *Actinomycetales*. *International Journal of Systematic Bacteriology* **31**, 420–431.

SUEOKA, N. 1961 Variations and heterogeneity of base composition of nucleic acids: a compilation of old and new data. *Journal of Molecular Biology* **3**, 31–40.

SUTTON, W. D. 1971 A crude nuclease preparation suitable for use in DNA reassociation experiments. *Biochimica et Biophysica Acta* **240**, 522–531.

ULITZER, S. 1972 Rapid determination of DNA base composition by ultraviolet spectroscopy. *Biochimica et Biophysica Acta* **272**, 1–11.

WETMUR, J. G. 1976 Hybridization and renaturation kinetics of nucleic acids. *Annual Review of Biophysics and Bioengineering* **5**, 337–361.

YAMADA, K. & KOMAGATA, K. 1970 Taxonomic studies on coryneform bacteria III DNA base composition of coryneform bacteria. *Journal of General and Applied Microbiology* **16**, 215–224.

ZADRAŽIL, S., ŠATAVA, J. & ŠORMOVÁ, Z. 1973 Isolation procedure for bacterial DNA based on gel permeation chromatography on a sepharose column. *Journal of Chromatography* **91**, 451–458.

Pyrolysis Mass Spectrometry in Bacterial Systematics

LESLEY A. SHUTE AND R. C. W. BERKELEY

Department of Microbiology, The Medical School, University of Bristol, Bristol, UK

J. R. NORRIS AND C. S. GUTTERIDGE

Cadbury Schweppes PLC Group Research, Lord Zuckerman Research Centre, University of Reading, Reading, UK

Characterisation and identification of microorganisms, whether done on fresh isolates or as a check on reference strains, constitutes a major role of diagnostic microbiology. Many systems and methods have been developed for this task, but in recent years increasing emphasis has been placed on the need to reduce the time involved and to introduce automated systems into routine microbiology. One of the methods that is likely to enable rapid characterisation of microorganisms and that also has potential for automation is pyrolysis mass spectrometry (Py–MS).

Pyrolysis is the breakdown of complex materials (e.g. bacteria) in an inert atmosphere to produce a series of volatile, lower-molecular-weight substances, using heat alone. In Py–MS, these fragments are subsequently detected and quantified using mass spectrometry. A similar technique, pyrolysis gas chromatography (Py–GC) uses gas chromatography as the detection system.

The first report of Py–MS applied to examine complex biological materials was that of Zemany (1952), who pyrolysed albumin and pepsin. Reiner (1965) showed that Py–GC could be used as an aid to bacterial identification and more interest was generated in the technique when it was used in the US Space Exploration program (Oyama 1963). Most of the earlier analytical pyrolysis work was done using Py–GC, but the problems with reproducibility encountered in early Py–GC systems led to the development

CHEMICAL METHODS IN BACTERIAL SYSTEMATICS

0-12-289675-0

of Py–MS as a technique that is not only more reproducible but faster and more easily automated than Py–GC.

Meuzelaar & Kistemaker (1973) made the major contribution to this field when they developed a pyrolysis mass spectrometer specifically for fingerprinting complex samples such as bacteria. This was later further developed (Meuzelaar *et al.* 1976) to produce a fully automated instrument, which has been used for discriminating a wide variety of bacterial types (Meuzelaar *et al.* 1976; Borst *et al.* 1978; Boon *et al.* 1980). The mass spectra produced by Py–MS can be treated as chemical profiles or "fingerprints" of the organisms analysed. These can be used for identification purposes by comparing them to spectra held in a data base, using computerized procedures. They can also be used for studying inter- and intragroup relationships, in combination with multivariate statistical techniques.

Two particular topics on which concentrated Py–MS studies have been carried out are the characterisation of mycobacteria (Wieten *et al.* 1981, 1982) and of selected yeast species (Windig *et al.* 1981b, Windig & Haverkamp 1982). In these areas, work has progressed beyond simple identification and differentiation of strains to factor analysis of mass spectra, so the differences due to the presence or absence of specific chemical components can be elucidated and correlated with other chemotaxonomic data.

This chapter outlines the principles of Py–MS and the operation of a commercially available pyrolysis mass spectrometer specifically designed for analysing complex biomaterials. Details of sample preparation, data handling and reproducibility are discussed. A study into four groups of closely related *Bacillus* strains is reported, to show a typical application of Py–MS to bacterial systematics.

Principles of Pyrolysis Mass Spectrometry

Approaches to Py–MS

A form of Py–MS can be carried out using standard mass spectrometers if the technique of direct probe mass spectrometry (DPMS) is used (Gutteridge & Puckey 1982), where samples are thermally degraded adjacent to the ion source of the mass spectrometer using a ballistically heated probe. Alternatively, a Curie-point pyrolyser can be linked to the mass spectrometer in the manner of Hughes *et al.* (1977), the pyrolysate passing through an empty glass chromatographic column and jet separator before entering the mass spectrometer.

These two approaches are useful in that they make the technique more accessible, since purpose-built Py–MS systems are at present expensive

(£80K+). However, reproducibility is not as good as that obtained with purpose-built Py–MS instruments and the sample analysis time is longer; thus, for routine analysis of large numbers of samples, an integrated Py–MS system is preferable.

Pyrolysis methods

A detailed account of pyrolysis techniques can be found in the reviews by Irwin & Slack (1978) and Irwin (1979). Only the technique of Curie-point (filament) pyrolysis is described here because it is the one used most commonly in integrated Py–MS systems.

In Curie-point pyrolysis, a thin ferromagnetic wire is placed in the work coil of a high-frequency oscillator. When the field is switched on, eddy currents cause the wire to heat inductively until the wire temperature reaches its Curie-point. The wire then becomes paramagnetic and stops heating until the temperature falls below the Curie-point, when it becomes ferromagnetic again. It thus acts as a thermostatic switch and the equilibrium temperature (Teq) stabilizes close to the Curie-point. The Curie-point is determined by the wire composition, pure iron having a Curie-point of 770°C and pure nickel of 358°C. Common alloys of the two have Curie-points within this range. Wires with Curie-points of 510 and 610°C are those used most commonly for the pyrolysis of microorganisms, since all types of organic materials will thermally degrade at these temperatures. The temperature rise time (TRT), i.e. the time taken to reach Teq is an important parameter that should be standardized, and it is advisable to use a pyrolysis system with a fast TRT to minimise the formation of nonvolatile chars (Lincoln 1965).

Mass spectrometry

After pyrolysis, the resulting mixture of volatile pyrolysis products (the pyrolysate) must be ionized to become amenable to mass analysis. Since pyrolysis is a fierce process, producing small fragments, further fragmentation of the pyrolysate is undesirable, so that maximum chemical specificity can be retained. For this reason "soft" ionization techniques are used, which minimize pyrolysate fragmentation.

The ionization procedure most commonly used in Py–MS is low-voltage electron impact ionization (EI). There are two other soft ionization techniques, chemical ionization (CI) (Hunt *et al.* 1976) and field ionization (FI) (Schulten *et al.* 1973), which are useful for pinpointing qualitative differences in pyrolysis products, especially when high resolution mass spectrometers are used. Unfortunately CI and FI are not as reproducible as EI and cannot be adapted easily to the production of quantitative data with a fast throughput of samples.

In EI, the pyrolysate is bombarded by electrons from an ion source, causing electrons to be removed from the pyrolysate fragments, producing positively charged ions. When low-voltage EI is used (14–16 eV) only singly charged fragments are produced. These are accelerated across a potential difference into the mass spectrometer where they are separated according to their mass/charge ratio (m/z). Thus as each fragment possesses only one charge, they are effectively separated by their mass.

Mass analysis is carried out by passing the ions through either an electric field, in the case of a quadrupole mass spectrometer, or a magnetic field, in the case of a magnetic sector mass spectrometer. Separation of the ions relies on the fact that ions of different masses will be deflected by different amounts and will pass through the field at different rates.

Following separation in the mass spectrometer, the ions are detected, either by ion counting or current monitoring. Ion counting is preferable and has been used in all purpose-built Py–MS systems because it allows detection of small signals even at high scanning speeds. At high amplifications, the bandwidth of current amplifiers is low and this limits the scanning speed that can be used. Once detected, the ions are recorded either by a signal

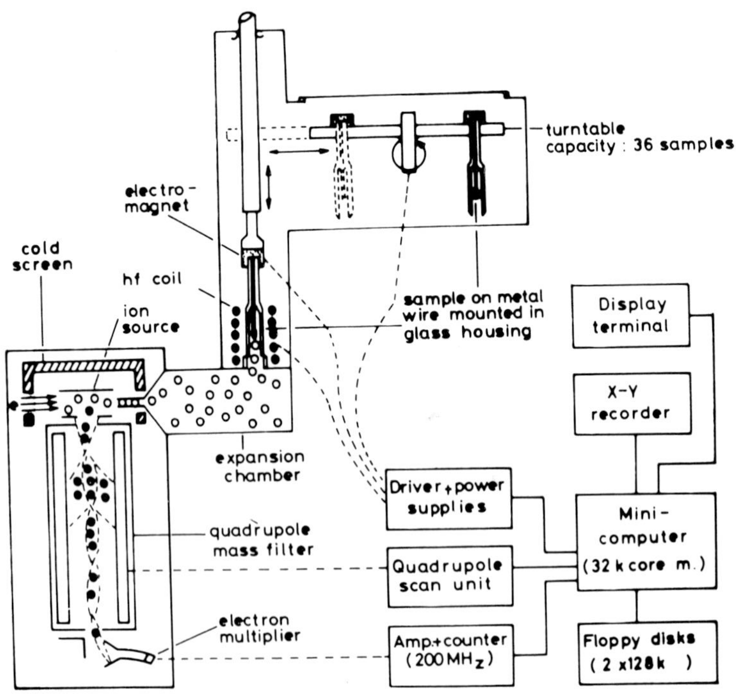

FIG. 1. Schematic representation of an automated Curie-point pyrolysis mass spectrometer based on a quadrupole mass analyser (Meuzelaar *et al.* 1976).

averager or by a computer. The pyrolysis mass spectra are then ready to be subjected to data analysis.

The first automated Py–MS system (Meuzelaar *et al.* 1976) incorporated a quadrupole (Fig. 1); much of the published Py–MS work has been done with this system. A manually operated version is now available commercially (Extranuclear Laboratories Inc., Pittsburgh, Pennsylvania, USA, Py–MS Model 5000-1). A similar Py–MS system, but utilising a magnetic mass analyser, will be described later. It is important to realise that the successful integration of pyrolysis with mass spectrometry depends on the method used to transfer the pyrolysate to the ionization region. In the fully integrated systems described here, the pyrolysate is generated at the front end of a heated expansion chamber, which is lined with gold to produce an inert surface. This reduces secondary reactions amongst the gaseous mixture and hence helps to maintain the specificity of the pyrolysis products. The pyrolysate is then leaked into the mass spectrometer source over a short period of time (15–30 sec) to be ionized.

Sample preparation

Samples for Curie-point pyrolysis are applied to wires held in glass reaction tubes. Both should be cleaned rigorously before use. After use, the wires are discarded, but the tubes can be cleaned and reused.

Filament cleaning can be done by a number of methods, e.g. ultrasonic cleaning in organic solvents, soaking in acids, or heating in a water-saturated hydrogen atmosphere. Our wires are cleaned by soaking in 10% hydrochloric acid for 30 min and rinsing in tap water, distilled water and acetone, followed by oven drying overnight at 200°C. Meuzelaar *et al.* (1982) use the reductive hydrogen cleaning technique. This has the advantage that it removes oxide layers, which may cause catalytic reactions, but it may also cause hydrogen adsorption by the wire, which can influence the pyrolytic reactions. Whichever method is chosen, the same one should be used throughout a study and kept as standard as possible, as the method of cleaning can influence the pyrolysis patterns (Windig *et al.* 1979). Glass reaction tubes are best cleaned by soaking in chromic acid and rinsing in tap water, distilled water and acetone, and oven drying at 200°C.

Preparation of samples for Curie-point pyrolysis varies, depending on whether the samples are taken from solid or liquid media. In both cases, the aim is to standardize the preparation technique as much as possible and to prepare samples free from medium components.

When grown on solid media, microbial growth may be applied directly to the wire, using a disposable plastic loop. Alternatively, a suspension of growth in distilled water or other solvents such as methanol can be made and

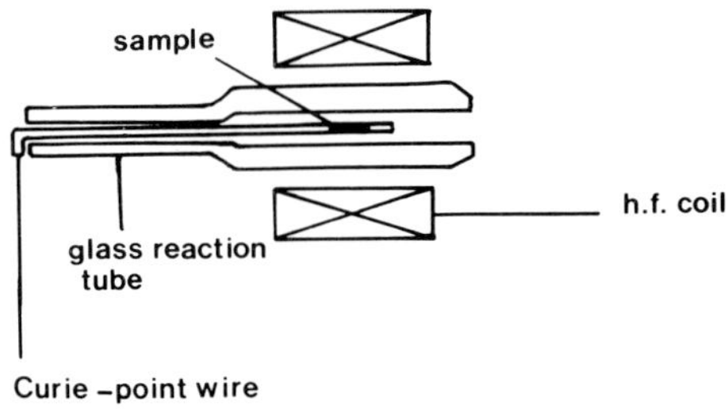

FIG. 2. Schematic representation of the Curie-point wire/reaction tube assembly located within the high-frequency coil.

applied to the wire by dropwise addition using a microsyringe, or by turning the wire in the suspension. Organisms that tend to grow into the agar surface can be grown on membrane filters (Oxborrow *et al.* 1976) to provide samples free of media components.

When liquid cultures are used, cells can be harvested by filtration or centrifugation and then washed in distilled water, resuspended in the desired solvent and applied to the wire by dropwise addition, 5–10 μl at a time, using a microsyringe. Alternatively, after washing, the cells can be freeze-dried and resuspended at a later date for analysis.

There is no evidence favouring one particular technique of preparing microbial cells for Py–MS and most workers have used colony material taken directly from agar, a simple and rapid approach, which is one of the advantages of this method of characterising microorganisms. However the sample is prepared, it is then applied about 3 mm from the end of the Curie-point wire. The wire is then retracted into the tube (Fig. 2) and secured with a plastic cap. The tube is then placed in the probe of the mass spectrometer and is ready for pyrolysis.

Sample size is usually between 5 and 50 μg. Over this range the data produced are linear with sample size, provided sample coating is uniform and the data are normalised to remove sample size differences during the data analysis phase. Samples in excess of 50 μg tend to be pyrolysed at different rates throughout the sample and are best avoided.

Pyromass 8-80

The Pyromass 8-80 (VG Gas Analysis Ltd., Middlewich, Cheshire, UK) is a Curie-point pyrolysis mass spectrometer based on the concepts pioneered by Meuzelaar and Kistemaker (1973) but incorporating a magnetic mass analyser (Figs. 3 and 4).

Power for pyrolysis is supplied by a Fischer Labortechnik, 1.5 kw, 1.1

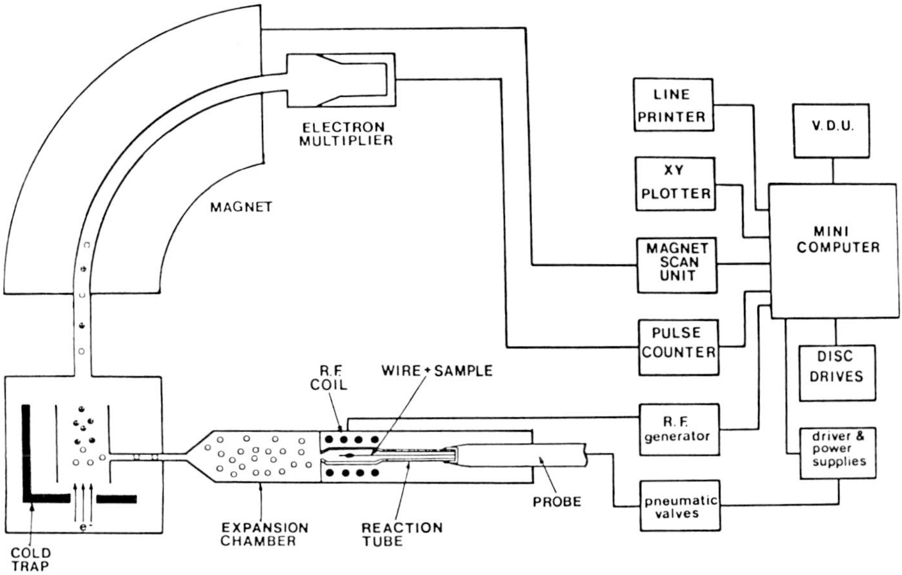

FIG. 3. Schematic representation of the Pyromass 8-80.

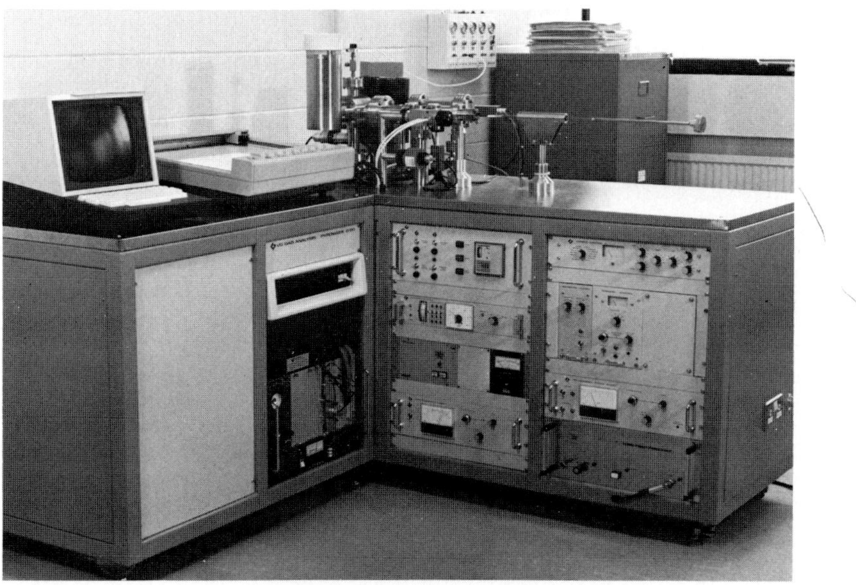

FIG. 4. The Pyromass 8-80.

MHz high-frequency generator. The reaction tube is inserted into the RF coil using the instrument probe. The reaction tube helps to trap relatively non-volatile pyrolysis products, which might condense on the walls and cause contamination of the ion source. On pyrolysis, the products pass into a 10-cm^3 gold-coated expansion chamber and leave it as a molecular beam, entering the ion source via a ceramic tube. The entire expansion assembly is maintained at a preset temperature (usually 150°C) to prevent condensation of the pyrolysate. The source is an electron-impact Nier type, surrounded by a liquid nitrogen cooled baffle to trap high molecular weight organic contaminants. Following ionization, an 8-cm, 80° single-focusing magnetic-sector instrument is used for mass analysis. The mass range is scanned exponentially from m/z 300 to m/z 12. Ions are detected by an electron multiplier coupled to a fast-response, x-10 preamplifier. The pulses generated are counted by a Camac-based system employing an LSI 11/23 minicomputer and the raw mass data are stored on a floppy diskette. Normalized data can be output as full or restricted range mass spectra by a line printer and/or an X–Y plotter. The time taken from insertion of the sample to completion of the spectrum printout is 5–6 min (depending on the amount of data collected).

The operational parameters of the instrument can be altered from the computer keyboard. For analysis of bacterial cultures, the operational parameters of the instrument were: pyrolysis temperature, 510°C for 2 sec; hold time, 60 sec; scan time, 1.3 sec $cycle^{-1}$; number of scans, 35; electron energy, 16 eV.

The machine is at present operated manually, though an automated system is being developed whereby a rack will be loaded with up to 65 samples, which will be processed sequentially, the operation being controlled by the computer.

Once the data have been recorded on the diskette, it is then ready for data analysis.

Data Handling

Good data handling is an essential prerequisite for the successful application of Py–MS to the characterisation of microorganisms. Analysis of pyrolysis data usually involves the use of a battery of multivariate analysis (chemometric, pattern recognition) procedures. Figure 5 is a flow diagram linking most of the methods that have been used where the aim of the analysis is to characterise, discriminate and identify samples (MacFie & Gutteridge 1982). The chemical interpretation of pyrolysis mass spectra and the analysis of mixtures use different techniques that are not discussed here (Windig *et al.*

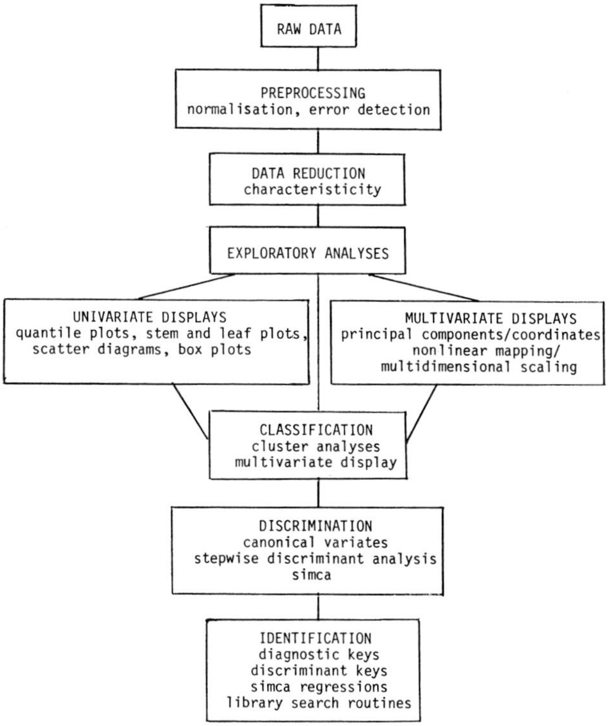

FIG. 5. Flow diagram indicating a comprehensive approach to pyrolysis–mass spectrometry data handling for classification, discrimination and identification.

1981a). In practice, although it would be desirable, it is not usually feasible to apply all of the techniques in Fig. 5 to a single set of data, and the methods used are always a function of the availability of particular programs and packages. Nevertheless, given the volume of data generated by Py–MS, it is safe to assume that all laboratories using the method will employ a data analysis strategy involving all of the major stages outlined on Fig. 5.

The general characteristics of Py–MS data govern the methods of analysis that can be used successfully. The data are quantitative and correlated both chemically and mathematically. Chemical correlations arise because each mass (m/z) in the spectrum has a multiple origin during the thermal degradation process. Mathematical correlations are introduced by the need to normalise the data (e.g. by expressing the mass intensities as a proportion of the total intensity) to remove differences in sample size. Also since Py–MS is a complex analytical technique, a certain amount of "noise" in the data is inevitable and all these factors combine to make the problem of handling the data a challenging one. The overall aim of data analysis, therefore, is to amplify differences between spectra that are reproducible and provide information useful for the discrimination of samples.

A full description of the concepts and mathematics of the methods used is beyond the scope of this chapter (see MacFie & Gutteridge 1982), but two key procedures are described so that the examples can be understood.

Graphical representation of multidimensional data

Most Py–MS data analyses include an attempt to represent the multidimensional data graphically. For a spectrum containing M masses, if $M = 2$, then a simple way of observing variation is to plot mass 1 against mass 2 for each sample. In normal Py–MS studies $M > 100$, and although it is still conceptually possible to consider each sample as a point in M-dimensional space, two- or three-dimensional scatter diagrams are selected that give the best approximations to the M-dimensional space. One method for achieving this is principal coordinates analysis (PCA) (Gower 1966), which operates on the matrix of the interdistances between the samples.

Discriminant analysis procedures

The linear discriminant analysis problem is to find a linear combination of the masses that best differentiates a preset (*a priori*) group structure. This can be extended to K orthogonal linear combinations, where $K < N_g - 1$, and N_g is the number of groups. For a statistically valid analysis, the following numerical constraint must be complied with (Dixon 1975):

$$N_v < N_s - N_g - 1$$

where N_v = number of masses (variables) and N_s = number of samples.

One form of discriminant analysis is canonical variates (CVA) (MacFie *et al.* 1978) in which successive axes are derived that are independent and seek to maximise the ratio of the variation between to that within the *a priori* groups, assuming that the structure of the variation within the groups is the same.

Stepwise discriminant analysis (Dixon 1975) is used to select a subset of the masses that gives a stable solution. This procedure works iteratively, including masses in the subset one at a time on the basis of maximising the ratio of between to within group variation (F ratio). Instead of axes, discriminant (classification) functions are calculated that permit samples to be classified in one of the groups. Alternatively, a generalised distance (Mahalanobis 1936) of the samples from each of the group means can be obtained. Recording the number of misclassified samples gives the apparent error rate. The classification functions obtained can be used to identify new samples.

In our laboratory the usual sequence of analyses, once the data have been transferred to the main frame computer, is:

(1) Load raw data; a selection of m/z values is used based on the range apparent on visual inspection of the data. A small number of intense low-molecular-weight fragments known to be non-reproducible (e.g. m/z 18, H_2O; m/z 28, N_2) are omitted routinely.
(2) Normalise data.
(3) Check for errors and outliers.
(4) Calculate characteristicity values (Eshuis *et al.* 1977).
(5) Average replicates (optional).
(6) Format reduced data file (usually to contain the 50 most "characteristic" masses).
(7) Load reduced data.
(8) Run principal coordinates and cluster analysis.
(9) Run canonical variates.
(10) Run stepwise discriminant analysis.
(11) Test for stability, using jackknife procedures (Gutteridge *et al.* 1980; O'Donnell *et al.* 1980a).
(12) Inspect feature–feature plots for groupings.
(13) Generate classification functions.
(14) Attempt identification of unknowns, using classification functions.

These analyses are carried out using two packages—HILDA and GENSTAT. HILDA (Highly Interactive Linear Discriminant Analysis) is a specially written Fortran-IV program containing procedures for:

(1) Normalisation.
(2) Error detection.
(3) Data reduction by calculation of characteristicity values.
(4) Exploratory univariate displays.
(5) Stepwise discriminant analysis.
(6) Identification using classification functions.

GENSTAT (Nelder 1979) is a general statistical package available for a wide range of main-frame computers. A number of other packages contain a wide range of multivariate analysis procedures and other workers in this field have used SPSS (Nie *et al.* 1975), BMDP (Dixon 1975) and ARTHUR (Kowalski 1975). In addition, some purpose-written software is available, e.g. FOMPYR (Eshuis *et al.* 1977) and NORMA (Huff *et al.* 1981).

Reproducibility of Py–MS

It is obvious that if Py–MS is to prove useful in bacterial systematics, mass spectra from the same organism, grown under the same condtions, must be reproducible. The short-term reproducibility of the Pyromass 8-80 was followed over the term of a Py–MS study of *Bacillus* strains and the mean dissimilarity calculated for duplicate samples was 4%. A small number of replicates (5/104) showed relatively poor reproducibility, giving dissimilarity values of >8%. The poor reproducibility of a few samples (termed outliers) is a common observation in pyrolysis studies, and is believed to be caused by inconsistencies in the coating of the Curie-point wires when samples are transferred direct from agar plates. Outliers are detected during the data analysis and are usually removed from the data base.

Our figure of 4% for short-term reproducibility compares favourably with that obtained by Windig *et al.* (1979) for short-term (6%) and long-term reproducibility (8%) of averaged spectra of glycogen and bovine serum albumin. The long-term reproducibility of the Pyromass 8-80 has not yet been assessed.

Factors affecting the long-term reproducibility of Py–MS have been studied by Meuzelaar (1978) and Windig *et al.* (1979). It is critical to standardize the growth conditions, sample preparation technique and the pyrolysis parameters used if reproducibility is to be maintained. The same ionization energy should be used throughout a Py–MS study because alterations produce greatly differing spectra. This may be an important factor in the maintenance of long-term reproducibility inasmuch as contamination of the ion source can induce changes in electron impact energies.

The ionizing energy of a pyrolysis mass spectrometer needs to be calibrated regularly. At present it is believed that prospects for interlaboratory reproducibility between instruments of the same basic design are good (Meuzelaar *et al.* 1982), although there is little published evidence to support that view.

Using Py–MS, the variation between separate bacterial species tends to be at least 15%, so it can be argued that a variation due to the method itself of 5–10% could be tolerated, as long as the data handling procedures are developed to cope with it. In normal operation, variability can be kept close to the 5% figure, which accounts for the ability of Py–MS to discriminate strains as well as species.

In applications of Py–MS where sample mass spectra are to be compared to spectra held in a data base, reproducibility is a critical factor. However, in Py–MS applications where samples are analysed along side reference samples, for example, in a quality control study, long-term reproducibility is less important. This application of Py–MS is termed operational fingerprinting (Meuzelaar *et al.* 1982) and is usually only used when the samples to be analysed are of the same type. As an example, operational fingerprinting would be useful for detecting misidentified bacteria when a batch of samples identified as the same species were analysed with reference samples of the species in question.

Thus with standardization of sample preparation methods, pyrolysis parameters and the type of instrument used, reproducible Py–MS results should be obtainable. The small variation in reproducibility observed should, with adequate data-handling procedures, be within tolerance for most biological applications.

Examples of Py–MS Studies

One of the advantages of Py–MS is that it can be applied to all types of microorganisms. This is illustrated by Figs. 6 (a), (b) and (c), which show mass spectra for three different organisms—a yeast, *Brettanomyces naardenensis*, and two bacteria—a strain of *Bacillus licheniformis* and an unclassified thermophilic *Archaebacterium* strain. The differences between these spectra are sufficiently obvious to be seen by eye, particularly in the case of the *Archaebacterium* strain, which has unique masses above m/z 150 that are believed to be fragmentation products of complex lipids. The differentiation of more closely related organisms requires computerised data analysis.

As an example of how Py–MS data can be used for these purposes, the results of a Py–MS study on four groups of closely related *Bacillus* strains (*B. subtilis, B. pumilus, B. licheniformis* and "*B. amyloliquefaciens*") are shown. The strains were all grown in duplicate on Nutrient Agar (Oxoid) at 35°C for 16 h and were examined microscopically before analysis to ensure that only nonsporulated cultures were used. Two samples were taken from each plate, giving four samples for each strain.

The relationships between strains of the same group were examined using principal coordinates analysis. This showed that the groups were not completely homogeneous, but tended to split into smaller clusters of very closely related strains, within the group. An example of this is shown in Fig. 7, which shows a principal coordinates plot of the "*B. amyloliquefaciens*" group.

To examine the relationship between strains from different groups and to assess the discrimination of the groups by Py–MS, canonical variates analysis was used. All four groups could be discriminated by using data from nonsporulated cultures, as shown in Fig. 8. These results agree with those of O'Donnell *et al.* (1980b), who carried out a study on the same group of strains using the alternative analytical pyrolysis technique of pyrolysis gas-liquid chromatography. Different results were obtained by using data from sporulated cultures (Shute *et al.* 1983). The discrimination of these four groups of strains was an exacting test of the discriminating abilities of the Pyromass 8-80, since they are very closely related and possess few characteristics by which they can be separated (Gordon *et al.* 1973).

The group separation was also examined using stepwise discriminant analysis. A total of 16 masses were needed to discriminate the group fully, to a

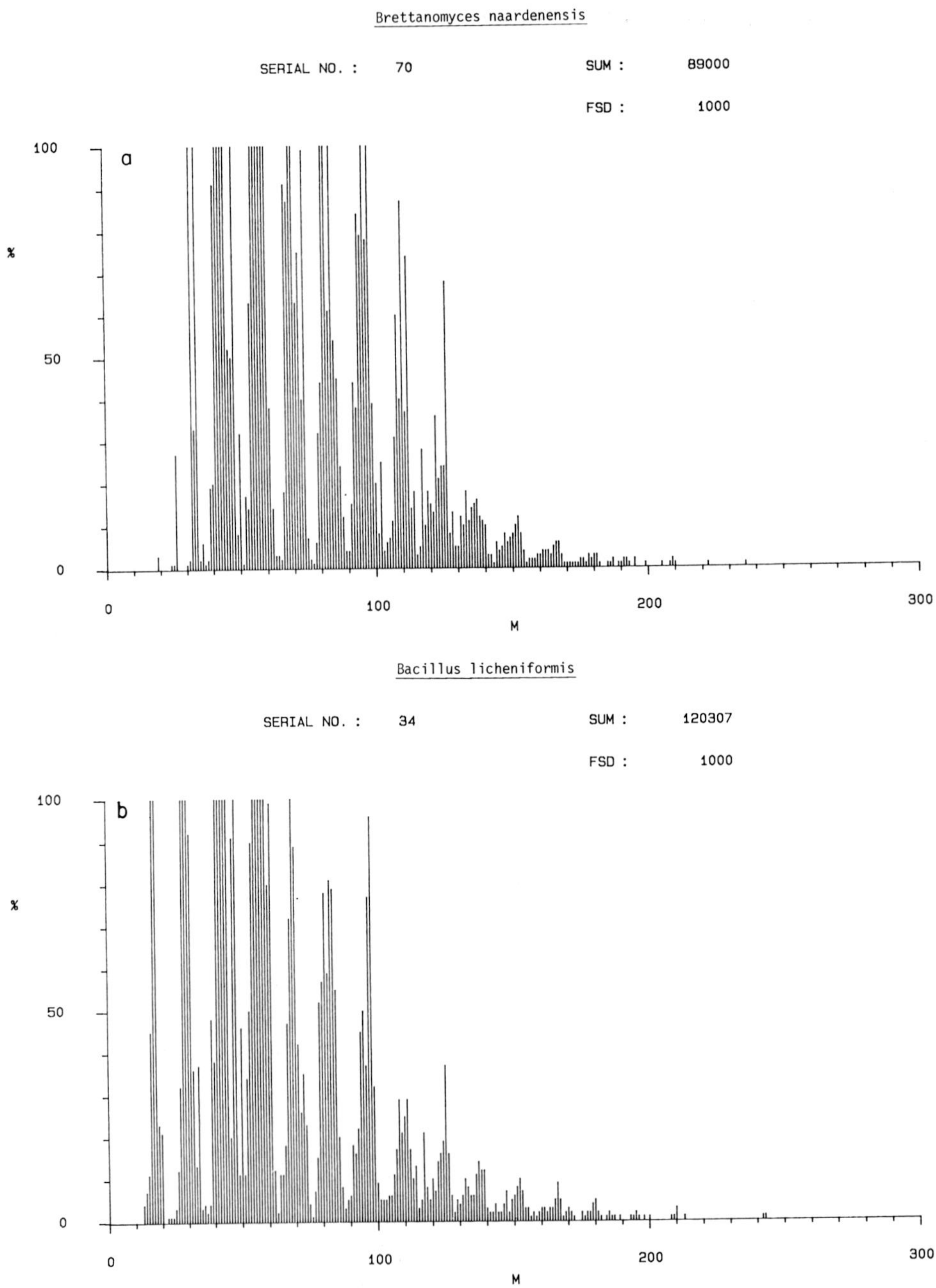

FIG. 6. Representative pyrolysis mass spectra of (a) *Brettanomyces naardenensis;* (b) *Bacillus licheniformis;* (c) a thermophilic *Archaebacterium* strain.

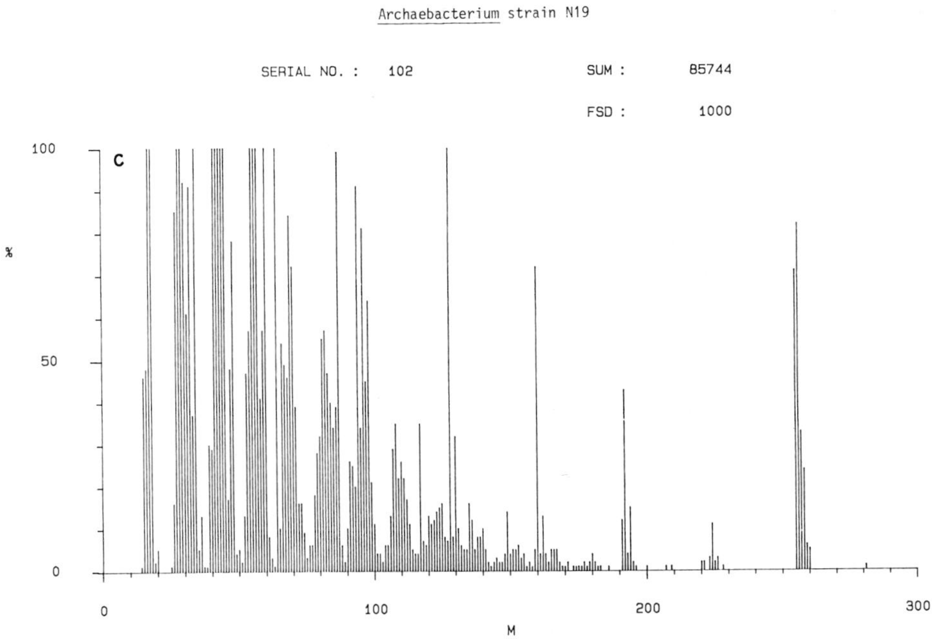

FIG. 6. (*Continued*)

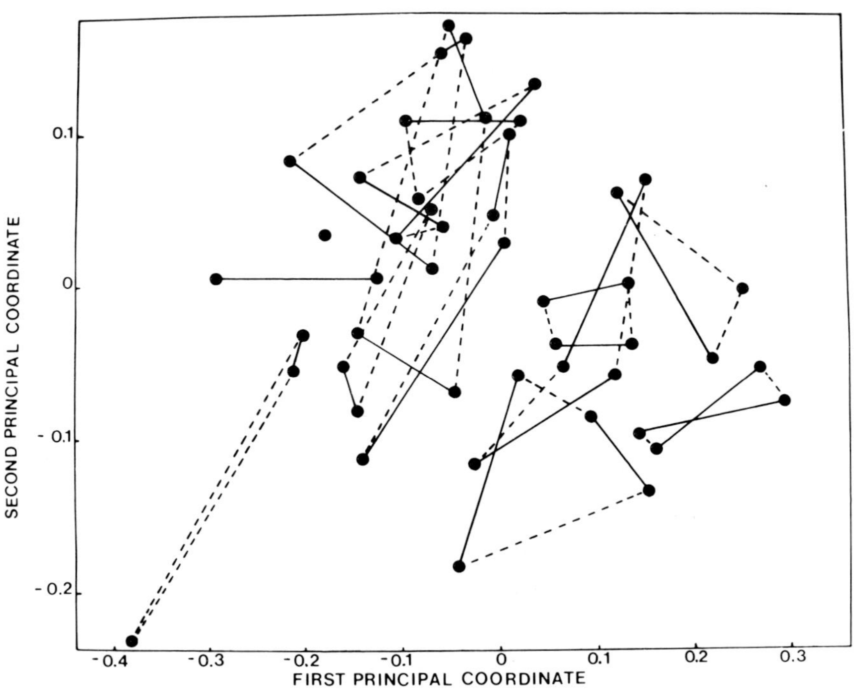

FIG. 7. Principal coordinates analysis of "*Bacillus amyloliquefaciens*" strains. Dotted lines join samples from replicate culture plates and complete lines samples from the same plate.

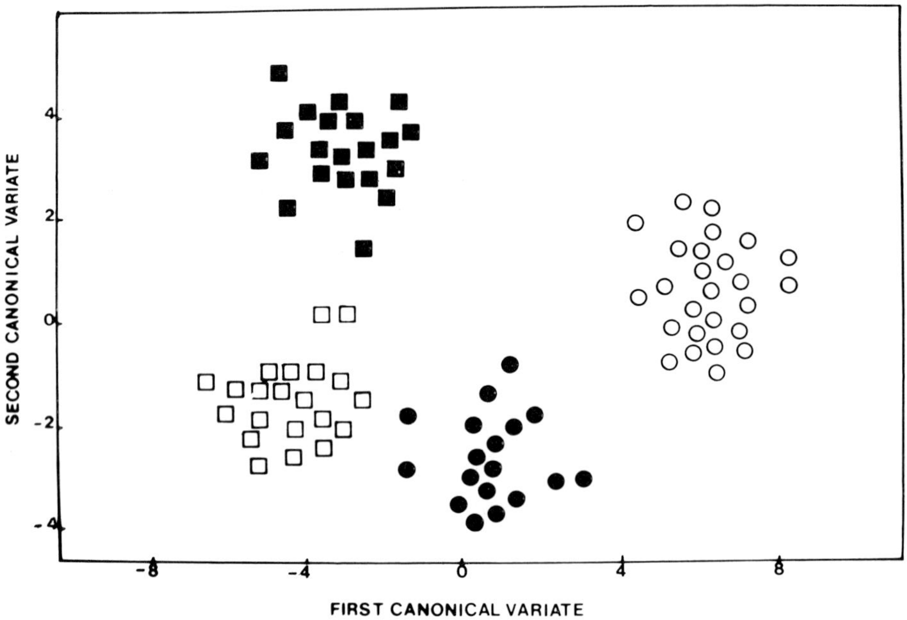

FIG. 8. Canonical variates analysis of four groups of *Bacillus* strains: ●, *B. subtilis;* ○, "*B. amyloliquefaciens*"; ■, *B. pumilus;* □, *B. licheniformis.*

level where only 4/86 samples were misclassified (Table 1). The masses chosen were used to identify eight samples, which were analysed as though from unknown strains, the samples being placed into their most probable group on the basis of these masses. Using this approach, all the unknowns analysed were correctly identified (Table 2).

TABLE 1

Classification of four groups of Bacillus *strains using stepwise discriminant analysis*

Group	Strain allocation				Misclassified
	S	P	L	A	
B. subtilis (S)	16	1	1	0	2/18
B. pumilus (P)	0	19	0	1	1/20
B. licheniformis (L)	0	0	22	0	—
"*B. amyloliquefaciens*" (A)	1	0	0	25	1/26
Total	17	20	23	26	4/86

TABLE 2

Identification of eight unknowns, using stepwise discriminant analysis

	Distance from group mean				
Actual identity	S	P	L	A	Identification
B. subtilis (S)	3.7	19.6	17.2	22 2	S
B. subtilis (S)	14.5	27.8	41.4	15.9	S
B. pumilus (P)	29.5	6.9	26.6	30.8	P
B. pumilus (P)	45.5	13.0	43.1	33.4	P
B. licheniformis (L)	24.6	24.4	7.2	38.4	L
B. licheniformis (L)	10.0	16.1	7.5	27.1	L
"*B. amyloliquefaciens*" (A)	39.8	20.8	32.5	11.6	A
"*B. amyloliquefaciens*" (A)	87.6	58.2	91.6	32.5	A

Discussion

The main advantages of Py–MS as a tool for the characterisation of microorganisms are that it is both rapid (< 10 min per sample) and widely applicable—if it can be grown it can be pyrolysed! Sample preparation is simple—in most cases, samples can be taken directly from culture plates. Sample sizes are sufficiently small to allow the analysis of single colonies—although this depends on the organism to be studied. The routine operation of a Py–MS system is simple and requires little operator expertise, although experience of mass spectrometry is essential to maintain the calibration of the instrument and to effect regular servicing. It is also amenable to automation and computerisation and is flexible enough to be used for a number of analytical tasks in a multifunction laboratory.

The disadvantages are related to the novelty and complexity of the method, which bridges three disciplines—microbiology, analytical chemistry and mathematics (statistics). Few laboratories have sufficient expertise available in all three disciplines to operate a pyrolysis mass spectrometer effectively. Also, the apparatus is expensive (currently £80–100K), and the data handling requires the use of a main-frame computer, or a large minicomputer, and a number of complex programs. However, it is likely that lower cost instrumentation, including on-line data handling using micro-computers, will be available in the future.

The key to the acceptance and use of Py–MS by microbiologists is that it should produce classifications that do not differ substantially from those obtained using conventional or other chemotaxonomic methods and that it should be stable enough to permit routine identification. So far, and this is supported by the example given in this chapter, this has proved to be the case, but there are only a few research groups using Py–MS at the present time and the number of published applications is small.

Thus far, Py–MS has mostly been used to investigate established taxonomies. The use of discriminant analysis using an *a priori* group structure is appropriate to this task, but the generation of novel taxonomies by using Py–MS will require a different data analysis approach, which is currently under development. The use of conventional numerical taxonomy techniques, i.e. calculation of similarity coefficients followed by cluster analysis has had only limited success with Py–MS data (MacFie & Gutteridge 1982), probably because of the complex nature of the data and the abundance of correlations.

Although overall correlation of Py–MS results with other taxonomic data is a desired requirement, because Py–MS examines, albeit indirectly, the total cellular composition of microorganisms, there may be situations where strains are differentiated irrespective of their taxonomic status. For example, a strain overproducing a secondary metabolite or depositing an intracellular storage compound may produce a spectrum dominated by the thermal degradation products of that compound.

The ability of Py–MS to differentiate at the strain level (Gutteridge & Norris 1983) is particularly valuable and one of the prime uses of the method may turn out to be in epidemiology, as some initial studies on hospital isolates of *Klebsiella* (Meuzelaar *et al.* 1982) have indicated. With this type of work, the concept of operational fingerprinting, i.e. comparing a batch of unknown strains with standard reference strains, is useful because it eliminates the need to establish libraries of spectra that have to remain stable for long periods of time.

In conclusion, the use of Py–MS for the rapid characterisation of microorganisms is a technique at an early stage of development but is one with a good deal of long-term potential. The ability to analyse small amounts of material with minimum sample preparation to obtain, in minutes, fingerprint data that can be used for identification is unparalleled by other characterisation methods and will be a powerful tool for diagnostic microbiology.

References

BOON, J. J., DE BOER, W. R., KRUYSSEN, F. J. & WOUTERS, J. T. M. 1980 Pyrolysis mass spectrometry of whole cell walls and isolated cell wall polymers of *Bacillus subtilis* var. *niger* WM. *Journal of General Microbiology* **122**, 119–127.

BORST, J., VAN DER SNEE-ENKELAAR, A. C. & MEUZELAAR, H. L. C. 1978 Typing of *Neisseria gonorrhoeae* by pyrolysis mass spectrometry. *Antonie van Leeuwenhoek* **44**, 253.

DIXON, W. J. 1975 *Biomedical Computer Programs*. Los Angeles: University of California Press.

ESHUIS, W., KISTEMAKER, P.G. & MEUZELAAR, H. L. C. 1977 In *Analytical Pyrolysis* eds. Jones, C. E. R. & Cramers, C. A., pp 151–156. Amsterdam: Elsevier.

GORDON, R. E., HAYNES, W. C. & HOA-NAY PANG, C. 1973 *The Genus Bacillus*. Washington D.C: U.S. Department of Agriculture.

GOWER, J. C. 1966 Some distance properties of latent root and vector methods used in multivariate analysis. *Biometrika* **53,** 325–338.

GUTTERIDGE, C. S. & NORRIS, J. R. 1983 Possible applications of pyrolysis mass spectrometry to biotechnology. In *Biotech 83* pp. 919–929. U.K.: Online Publications Ltd.

GUTTERIDGE, C. S. & PUCKEY, D. J. 1982 Discrimination of some Gram negative bacteria by direct probe mass spectrometry. *Journal of General Microbiology* **128,** 721–730.

GUTTERIDGE, C. S., MACKEY, B. M. & NORRIS, J. R. 1980 A pyrolysis gas-liquid chromatography study of *Clostridium botulinum* and related organisms. *Journal of Applied Bacteriology* **49,** 165–174.

HUFF, S. M., MEUZELAAR, H. L. C., POPE, D. L. & KJELDSBERG, C. R. 1981 Characterization of leukemic and normal white blood cells by Curie-point pyrolysis mass spectrometry. 1. Numerical evaluation of the results of a pilot study. *Journal of Analytical and Applied Pyrolysis* **3,** 95–111.

HUGHES, J. C., WHEALS, B. B. & WHITEHOUSE, M. J. 1977 Simple technique for the pyrolysis mass spectrometry of polymeric materials. *Analyst* **102,** 143–144.

HUNT, D. F., STAFFORD, J. C., CROW, F. W. & RUSSELL, J. W. 1976 Pulsed positive negative ion chemical ionization mass spectrometry. *Analytical Chemistry* **48,** 2098–2105.

IRWIN, W. J. 1979 Analytical pyrolysis—an overview. *Journal of Analytical and Applied Pyrolysis* **1,** 1–25, 89–122.

IRWIN, W. J. & SLACK, J. A. 1978 Analytical pyrolysis in biomedical studies, a review. *Analyst* **103,** 673–704.

KOWALSKI, B. R. 1975 Measurement analysis by pattern recognition. *Analytical Chemistry* **47,** 1152A–1162A.

LINCOLN, K. A. 1965 Flash pyrolysis of solid fuel materials by thermal radiation. *Pyrodynamics* **2,** 133–143.

MACFIE, H. J. H. & GUTTERIDGE, C. S. 1982 Comparative studies on some methods for handling quantitative data generated by analytical pyrolysis. *Journal of Analytical and Applied Pyrolysis* **4,** 175–204.

MACFIE, H. J. H., GUTTERIDGE, C. S. & NORRIS, J. R. 1978 Use of canonical variates analysis in differentiation of bacteria by pyrolysis gas-liquid chromatography. *Journal of General Microbiology* **104,** 67–74.

MAHALANOBIS, P. C. 1936 On the generalized distance in statistics. *Proceedings of the National Institute of Sciences of India* **2,** 49–55.

MEUZELAAR, H. L. C. 1978 Pyrolysis mass spectrometry: prospects for interlaboratory standardization. *Proceedings 26th Annual ASMS Conference on Mass Spectrometry and Allied Topics* pp. 29–41.

MEUZELAAR, H. L. C. & KISTEMAKER, P. G. 1973 A technique for fast and reproducible fingerprinting of bacteria by pyrolysis mass spectrometry. *Analytical Chemistry* **45,** 587–590.

MEUZELAAR, H. L. C., KISTEMAKER, P. G., ESHUIS, W. & BOERBOOM, H. A. J. 1976 Automated pyrolysis mass spectrometry: Application to the differentiation of micro-organisms. *Proceedings 26th Annual Conference on Mass Spectrometry and Allied Topics* pp. 29–41.

MEUZELAAR, H. L. C., HAVERHAMP, J. & HILEMAN, F. D. 1982 *Pyrolysis Mass Spectometry of Recent and Fossil Biomaterials. Compendium and Atlas.* Amsterdam: Elsevier.

NELDER, J. A. 1979 *Genstat Reference Manual.* Scientific and Social Service Program Library, University of Edinburgh.

NIE, N. H., HULL, C. H., JENKINS, J. G., STEINBRENNER, K. & BENT, D. H. 1975 *Statistical Package for the Social Sciences (SPSS),* 2nd ed. New York: McGraw Hill.

O'DONNELL, A. G., MACFIE, H. J. H. & NORRIS, J. R. 1980a An investigation of the relationships between *B. cereus, B. thuringiensis* and *B. mycoides* using pyrolysis gas-liquid chromatography. *Journal of General Microbiology* **32,** 306–309.

O'Donnell, A. G., Norris, J. R., Berkeley, R. C. W., Claus, D., Kaneko, T., Logan, N. A. & Nozaki, R. 1980b Characterization of *Bacillus amyloliquefaciens* by pyrolysis gas-liquid chromatography, deoxyribonucleic acid–deoxyribonucleic acid hybridization, biochemical tests and API systems. *International Journal of Systematic Bacteriology* **30,** 448–459.

Oxborrow, G. S., Fields, N. D. & Puleo, J. R. 1976 Preparation of pure microbiological samples for PGLC studies. *Applied and Environmental Microbiology* **32,** 306–309.

Oyama, V. I. 1963 Use of gas chromatography for the detection of life on Mars. *Nature, London* **200,** 1058–1059.

Reiner, E. 1965 Identification of bacterial strains by pyrolysis gas-liquid chromatography. *Nature, London* **200,** 1272–1274.

Schulten, H. R., Beckey, H. D., Meuzelaar, H. L. C. & Boerboom, A. J. H. 1973 High resolution field ionization mass spectrometry of bacterial pyrolysis products. *Analytical Chemistry* **45,** 191–195.

Shute, L. A., Gutteridge, C. S., Norris, J. R. & Berkeley, R. C. W. 1983 Curie-point pyrolysis mass spectrometry applied to characterization and identification of selected *Bacillus* species. *Journal of General Microbiology* **130,** 343–355.

Wieten, G., Haverkamp, J., Meuzelaar, H. L. C., Engel, H. B. W. & Berwald, L. G. 1981 Pyrolysis mass spectrometry: a new method to differentiate between the mycobacteria of the 'Tuberculosis' complex and other mycobacteria. *Journal of General Microbiology* **122,** 109–118.

Wieten, G., Haverkamp, J., Berwald, L. G., Groothuis, D. G. & Draper, P. 1982 PMS: its application to mycobacteriology, including *Mycobacterium leprae. Annales de Microbiologie* **133B,** 15–27.

Windig, W. & Haverkamp, J. 1982. Pyrolysis mass spectrometry of selected yeast species: *I. Rhodospiridium; II Spirodiobolus* and relationships. *Studies in Mycology* **22,** 56–64.

Windig, W., Kistemaker, P. G., Haverkamp, J. & Meuzelaar, H. L. C. 1979 The effects of sample preparation, pyrolysis and pyrolysate transfer conditions on pyrolysis mass spectra. *Journal of Analytical and Applied Pyrolysis* **1,** 39–52.

Windig, W., Kistemaker, P. G. & Haverkamp, J. 1981a Chemical interpretation of differences in pyrolysis-mass spectra of simulated mixtures of biopolymers by factor analysis with graphical rotation. *Journal of Analytical and Applied Pyrolysis* **3,** 199–212.

Windig, W., DeHoog, G. S. & Haverkamp, J. 1981b Chemical characterization of yeasts and yeast-like fungi by factor analysis of their pyrolysis-mass spectra. *Journal of Analytical and Applied Pyrolysis* **3,** 213.

Zemany, P. D. 1952 Identification of complex organic materials by mass spectrometric analysis of their pyrolysis products. *Analytical Chemistry* **24,** 1709–1713.

Bacterial Taxonomy Based on Electrophoretic Whole-Cell Protein Patterns

P. J. H. Jackman

Department of Bacteriology, Institute of Dermatology, London, UK

The major expression of the microbial genome results in the synthesis of about 2000 protein molecules that constitute the microbial cell. These molecules form an information source of immense potential and richness for the characterization of microorganisms. Many techniques have been developed for the separation and characterization of proteins and one of these, polyacrylamide gel electrophoresis (PAGE), is a particularly convenient and valuable tool in microbial systematics. Electrophoretic methods of suitable resolution and reproducibility provide a basis for microbial systematics that is objective, rapid and relevant to many laboratories. The application of whole-cell protein PAGE electrophoresis in microbial classification, identification and typing has been reviewed by Kersters & De Ley (1980) and Jackman (1984). Good correlation has been found between the results of DNA–DNA hybridization and PAGE studies in a number of genera including *Agrobacterium* (Kersters & De Ley 1975), *Brucella* (Morris 1973), *Propionibacterium* (Gross *et al.* 1978), *Pseudomonas* (Owen & Jackman 1982) and *Zymomonas* (Swings *et al.* 1976).

A Standard PAGE Technique for Microbial Systematics

Jackman (1984) has proposed a standard PAGE technique for microbial systematics, which is described in detail here. The technique, which has been adopted by the National Collection of Type Cultures (NCTC) as the foundation of a system for the computer identification of medically important bacteria, may be applicable elsewhere. Indeed, it would be advantageous for research workers to adopt a common technique in order to enhance the

CHEMICAL METHODS IN
BACTERIAL SYSTEMATICS

0-12-289675-0

prospect of interlaboratory reproducibility and comparability of patterns, to help focus research on the reproducibility and analysis of patterns, and to aid in the exchange of computer programs.

A priori considerations

Taxonomic studies using PAGE must be planned with care and pilot experiments may be necessary. It is important to have some idea of the variation that exists between the test strains since one-dimensional, whole-cell patterns can generally only be used to discriminate at the subgeneric level. This is certainly the case in genera that encompass a broad range of variation. If strains of several genera are under study, other taxonomic techniques should be employed to allocate strains to genera, and PAGE patterns may then be used to discriminate between strains within each genus.

Preliminary experiments on a small number of strains should be designed to find the optimal cell breakage and protein sample preparation method (see later section for suggestions). Once the sample preparation method has been chosen, the protein patterns of a range of test strains can be determined on a single gel. If the patterns prove to be very similar and more discrimination is required, as in epidemiological typing, then other PAGE methods such as two-dimensional separation, studies of isoenzymes or peptide mapping can be considered. It must be conceded that some minor differences visible between patterns adjacent on a single gel may not be densitometrically resolved or sufficiently reproducible between gels to enable computer analysis to be used. If the patterns are very diverse, some preselection of strains by other taxonomic criteria is indicated.

The next step is to carry out a quantitative reproducibility study. This is best done by comparing replicates of a single sample separated in a number of independently run gels. A numerical analysis of the replicates should give a mean similarity of at least 80%. Below this level the structure of a resulting dendrogram would differ little from one based on random data (Sneath & Johnson 1972). Experience using the standard PAGE method indicates that reproducibility can be about 95% (Whiley *et al.* 1982).

Culture of microorganisms

A medium that supports the optimum growth of the test strains should be used and all culture conditions kept as constant as possible. Kersters & De Ley (1975) found a strain of *Agrobacterium* to have fewer protein bands when grown on a simple defined medium compared with the pattern on a complex medium.

Sample preparation

It is difficult to standardise protein sample preparation inasmuch as organisms vary greatly in their susceptibility to cell-breakage methods. Two factors are often important: first, the use of the minimum amount of cell mass to facilitate the study of many strains and second, safety considerations when working with pathogens.

It is most convenient if cells can be broken by boiling in SDS because this kills and disrupts them in a single step and can be used with small quantities of cells. Cells (0.1 g wet weight) are harvested from solid media, care being taken not to disturb the agar surface, and suspended in 1.0 ml of sample treatment buffer (0.0625 *M* Tris–HCl, pH 6.8; 2% SDS; 5% 2-mercaptoethanol; 10% glycerol). The suspension is then heated at 100°C for 10 min, centrifuged at 10,000 *g* for a further 10 min and stored at −20°C until required. In many cases, all sample preparation steps can be performed in a single 1.5-ml Eppendorf tube. When cells cannot be broken in this way, other chemical methods such as lysis by lysozyme may be tried; should these fail, physical methods such as sonication, shaking with glass beads or pressure techniques can be tried prior to SDS treatment (Coakley *et al.* 1977). The physical methods, however, generally require a larger amount of cell mass and are often inconvenient with pathogens.

Electrophoresis

The electrophoresis method described is a modification from that of Laemmli (1970). The exact type of electrophoresis apparatus used may depend on what is already available, but the vertical slab apparatus manufactured by Hoefer Scientific Instruments (San Francisco, California, USA) can be recommended. The Hoefer multiple slab unit (Fig. 1) can be used to run as many as 12 gels, each with 20 samples at one time, under uniform electrophoretic conditions. The separation gel should be long enough to accommodate comfortably a 100-mm protein pattern; generally about 120 mm is sufficient. Gels 0.75 or 1.5 mm thick are recommended. The separation gel contains 10% total acrylamide with 2.7% crosslinking *N,N'*-bismethyleneacrylamide, 0.375 *M* Tris–Cl, and 0.1% SDS, and should be adjusted to pH 8.8. The stacking gel contains 5% total acrylamide with 2.7% crosslinking *N,N'*-bismethyleneacrylamide, 0.125 *M* Tris–Cl, and 0.1% SDS, and should be 10 mm long and at pH 6.8. Stock solutions should be made up in large quantities so that many gels can be prepared from the same batch of chemicals. The quantities given below are sufficient for 300, 0.75 mm-thick gels using the Hoefer apparatus.

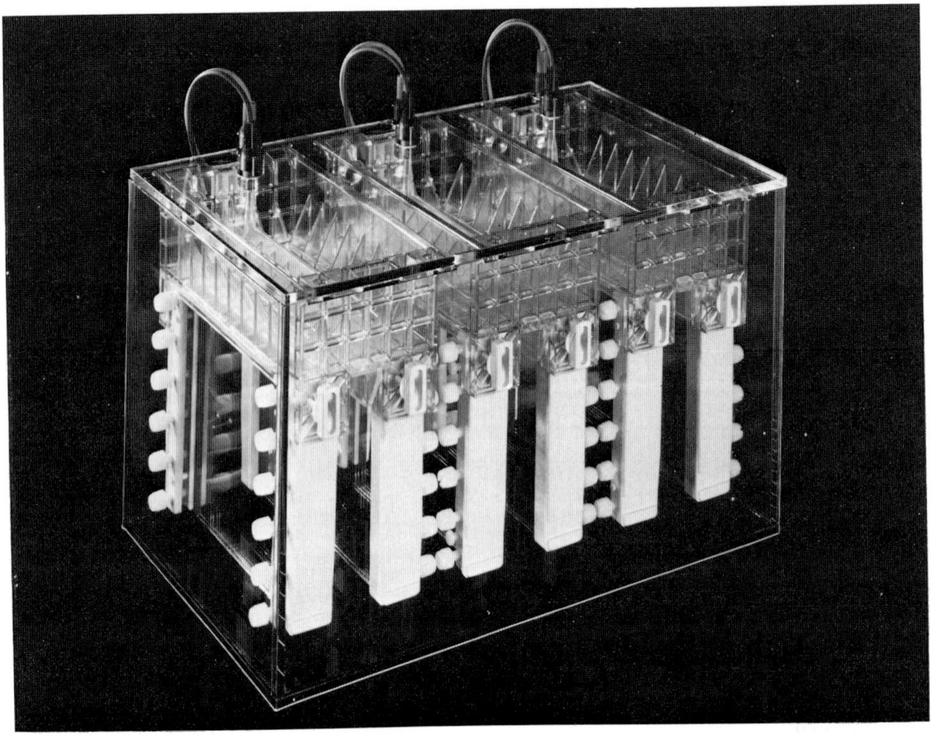

FIG. 1. Multigel electrophoresis apparatus. Courtesy of Hofer Scientific Instruments Ltd, 650 Fifth Street, Box 77387, San Francisco, USA.

Reagents for SDS–PAGE

2X Treatment buffer	
Stacking gel buffer	25 ml
SDS	4 g
Glycerol	20 ml
2-Mercaptoethanol	10 ml
Distilled water	to 100 ml
Monomer solution	
Acrylamide	584 g
N,N'-bismethyleneacrylamide	16 g
Distilled water (stir and filter)	to 2 l
Stacking gel buffer	
Tris (adjust to pH 6.8 with HCl)	30 g
Distilled water	to 500 ml
Separation gel buffer	
Tris (adjust to pH 8.8 with HCl)	363 g
Distilled water	to 2l

10% SDS	
SDS	10 g
Distilled water	to 100 ml

All of these solutions should be stored at 4°C except the 10% SDS which should be kept at room temperature.

Tank buffer	
Tris	48 g
Glycine	230 g
SDS	16 g
Distilled water	to 16 l
Stain stock	
Coomassie Blue R-250	20 g
Distilled water (stir and filter)	to 1 l
Stain	
Stain stock	250 ml
Methanol	1000 ml
Acetic acid	200 ml
Distilled water	to 2 l
Destain 1	
Methanol	1000 ml
Acetic acid	200 ml
Distilled water	to 2 l
Destain 2	
Methanol	500 ml
Acetic acid	700 ml
Distilled water	to 10 l

Chemicals are from Sigma, except SDS, glycine, methanol, acetic acid which are from BDH.

Preparation of gels

Add a small paper label with a unique gel number to the bottom left corner of each gel cassette prior to pouring. A PVC blank comb should be used to form the top surface of the separation gel 10 mm below the bottom of a PTFE well-forming comb to give 10 mm of stacking gel. For 12 0.75 mm thick gels with the Hoefer apparatus, the quantities are as follows:

Monomer solution (4°C)	67 ml
Separation gel buffer (4°C)	50 ml
Distilled water (4°C)	to 197 ml

Deaerate under vacuum for 10 min in a side-arm 500-ml conical flask with stirring, then add:

Ammonium persulphate	0.45 ml of a freshly made 10% solution
10% SDS (room temperature)	2 ml
TEMED (4°C)	60 μl

Swirl to mix and pour gels using a 50-ml syringe with a short length of plastic tubing. After 20 min remove the blank combs and overlay with working-strength separation gel buffer. The amounts of deaeration should be adjusted to give a polymerization time of 10 ± 2 min. After overlaying, the gels should not be used for 24 h to allow complete polymerization to occur. When ready, insert well-forming combs and cast the stacking gels:

Monomer solution (4°C)	10 ml
Stacking gel buffer (4°C)	15 ml
Distilled water (4°C)	to 59 ml

Deaerate under vacuum for 10 min with stirring, then add:

Ammonium persulphate	0.30 ml of a freshly made 10% solution
10% SDS	(room temperature) 0.6 ml
TEMED (4°C)	40 μl

Swirl to mix and pour gels. After 30 min the combs are carefully removed and the wells are rinsed with distilled water and then filled with tank buffer.

Samples are best applied using a 10- or 50-μl syringe with repeater. The outer 20 mm of gel should not be used for samples, to avoid distortion. A protein sample of 10 μl of a 10-mg/ml solution is often satisfactory, using the gels described. The central well must always contain an aliquot of a reference strain sample, which can then be used to measure and control reproducibility. Bromophenol blue should be added to this reference sample (final concentration 0.001%) to act as a visible marker during electrophoresis. A reference sample, containing a mixture of pure proteins of known molecular weight, (e.g. low-molecular-weight kit, Pharmacia) on every gel is also recommended for each gel. The electrode buffer is 0.25 *M* Tris, 0.192 glycine, and 0.1% SDS and is at pH 8.3. The samples are then run at 15 mA per 0.75 mm gel until the bromophenol blue has migrated 100 ± 1 mm through the separation gel, which generally takes about 4 h. The gels are then stained overnight in Coomassie brilliant blue R-250 (0.125%) in 50% methanol–10% acetic acid. Gels are destained in Destain 1 for 1 h and then in changes of Destain 2 until a clear background is obtained. After destaining, pattern length must be brought back to 100 ± 1 mm by immersing the gel in appropriate strength methanol–10% acetic acid. Gels are then dried under vacuum on a Hoefer drier for ease of handling, scanning and long-term storage.

Analysis of patterns

A set of computer programs for numerical taxonomic analysis of protein patterns are available including listings (Jackman 1983; Jackman *et al.* 1983). The analysis requires a scanning densitometer linked to a computer. If the densitometer lacks a computer interface, an analogue-to-digital converter (ADC) can be used to digitise the chart recorder output and send the absorbance values to a computer.

Program description

The program suite consists of five programs written in BASIC for the Commodore PET 3032 microcomputer and 3040 disk unit.

TRECORD 1.00

Program TRECORD 1.00 accepts voltage values from an ADC at constant time intervals and records them as a sequential file on a floppy disk. The voltage values (*Y%* values) then represent optical absorbance at regular intervals along the gel. Using an eight-bit ADC, the *Y%* values will be integers between 0 and 255. No subroutine for accepting values from an ADC is given since this will vary from model to model and is usually supplied by the manufacturer. Routines for use with the Ortholog 181 (Ortholog Ltd., Middlesex, UK) are available from the author. The frequency of sampling should be adjusted with the scanning speed of the densitometer to give a suitable number of points per mm of scan. Ten values per mm of pattern gives a suitable resolution for permanent record. If required, the trace may be reduced for subsequent analysis by averaging or interpolation (Jackman *et al.* 1983).

PRECORD 1.00

Program PRECORD 1.00 detects distinct peaks in traces generated by the program TRECORD 1.00. In order for these data to be used successfully, two factors need to be optimised. First, noise in the signal from the ADC gives rise to spurious peaks if not minimised. One successful method of noise reduction is signal averaging. Such a program for use with the Commodore PET 3032 and Ortholog 181 ADC is available from the author. The

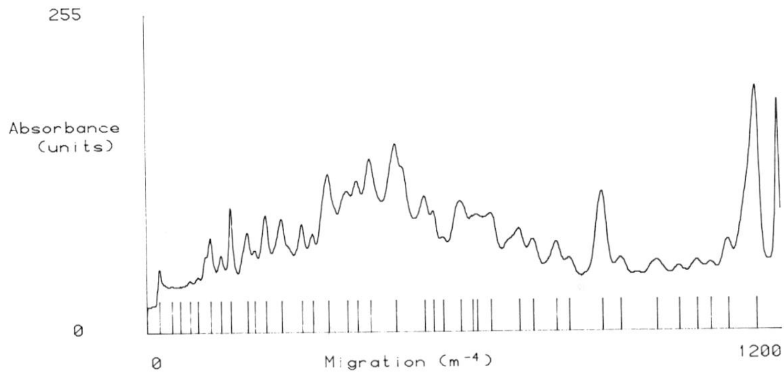

FIG. 2. Peak detection in absorbance trace using PRECORD 1.00.

second factor is the resolution of band positions. The best results are obtained if the trace is reduced to as few values as possible whilst still maintaining all peaks separate. This gives the fastest calculation of similarity without distorting the data. Use of too high a resolution trace may reduce reproducibility to an unacceptably low level since PMATRIX 1.00 requires exact matching of bands. PRECORD 1.00 uses a simple algorithm to detect the characteristic shape of peak tops with a refinement to detect flat peak tops up to any desired width. The number and position of the peaks detected are recorded on disk. An example of peak detection is shown in Fig. 2.

PMATRIX 1.00

Program PMATRIX 1.00 uses the PEAKFILES generated by PRECORD 1.00 to form a similarity matrix using the Dice coefficient (Dice 1945).

$$S_d = \frac{2m}{a+b}100$$

where:

m = number of matching bands in patterns A and B, and a,b = number of bands in patterns A and B.

The same routine to that in TMATRIX 1.00 is provided to search for the best-fit between misaligned patterns. The similarity matrix formed by PMATRIX 1.00 may be clustered with the program TCLUSTER 1.00.

TMATRIX 1.00

Program TMATRIX 1.00 forms a similarity matrix using the correlation coefficient to calculate the similarity between traces (Sneath & Sokal 1973).

The program will accommodate up to 100 traces each of 100 points on a PET 3032.

Background subtraction

There is often considerable background absorbance in whole-cell patterns that is due mainly to overlapping of peaks (Feltham & Sneath 1979). In this program, the background trend may be removed. For each trace, the program constructs a grossly averaged trace consisting of just ten points. A fraction of this grossly averaged trace, typically 0.5–0.8, is then subtracted from the original. Taxonomies based on such traces show increased discrimination between strains owing to the increased emphasis on the pattern of peaks in comparison with the overall background trend, which tends to be similar in all traces (Jackman 1981).

Best-fit on the X*-axis*

Some misalignment will usually exist between patterns from different electrophoresis runs however much attention is given to technical detail in PAGE. Misalignment in the X-axis considerably reduces calculated similarity between otherwise identical traces (Jackman 1981). This program has an option to search for the best-fit between traces by recalculating the similarity between each pair of traces at successive points of misalignment on the overlapping portion of the traces using a similar idea to that of Feltham & Sneath (1979).

TCLUSTER 1.00

Program TCLUSTER 1.00 clusters the similarity matrix generated by TMATRIX 1.00 using the unweighted pair group method using averages algorithm (UPGMA, Sneath & Sokal 1973) (Fig. 3). The instructions for drawing a dendrogram should be output to a printer if the dendrogram is to be drawn by hand. If an X–Y plotter is available, the dendrogram may be drawn automatically. A program for plotting dendrograms and shaded matrices using a Tektronic 4662 plotter interfaced to a Commodore PET 3032 is available from the author.

Computer time

The current disadvantage of microcomputers is their slowness in comparison with main frames, although this should be improved in the future. TMATRIX 1.00 takes about 11 h per point of alignment to produce a matrix

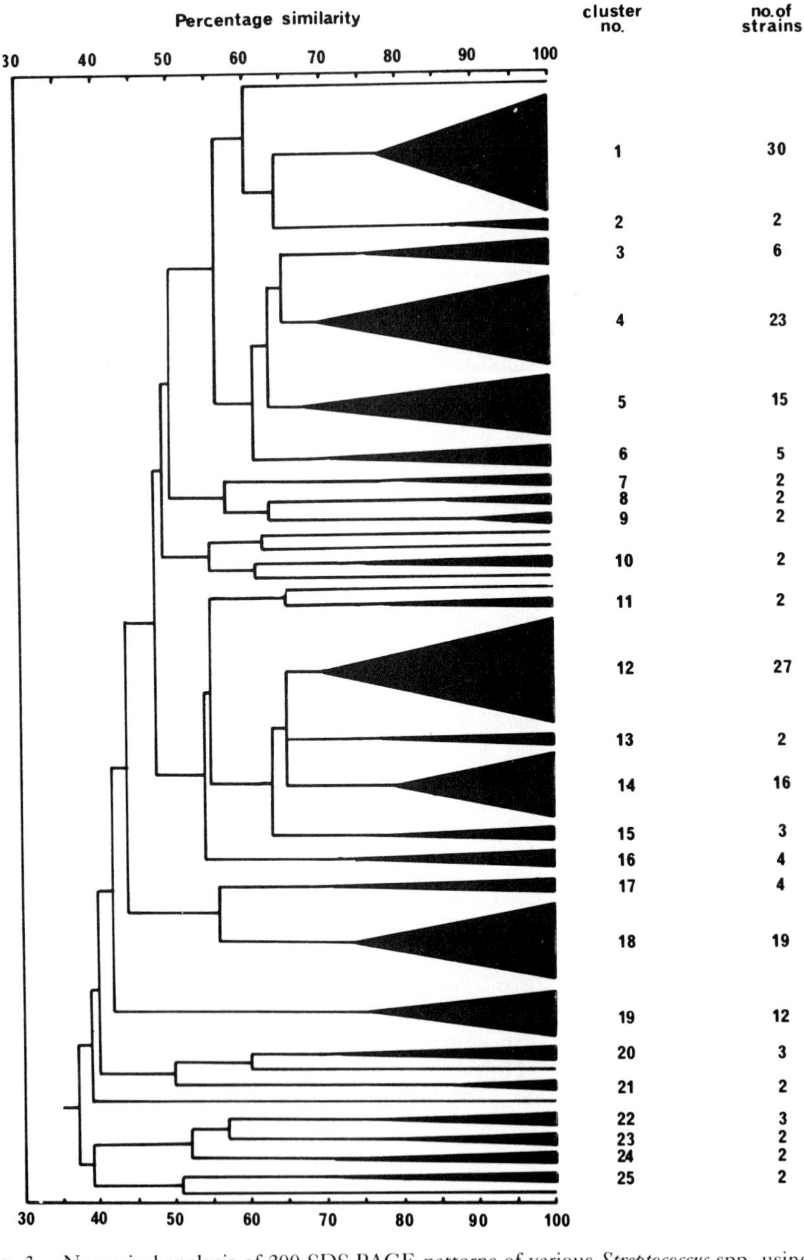

FIG. 3. Numerical analysis of 200 SDS-PAGE patterns of various *Streptococcus* spp. using the correlation coefficient and clustered using the unweighted pair-group method using averages (UPGMA).

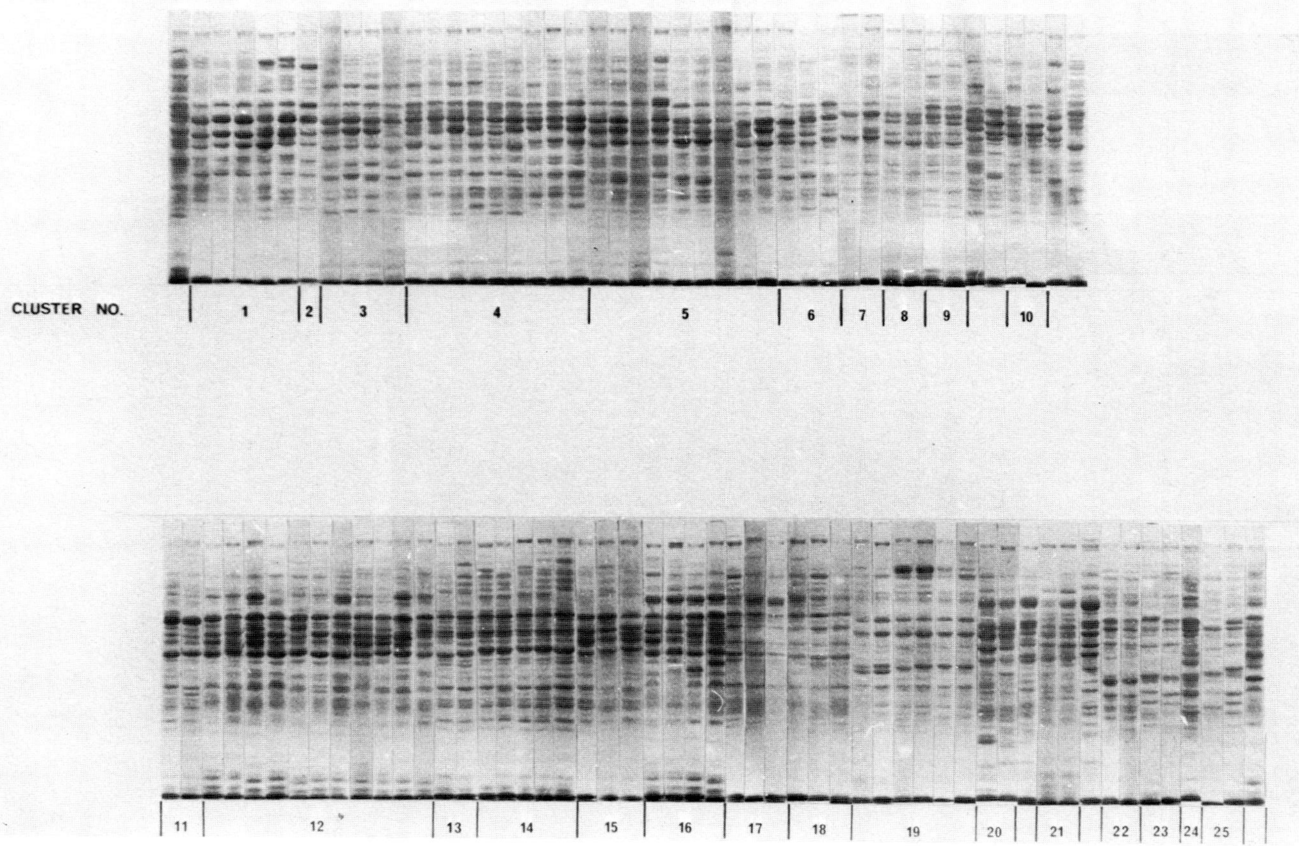
CLUSTER NO.
1
2
3
4
5
6
7
8
9
10
11
12
13
14
15
16
17
18
19
20
21
22
23
24
25

from 100 traces each of 100 points on a Commodore PET 3032. Clearly with some data it will be necessary to run overnight or longer. A refined program (TMATRIX 2.00), specifically for the Commodore PET 3032, incorporating a machine code correlation coefficient routine is available from the author. TMATRIX 2.00 is some 15-fold faster than TMATRIX 1.00 and in addition can accommodate up to 200 traces each of 100 points.

Reproducibility of Protein Patterns

Reproducibility of band mobilities is the single most important factor in the use of protein patterns in numerical taxonomy. The first aspect of reproducibility that must be considered is related to factors affecting the separation of a protein sample determined by electrophoresis such as run conditions and gel structure. Reproducibility of the electrophoretic technique is best measured by the numerical analysis of aliquots of a single protein sample separated on several independently run gels.

In the study of Whiley *et al.* (1982), 54 aliquots of a single protein sample were separated on three independently run gels using 10% SDS gels. The best measure of reproducibility is probably the mean of the matrix of similarities between those patterns which, in this study, used the best-fit procedure described; the correlation coefficient was 95%. Jackman (1981, 1982), in a study of gradient PAGE patterns, applied the same methods of pattern analysis and found a reproducibility of 82%. Values below 80%, representing an average test error of more than about 10%, are considered unacceptably low (Sneath & Johnson 1972). It should be noted that reproducibility is higher within a gel and in gels run together than in samples separated in different electrophoretic runs.

The second aspect of reproducibility concerns factors that can vary from sample to sample such as age, differences in sample preparation, and differences in microbial culture conditions. The effect of some of these variations was measured by Jackman (1981) and was found to be within the same range as the reproducibility of the analytical method. Four independently cultured and prepared samples gave identical patterns. A *Corynebacterium* strain cultured for 4 days on medium lacking either Tween 80 or horse serum showed patterns varying from those obtained from cells grown on the complete medium only in the relative intensity of a few bands. Strains cultured for 1, 2, 4 and 7 days showed some variation inasmuch as there was a loss of several bands in the 2- and 7-day samples. Long-term storage of protein samples, i.e. more than 6 months, causes a general deterioration and increase in protein denaturation but has no particular effect on band mobilities. It would be expected that gross variations in culture, or in cell-breakage method, would have some influence on protein patterns, but these aspects have not been fully explored. Variations in total protein concentration and in background staining do affect the pattern in the *Y*-axis but have only a small effect on similarities if the similarity coefficients described here are employed.

Resolution of Protein Patterns

A priori, the resolution of the electrophoretic system used would seem to be of great importance. The densitometer employed in a study may place the main restriction on resolution. Under optimal conditions, bands down to about 0.5 mm apart can be resolved on separate peaks, although bands closer together sometimes appear as shoulders (Jackman 1982). High-resolution traces composed of many points may also be unsatisfactory in some cases because of constraints on computing time. The resolution of the recording system may also be impaired by the presence of noise in the signal from the densitometer or ADC. A proportion of the fainter bands will produce peaks that fall into the same range of amplitude and width as noise peaks. Considerations of resolution or noise seem to be of limited importance, however, where a correlation coefficient is used, since similarities calculated on 5 points/mm traces have been found to be identical to those calculated on 1 point/mm point traces (Jackman 1982). Such considerations do become very important where a coefficient is based on the peak positions as additional peaks may be introduced by noise or peaks lost by merging owing to low resolution scanning. Possibly the main effect of electrophoretic system resolution that should be considered is on the reliability of classifications or identifications based on protein patterns. The higher the resolution, the less is the probability of unrelated strains showing a high similarity by chance. The phenomenon of "intruders" noted by Kersters & De Ley (1975) may be related to the moderate resolution of their electrophoretic system, although some degree of this phenomenon would be expected in any numerical study.

Conclusions

PAGE patterns seem to possess many fine attributes that can be exploited for measuring systematic relationships between microorganisms. They represent a large, randomly selected number of perhaps single-gene characters. Pattern can be determined relatively easily and quickly and the same technique can be applied to any group of microorganisms. The recording and objective comparison of patterns can be performed by computers. Taxonomic relationships based on PAGE patterns show a high correlation with both those from numerical taxonomy based on classical microbiological tests and those derived from DNA–DNA hybridization measurements. Considering the ease of protein pattern comparison relative to DNA–DNA hybridization it seems likely that electrophoretic protein patterns could be used for rapid screening of isolates prior to hybridization of selected strains and indeed supplant DNA–DNA hybridization studies altogether in many applications in microbial systematics. Despite these attributes, PAGE pattern numerical taxonomy is still comparatively rarely undertaken, and some limitations must be recognised.

The reproducibility level in protein pattern numerical taxonomy is of critical importance and improvements in reproducibility should be pursued vigorously. Lack of reproducibility is paradoxically exacerbated by high-resolution electrophoretic systems. Proteins bands in gradient PAGE or iso-electric focusing may frequently be only 0.2–0.5 mm wide, so a very high accuracy of matching of homologous bands is necessary to achieve a high calculated similarity. It is proposed that a standardised system based on that of Laemmli (1970), which uses electrophoresis in 10% homogeneous gels in the presence of SDS, might prove suitable for most practical systematic studies.

The resolution of the densitometry and the signal-to-noise ratio of the recording system should be at least equal to the reproducibility of the electrophoretic system and will be the critical factor if a similarity coefficient based on band position is to be employed. Reproducibility can be improved by the use of best-fit procedures in the X-axis. However, if computing time is restricted, attention to the technical aspects of the electrophoresis system may prove equally productive. The use of microcomputers for numerical taxonomy based on protein patterns has been shown to be perfectly feasible and increases the range of laboratories that could carry out the procedures. It is essential, however, to optimise the speed of the programs and machine code programing may be desirable under some circumstances. The use of protein patterns in classification, identification and typing of bacterial strains undoubtedly has a very high potential, and it is hoped that the introduction of a standard technique will help to realise that potential in the future.

The Medical Research Council, the Dunhill Trust, the Royal Society, and the Mason Medical Foundation are thanked for financial support of this work. R. A. Whiley is thanked for provision of Fig. 3.

References

Coakley, W. T., Bater, A. J. & Lloyd, D. 1977 Disruption of micro-organisms. *Advances in Microbial Physiology* **16**, 279–341.

DICE, L. R. 1945 Measures of the amounts of ecologic association between species. *Ecology* **26**, 297–302.

FELTHAM, R. K. A. & SNEATH, P. H. A. 1979 Quantitative comparison of electrophoretic traces of bacterial proteins. *Computers and Biomedical Research* **12**, 247–263.

GROSS, C. S., FERGUSON, D. A. & CUMMINS, C. S. 1978 Electrophoretic protein patterns and enzyme mobilities in anaerobic coryneforms. *Applied and Environmental Microbiology* **35**, 1102–1108.

JACKMAN, P. J. H. 1981 Taxonomy of aerobic axillary coryneforms based on electrophoretic protein patterns. Ph.D. Thesis, University of London.

JACKMAN, P. J. H. 1982 Classification of *Corynebacterium* species from axillary skin by numerical analysis of electrophoretic protein patterns. *Journal of Medical Microbiology* **15**, 485–492.

JACKMAN, P. J. H. 1983 A program in BASIC for numerical taxonomy of micro-organisms based on electrophoretic band positions. *Microbios Letters* **23**, 119–124.

JACKMAN, P. J. H. 1984 Characterisation of micro-organisms by electrophoretic protein patterns. In *New Methods for the Detection and Characterisation of Micro-organisms* ed. Gutteridge, C. S. Chichester: John Wiley and Sons. (In press.)

JACKMAN, P. J. H., FELTHAM, R. K. A. & SNEATH, P. H. A. 1983 A program in BASIC for numerical taxonomy of micro-organisms based on electrophoretic protein patterns. *Microbios Letters* **23**, 87–93.

KERSTERS, K. & DE LEY, J. 1975 Identification and grouping of bacteria by numerical analysis of their electrophoretic protein patterns. *Journal of General Microbiology* **87**, 333–342.

KERSTERS, K. & DE LEY, J. 1980 Classification and identification of bacteria by electrophoresis of their proteins. In *Microbiological Classification and Identification* eds. Goodfellow, M. & Board, R. G., pp. 273–297. London: Academic Press.

LAEMMLI, U. K. 1970 Cleavage of structural proteins during the assembly of the head of bacteriophage T4. *Nature, London* **227**, 680–685.

MORRIS, J. A. 1973 The use of polyacrylamide electrophoresis in taxonomy of *Brucella*. *Journal of General Microbiology* **76**, 231–237.

OWEN, R. J. & JACKMAN, P. J. H. 1982 The similarities between *Pseudomonas paucimobilis* and allied bacteria derived from analysis of deoxyribonucleic acids and electrophoretic protein patterns. *Journal of General Microbiology* **128**, 2945–2954.

SNEATH, P. H. A. & JOHNSON, R. 1972 The influence on numerical taxonomic similarities of errors in microbiological tests. *Journal of General Microbiology* **72**, 377–391.

SNEATH, P. H. A. & SOKAL, R. R. 1973 *Numerical Taxonomy: The Principles and Practice of Numerical Classification*. San Francisco: W. H. Freeman.

SWINGS, J., KERSTERS, K. & DE LEY, J. 1976 Numerical analysis of electrophoretic protein patterns of *Zymomonas* strains. *Journal of General Microbiology* **93**, 266–271.

WHILEY, R. A., HARDIE, J. M. & JACKMAN, P. J. H. 1982 SDS-polyacrylamide gel electrophoresis of oral streptococci. *In Proceedings of VIII International Symposium on Streptococci and Streptococcal Diseases* eds. Holm, S. E. & Christensen, P., pp. 61–62. Chertsey: Reedbooks Ltd.

Integrated Lipid and Wall Analysis of Actinomycetes

A. G. O'DONNELL, D. E. MINNIKIN AND M. GOODFELLOW

Departments of Organic Chemistry and Microbiology, University of Newcastle upon Tyne, Newcastle upon Tyne, UK

Integrated Chemotaxonomy

The impact of chemical studies on bacterial classification and identification is reflected in the number of genus and species descriptions that now contain details of chemical composition. Indeed, for certain actinomycete taxa, data on the guanine plus cytosine (GC%), lipid and cell wall diamino acid composition provide the most reliable way of identifying unknown isolates to the generic level (Goodfellow & Cross 1984; Minnikin & O'Donnell 1984). Developments in gas chromatography (GC) and high performance liquid chromatography (HPLC) and the increasing use of these techniques in bacterial systematics have made possible the introduction of simple and rapid chemical extraction and analysis procedures. The time taken to classify and identify bacteria successfully is often governed, therefore, by the length of incubation necessary to obtain sufficient biomass. Furthermore, the fact that different characters have different discriminating ability makes it advantageous to maximise the amount of chemotaxonomic information obtained from samples of biomass.

This chapter describes the development of an integrated chemotaxonomic procedure that enables profiles of lipids, wall amino acids and sugars to be obtained from the same biomass (30–50 mg dry weight). Clearly, methods used early in the procedure have been selected so that they do not interfere with or degrade components of interest extracted later in the system. All of the procedures are flexible and simple to use and should be readily adaptable for use in other laboratories. Outlines of the schemes used are shown in Figs. 1 and 2.

0-12-289675-0

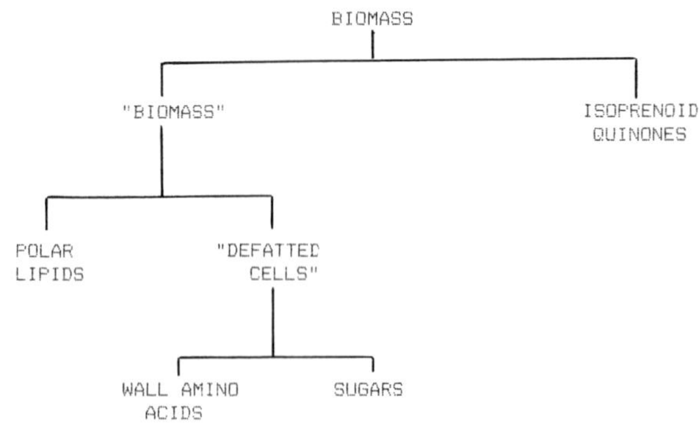

FIG. 1. Integrated extraction procedure for the analysis of isoprenoid quinones, polar lipids, wall amino acids and sugars.

Lipid analysis

Fatty acids

Taxonomically important lipids can be divided into three main types: long-chain compounds, isoprenoid quinones and polar lipids. The analysis of long-chain fatty acids and isoprenoid quinones are considered elsewhere (see Chapters 8, 9 and 13, this volume) and will only be discussed here in relation to their place in the integrated chemotaxonomic procedures outlined in Figs. 1 and 2. Actinomycete long-chain fatty acids can be divided into two main groups: those ranging in size from 12 to 20 carbons and the mycolic acids (see Chapter 12, this volume), which can have from 20 to ~90 carbons. The former group of acids correspond to the apolar portion of phospholipids, the most common being the nonhydroxylated straight-chain, monounsaturated, 10-methyl branched and iso and anteiso methyl branched acids (Minnikin & Goodfellow 1981a).

Fatty acids are usually analysed as their methyl ester derivatives, the latter being formed during acid (Table 1) or alkaline (Table 2) methanolysis. For certain representatives of genera such as *Actinomyces* (Kroppenstedt & Kutzner

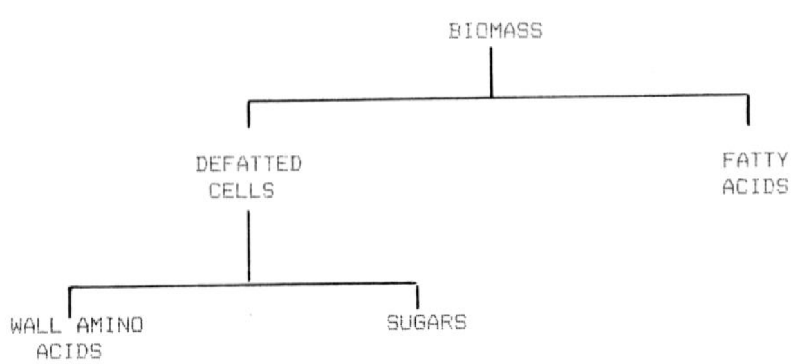

FIG. 2. Integrated procedure for the analysis of fatty acids, wall amino acids and sugars.

TABLE 1

*Acid methanolysis procedure**

Dry biomass (10–50 mg)

↓ ← Add 3 ml methanol–toluene–H_2SO_4 (6:3:0.2, v/v)

Heat overnight at 75°C

↓ ← Cool and add petroleum ether (bp 60–80°C)

Mix and centrifuge

↓

Pass upper layer through 1-cm column of ammonium hydrogen carbonate

↓ ← Wash lower layer with 1 ml petroleum ether

Pass upper layer through 1-cm column of ammonium hydrogen carbonate

↓

Fatty acid methyl esters (FAMES)

*From Minnikin *et al.* (1980).

1978) and *Lactobacillus* (Lechevalier 1977), which contain cyclopropane fatty acids, an alkaline extraction is preferred since, unlike acid methanolysis, it does not attack cyclopropane rings (Moss 1978; Christie 1982). To maximise the amount of information obtained from the same biomass sample (Fig. 2), an alkaline degradation is used since the chemically more exacting acid methanolysis makes it difficult to purify the defatted biomass further for cell wall analysis (Fig. 2). The chemotaxonomic method outlined in Fig. 2 omits the analysis of isoprenoid quinones and polar lipids and thereby provides a valuable alternative to the more detailed procedure shown in Fig. 1; it also has the advantage of requiring less starting biomass (10–20 mg dry weight).

Isoprenoid quinones and polar lipids

Bacterial plasma membranes are composed of amphipathic polar lipids associated with specific membrane proteins. Amphipathic polar lipids usally consist of hydrophilic head groups linked to two hydrophobic fatty acid chains. Phospholipids are the most common polar lipids, but glycolipids and acylated ornithine or lysine amides also fall into this category. These amphipathic lipids combine to form a bilayer, the hydrophobic interior of which

TABLE 2

*Alkaline methanolysis procedure**

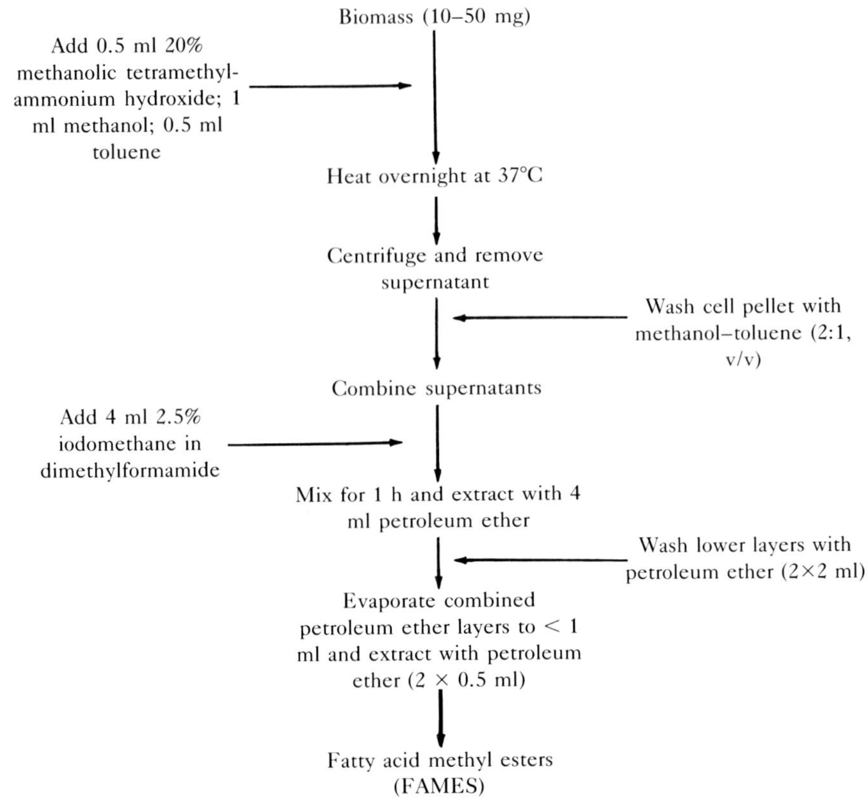

*From Minnikin *et al.* (1984a).

may provide a suitable environment for the location of isoprenoid quinones. To date, menaquinones are the only type of respiratory isoprenoid quinones found in actinomycetes and the variations in the number of isoprene units and hydrogenated double bonds make these membrane constituents of considerable chemotaxonomic value (see Chapters 8 and 13, this volume).

Isolated isoprenoid quinones (Table 3) are purified on 10 × 10-cm pieces of plastic-backed, silica-gel sheets (Merck 5735), using a developing mixture of petroleum ether (bp 60–80°C)–acetone (95:5 v/v). Purified isoprenoid quinones are detected by using UV light (254 nm.). Zones containing isoprenoid quinones are then cut from the plate and extracted using diethyl ether. To protect extracted isoprenoid quinones from losses due to the ef-

TABLE 3

*Combined extraction of isoprenoid quinones and polar lipids**

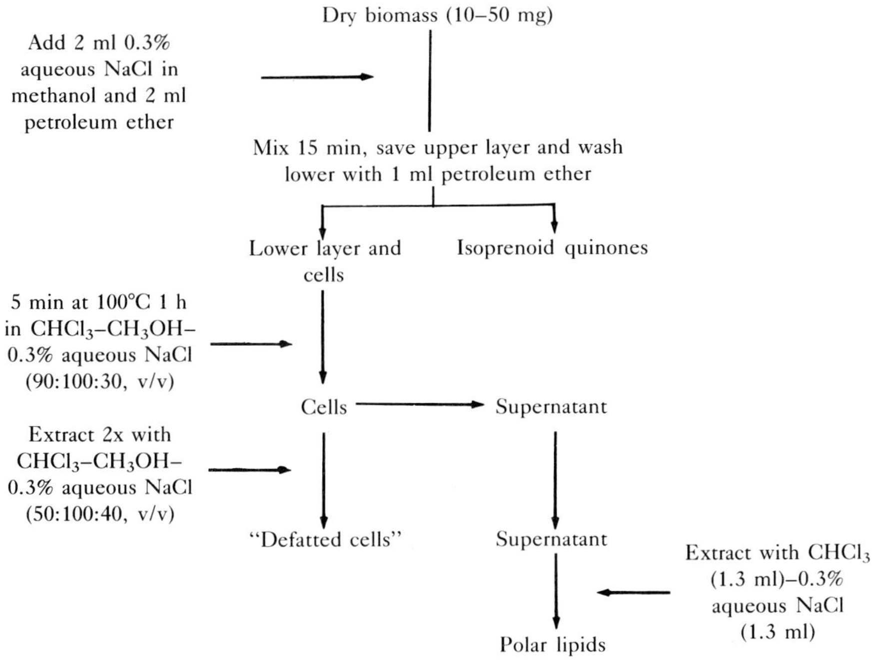

*From Minnikin *et al.* (1984b).

fects of air and light, samples are transferred to small glass vials, dissolved in petroleum ether (bp 60–80°C), wrapped in silver foil and stored at −20°C prior to analysis by HPLC, mass spectrometry or reverse-phase thin layer chromatography (see Chapters 8 and 13, this volume).

The lower layers remaining after isoprenoid quinone extraction are examined using a modification (Card 1973; Minnikin *et al.* 1984b) of the method of Bligh & Dyer (1959) to give a polar lipid extract (Table 3). Polar lipid patterns (Fig. 3) are obtained on 6.6 × 6.6-cm pieces of aluminum-backed silica-gel sheets (Merck 5554), using two-dimensional TLC (O'Donnell *et al.* 1982a; Embley *et al.* 1983; Minnikin *et al.* 1984b); solvent systems are shown on Fig. 3. All lipids are detected by spraying with 5% ethanolic molybdophosphoric acid followed by charring at 180°C. Specific spray reagents for lipid phosphate (Dittmer & Lester 1964), α-glycols (periodate-Schiff) (Shaw 1968), sugars (α-naphthol) (Jacin & Mishkin 1965) and amino groups (0.2% ninhydrin in water-saturated butanol) (Consden & Gordon 1948) are also used.

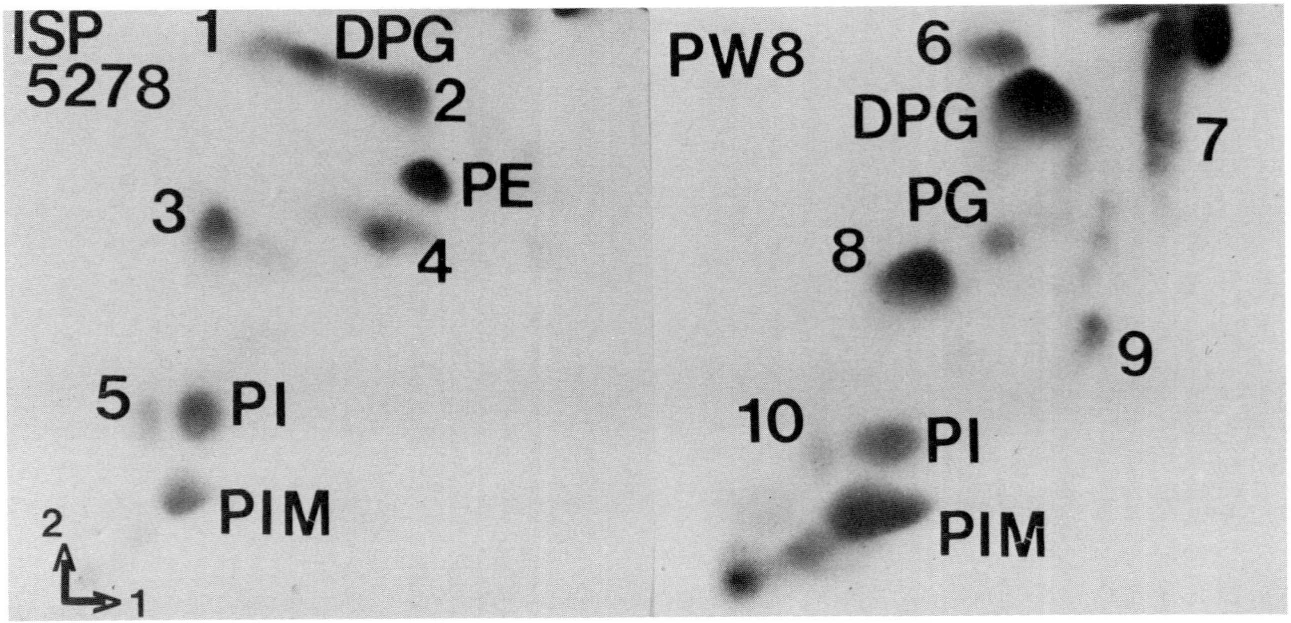

FIG. 3. Two-dimensional thin-layer chromatograms of polar lipids from *Streptomyces umbrinus* ISP 5278 and *Corynebacterium diphtheriae* PW8. Chloroform–methanol–water (65:25:4) by volume was used in the first direction and chloroform–acetic acid–methanol–water (80:15:12:4) by volume in the second direction. Abbreviations: DPG, diphosphatidylglycerol; PG, phosphatidylglycerol; PI, phosphatidylinositol; PIM, monoacylphosphatidylinositol dimannoside; PE, phosphatidylethanolamine. Spots 1–5,8,10 incompletely characterised phospholipids. Spots 6,7,9 incompletely characterised glycolipids.

TABLE 4

*Wall amino-acid analysis**

Defatted cells

↓ ← Wash 2× in 2 ml methanol, add 2 ml Tween 80–0.3 *N* KOH (0.1%, v/v), autoclave for 10 min, wash 2× with distilled water

Cell pellet

↓ ← Add 3 ml 10% trichloroacetic acid, heat at 100°C 15 min, wash 3× in distilled water, hydrolyse at 120°C in 6 *N* HCl

Hydrolysate

→ Retain small amount for TLC

↓ ← Filter and extract with petroleum ethers, dry under reduced pressure, esterify in 0.5 ml 20% (v/v) acetyl chloride–isobutanol (1 h at 120°C)

Isobutyl esters

↓ ← Dry under reduced pressure, add 100 μl heptafluorobutyric anhydride and 50 μl acetonitrile, heat 15 min at 120°C

Evaporate under N_2 until just dry, dissolve in 150–200 μl ethyl acetate

↓

Analyse by gas chromatography

*Modified from O'Donnell *et al.* (1982).

Cell wall analysis

Wall amino acids

The introduction of wall amino acid analysis led to major revisions in the classification of many Gram-positive taxa (see Chapters 10 and 11, this volume; Schleifer & Kandler 1972; Keddie & Bousfield 1980). The analysis of wall amino acids by gas chromatography (Moss *et al.* 1971; Ogata *et al.* 1975; O'Donnell *et al.* 1982b) is fast, uses small amounts of biomass and produces quantitative data. In the procedures outlined in Figs. 1 and 2, lipid-free

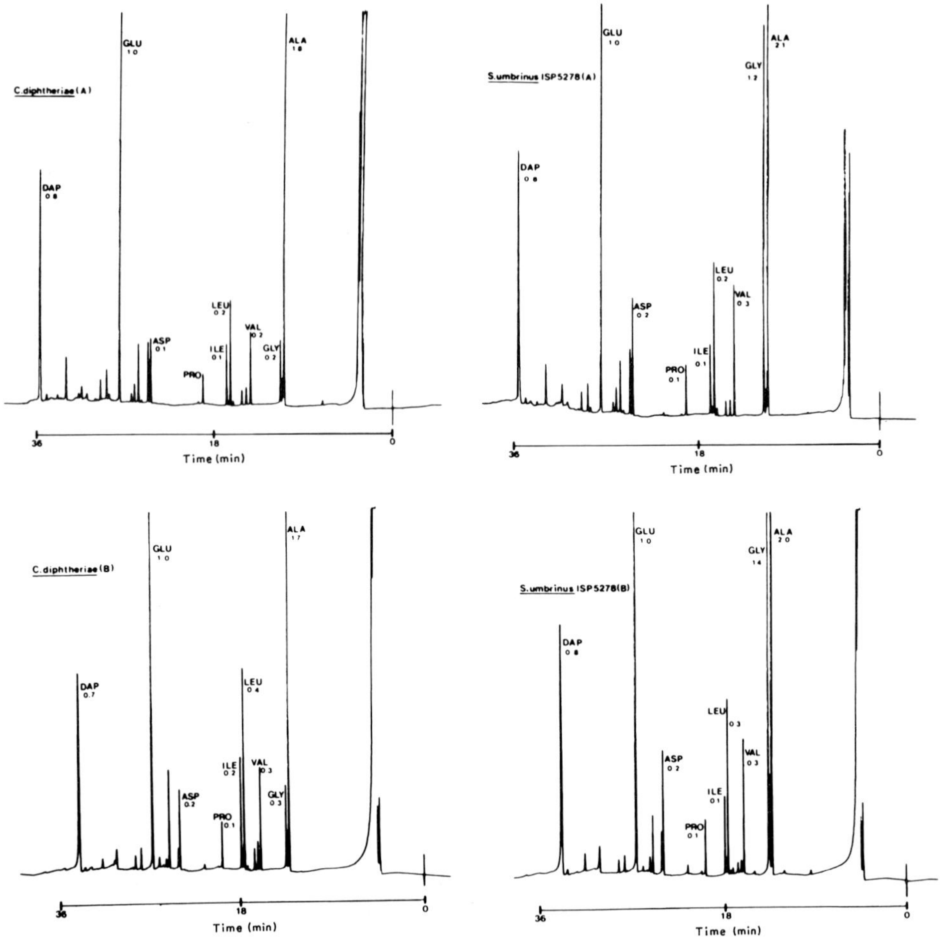

FIG. 4. Gas chromatogram of wall amino acid derivatives from alkaline methanolysis (A) and polar lipid (B) extracted biomass of *Corynebacterium diphtheriae* PW8 and *Streptomyces umbrinus* ISP5278. Abbreviations: ALA, alanine; GLY, glycine; VAL, valine; LEU, leucine; ILE, isoleucine; PRO, proline; ASP, aspartic acid; GLU, glutamic acid; DAP, diaminopimelic acid. Molar ratios relative to glutamic acid (1.0) are shown next to each peak. Analysis conditions: 50 m OV-1, 5 min at 100°C, then 4°C min^{-1} to 230°C.

biomass is used as starting material. Only 25–50% of this defatted biomass is necessary for wall amino acid analysis, the remainder being kept for sugar analysis. Following purification, hydrolysis and derivatization (O'Donnell *et al.* 1982b) (Table 4) the resultant amino acid derivatives are dissolved in 150–200 μl of ethyl acetate and analysed using a Shimadzu GC Mini-2, on a

50-m OV-1 bonded, fused silica capillary column with a split ratio of 50:1, an injection of 0.5–1 μl being sufficient for gas chromatography. Analysis times, using the temperature program shown in Fig. 4, are ~40 min. Response factors, calculated using an equimolar mixture of amino acids, are used by a Shimadzu CE1B electronic integrator to record automatically the concentration of each component relative to glutamic acid. Examples of the GC analysis of wall amino acid derivatives using both alkaline methanolysis defatted cells and polar lipid and isoprenoid quinone free biomass are shown in Fig. 4.

Determination of diaminopimelic acid isomers

The analysis system described above does not distinguish between the various isomeric forms of diaminopimelic acid (DAP); hydrolysates containing DAP are screened, therefore, by TLC on cellulose plates (Staneck & Roberts 1974). Hydrolysates (Table 4) are applied as short (0.5–1.0 cm) bands to cellulose plates (Kodak 13255) and run in a developing mixture containing methanol–water–HCl–pyridine (40:13:2:5, v/v) for 3 h. After drying, the cellulose plates are lightly sprayed with 0.2% ninhydrin in acetone and heated for 5 min at 100°C. Diaminopimelic acid isomers give a characteristic green colour and are identified by comparison with DAP standards.

Sugar analysis

For the analysis of sugars, the pellet, after polar lipid or fatty acid extraction, is subject to an overnight acid methanolysis in 2 ml of 4% (v/v) acetyl chloride in methanol and 0.1 ml of toluene, at 75°C. This releases bound lipids such as mycolic acids and hydroxy acids as methyl esters and sugars as O-methyl glycosides. The lipid and sugar derivatives are separated by extracting the supernatant with an equal volume of petroleum ether (bp 60–80°C) (Table 5). The O-methyl glycosides are dissolved in 50 μl of ethyl acetate and 100 μl of trifluoroacetic anhydride and heated at 150°C. The resultant trifluoroacetate derivatives are then analysed by gas chromatography on a 50-m OV-1 capillary column with temperature programing (10 min at 100°C, then 2°C/min to 230°C) (Fig. 5). Such an analysis system (Table 5) gives adequate resolution of common actinomycete wall sugars such as arabinose and galactose and provides a relatively simple method for detecting the presence of an arabinogalactan. However, the difficulties in quantifying trifluoracetate sugar derivatives because of multiple peaks resulting from different ring sizes and anomeric forms, makes the analysis of alternative derivatives, such as acetates of alditols and aldononitriles (Albersheim *et al.* 1967; Drucker 1981), potentially more valuable.

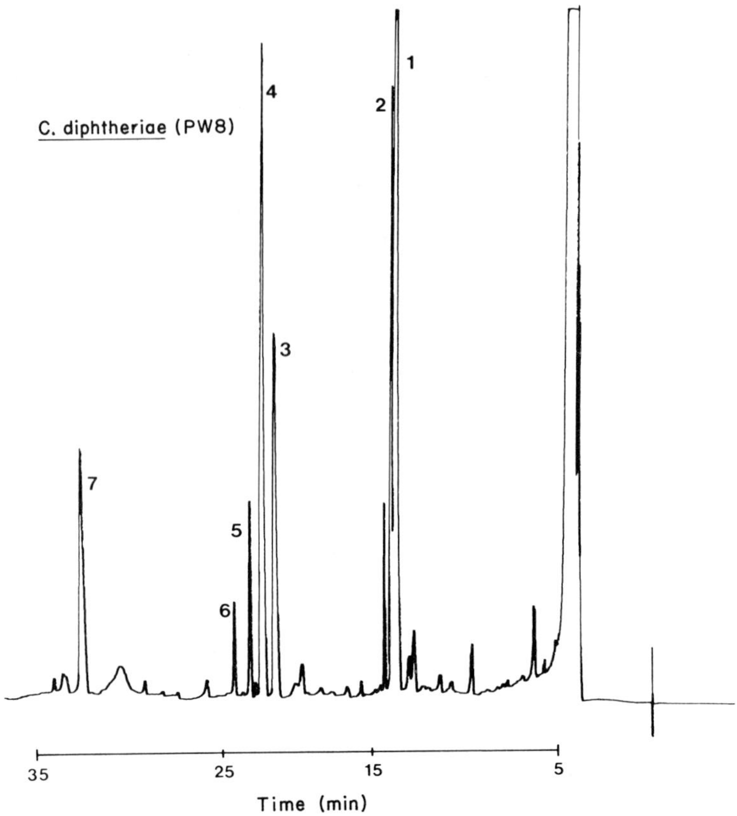

FIG. 5. Gas chromatogram of wall sugar derivatives from *Corynebacterium diphtheriae* PW8. Peaks 1 and 2, arabinose; 3,4, and 7, galactose; 5 and 6, glucose. Analysis conditions as described in the text.

Summary

The continued use of chemical markers in bacterial systematics depends on the development of simple but comprehensive extraction and analysis procedures. Although the methods described above were developed for the routine chemical characterisation of actinomycetes and related organisms, they have been used in chemotaxonomic investigations of the genera *Bacillus* (Minnikin & Goodfellow 1981b), *Renibacterium* (Embley *et al.* 1983) and *Staphylococcus* (Nahaie *et al.* 1984). With certain actinomycetes, such as the slow growing mycobacteria, the development of small-scale analytical methods that give large amounts of chemotaxonomic information from relatively small quantities of biomass (see Chapter 12, this volume) is particularly

TABLE 5

Extraction and derivatization of wall sugars

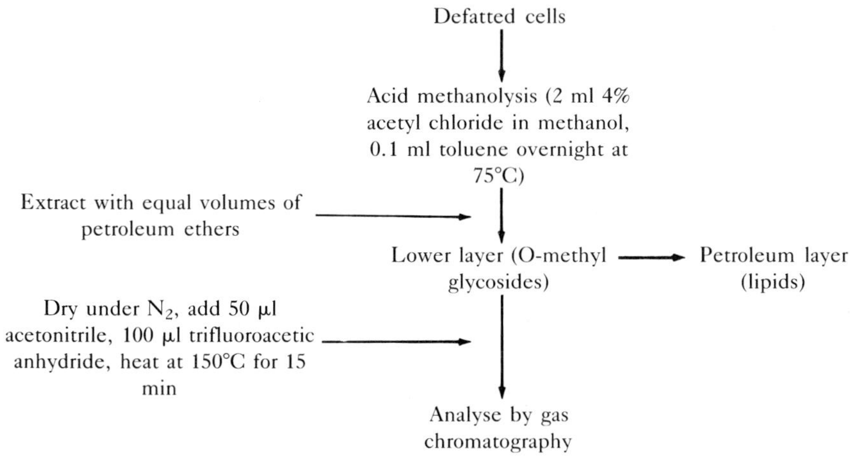

advantageous and makes chemical analysis a powerful technique in bacterial classification and in the routine identification of bacterial isolates.

The authors are indebted to the Medical Research Council (Grant G979/134/S) and to the Science and Engineering Research Council (Grant GR/C/18830) for support.

References

ALBERSHEIM, P., NEVINS, D. J., ENGLISH, P. D. & KARR, A. 1967 A method for the analysis of sugars in plant cell-wall polysaccharides by gas-chromatography. *Carbohydrate Research* **5**, 340–345.

BLIGH, E. G. & DYER, W. J. 1959 A rapid method of total lipid extraction and purification. *Canadian Journal of Biochemistry and Physiology* **37**, 911–917.

CARD, G. L. 1973 Metabolism of phosphatidylglycerol, phosphatidylethanolamine and cardiolipin of *Bacillus stearothermophilus*. *Journal of Bacteriology* **114**, 1125–1137.

CHRISTIE, W. W. 1982 *Lipid Analysis*. Oxford: Pergamon Press.

CONSDEN, R. & GORDON, A. H. 1948 Effect of salt on partition chromatograms. *Nature, London* **162**, 180–181.

DITTMER, J. C. F. & LESTER, R. L. 1964 A simple specific spray for the detection of phospholipids on thin-layer chromatograms. *Journal of Lipid Research* **5**, 126–127.

DRUCKER, D B. 1981 *Microbiological Applications of Gas Chromatography*. Cambridge: Cambridge University Press.

EMBLEY, T. M., GOODFELLOW, M., MINNIKIN, D. E. & AUSTIN, B. 1983 Fatty acid, isoprenoid quinone and polar lipid composition in the classification of *Renibacterium salmoninarum*. *Journal of Applied Bacteriology* **55**, 31–37.

GOODFELLOW, M. & CROSS, T. 1984 Classification. In *The Biology of the Actinomycetes* eds. Goodfellow, M., Mordarski, M., Williams, S. T., pp. 7–164. London: Academic Press.

JACIN, H. & MISHKIN, A. R. 1965 Separation of carbohydrates on borate-impregnated silica gel G plates. *Journal of Chromatography* **18**, 170–173.

KEDDIE, R. M. & BOUSFIELD, I. J. 1980 Cell wall composition in the classification and identification of coryneform bacteria. In *Microbiological Classification and Identification* eds. Goodfellow, M. & Board, R. G., pp. 167–188. London: Academic Press.

KROPPENSTEDT, R. M. & KUTZNER, H. J. 1978 Biochemical taxonomy of some problem actinomycetes. *Zentralblatt für Bakteriologie Parasitenkunde, Infektionskrankheiten und Hygiene.* I. *Abteilung, Supplement* **6**, 125–133.

LECHEVALIER, M. P. 1977 Lipids in bacterial taxonomy—a taxonomist's viewpoint. *CRC Critical Reviews in Microbiology* **5**, 109–210.

MINNIKIN, D. E. & GOODFELLOW, M. 1981a Lipids in the classification of actinomycetes. *Zentralblatt für Bakteriologie, Mikrobiologie und Hygiene.* I. *Abteilung, Supplement* **11**, 100–109.

MINNIKIN, D. E. & GOODFELLOW, M. 1981b Lipids in the classification of *Bacillus* and related taxa. In *The Aerobic Endospore-forming Bacteria* eds. Berkeley, R. C. W. & Goodfellow, M., pp. 59–90. London: Academic Press.

MINNIKIN, D. E. & O'DONNELL, A. G. 1984 Actinomycete envelope and peptidoglycan composition. In *The Biology of the Actinomycetes* eds. Goodfellow, M., Mordarski, M., & Williams, S. T., pp. 337–388. London: Academic Press.

MINNIKIN, D. E., HUTCHINSON, I. G., CALDICOTT, A. B. & GOODFELLOW, M. 1980 Thin-layer chromatography of methanolysates of mycolic acid containing bacteria. *Journal of Chromatography* **188**, 221–233.

MINNIKIN, D. E., MINNIKIN, S. M., O'DONNELL, A. G. & GOODFELLOW, M. 1984a Extraction of mycobacterial mycolic acids and other long chain compounds by an alkaline methanolysis procedure. *Journal of Microbiological Methods* **2**, 243–249.

MINNIKIN, D. E., O'DONNELL, A. G., GOODFELLOW, M., ALDERSON, G., ATHALYE, M., SCHAAL, A. & PARLETT, J. H. 1984b An integrated procedure for the extraction of bacterial isoprenoid quinones and polar lipids. *Journal of Microbiological Methods* **2**, 233–241.

MOSS, C. W. 1978 New methodology for identification of nonfermenters: Gas-liquid chromatographic chemotaxonomy. In *Glucose Non-fermenting Gram-Negative Bacteria in Clinical Microbiology* ed. Gilardi, G. L., pp. 171–201. West Palm Beach: CRC Press.

MOSS, C. W., DIAZ, F. J. & LAMBERT, M. A. 1971 Determination of diaminopimelic acid, ornithine and muramic acid by gas chromatography. *Analytical Biochemistry* **44**, 458–461.

NAHAIE, M. R., GOODFELLOW, M., MINNIKIN, D. E. & HÁJEK, V. 1984 Polar lipid and isoprenoid quinone composition in the classification of *Staphylococcus*. *Journal of General Microbiology* **130**, 2427–2437.

O'DONNELL, A. G., GOODFELLOW, M. & MINNIKIN, D. E. 1982a Lipids in the classification of *Nocardioides:* Reclassification of *Arthrobacter simplex* (Jensen) Lochhead in the genus *Nocardioides* (Prauser) emend. O'Donnell *et al.* as *Nocardioides simplex* comb.nov. *Archives of Microbiology* **133**, 323–329.

O'DONNELL, A. G., MINNIKIN, D. E., GOODFELLOW, M. & PARLETT, J. H. 1982b The analysis of actinomycete wall amino acids by gas chromatography. *FEMS Microbiology Letters* **15**, 75E–78E.

OGATA, S., TAHARA, Y. & HONGO, M. 1975 Chemical composition of cell wall peptidoglycan from *Clostridium saccharoperbutylacetonicum* studied with phage endolysins and gas chromatography. *Journal of General and Applied Microbiology* **21**, 65–74.

SCHLEIFER, K. H. & KANDLER, O. 1972 Peptidoglycan types of bacterial cell walls and their taxonomic implications. *Bacteriological Reviews* **36**, 407–477.
SHAW, N. 1968 The detection of lipids on thin-layer chromatograms with the periodate–Schiff reagent. *Biochimica et Biophysica Acta* **164**, 435–436.
STANECK, J. L. & ROBERTS, G. D. 1974 Simplified approach to identification of aerobic actinomycetes by thin-layer chromatography. *Applied Microbiology* **28**, 226–231.

Whole-Cell and Lipopolysaccharide Fatty Acids and Sugars of Gram-Negative Bacteria

E. Jantzen and K. Bryn

National Institute of Public Health, Oslo, Norway

The large and heterogeneous collection of microorganisms that fail to retain the Gram stain seems to share certain fundamental properties of cell-wall assembly: the presence of a chemically complex outer membrane and a thin peptidoglycan layer. In comparison, the typical Gram-positive cell lacks the outer membrane, and the cell wall is mainly composed of a thick layer of peptidoglycan (Inouye 1979; Rogers *et al.* 1980).

Chemically, the unique outer membrane of Gram-negative cells is mainly composed of phospholipids, certain proteins, as well as characteristic lipopolysaccharides (LPS). Although the general architecture of LPS is similar for most Gram-negative bacteria, including cyanobacteria, the composition and internal arrangement of sugars and fatty acids varies strikingly among the constituent taxa (Wilkinson 1977; Rietschel & Lüderitz 1980).

Together with the generally endotoxic LPS, the more hydrophilic and loosely bound capsular polysaccharides play an important role in infection and immunity, and have accordingly been extensively studied by chemical and immunological methods. It has clearly been demonstrated that data on both cellular sugars and fatty acids are of high value in the precise description of bacterial species (Jantzen *et al.* 1978; Jantzen 1984).

Sensitive and reliable chromatographic techniques, such as gas chromatography (GC), have made it possible to obtain profiles of cellular constituents from small amounts of cell material. This kind of direct approach for analysis of bacterial marker substances represents a relatively new and powerful methodology useful not only in fundamental studies of bacterial systematics, but also in practical diagnostic work.

CHEMICAL METHODS IN
BACTERIAL SYSTEMATICS

0-12-289675-0

Lipids are present in almost all bacterial cell walls, and considerable interest has been focused on their implications in bacterial systematics (Goldfine 1972; Shaw 1974; Lechevalier 1977; Drücker 1981; Asselineau and Asselineau 1984). In 1963 Abel *et al.* demonstrated the feasibility of utilizing GC analysis of whole-cell fatty acids in bacterial classification. Since that time a large number of publications on bacterial fatty acids have appeared, and the most important Gram-negative bacteria commonly involved in human infections have now been examined for their fatty-acid composition (Drücker 1981; Moss, 1981; Jantzen 1984). More than 50 different fatty acids are known to occur in bacteria, and a typical Gram-negative organism may contain about 12–20 fatty acids in quantities ranging from trace level to 60–70% of the total. Isolates of Gram-negative organisms can usually be identified to the genus level solely by their fatty-acid patterns.

Lipopolysaccharides of the outer membrane represent the well-known, heat-stable O antigens, which have long been utilized in serological diagnosis. Detailed structural analysis of LPS from a variety of bacteria has revealed some common structural features (Lüderitz *et al.* 1971; Wilkinson 1977). The lipid part, lipid A, usually consists of a disaccharide of glucosamine with unhydroxylated and hydroxylated fatty acids in ester and amide linkages (Wollenweber *et al.* 1982). The polysaccharide part is nearly always linked to lipid A by a ketosidic bond of 2-keto-3-deoxyoctonate (KDO), which may in some rare cases be absent or replaced by another ketosidic sugar (Jann *et al.* 1973). Quite often a heptose (L-*glycero*-D-*manno*-heptose) is found in the inner part of the polysaccharide. The number or kinds of sugars may vary (even within a genus or species), as well as their sequences and types of linkages. Lipopolysaccharides of some bacteria, e.g. species of *Enterobacteriaceae,* have a so-called O-specific side chain where repeating units of specific oligosaccharides contribute to the chemical and serological diversity. Thus LPS contains both fatty acids and sugars of potential value in bacterial systematics (Rietschel & Lüderitz 1980).

Polysaccharides other than LPS may also contribute to the chemical and immunological character of bacterial cell surfaces. In most cases, the occurrence of a specific polysaccharide is a characteristic of a single strain and hence may be valuable for tracing organisms causing epidemics. The presence of capsular polysaccharides, however, is also more dependent on cultural conditions than LPS, inasmuch as the former may disappear after growth on artificial media (Ellwood & Tempest 1972). In other cases, such polysaccharides are stable characteristics of species or groups, e.g. the capsular polysaccharides of urinary tract-infective strains of *Escherichia coli* (Jann & Jann 1983) or the group-specific polysaccharides of the clinically important *Neisseria meningitidis* (Jennings *et al.* 1973).

Methods

A standard gas chromatograph equipped with a flame ionization detector and temperature programmer is required. Glass or fused-silica capillary columns are highly recommended. An automatic integrator is also very helpful, especially when complex profiles are analyzed.

Harvesting of bacteria is usually done by washing bacterial colonies off agar plates with a right-angled glass rod and a small amount of distilled water. Cell material from a single plate is enough for several GC analyses. The cells are centrifuged at 2500 rpm, washed twice in distilled water, processed further or freeze-dried for storage.

Analysis of whole-cell fatty acids

An outline of the described methods of analysis of whole-cell fatty acids is given in Fig. 1.

Two methods of bond cleavage may be required for quantitative release of

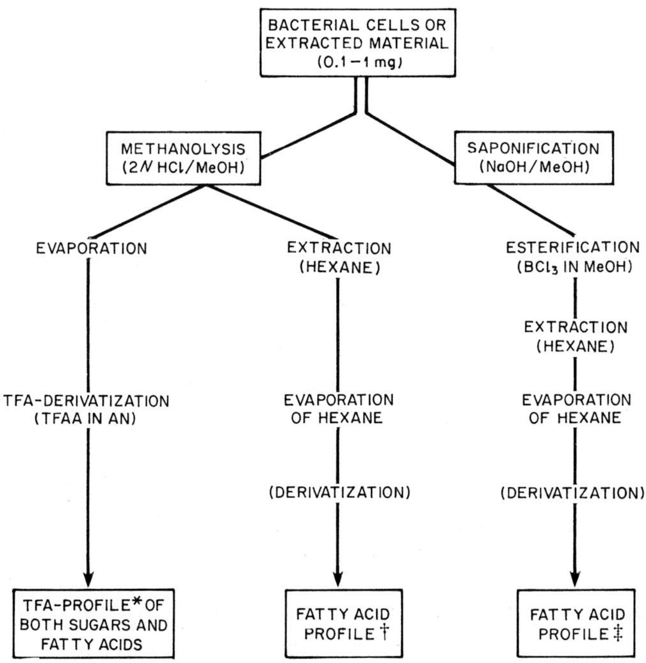

FIG. 1. Outline for analysis of cellular fatty acids and sugars. *, low yield of short fatty acids; †, cyclopropane fatty acids are partly degraded; ‡, low yield of amide-linked hydroxy fatty acids.

fatty acids (Jantzen *et al.* 1978). An acidic reagent provides the best yield of amide-bond constituents, e.g. hydroxy fatty acids of LPS, but destroys cyclopropane fatty acids, which otherwise are unaffected by alkaline procedures (Moss *et al.* 1974). Many bacteria have only one kind of such substituted fatty acids and hence only one of the following two methods would then be required. Most sugars are destroyed by alkaline conditions.

Method A

Freeze-dried bacteria (1–10 mg) or bacterial cell material from an agar plate is suspended in anhydrous 2 *M* HCl in methanol (2 ml) in Teflon-lined, screw-cap vials. Air is expelled by nitrogen, and the samples are heated at 85°C for 18 h. After cooling, the methanolysate is concentrated by a stream of nitrogen (to 1 ml), a NaCl solution is added (50% saturated; 2ml) and fatty acid methyl esters are extracted with hexane (2 × 3 ml). The hexane is then evaporated and the volume adjusted to 100 μl. Samples can be stored in sealed glass capillaries at −20°C.

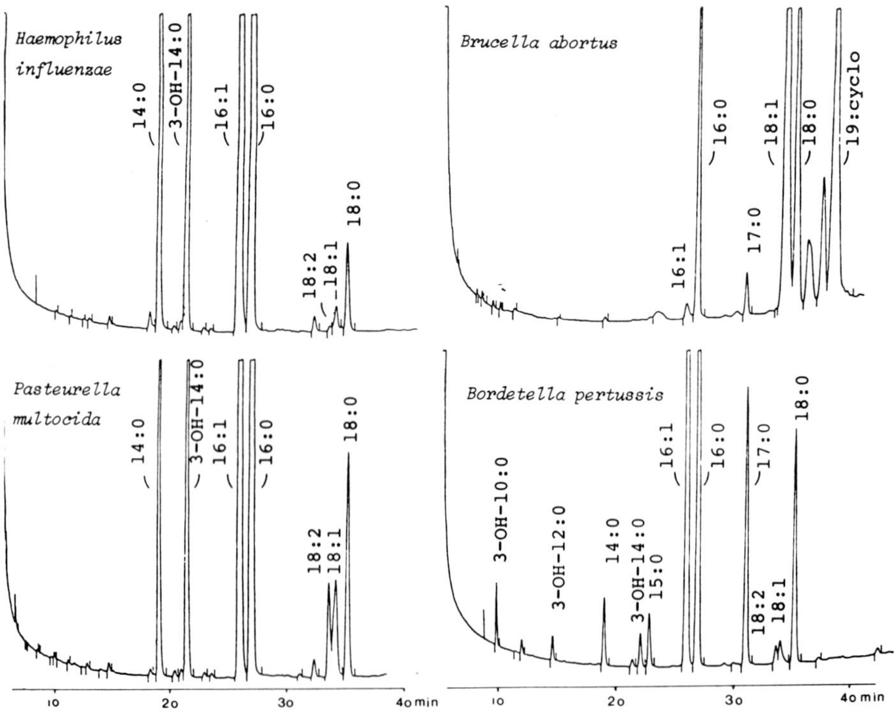

FIG. 2. Typical gas chromatographic fatty acid profiles of Gram-negative bacteria.

Method B

Bacterial cells are suspended in 0.5 ml of a stock solution (4 g of NaOH in 100 ml of methanol–toluene 60:40 + 2 drops of a phenolphthalein solution) in a Teflon-lined screw-cap vial. The air is expelled by nitrogen and the reaction mixture is heated for 15 min at 100°C. After cooling, the solution is acidified by 2 *M* HCl in methanol (until colourless + 1 drop, pH 2). Methyl ester formation is then completed by adding 0.5 ml of 14% BCl_3 in methanol and heated for 5 min at 100°C. Extraction is performed as described for Method A.

Some typical GC profiles of cellular fatty acids are shown in Fig. 2.

Analysis of whole-cell sugars

An outline of the described methods of analysis of whole-cell sugars is given in Fig. 1.

It is possible to analyse whole-cell sugars (originating from nucleic acids, capsular polysaccharides and LPS) after acidic methanolysis as described in method A above (Jantzen *et al.* 1974b). By omitting the extraction step with hexane, a combined sugar and fatty-acid GC profile is obtained. If analysis of sugars alone is wanted, the hexane extraction of fatty-acid methyl esters is preferably done in the absence of salt. Sugar moieties in stable phosphate linkages are not detected by this method.

A rather high background of nonsugar material may contribute to the complexity of the GC profiles, and repeated injections of whole-cell methanolysates reduce the lifetime of GC columns. Modern, bonded-phase capillary columns, which are possible to clean and renew, may solve part of this problem. Otherwise an extraction step (see next section) is recommended, and more reproducible GC profiles may then be obtained.

Analysis of fatty acids and sugars of LPS

Bacteria from an agar plate are harvested with 3 ml water, and LPS (and polysaccharides) is extracted at 68°C with the hot phenol method of Westphal *et al.* (1952). Dialysed and lyophilized water phases, containing a mixture of nucleic acids, polysaccharides and LPS, can be analysed as such (Fig. 3; Bryn & Jantzen, 1982). Purification of LPS is commonly done by repeated ultracentrifugation for 4 h at 105,000 *g*, which selectively sediments the LPS, leaving the polysaccharides enriched in the supernatant. The mixture of LPS constituent sugars and fatty acids released by methanolysis (Method A) is concentrated to dryness by nitrogen or on a rotary evaporator before derivatization and gas chromatography.

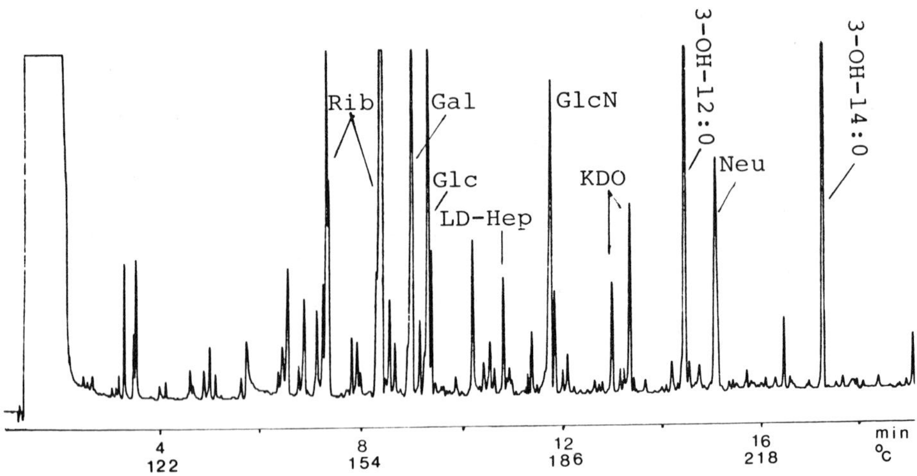

FIG. 3. Gas chromatographic profile of phenol–water-extracted carbohydrates from *Neisseria meningitidis*. See Table 6 for abbreviations.

Derivatization and gas chromatography

Sugars (as methyl glycosides) and hydroxy fatty acids (as methyl esters) are derivatized by heating to boiling with 50% trifluoroacetic (TFA) anhydride in acetonitrile, and the TFA concentration is adjusted to 10% before GC. The GC analyses are carried out as previously described (Jantzen *et al.* 1978; Bryn & Jantzen 1982). The three most frequently used columns in the authors' laboratory are 10% methylsilicone (SE-30) on Gas Chrom Q (200 × 0.2 cm), 10% cyanopropylsilicone (SILAR 10C) on Gas Chrom Q (200 × 0.2 cm) and a fused-silica capillary column coated with SE-30 methylsilicone (25 m × 0.02 cm). The injection port temperature is 250°C and the detector temperature 300°C. The two packed columns are operated from 120°C to 250°C, increasing at 2°C/min and with a flow rate (nitrogen) of 18 ml/min. The capillary column is operated in splitless mode with a carrier gas (helium) flow rate of 1.5 ml/min. The oven temperature is programmed from 90°C (LPS constituents) or 120°C (fatty acid methyl esters) to 270°C at a rate of 8°C/min.

Peak identification

For identification of the individual fatty acids and sugars, comparison of the GC retention times with those of standards is the primary method, at best by use of two columns of different polarity. Hydroxyl groups on fatty acids are recognized after TFA derivatization, which sharpens the GC peak and reduces the retention times in a regular manner. Cyclopropane fatty acids are recognized by their acid lability, resulting in characteristic patterns of degraded products (Jantzen *et al.* 1978). Unsaturated fatty acids lose their double bond upon hydrogenation, and they are converted to the corresponding saturated fatty acids. Monosaccharides give a typical peak pattern upon methanolysis, with rather constant proportion between peaks (Bryn & Jantzen 1982). Fatty-acid methyl esters and TFA-derivatized sugars are easily distinguished by their widely different partition coefficients in the two-phase system hexane–acetonitrile (Jantzen *et al.* 1974a). This two-phase system can also be used to discriminate between saturated and unsaturated fatty acids, as well as between different locations of hydroxyl groups. The final proof of the identity of a certain GC peak is by use of mass spectrometry.

Fatty Acids

Most of the accumulated data on bacterial lipids concern fatty-acid composition and certain taxonomic features are currently encountered. In contrast to plants and animals, the prokaryotic cell generally lacks polyunsaturated fatty acids, but contains several types not commonly observed in the plant and animal kingdoms. Thus the Gram-negative bacterial groups that hitherto have been examined for fatty acids contain one or more of the following categories: straight chain, monounsaturated, cyclopropane substituted, 2- and 3-hydroxylated and methyl-branched (iso and anteiso) fatty acids. Two substitutions on a single fatty acid (e.g. methyl-branched or monoenoic hydroxy fatty acids) may also be observed, but are uncommon. Usually only a few of these main categories of fatty-acid types predominate within a certain taxon, and hence a simple recording of the fatty acid types of bacterial isolates can often be of considerable value as a primary step in diagnosis.

Straight chain acids

Straight chain saturated and monounsaturated fatty acids of the same kinds as are present in mammalian tissue have been found in almost all bacterial groups as part of membrane lipids. Fatty-acid chain lengths may vary considerably between bacterial species, from C_8 to C_{26}. The length of the alkyl chain of a fatty acid influences the membrane fluidity, and growth at higher temperatures promotes the synthesis of fatty acids with longer alkyl chains, thereby rendering the membrane fluidity constant (Melchior 1982). Introduction of a double bond in the alkyl chain is another way for bacteria to adjust the fluidity of membranes (Abbas & Card 1980). The double bond is formed by dehydrogenation of a carbon–carbon bond and is usually located near the centre of the molecule (either ω-7 or ω-9). Positional isomers of monounsaturated fatty acids are resolved by high resolution GC but their potential in taxonomy has still to be exploited (Uchida & Mogi 1972; Sugimoto *et al.* 1983). Occurrence of diunsaturated fatty acids in bacteria has been described (Roussel & Asselineau 1980), but in most cases it is not clear whether they are biosynthesized by the bacteria or taken up from the growth media.

Hydroxylated fatty acids

Hydroxylated fatty acids occurring in bacteria usually have the OH group either in position 2 or 3 (2-OH-FA and 3-OH-FA). The former may occur to some extent in animal tissue as D isomers (Tatsumi *et al.* 1974) but are more typical of some Gram-negative bacteria where the L isomers occur (Wollenweber *et al.* 1982). The 3-OH-FA acids, which always occur as the D isomers,

TABLE 1

*Gram-negative bacteria grouped according to their content of hydroxy-, methyl-, and cyclopropane-substituted fatty acids**

FA-group	Taxon			OH-FA	BR-FA	CYCLO-FA
1	*Acinetobacter* *Actinobacillus* *Bordetella pertussis* *Branhamella* *Campylobacter*-1	*Cardiobacterium* *Eikenella* *Francisella* *Fusobacterium* *Haemophilus*	*Kingella* *Leptotrichia* *Moraxella* *Neisseria* *Pasteurella* *Pseudomonas*-1	+	−	−
2	*"Achromobacter"*-1 *Alcaligenes* *Bordetella bronchiseptica* *Bordetella parapertussis*	*Campylobacter*-2 *Enterobacteriaceae* *Flavobacterium/Cytophaga*-1 *Pseudomonas*-2		+	−	+
3	*Bacteroides* *Flavobacterium/Cytophaga*-2		*Flexibacter* *Pseudomonas*-3	+	+	−
4	*Legionella*			−	+	−
5	*"Achromobacter"*-2 *Brucella*			−	−	+
6	*Gardnerella* *Treponema*			−	−	−

*Abbreviations: FA, Fatty acid; OH-FA, Hydroxylated fatty acid; BR-FA, Branched (iso or anteiso) fatty acid; CYCLO-FA, Cyclopropane substituted fatty acid. Substituted fatty acids present at trace levels have been neglected.

"Achromobacter"-1: *"A. xylosoxidans"*; *"Achromobacter"*-2: *A.* sp. (King group Vd); *Campylobacter*-1: *C. fetus; C. sputorum*, *Campylobacter*-2: *C. coli, C. jejuni; Flavobacterium/Cytophaga*-1; *Flavobacterium* sp., *F. capsulatum; Flavobacterium/Cytophaga*-2 (see Table 4); *Pseudomonas*-1: *P. paucimobilis; Pseudomonas*-2 (see Table 3) and *Pseudomonas*-3 (see Table 4).

have been found in lipid A of nearly all examined LPS (Wilkinson 1977; Rietschel & Lüderitz 1980) and in some cases also in amino acid-containing lipids (Lechevalier 1977). The biosynthesis of 2- and 3-hydroxy fatty acids may differ, inasmuch as 2-OH-FA are intermediates of α oxidation of fatty acids, and 3-OH-FA of the ordinary fatty acyl elongation pathway for biosynthesis of fatty acids (Osborn 1979). As shown in Table 1, hydroxylated fatty acids are found in almost all Gram-negative species.

Cyclopropane fatty acids

Cyclopropane fatty acids, not occurring in mammalian cells, are found in major quantities in several genera of both Gram-negative (Table 1) and Gram-positive bacteria. They are formed by methyl addition to a preformed monounsaturated fatty acid (Cronan *et al.* 1979). Genera that generally have cyclopropane fatty acids may occasionally, because of the one-step enzymatic character of biosynthesis, comprise strains or species devoid of these acids, e.g. *Bordetella pertussis* and *Brucella canis* (Dees *et al.* 1981; Jantzen *et al.* 1982). Cyclopropane fatty acid formation is an aging phenomenon, and the quantity of a certain cyclopropane fatty acid is a less stable characteristic of a species than the sum of this acid and its monounsaturated precursor.

Branched-chain fatty acids

Fatty acids with branched alkyl chains predominate among several bacterial groups. Methyl branches are located in the ultimate (iso) or penultimate (anteiso) positions. Such branched fatty acids are usual constituents of lipids of Gram-positive bacteria, but they have also been found in some Gram-negative organisms (Table 1). Occasionally the methyl branch may occur in other locations, e.g. in 10-methyl-stearic acid of mycobacterial species. The branched fatty acids may serve the same purpose of regulation of membrane fluidity as the unsaturated fatty acids, and these two kinds of fatty acids seem not to occur concurrently.

Other kinds of fatty acids

Two different substitutions may occasionally be found in the same fatty acid. As seen in Table 4 (FA-group 4), methyl-branched hydroxy fatty acids are characteristic of the genera *Bacteroides*, *Cytophaga*, *Flavobacterium*, and *Flexibacter*. A dihydroxy fatty acid has been found in an LPS preparation of *Legionella pneumophila* (Mayberry 1981). Unsaturated OH-FA seem to be rare, but 2-hydroxy monoenoic acids have been detected in a strain of *Spirillum* (Moss & Lambert 1976) and in *Pseudomonas solanacearum* (Drigues *et al.*

1981). Iso and anteiso monoenoic acids characterize *Desulfovibrio desulfurans* (Boon *et al.* 1977) and iso-17:1 is found in certain *Pseudomonas* and *Flavobacterium/Cytophaga* strains (Oyaizu & Komagata 1981, 1983) but are otherwise rare.

The high-molecular-weight 3-OH-α-branched fatty acids (mycolic acids), which are found in the complex lipids of *Mycobacterium, Corynebacterium* and *Nocardia* strains, have not been found in Gram-negative bacteria. Haeffner *et al.* (1977) described the simpler but related 2-methyl-3-hydroxy fatty acids of LPS isolated and found in small amounts from an LPS preparation of *Bordetella pertussis,* but this finding could not be verified (Jantzen *et al.* 1982; Kawai & Moribayashi 1982).

Some bacterial lipids, e.g. plasmalogens, which commonly occur among anaerobic bacteria (Kamio *et al.* 1969) have ether-linked β-unsaturated fatty acid constituents. These linkages are cleaved on methanolysis and fatty aldehydes (as methyl acetals) are released. The latter, like the long-chain fatty alcohols of wax esters (Bryn *et al.* 1977), occur in the fatty-acid fraction where they can cause confusion in fatty acid determinations. The distribution of these two fatty-acid-like substance classes among bacteria is of taxonomic interest, but their analysis requires special methods, and will not be considered here.

Fatty Acid Patterns

In Table 1 a number of Gram-negative bacteria have been grouped according to their content of the three types of substituted fatty acids. Fatty-acid groups 1 and 2 predominate and they comprise most of the organisms causing human infections. The characteristic fatty acid patterns of the six FA groups are shown in Tables 2, 3 and 4. Hydroxy acids are found in most but not all Gram-negative species. Some bacteria may contain only small amounts of hydroxylated fatty acids in their LPS. For instance, both *Brucella* and *Legionella* have been reported to contain hydroxy fatty acids (OH-FA) in their LPS (Moreno *et al.* 1979; Mayberry 1981) but the quantities are too small to be recognized in whole-cell profiles. Similarly, certain species apparently lack an active cyclopropane-forming enzyme (e.g. *Bordetella pertussis, Brucella canis, Campylobacter fetus* and *Campylobacter sputorum*) and phylogenetically closely related taxa will accordingly, in some cases, be split into different groups. On the other hand, the diversity shown by *Flavobacterium* and *Pseudomonas* (Table 1) is due to the well-known phylogenetic heterogeneity of these taxa.

Table 2

*Fatty acid patterns of fatty acid group 1**

Taxon	OH-FA	FA	Unsat. FA	References
Acinetobacter	2–12, 3–12, 3–14	12, 14, 15, <u>16,</u> 17, 18	<u>16,</u> 17, 18	Jantzen *et al.* 1975
Actinobacillus	3–14	14, <u>16,</u> 18	<u>16,</u> 18	Jantzen *et al.* 1980
Bordetella pertussis	3–10, 3–12, 3–14, 3–16	12, 14, 15, <u>16,</u> 17, 18	<u>16,</u> 18	Jantzen *et al.* 1982 Kawai & Moribayashi 1982
Campylobacter-1	3–14	12, 14, <u>16,</u> 18	16, <u>18</u>	Hebert *et al.* 1982
Cardiobacterium	3–14, 3–16	12, 14, <u>16,</u> 18	16, <u>18</u>	K. Bøvre & N. Hagen unpublished
Eikenella	3–12, 3–14, 3–16	12, 14, 15, <u>16,</u> 18	16, <u>18</u>	K. Bøvre & N. Hagen unpublished
Francisella	2–10, 3–16, <u>3–18</u>	10, 14, 16, 18, 20, 22, 24, 26	16, 18, 20, 22, 24, 26	Jantzen *et al.* 1979
Fusobacterium	3–12, 3–14, 3–16	12, <u>14,</u> 15, <u>16,</u> 18	16, 18	Jantzen & Hofstad 1981
Haemophilus	3–14	14, <u>16,</u> 18	<u>16,</u> 18	Jantzen *et al.* 1980
Kingella	3–12, 3–14	12, <u>14,</u> 16, 18	16, 18	K. Bøvre & N. Hagen unpublished
Leptotrichia	3–14	12, 14, 15, <u>16,</u> 18	16, <u>18</u>	Hofstad & Jantzen 1982
Moraxella				
1	3–12, 3–14	12, 14, 16, 18, 18alc	16, 17, <u>18</u>	
2	3–12, 3–14	14, 16, 18, 18alc	16, 17, <u>18</u>	
3	3–12, 3–14	12, 14, 15, <u>16,</u> 18	16, 17, <u>18</u>	Jantzen *et al.* 1974b
4	3–12, 3–14	12, 14, 15, <u>16,</u> 17, 18, 18alc	16, 17, <u>18</u>	
5	3–14, 3–16	14, 15, <u>16,</u> 17, 18	16, <u>18</u>	
Neisseria				
1	3–12, 3–14	12, 14, 15, <u>16,</u> 18	<u>16,</u> 18	Lambert *et al.* 1971 Jantzen *et al.* 1974b
2	3–12, 3–14, 3–16	12, 14, 15, <u>16,</u> 18	<u>16,</u> 18	
Pasteurella	3–14	14, <u>16,</u> 18	<u>16,</u> 18	Jantzen *et al.* 1980
Pseudomonas-1	<u>2–12</u>	16, 18	16, <u>18</u>	Oyaizu & Komagata 1983

*Abbreviations: FA, Fatty acid; OH-FA, hydroxylated fatty acid; Unsat., monounsaturated fatty acid, no distinction between isomers has been made. Number indicates chain length; 2- and 3- indicate position of hydroxyl group. Underlined fatty acids are major constituents (above 20%); alc, fatty alcohol.

Campylobacter-1: *C. fetus*, *C. sputorum*. *Moraxella:* 1: *Branhamella catarrhalis*, *Moraxella bovis*, *M. caviae*, *M. lacunata*, *M. nonliquefaciens*, *M. ovis*. 2: *M. osloensis*. 3: *M. phenylpyruvica*. 4: *M. atlantae*. 5: *M. urethralis* and *Pseudomonas*-1: *P. paucimobilis*.

TABLE 3

*Fatty-acid patterns of fatty acid group 2**

Taxon	OH-FA	Cyclo-FA	FA	Unsat. FA	References
"Achromobacter"-1	2–12, 2–14, 3–14, 3–16	17, 19	14, 15, 16, 17, 18, 19	16, 18	Dees & Moss 1978
Alcaligenes	2–12, 2–14, 3–14, 3–16	17, 19	12, 14, 16, 18	16, 18	Dees & Moss 1975
Bordetella bronchiseptica	2–12, 3–14, 3–16	17, 19	14, 16, 17, 18	16, 18	Jantzen *et al.* 1982; Kawai & Moribayashi 1982
B. parapertussis	3–10, 3–14, 3–16	17, 19	14, 16, 18	16, 18	Jantzen *et al.* 1982; Kawai & Moribayashi 1982
Campylobacter-2	3–14	19	14, 16	16, 18	Hebert *et al.* 1982
Enterobacteriaceae	3–13, 2–14, 3–14	17, 19	12, 13, 14, 15, 16, 17, 18	16, 18	Jantzen & Lassen 1980, and unpublished data
Flavobacterium/Cytophaga-1	3–10, 2–14, 3–14, 2–16, 3–16	17, 19	12, 14, 16, 18	16, 18	Oyaizu & Komagata 1981
Pseudomonas-2					
1	3–10, 2–12, 3–12	17, 19	12, 14, 15, 16, 18	16, 18	
2	3–14, 2–16, 3–16	17, 19	12, 14, 15, 16, 17, 18, 19	16, 18	
3	3–10, 2–14, 2–16, 2–18	17	12, 14, 15, 16, 17, 18, 19	16, 18	Oyaizu & Komagata 1983; Moss 1978
4	3–12, 2–14 3–14	17, 19	12, 14, 15, 16, 17, 18, (b15)	16, 18	
5	3–8	17, 19	8, 12, 16, 17, 18, 19	16, 18	
6	3–10, 3–12, 3–14	17	12, 14, 16, 19	16, 19	

*See footnotes of Tables 1 and 2 for abbreviations.

"Achromobacter"-1: *"A. xylosoxidans"*; *Campylobacter*-2: *C. coli*, *C. jejuni*; *Flavobacterium/Cytophaga*-1: *F. capsulatum*, *Flavobacterium* sp., *Pseudomonas*-2: 1: *P. aeruginosa*, *P. alcaligenes*, *P. fluorescens*, *P. mendocina*, *P. putida*, and *P. stutzeri*; 2: *P. caryophylli*, *"P. cepacia"*, and *P. gladioli*; 3: *P. acidovorans* and *P. testosteroni*; 4: *P. diminuta*; 5: *P. palleronii*; and 6: *P. avenae*.

TABLE 4

*Fatty acid patterns of fatty acid groups 3, 4, 5 and 6**

Taxon	OH-FA†	Br-FA	Cyclo-FA	FA	Unsat. FA†	References
FA-Group 3						
Bacteroides	3-b15, 3-15, 3-16, <u>3-b17,</u> 3-17	14, <u>15</u>	—	14, 15, 16	—	Mayberry 1980; Miyagawa & Suto 1980
Flavobacterium/ Cytophaga-2						
1	3-b11, 3-b13	14, 16	—	14, 15, 16, 17	16, b17, 18	
2	2-b15, 3-b15, 3-b16, 3-b17	<u>15</u>	—	14, 15, 16	16, b17, 18	
3	2-b15, 3-b15, 3-16, 3-b17	<u>15</u>	(17)	12, 14, 15, 16, 18	16, b17 18	Oyaizu & Komagata 1981
4	2-b15, 3-b15, 3-b16, 3-b17	<u>15</u>	—	12, 14, 15	16, b17 18	
Flexibacter	2-b14, 3-b14, 3-14, 2-b16, 3-16	12, 15, 16	—	14, 15, 16, 18	b16, 16	Fautz *et al.* 1979
Pseudomonas-3	3-10, 2-b11, 3-b13	13, <u>15</u>	—	14, 15, 16,	16, b17	Oyaizu & Komagata 1983

(*continued*)

TABLE 4 (*Continued*)

Taxon	OH-FA	Br-FA	Cyclo-FA	FA	Unsat. FA	References
FA-Group 4						
Legionella	—	14, 15, 16, 17	—	14, 15, 16, 17, 18, 19	14, b16. 16	Moss *et al.* 1977; Mayberry 1981
FA-Group 5						
"Achromobacter"-2	—	—	19	16, 17, 18, 19, 20	16, 17, 18	Dees & Moss 1978
Brucella	—	—	19	16, 17, 18	16, 17, 18	Jantzen *et al.* 1982; Dees *et al.* 1981
FA-Group 6						
Gardnerella	—	—	—	14, 16, 18	16, 18	Csango *et al.* 1982
Treponema	—	—	—	14, 15, 16, 17, 18	16, 18 20	Matthews *et al.* 1979; Livermore & Johnson 1974

*See footnotes of Tables 1 and 2 for abbreviations; †b, branched.

"Achromobacter"-2: *"Achromobacter"* sp. (King group Vd) *Flavobacterium*/*Cytophaga*-2: 1: *"F. flavescens," "F. fuscum,"* and *"F. lutescens"*; 2: *"F. breve," F. ferrugineum, F. heparinum, F. meningosepticum,* and *"F. sewanense"*; 3: *F. aquatile* and *"C. johnsonae"*; 4: *F. odoratum, Flavobacterium* sp. and *Pseudomonas*-3: *"P. maltophila"* and *P. pictorum*.

TABLE 5

Proportion (%) of lipopolysaccharide fatty acids in whole-cell fatty acid profiles*

Taxon	Number of examined strains	Percent LPS fatty acids	LPS marker
Salmonella	5	18	3-OH-14:0
Yersinia	6	15	3-OH-14:0
"Acinetobacter"	11	20†	2-OH-12:0, 3-OH-12:0
Neisseria	6	17	3-OH-12:0, 3-OH-14:0
Moraxella	9	18‡	3-OH-12:0
Kingella	1	13	3-OH-14:0
Bordetella	3	15	2-OH-12:0, 3-OH-14:0
Actinobacillus	1	22	3-OH-14:0

*Determined by quantitation of one or more LPS-specific fatty acids in whole-cell preparations and in corresponding extracted LPS.
†Coefficient of variation 17%.
‡Range observed (6–41%) is higher within this taxon than normally observed.

Lipopolysaccharide Fatty-Acid Patterns

Generally, LPS fatty acids seem to constitute about 15–20% of total cellular fatty acids of Gram-negative bacteria, as determined in the authors' laboratory (Table 5). On the basis of these data, one can estimate the total weight of LPS present in bacterial cells to be of the same order as for extractable lipids (5–15%), or approximately five times higher than yields of extracted LPS; 1–2% are frequently reported in the literature. Fatty acids occurring in LPS in amounts lower than about 5% are difficult to detect in whole-cell analyses.

Patterns of LPS fatty acids vary between bacterial groups (Table 6) but several common characteristics are found. Usually two molecules of a 3-OH-FA occur in amide linkage in lipid A, and the same or other 3-OH-FA may additionally be ester linked with a total number of FA molecules of about six. Thus the predominant 3-OH-FA will usually constitute at least 30%, often 60–65%. Usually 3-OH-FA with chain lengths longer than the major one do not occur or when present, only in small amounts. Although 2-OH-FA are also found, they have a rather limited occurrence, e.g. in LPS of *Bordetella bronchiseptica*, *Acinetobacter*, *Pseudomonas*, *Salmonella*, and *Serratia* (Table 6). In a few cases 3-OH-FA have been found to be nearly absent *Brucella*, *Legionella*, or alternatively totally predominant in LPS (85–95% in *Moraxella phenylpyruvica;* K. Bryn, unpublished). The number of different LPS fatty acids may occasionally be very low (1–3) or unusually high (10–15). Quantitative differences may be important taxonomically, and fairly often rather different LPS fatty-acid patterns are found within a genus (e.g.

TABLE 6

Characteristic LPS constituents of some Gram-negative bacteria related to human infections

Taxon	KDO*	Hep†	Typical sugars‡	2-OH-FA**	3-OH-FA**	FA**	References
Enterobacteriaceae			A large variety of different				Lüderitz *et al.* 1971;
Enterobacter	+	LD	pentoses, hexoses, de-	—	14	14, 16	Wilkinson 1977; Bryn &
Escherichia	+	LD	oxy, dideoxy, O-Me,	—	14	12, 14, 16	Rietschel 1978
Klebsiella	+	LD	amino, diamino sugars,	14	14	12, 14, 16	
Proteus	+	LD	uronic acids, etc. char-	—	14	14, 16	
Salmonella	+	LD	acterize this group	14	14	12, 14, 16	
Serratia	+	LD		14	14	12, 14, 16	
Shigella	+	LD		—	14	12, 14, 16	
Yersinia	+	LD, DD		—	13, 14	12, 16	
Campylobacter jejuni	+	LD	Gal, Glc, GalN	—	14	16	Næss & Hofstad 1982
Vibrio cholera	−	LD	QuinN, AraN, Fru	—	12, 14	14, 16	Broady *et al.* 1981; Kabir 1982
Actinobacillus actinomycetem comitans	+	LD, DD	Fuc, Gal, Glc, Rha	—	14	14	Kiley & Holt 1980
Haemophilus influenzae	+	LD	Gal, Glc, Man	—	14	14	Flesher & Insel 1978
Pasteurella multocida	+	LD, DD	Gal, Glc	—	14	14	Erler *et al.* 1981

Brucella abortus	+	—		—	18?	16, 18	Moreno *et al.* 1979
Brucella melitensis	+	LD, DD	Ara, Gal, Rha	n.d.	n.d.	n.d.	Lacave *et al.* 1969
Bordetella bronchiseptica	+	LD	Glc, GlcU	12	12, 14	14, 16	LeDur *et al.* 1980; Jantzen *et al.* 1982
Bordetella parapertussis	+	LD	Glu, GlcU	—	10, 14	14, 16	
Bordetella pertussis	+	LD	Glc, GlcU	—	10, 12, 14	14, 16	
Neisseria gonorrhoeae	+	LD	Gal, Glc, Man, Rha, GalN	—	12, 14	12, 14, 16	Perry *et al.* 1975
Neisseria meningitidis	+	LD	Gal, Glc	—	12, 14	12, 14, 16	Jennings *et al.* 1980; Bryn *et al.* 1983
Pseudomonas aeruginosa	+	LD	Glc, Rha, GalN, FucN, QuinN	12	10, 12	12, 16	Wilkinson & Galbraith 1975
Bacteroides fragilis	–	—	Gal, Glc, Rha	—	14, 15, 16, i17, 17	i15	Kasper *et al.* 1983

*KDO, 3-deoxy-D-manno-octulosonic acid.

†Hep, heptose, the two isomers L-glycero-D-manno-heptose (LD) and D-glycero-D-manno-heptose (DD).

‡Glucosamine occurs as a Lipid A constituent in all species. Strains of *Enterobacteriaceae* normally have LPS with an O-specific chain; hence their structural complexity and intragroup variation in sugar patterns seem to be exceptional among Gram-negative bacteria.

Ara, Arabinose; AraN, Arabinosamine; Fru, Fructose; Fuc, Fucose; FucN, Fucosamine; Gal, Galactose; GalN, Galactosamine; Glc, Glucose; GlcU, Glucuronic acid; Man, Mannose; Rha, Rhamnose; QuinN, Quinovosamine.

**FA, Fatty acid; 2-OH and 3-OH indicates a hydroxyl group and its position; n.d., not determined.

Bordetella, Moraxella) or even within a species (e.g. *Acinetobacter calcoaceticus* K. Bryn, unpublished).

Some examples of LPS profiles obtained by capillary gas chromatography are given in Fig. 4.

Typical Sugars

A number of different glyco-conjugates have been described as part of the bacterial cell wall (including more loosely bound capsular material). Several of the corresponding monosaccharide constituents have been found to occur only in bacteria and their distribution may be limited to certain taxa.

The analysis of bacterial sugars is often hampered by the extreme variation in linkage stabilities toward acid. The strong hydrolytic conditions necessary to release the most firmly bound sugars, e.g. amino sugars and glucuronic acid, may be detrimental for the more acid-labile sugars, e.g. dideoxyhexoses and 2-keto-3-deoxyoctonate.

Unsubstituted sugars

The widely distributed hexoses glucose and galactose are also common in bacteria, whereas mannose is less frequent. Of the pentoses, ribose is always present in RNA and may also be part of LPS or polysaccharides, as may arabinose. Since the latter does not occur in mammalian tissue, it may serve as a marker of bacterial infection. Almost unique for bacteria are also aldoheptoses. Two or more moieties of L-glycero-D-manno-heptose (one or more being phosphorylated) occur as a cluster in LPS of many Gram-negative bacteria (Lüderitz *et al.* 1971). Some species (e.g. *Actinobacillus actinomycetemcomitans*) or serotypes (e.g. of *Yersinia enterocolitica*) also have the D-glycero analogue in their LPS, and heptoses may occasionally occur in major quantities in other bacterial polysaccharides (Jann & Jann 1977).

Deoxy sugars

Some precautions must be taken in the analysis of 3,6-dideoxyhexoses because of their acid lability. Five such rather nonpolar sugars (abequose, ascarylose, paratose, colitose, and tyvelose) have been detected in LPS (Lüderitz *et al.* 1971), or more rarely in other bacterial polysaccharides (Jann & Jann 1977). The 6-deoxyhexoses rhamnose and fucose are found in LPS and in polysaccharides of many bacteria, whereas 6-deoxy-L-altrose is more rare, occurring, e.g. in LPS of certain strains of *Yersinia* (Beer & Seltmann 1973); 6-deoxy-D-manno-heptose has also been identified in strains of the latter (Samuelsson *et al.* 1974).

O-methyl sugars

A number of different O-methylated sugars have been detected in LPS of certain bacteria, and with an increasing number of methyl groups on the sugar moiety their polarity decreases. Together with deoxy sugars these moieties can render the LPS nonpolar enough to be recovered in the phenol phase upon extraction. This is typically the case with some *Rhodopseudomonas* strains where 4-O-methyl-D-xylose and 3-O-methyl-6-deoxy-D-talose contribute to the lipophilic character of the LPS (Weckesser *et al.* 1974).

Amino sugars

Glucosamine is a universal constituent (as *N*-acetyl) of the peptidoglycan of both Gram-positive and Gram-negative bacteria. In addition, this sugar is a nearly universal constituent of lipid A of Gram-negative species where it is acylated with long chain fatty acids (Wollenweber *et al.* 1982). In the *N*-acetylated form, glucosamine may also occur in the polysaccharide part of LPS, or in capsular polysaccharides. Galactosamine occurs much less frequently in bacteria but is *N*-acetylated in LPS of, for example, *Yersinia pseudotuberculosis* (Samuelsson *et al.* 1974) and in the group-specific polysaccharide of *Neisseria meningitidis* 29E (Bhattacharjee *et al.* 1978). 4-Aminoarabinose is a very acid-labile constituent of a number of strains of *Salmonella* (Galanos *et al.* 1979), and it may have been overlooked in other bacteria. Several other amino sugars are found in bacteria, including 6-deoxyglucosamine and a 2,3-diamino-2,3-dideoxyhexose (Weckesser *et al.* 1974), which require drastically strong acidic conditions for release. The latter may, therefore, be overlooked when ordinary procedures of bond splitting are used.

Sugar acids

2-Keto-3-deoxyoctonate, commonly denoted KDO or dOclA, is an almost universal eight-carbon sugar acid of LPS of Gram-negative species. It is usually found as a cluster of a few moieties (partly phosphorylated) in the linkage region between the polysaccharide chain and lipid A. The ketosidic linkage is extremely labile toward acid, which may also partly convert KDO to its anhydro form, which can be determined by GC (Bryn & Jantzen 1982; the first KDO GC peak is shown in Fig. 4), thus overcoming some of the problems of colorimetric KDO determinations (Unger 1981). 2-Keto-3-deoxyoctonate occurs as a linkage sugar to a phospholipid anchor in minor quantities in some capsular polysaccharides of *E. coli* (Schmidt & Jann 1982) or in major amounts in polysaccharides of some urinary-tract-infective strains. Also, the group-specific polysaccharide of *Neisseria meningitidis* 29E is characterized by high contents of KDO (Bhattacharjee *et al.* 1978).

The nine-carbon sugar acid, *N*-acetylneuraminic acid, is a widely occurring constituent of mammalian glycoproteins and some glycolipids, and in bacteria it occurs in major amounts in polysaccharides of, for example, *E. coli* K1 and of some serogroups of *N. meningitidis* (Jann & Jann 1977). Glucuronic acid may be found as part of LPS of, for example, *Bordetella pertussis* (LeDur *et al.* 1980) and otherwise in polysaccharides of some serotypes of *E. coli* and in some *Klebsiella* serotypes, which also contain galacturonic acid (Jann & Jann 1977).

Sugar Patterns

Studies on the occurrence and distribution of sugars among bacterial taxa are still incomplete, and few comprehensive surveys of taxonomic value have been presented. Most reports deal with extensive chemical examinations of LPS or capsular polysaccharides from a limited number of strains of a species. Thus the representative nature of such strains can be questioned. Ideally, many strains should be screened using a fairly simple and sensitive technique in order to discover taxonomically useful sugars. One approach is based on GC profiles of TFA-derivatized methyl glycosides obtained after methanolysis. Because of the high sensitivity (cells from a single agar plate are sufficient) and the excellent separation of sugars achieved by gas chromatography, this method fulfills many of the requirements of a useful screening technique for bacterial sugars (Jantzen *et al.* 1974a).

In a survey of 64 strains of *Acinetobacter*, *Moraxella* and *Neisseria*, Jantzen *et al.* (1976) detected KDO in all strains. L-glycero-D-manno-heptose was only found in the genus *Neisseria* and in "*Moraxella*" *urethralis*. The distribution of neuraminic acid served to distinguish strains of *N. meningitidis* groups B and C from other meningococcal serogroups. In *Acinetobacter* there was evidence of several capsular types (Jantzen *et al.* 1976). Thus 14 of 17 strains examined had elevated amounts of one or more of the sugars glucose, galactose, glucosamine, mannose, or rhamnose, compared to typical acapsular strains. High concentrations of mannose were also detected in the two strains of *Moraxella atlantae*. Some strains of *Moraxella nonliquefaciens* contained significant amounts of neuraminic acid. Interestingly, these strains have recently been shown to possess a polysaccharide very similar to the *Neisseria meningitidis* group B polyneuraminic acid (Bøvre *et al.* 1983). Thus by screening a large number of strains, the taxonomic potential of certain polysaccharides can be uncovered. However, the complexity of the whole-cell GC profiles may cause doubtful results. Simplification of the profiles is obtained by introducing a carbohydrate extraction step and capillary column GC (Bryn & Jantzen 1982);

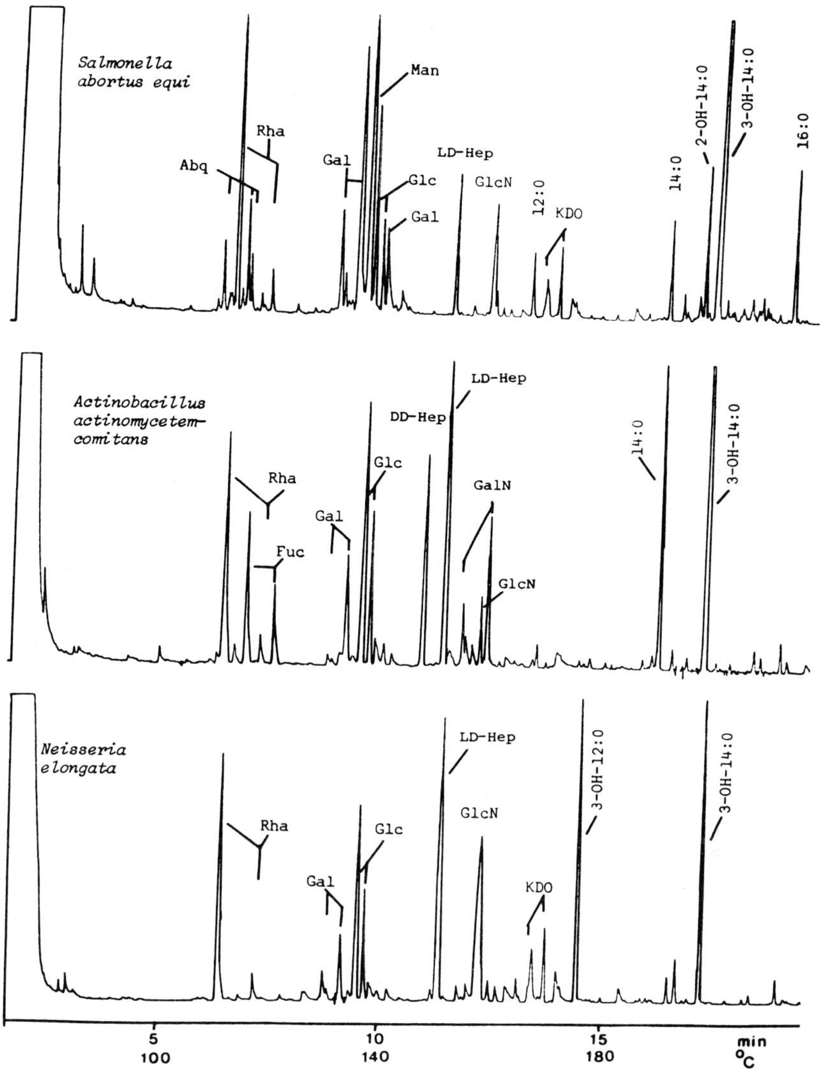

FIG. 4. Example of gas chromatographic lipopolysaccharide profiles of Gram-negative bacteria. (From Bryn & Jantzen 1982.)

this reduces the background, gives more reliable results and represents an improvement in the method. This extended method also furnishes sugar profiles from bacterial growth taken from a single agar plate, and it proved useful for the mapping of carbohydrates of *Neisseria meningitidis* strains (Bryn *et al.* 1983).

Of the glycoconjugates occurring in the Gram-negative cell envelope, LPS have been especially thoroughly examined, and strains belonging to the family *Enterobacteriaceae* have served as model organisms in this respect. The LPS structures have been elucidated by close cooperation between bacteriologists, immunologists and chemists (Lüderitz *et al.* 1971). Presently, the serological diagnosis of *Salmonella* is based on LPS (or O antigens) of known structures and the serological specificities have been traced to specific combinations of sugars (Lüderitz *et al.* 1971). Some typical GC profiles of LPS are shown in Fig. 4. Generally, the LPS found in strains of the *Enterobacteriaceae* contain glucosamine, KDO, L-glycero-D-manno-heptose, glucose and galactose in the inner parts of LPS. An outer part, the O chain, is more specific and may contain other sugars characteristic of a certain species or serotype. Since the O chain is composed of several repeating units, these sugars may predominate quantitatively, and many rare sugars have also been described (Lüderitz *et al.* 1971; Wilkinson 1977).

In other bacterial groups, O chains in the LPS are less common, e.g. LPS of both *Neisseria* (Jennings *et al.* 1980) and *Bordetella* (LeDur *et al.* 1980) are of the rough (R) type. Lipopolysaccharide sugars may therefore be characteristic of either genera or species, depending on the taxon (Table 6). However, for many genera, LPS from only a few species or strains have been examined by chemical techniques and data may be completely lacking in some taxa.

Similar to the chemotaxonomic investigations of LPS, studies of capsular polysaccharides cover only a restricted number of bacterial species. The *Enterobacteriaceae* is especially well mapped (Jann & Jann, 1977). The so-called K antigens are essential in the serodiagnosis of, e.g. *Escherichia coli* (Jann & Jann 1983) and other taxa. Also in *Neisseria meningitidis* the capsular polysaccharides are well known structurally. Twelve different polysaccharides form the basis of the serological diagnosis of this bacterium. Recent studies by capillary GC were performed on crude extracts of 90 serogrouped strains of *N. meningitidis* of clinical and carrier origin (Bryn *et al.* 1983); an example is given in Fig. 3. In addition to the expected sugars from LPS and capsular polysaccharides, many of the strains were found to contain constituents of other polysaccharides including undescribed moieties. Thus some of the meningococcal strains that could not be grouped by standard sera showed considerable amounts of certain sugars of possible capsular origin. This again demonstrates the value of screening a large number of bacterial strains for sugar patterns, and much work remains to be done in this field.

References

ABBAS, C. A. & CARD, G. L. 1980 The relationships between growth temperature, fatty acid composition and the physical state and fluidity of membrane lipids in *Yersinia enterocolitica*. *Biochimica et Biophysica Acta* **602**, 469–476.

ABEL, K., DESMERTZING, H. & PETERSON, J. I. 1963 Classification of microorganisms by analysis of chemical composition. I. Feasibility of utilizing gas chromatography. *Journal of Bacteriology* **85**, 1039–1044.

ASSELINEAU, C. & ASSELINEAU, J. 1984 Fatty acids and complex lipids. In *Gas Chromatography/Mass Spectrometry Applications in Microbiology* eds. Odham, G., Larsson, L. & Mårdh, P. E. pp. 57–103. New York: Plenum.

BHATTACHARJEE, A. K., JENNINGS, H. J. & KENNY, C. P. 1978 Structural elucidation of the 3-deoxy-D-manno-octulosonic acid containing meningococcal 29-e capsular polysaccharide antigen using carbon-13 nuclear magnetic resonance. *Biochemistry* **17**, 645–651.

BEER, W. & SELTMANN, G. 1973 Zuckerzusammensetzung der Lipopolysaccharide einiger *Yersinia enterocolitica*-Stamme. *Zeitschrift für Allgemeine Mikrobiologie* **13**, 167–169.

BOON, J. J., DE LEEUW, J. W., VAN DER HOEK, G. J. & VOSJAN, J. H. 1977 Significance and taxonomic value of *iso* and *anteiso* monoenoic fatty acids and branched β-hydroxy acids in *Desulfovibrio desulfuricans*. *Journal of Bacteriology* **129**, 1183–1191.

BROADY, K. W., RIETSCHEL, E. T. & LÜDERITZ, O. 1981 The chemical structure of the lipid A component of lipopolysaccharides from *Vibrio cholerae*. *European Journal of Biochemistry* **115**, 463–468.

BRYN, K. & JANTZEN, E. 1982 Analysis of lipopolysaccharides by methanolysis, trifluoroacetylation, and gas chromatography on a fused-silica capillary column. *Journal of Chromatography* **240**, 405–413.

BRYN, K. & RIETSCHEL, E. T. 1978 L-2-Hydroxytetradecanoic acid as a constituent of *Salmonella* lipopolysaccharides (Lipid A). *European Journal of Biochemistry* **86**, 311–315.

BRYN, K., JANTZEN, E. & BØVRE, K. 1977 Occurrence and patterns of waxes in *Neisseriaceae*. *Journal of General Microbiology* **102**, 33–43.

BRYN, K., FRØHOLM, L. O., HOLTEN, E. & BØVRE, K. 1983 Gas-chromatographic screening of capsular polysaccharides of *Neisseria meningitidis*. *NIPH Annals* **6**, 91–101.

BØVRE, K., BRYN, K., CLOSS, O., HAGEN, N & FRØHOLM, L. O. 1983 Surface polysaccharide of *Moraxella nonliquefaciens* identical to *Neisseria meningitidis* group B capsular polysaccharide. A chemical and immunological investigation. *NIPH Annals* **6**, 65–73.

CRONAN, J. R., RANDALL, R., TAYLOR, F. R. & JACKSON, M. B. 1979 Properties and biosynthesis of cyclopropane fatty acids in *Escherichia coli*. *Journal of Bacteriology* **138**, 118–121.

CSANGO, P. A., HAGEN, N. & JAGARS, G. 1982 Method for isolation of *Gardnerella vaginalis* (*Haemophilus vaginalis*). Characterization of isolates by gas chromatography. *Acta Pathologica et Microbiologica Scandinavica, Section B: Microbiology* **90**, 89–93.

DEES, S. B. & MOSS, C. W. 1975 Cellular fatty acids of *Alcaligenes* and *Pseudomonas* species from clinical specimens. *Journal of Clinical Microbiology* **1**, 414–419.

DEES, S. B. & MOSS, C. W. 1978 Identification of *Achromobacter* species by cellular fatty acids and production of keto acids. *Journal of Clinical Microbiology* **8**, 61–66.

DEES, S. B., HOLLIS, D. G., WEAVER, R. E. & MOSS, C. W. 1981 Cellular fatty acids of *Brucella canis* and *Brucella suis*. *Journal of Clinical Microbiology* **14**, 111–112.

DRIGUES, P., LAFFORGUE, D. & ASSELINEAU, J. 1981 Etude des phospholipides de *Pseudomonas solanacearum*. Presence d'acides gras alpha hydroxyles mono ethyleniques. *Biochimica et Biophysica Acta* **666**, 504–507.

DRÜCKER, D. P. 1981 *Microbial Application of Gas Chromatography*. Cambridge: Cambridge University Press.

ELLWOOD, D. C. & TEMPEST, D. W. 1972 Effects of environment on bacterial wall contents and composition. *Advances in Microbial Physiology* **7**. 83–117.

ERLER, W., FLOSSMANN, K. D., FLEIST, H. & JACOB, B. 1981 Die Strukturelemente in den

Lipopolysacchariden des Chemotyps IV von *Pasteurella multocida*. *Zeitschrift für Allgemeine Mikrobiologie* **21**, 569–579.

FAUTZ, E., ROSENFELDER, G. & GROTJAHN, L. 1979 *Iso*-branched 2- and 3-hydroxy fatty acids as characteristic lipid constituents of some gliding bacteria. *Journal of Bacteriology* **140**, 852–858.

FLESHER, A. R. & INSEL, R. A. 1978 Characterization of lipopolysaccharides of *Haemophilus influenzae*. *Journal of Infectious Diseases* **138**, 719–730.

GALANOS, C., FREUDENBERG, M. A., LÜDERITZ, O., RIETSCHEL, E. T. & WESTPHAL, O. 1979 Chemical, physiochemical and biological properties of bacterial lipopolysaccharides. *Progress of Clinical Biological Research* **29**, 321–332.

GOLDFINE, H. 1972 Comparative aspects of bacterial lipids. *Advances in Microbial Physiology* **8**, 1–57.

HAEFFNER, N., CHABY, R. & SZABO, L. 1977 Identification of 2-methyl-3-hydroxydecanoic and 2-methyl-3-hydroxytetradecanoic acids in the "lipid x" fraction of the *Bordetella pertussis* endotoxin. *European Journal of Biochemistry* **77**, 535–544.

HEBERT, G. A., HOLLIS, D. G., WEAVER, R. E., LAMBERT, M. A., BLASER, M. J. & MOSS, C. W. 1982 30 years of campylobacters: biochemical characteristics and a biotyping proposal for *Campylobacter jejuni*. *Journal of Clinical Microbiology* **15**, 1065–1073.

HOFSTAD, T. & JANTZEN, E. 1982 Fatty acids of *Leptotrichia buccalis:* taxonomic implications. *Journal of General Microbiology* **128**, 151–153.

INOUYE, M. 1979 What is the outer membrane? In *Bacterial Outer Membranes* ed. Inouye, M., pp. 1–12. New York: John Wiley & Sons.

JANN, K. & JANN, B. 1977 Bacterial polysaccharide antigens. *Surface Carbohydrates of the Procaryotic Cell* ed. Sutherland, I., pp. 247–287. London & New York: Academic Press.

JANN, K. & JANN, B. 1983 The K antigens of *Escherichia coli*. *Progress in Allergy* **33**, 53–79.

JANN, B., JANN, K. & BEYART, G. O. 1973 2-Amino-2,6-dideoxy-D-glucose (D-quinovosamine): a constituent of the lipopolysaccharides of *Vibrio cholerae*. *European Journal of Biochemistry* **37**, 531–534.

JANTZEN, E. 1984 Analysis of cellular components in bacterial classification and diagnosis. In *Gas Chromatography/Mass Spectrometry Applications in Microbiology* ed. Odham, G., Larsson, L. & Mårdh, P.-A., pp. 257–302. New York & London: Plenum Press.

JANTZEN, E. & HOFSTAD, T. 1981 Fatty acids of *Fusobacterium* species: taxonomic implications. *Journal of General Microbiology* **123**, 163–171.

JANTZEN, E., BRYN, K. & BØVRE, K. 1974a Gas chromatography of bacterial whole cell methanolysates. IV. A procedure for fractionation and identification of fatty acids and monosaccharides of cellular structures. *Acta Pathologica et Microbiologica Scandinavica, Section B: Microbiology* **82**, 753–766.

JANTZEN, E., BRYN, K., BERGAN, T. & BØVRE, K. 1974b Gas chromatography of bacterial whole cell methanolysates. V. Fatty acid composition of neisseriae and moraxellae. *Acta Pathologica et Microbiolologica Scandinavica, Section B: Microbiology* **82**, 767–779.

JANTZEN, E., BRYN, K., BERGAN, T. & BØVRE, K. 1975 Gas chromatography of bacterial whole cell methanolysates. VII. Fatty acid composition of *Acinetobacter* in relation to the taxonomy of *Neisseriaceae*. *Acta Pathologica et Microbiologica Scandinavica, Section B: Microbiology* **83**, 569–580.

JANTZEN, E., BRYN, K. & BØVRE, K. 1976 Monosaccharide patterns of *Neisseriaceae*. *Acta Pathologica et Microbiologica Scandinavica, Section B: Microbiology* **84**, 177–188.

JANTZEN, E., BRYN, K., HAGEN, N., BERGAN, T. & BØVRE, K. 1978 Fatty acids and monosaccharides of *Neisseriaceae* in relation to established taxonomy. *NIPH Annals* **1**, 59–71.

JANTZEN, E., BERDAL, B. P. & OMLAND, T. 1979 Cellular fatty acid composition of *Francisella tularensis*. *Journal of Clinical Microbiology* **10**, 928–930.

JANTZEN, E., BERDAL, B. P. & OMLAND, T. 1980 Cellular fatty acid composition of *Haemophilus* species, *Pasteurella multocida, Actinobacillus actinomycetemcomitans* and *Haemophilus vaginalis* (*Corynebacterium vaginale*). *Acta Pathologica et Microbiologica Scandinavica, Section B: Microbiology* **88,** 89–93.

JANTZEN, E., KNUDSEN, E. & WINSNES, R. 1982 Fatty acid analysis for differentiation of *Bordetella* and *Brucella* species. *Acta Pathologica et Microbiologica Scandinavica, Section B: Microbiology* **90,** 353–359.

JANTZEN, E. & LASSEN, J. 1980 Characterization of *Yersinia* species by analysis of whole-cell fatty acids. *International Journal of Systematic Bacteriology* **30,** 421–428.

JENNINGS, H. J., HAWES, G. B., ADAMS, G. A. & KENNY, C. P. 1973 The chemical composition and serological reactions of lipopolysaccharides from serogroups A, B, X, and Y of *Neisseria meningitidis. Canadian Journal of Biochemistry* **51,** 1347–1354.

JENNINGS, H. J., BHATTACHARJEE, A. K., KENNE, L., KENNY, C. P. & CALVER, G. 1980 The R-type lipopolysaccharides of *Neisseria meningitidis. Canadian Journal of Biochemistry* **58,** 128–136.

KABIR, S. 1982 Characterization of the lipopolysaccharide from *Vibrio cholerae* 395 (Ogawa). *Infection and Immunity* **38,** 1263–1272.

KAMIO, Y., KANEGASAKI, S. & TAKAHASHI, H. 1969 Occurrence of plasmalogens in anaerobic bacteria. *Journal of General and Applied Microbiology* **15,** 439–451.

KASPER, D. L., WEINTRAUB, A., LINDBERG, A. A. & LÖNNGREN, J. 1983 Capsular polysaccharides and lipopolysaccharides from two *Bacteroides fragilis* reference strains: chemical and immunochemical characterization. *Journal of Bacteriology* **153,** 991–997.

KAWAI, J. & MORIBAYASHI, A. 1982 Characteristic lipids of *Bordetella pertussis:* simple fatty acid composition, hydroxy fatty acids, and an ornithine-containing lipid. *Journal of Bacteriology* **151,** 996–1005.

KILEY, P. & HOLT, S. C. 1980 Characterization of the lipopolysaccharide from *Actinobacillus actinomycetemcomitans* Y4 and N27. *Infection and Immunity* **30,** 862–873.

LACAVE, C., ASSELINEAU, J., SERRE, A. & ROUX, J. 1969 Comparison de la composition chimique d'une fraction polysaccharidique isolees de *Brucella melitensis. European Journal of Biochemistry* **9,** 189–198.

LAMBERT, M. A., HOLLIS, D. G., MOSS, C. W., WEAVER, R. E. & THOMAS, M. L. 1971 Fatty acid composition of nonpathogenic *Neisseria. Canadian Journal of Microbiology* **17,** 1491–1502.

LECHEVALIER, M. P. 1977 Lipids in bacterial taxonomy—a taxonomist's viewpoint. *Critical Review of Microbiology* **5,** 109–210.

LEDUR, A., CAROFF, M., CHABY, R. & SZABO, L. 1980 A novel type of endotoxin structure present in *Bordetella pertussis.* Isolation of two different polysaccharides bound to Lipid A. *European Journal of Biochemistry* **84,** 579–589.

LIVERMORE, B. P. & JOHNSON R. C. 1974 Lipids of the *Spirochaetales, Spirochaeta, Treponema,* and *Leptospira. Journal of Bacteriology* **120,** 1268–1273.

LÜDERITZ, O., WESTPHAL, O., STAUB, A. M. & NIKAIDO, H. 1971 Isolation and chemical and immunological characterization of bacterial lipopolysaccharides. In *Microbial Toxins* eds. Weinbaum, G., Kadis, S. & Ajl, S., Vol. IV, pp. 145–224. New York & London: Academic Press.

MATTHEWS, H., YANG, T.-K. & JENKIN, H. 1979 Unique lipid composition of *Treponema pallidum* (Nichols virulent strain). *Infection and Immunity* **24,** 713–719.

MAYBERRY, W. R. 1980 Cellular distribution and linkage of D-(-)-3-hydroxy fatty acids in *Bacteroides* species. *Journal of Bacteriology* **144,** 200–204.

MAYBERRY, W. R. 1981 Dihydroxy and monohydroxy fatty acids in *Legionella pneumophila. Journal of Bacteriology* **147**, 373–381.

MELCHIOR, D. L. 1982 Lipid phase transitions and regulation of membrane fluidity in prokaryotes. *Current Topics in Membranes and Transport* **17**, 263–316.

MIYAGAWA, E. & SUTO, T. 1980 Cellular fatty acid composition in *Bacteroides oralis* and *Bacteroides ruminicola*. *Journal of General and Applied Microbiology* **26**, 331–343.

MORENO, E., PITT, M. W., JONES, L. M., SCHURIG, G. G. & BERMAN, D. T. 1979 Purification and characterization of smooth and rough lipopolysaccharides from *Brucella abortus*. *Journal of Bacteriology* **138**, 361–369.

MOSS, C. W. 1978 New methodology for identification of non-fermenters: gas-liquid chromatographic chemotaxonomy. In *Glucose Non-fermenting Gram-negative Bacteria in Clinical Microbiology* ed. Gilardi, G. L., pp. 171–201. Boca Raton, USA: CRC Press.

MOSS, C. W. 1981 Gas-liquid chromatography as an analytical tool in microbiology. *Journal of Chromatography* **203**, 337–347.

MOSS, C. W. & LAMBERT, M. A. 1976 An uncommon fatty acid of *Spirillum*. *Microbios Letters* **3**, 207–212.

MOSS, C. W., LAMBERT, M. A. & MERWIN, W. 1974 Comparison of rapid methods for analysis of bacterial fatty acids. *Applied Microbiology* **28**, 80–85.

MOSS, C. W., WEAVER, R. E., DEES, S. B. & CHERRY, W. B. 1977 Cellular fatty acid composition of isolates from legionnaires disease. *Journal of Clinical Microbiology* **6**, 140–143.

NÆSS, V. & HOFSTAD, T. 1982 Isolation and chemical composition of lipopolysaccharide from *Campylobacter jejuni*. *Acta Pathologica et Microbiologica Scandinavica, Section B: Microbiology* **90**, 135–139.

OSBORN, M. J. 1979 Biosynthesis and assembly of lipopolysaccharide of the outer membrane. In *Bacterial Outer Membranes* ed. Inouye, M., pp. 15–35. New York: John Wiley & Sons.

OYAIZU, H. & KOMAGATA, K. 1981 Chemotaxonomic and phenotypic characterization of the strains of species in the *Flavobacterium-Cytophaga* complex. *Journal of General and Applied Microbiology* **27**, 57–107.

OYAIZU, H. & KOMAGATA, K. 1983 Grouping of *Pseudomonas* species on the basis of cellular fatty acid composition and the quinone system with special reference to the existence of 3-hydroxy fatty acids. *Journal of General and Applied Microbiology* **29**, 17–40.

PERRY, M. B., DAOUST, V., DIENA, B. B., ASHTON, F. E. & WALLACE, R. 1975 The lipopolysaccharides of *Neisseria gonorrhoeae* colony types 1 and 4. *Canadian Journal of Biochemistry* **53**, 623–629.

RIETSCHEL, E. T. & LÜDERITZ, O. 1980 Struktur von Lipopolysaccharid und Taxonomie Gram-negativer Bakterien. *Forum Microbiologie 1/80*. 12–20.

ROGERS, H. J., PERKINS, H. R. & WARD, J. B. 1980 *Microbial Cell Walls and Membranes*. London & New York: Chapman & Hall.

ROUSSEL, J. & ASSELINEAU, J. 1980 Fatty acid composition of the lipids of *Pseudomonas mildenbergii*. Presence of a fatty acid containing two conjugated double bonds. *Biochimica et Biophysica Acta* **619**, 689–692.

SAMUELSSON, K., LINDBERG, B. & BRUBAKER, R. R. 1974 Structure of O-specific side chains of lipopolysaccharides from *Yersinia pseudotuberculosis*. *Journal of Bacteriology* **117**, 1010–1016.

SCHMIDT, M. A. & JANN, K. 1982 Phospholipid substitution of capsular (K) polysaccharide antigens from *Escherichia coli* causing extra-intestinal infections. *FEMS Microbiology Letters* **14**, 69–74.

SHAW, N. 1974 Lipid composition as a guide to the classification of bacteria. *Advances of Applied Microbiology* **17**, 63–108.

SUGIMOTO, C., MIYAGAWA, E., NAKAZAWA, M., MITANI, K. & ISAYAMA, Y. 1983 Cellular fatty acid composition comparisons of *Haemophilus equigenitalis* and *Moraxella* species. *International Journal of Systematic Bacteriology* **33**, 181–187.

TATSUMI, K., KISHIMOTO, Y. & HIGNITE, C. 1974 Stereochemical aspects of synthetic and naturally occurring 2-hydroxy fatty acids. Their absolute configurations and assays of optical purity. *Archives of Biochemistry and Biophysics* **165**, 656–664.

UCHIDA, K. & MOGI, K. 1972 Cellular fatty acid spectra of *Pediococcus* species in relation to their taxonomy. *Journal of General and Applied Microbiology* **18**, 109–129.

UNGER, F. M. 1981 The chemistry and biological significance of 3-deoxy-D-*manno*-2-octulosonic acid (KDO). *Advances in Carbohydrate Chemistry and Biochemistry* **38**, 323–388.

WECKESSER, J., DREWS, G., MAYER, H. & FROMME, I. 1974 Lipopolysaccharide aus Rhodospirillaceae, Zusammensetzung und taxonomische Relevanz. *Zentralblatt für Bakteriologie, Parasitenkunde, Infektionskrankheiten und Hygiene. I. Abteilung, Originale A* **228**, 193–198.

WESTPHAL, O., LÜDERITZ, O. & BISTER, F. 1952. Über die Extraktion von Bakterien mit Phenol/Wasser. *Zeitschrift für Naturforschung* **7B**, 148–155.

WILKINSON, S. G. 1977 Composition and structure of bacterial lipopolysaccharides. In *Surface Carbohydrates of the Procaryotic Cell* ed. Sutherland, I. W., pp. 97–175. New York & London: Academic Press.

WILKINSON, S. G. & GALBRAITH, L. 1975 Studies of lipopolysaccharides from *Pseudomonas aeruginosa*. *European Journal of Biochemistry* **52**, 331–343.

WOLLENWEBER, H.-W., BROADY, K. W., LÜDERITZ, O. & RIETSCHEL, E. T. 1982 The chemical structure of Lipid A. Demonstration of amide-linked 3-acyloxyacyl residues in *Salmonella minnesota* Re lipopolysaccharide. *European Journal of Biochemistry* **124**, 191–198.

Fatty Acid and Menaquinone Analysis of Actinomycetes and Related Organisms

R. M. KROPPENSTEDT

German Collection of Microorganisms, Gesellschaft für Biotechnologische Forschung Braunschweig, Federal Republic of Germany

The application of chemical methods has not only led to improvements in the classification and identification of actinomycetes but has also provided an insight into the phylogenetic relationships of these bacteria (Stackebrandt *et al.* 1983a,b; see Chapters 2 and 3). Valuable data were derived from the analysis of cell-wall sugars and amino acids (Lechavalier & Lechevalier 1970a,b; see Chapters 7, 8, 10 and 11), fatty acids including mycolic acids (Minnikin & Goodfellow 1980; Goodfellow & Wayne 1982), phospho- and glycolipids (Lechevalier *et al.* 1977; Minnikin *et al.* 1978), isoprenoid quinones (Collins & Jones 1981; Chapter 13) and from ribosomal RNA sequencing studies (see Chapter 2). In the present contribution techniques used to determine the fatty acid profiles and menaquinone patterns of actinomycetes are considered together with their impact on actinomycete classification.

Fatty Acids

Fatty acids can be released from lipids by transesterification. The separation and identification of complex mixtures of fatty-acid methyl esters can be achieved either by using two glass capillary columns, each with a different separation behaviour (Kroppenstedt 1977, 1982b) or by employing a single glass capillary column connected to a mass spectrometer (Drucker 1981). This latter technique is usually beyond the scope of most microbiological laboratories. When only one capillary column is used, the unknown complex mixture must be analysed before and after hydrogenation if a reliable identification is to be achieved (Ballio & Barcellona 1968). Several laborious derivatisation and precolumn separation procedures must be employed when

CHEMICAL METHODS IN
BACTERIAL SYSTEMATICS

0-12-289675-0

short packed columns are used (Hofheinz & Grisebach 1965; Brian & Gardner 1968). Many microbiological laboratories are now equipped with gas chromatographs for packed columns that can easily be converted for the use of fused-silica, wide-bore capillary columns. With only one polar or nonpolar capillary column it is possible to identify an unknown fatty-acid pattern with one already described (Fig. 2). Even with short packed columns it is possible to recognize the major fatty-acid patterns exhibited by actinomycetes (Table 3).

Most fatty acids are common to eubacteria including actinomycetes, though a few are restricted to the latter. In addition, actinomycetes, contrary to many other bacteria, synthesize their fatty acids via the aerobic pathway (Scheuerbrandt & Bloch 1961). Oleic acid (*cis*-9,10-octadecenoic acid) can be detected in these organisms (Kroppenstedt & Kutzner 1978). The position of the methylene bridge in the cyclopropane fatty acids varies correspondingly, only those derived from oleic acid (dihydrosterculic acid = *cis*-9,10-methyleneoctadecanoic acid) occurring in actinomycetes. Fatty acids found exclusively in actinomycetes and related organisms are those with a methyl group in position 10; the best example is tuberculostearic acid (10-methyloctadecanoic acid) first detected in *Mycobacterium tuberculosis*.

Menaquinones

The first studies on menaquinones in actinomycetes employed mass spectrometry (Yamada *et al.* 1976a,b, 1977a,b; Collins *et al.* 1977) but recently exact quantitative data have been obtained using high performance liquid chromatography (HPLC) (Kroppenstedt *et al.* 1981; Collins 1982a; Tamaoka *et al.* 1983). In most cases, menaquinone mixtures can be separated and single menaquinones identified but there are instances where the identification of single peaks in complex mixtures is difficult. Thus *Thermomonospora* and some *Actinomadura* species synthesize mixtures of highly saturated menaquinones; in such cases, an additional stationary phase with a different separation behaviour must be applied (Kroppenstedt 1982a). Alternatively, HPLC eluates of single peaks can be collected and identified by mass spectrometry (Tamaoka *et al.* 1983).

Most bacteria contain either menaquinones, ubiquinones or both (see Chapter 13). Gram-positive bacteria including actinomycetes synthesize only menaquinones (Collins & Jones 1981). The menaquinones of actinomycetes and related bacteria, however, differ in their "degree of saturation"; they are partly saturated and comprise eight to ten isoprene units. Two exceptions, "*Gordona aurantiaca*" and "*Corynebacterium*" *paurometabolum*, have fully unsaturated isoprenylic side chains (Goodfellow *et al.* 1978; Collins & Jones

1982; Meisel 1982; Lemmer & Kroppenstedt 1984). This actinomycete menaquinone type is also found in *Micrococcus* (Yamada *et al.* 1976b), a taxon now known to contain strains closely related to arthrobacters (Stackebrandt & Woese 1981).

Material and Methods

Organisms

About 600 strains representing most of the genera of the order *Actinomycetales* were analysed for fatty acids and around 300 for their menaquinone composition. All of the microorganisms were obtained from the German Collection of Microorganisms (DSM Catalogue of Strains 1983). Strains not included in the *Approved Lists of Bacterial Names* (Skerman *et al.* 1980) or subsequent validation lists are cited in quotation marks.

Cultivation

The test strains were grown in a liquid medium (70 ml) in 300-ml baffled flasks on a rotary shaker at 140 rpm. Representatives of the genera *Actinomyces* and *Arachnia* were grown anaerobically. The growth medium and incubation regimes were chosen to support good growth of the test strains and in all cases cells were harvested in the early stationary phase after being checked for purity. Autoclaved cells were centrifuged, washed twice with distilled water and lyophilized. Freeze-dried samples were either used immediately or kept at room temperature in the dark until required. Little is known about the effect of cultivation regimes on menaquinone composition but the influence of such conditions on fatty acid profiles in actinomycetes has been demonstrated (Kroppenstedt 1977).

Fatty acid transesterification, extraction, and purification

Separation and identification

Two methods can be used for the transesterification of fatty-acid glycerol esters to fatty-acid methyl esters (FAMES). The method of Brian & Gardner (1967) is simple and rapid but gives mixtures that may contain nonfatty-acid substances. Transesterification of the fatty acids is carried out by refluxing 30–50 mg of dried cells in 10 ml of methanol–BCl_3 (or MeOH–BF_3) for 5 min or by keeping the reaction mixture at room temperature overnight. The sample is then diluted with 15 ml of H_2O to decompose the BCl_3, and the FAMES are extracted twice with 4 ml of chloroform. The sample is centrifuged, and the lower chloroform phase is collected, washed with 1 ml of H_2O, dried with Na_2SO_4 and evaporated on a rotary evaporator. The residue is redissolved in 100 μl of chloroform and 0.1 to 1.0 μl is used for gas liquid chromatography. The methanolysates are suitable for separation on packed columns and supported, open tubular columns.

The second method (Minnikin *et al.* 1975, 1980) is more laborious but gives clean FAME fractions that can be separated on narrow-bore, thin-film, capillary columns. Thus lyophilized cells (50 mg) are mixed with dry methanol (2.5 ml), toluene (2.5 ml) and concentrated sulphuric acid (0.2 ml) in a tube (10 ml) closed with a PTFE-lined screw cap, and methanolysis is allowed to proceed overnight at 75°C. The methanolysates are then cooled to room temperature. For the extraction of the FAMES, 1 ml of light petroleum (bp 60–80°C) is added, and the tube is shaken thoroughly. The organic phase (upper layer) is collected and an additional extraction with 1 ml of light petroleum is carried out. Should separation not occur in the first extraction, an additional ml of light petroleum is added and the sample is kept at −10°C in a deep freezer for about 10 min. Collected extracts are passed through columns, using Pasteur pipettes (7 × 150 mm) filled with ammonium hydrogen carbonate to eliminate residual acid. The eluates are reduced to dryness on a rotary evaporator and redissolved in 200–300 μl of light petroleum.

Methanolysates are applied with a syringe to 20 × 20-cm, silica-gel thin-layer plates (Merck 7748 Kieselgel 60 $PF_{254/366}$, thickness 0.3–0.4 mm) as a streak (20 mm). The plates are developed in light petroleum (60–80°C), diethyl ether (85:15, v/v) until the mobile phase has nearly reached the top of the plate (19 cm). Chromatograms are then air-dried and viewed under UV light (366 nm), which causes FAMES to appear as light blue spots against a dark blue background (Fig. 1). Nonhydroxylated FAMES have R_f values of 0.75–0.85 in this mobile phase, and 2-hydroxy and 3-hydroxy FAMES have R_f values of 0.22 and 0.16, respectively (Kroppenstedt 1982b). Long-chain 3-hydroxy fatty acids (mycolic acids) have R_f values from 0.25 to 0.50 according to their chain length and the occurrence of additional functional groups (Fig. 1). The individual fatty acid spots on the thin-layer plates are scraped off by sucking the silica gel into Pasteur pipettes packed with silanised glass wool. The FAMES are eluted from the columns with 5 ml of diethyl ether. The ether extract is evaporated to dryness and taken up in 100 μl of light petroleum of which 0.1–1.0 μl is used for gas liquid chromatography.

In the present study, the two gas chromatographs used were equipped with various columns that were used under optimal conditions: (a) a double column Varian 2700 GC with an all-glass system and two separate flame ionisation detectors; one injector and detector was supplied with a device for the use of capillary columns and the other was run with a packed glass column; (b) a Hewlett Packard 5700 GC with a modified WGA splitter

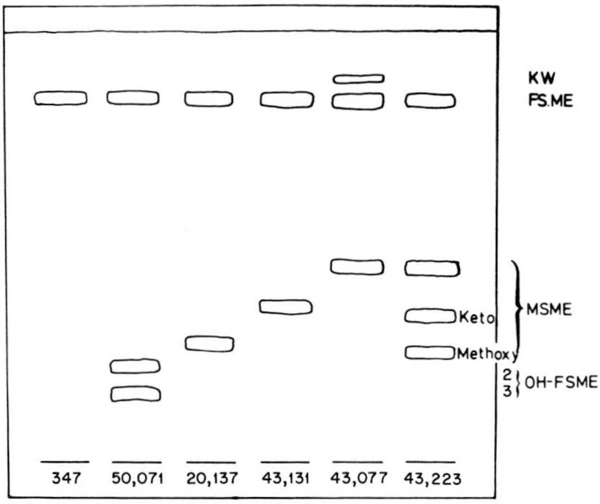

FIG. 1. Separation of hydroxylated and nonhydroxylated fatty-acid methyl esters by adsorption chromatography on precoated TLC plates: silica gel 60, layer thickness 0.25 mm (Merck, no 5721), 20 × 20 cm, activated for 30 min at 120°C; mobile phase light petroleum (60–80°C)–diethyl ether (85:15, v/v), detection by spraying with a 3.5% ethanolic solution of molybdophosphoric acid spray reagent (Merck, no 531) followed by heating at 120°C for 15 min.

Abbreviations		R_f values	DSM number	Species
KW	Hydrocarbons			
FSME	Fatty acid methyl esters	0.85	347	*Bacillus subtilis*
2-OH-FSME	2-Hydroxy-FAMES	0.22	50071	*Pseudomonas aeruginosa*
3-OH-FSME	3-Hydroxy-FAMES	0.16		
MSME	Mycolic acid methyl esters (2-Branched-3-OH-FAMES, MAMES)	0.28 to 0.45		
	From: Corynebacteria (C_{22}–C_{38})	0.28	20131	*Corynebacterium lilium*
	Nocardiae (C_{45}–C_{60})	0.36	43131	*Nocardia asteroides*
	Mycobacteria (C_{60}–C_{90})	0.45	43077	*Mycobacterium fortuitum*
Keto-MSME	Keto-MAMES	0.25	43223	*Mycobacterium intracellulare*
Methoxy-MSME	Methoxy-MAMES	0.34		

system in one of the two injection blocks and an all-glass connection at one of the two flame ionisation detectors. The detectors were supplied with an auxillary makeup gas device. With this modification the GC could also be used for capillary gas chromatography.

Several columns were employed, i.e. packed columns, support-coated, open tubular columns (SCOT), and wall-coated, open tubular columns (WCOT) (Kroppenstedt & Kutzner 1976, 1978; Kroppenstedt 1982b). With packed columns, fatty acid mixtures giving very simple patterns (Fig. 2a,b; Table 3; FA-type 1) can be resolved. At least a 5-m polar column is required for a sufficient resolution of branched iso and anteiso C_{15} fatty acids (Ackman 1972). A capillary column is essential to separate the isomers of unsaturated fatty acids such as oleic acid and *cis*-vaccenic acid (Panos & Henrikson 1968). For the separation and identification of fatty acids of actinomycetes (strains in some genera have very complex mixtures of more than 20 different fatty acids) two capillary columns with stationary phases of different polarity are needed for comparative analyses (Kroppenstedt 1977).

Of the many columns used by the author, the following can be recommended as a good compromise between high resolution of the fatty-acid mixtures and stability of the stationary phases: two wide-bore, fused-silica, capillary columns (25 m × 0.32 mm) with stationary phase FFAP (polar) and OV 101 (nonpolar). The identification of the fatty acids is based on the retention times obtained from standard mixtures, e.g., BC-Mix Applied Science Laboratories; bacterial acid methyl esters mixture (4-5436) Supelco, Inc., Standard 10 B and Standard 1xB from Nu-Chek Copenhagen, and

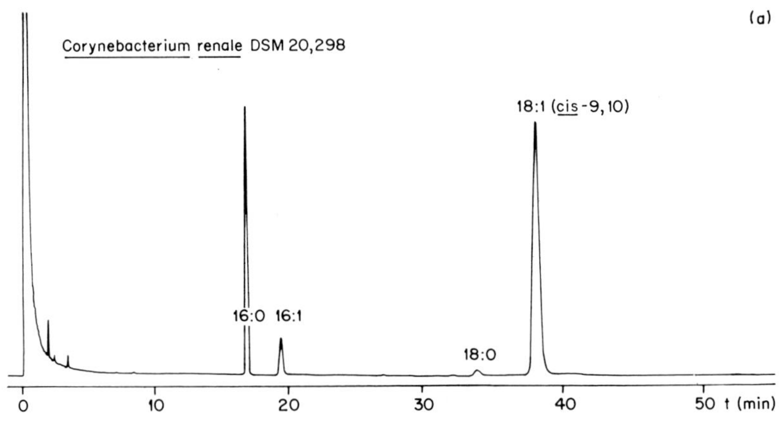

FIG. 2. Chromatograms of fatty-acid methyl esters of some *Actinomycetales* separated on a support-coated, open-tubular column (SCOT) with DEGS. Column (30 m × 0.30 mm i.d.), temperature 170°C, helium flow rate 2 ml/min, inlet pressure 0.5 kp/cm^2, injection volume 0.2 μl, and split rate 1:20. Abbreviations for the fatty acids are explained by the examples: 15:0 for straight-chain pentadecanoic acid; *i*-15:0 for iso-branched 13-methyltetradecanoic acid; *ai*-15:0 for anteiso-branched 12-methyltetradecanoic acid; 18Me(10) for 10-methyl-branched 10-methyloctadecanoic acid (tuberculosteric acid) and 18:1(*cis*-9,10) for the monounsaturated *cis*-9,10-octadecenoic acid (oleic acid). (a) *Corynebacterium renale* (saturated–unsaturated) FA-type 1a. (b) *Rhodococcus rhodochrous* (saturated–unsaturated–10-methyl branched) FA-type 1b. (c) "*Streptomyces verne*" (iso–anteiso-branched) FA-type 2c. (d) *Actinomadura pelletieri* (iso-branched–10-methyl-branched) FA-type 3a.

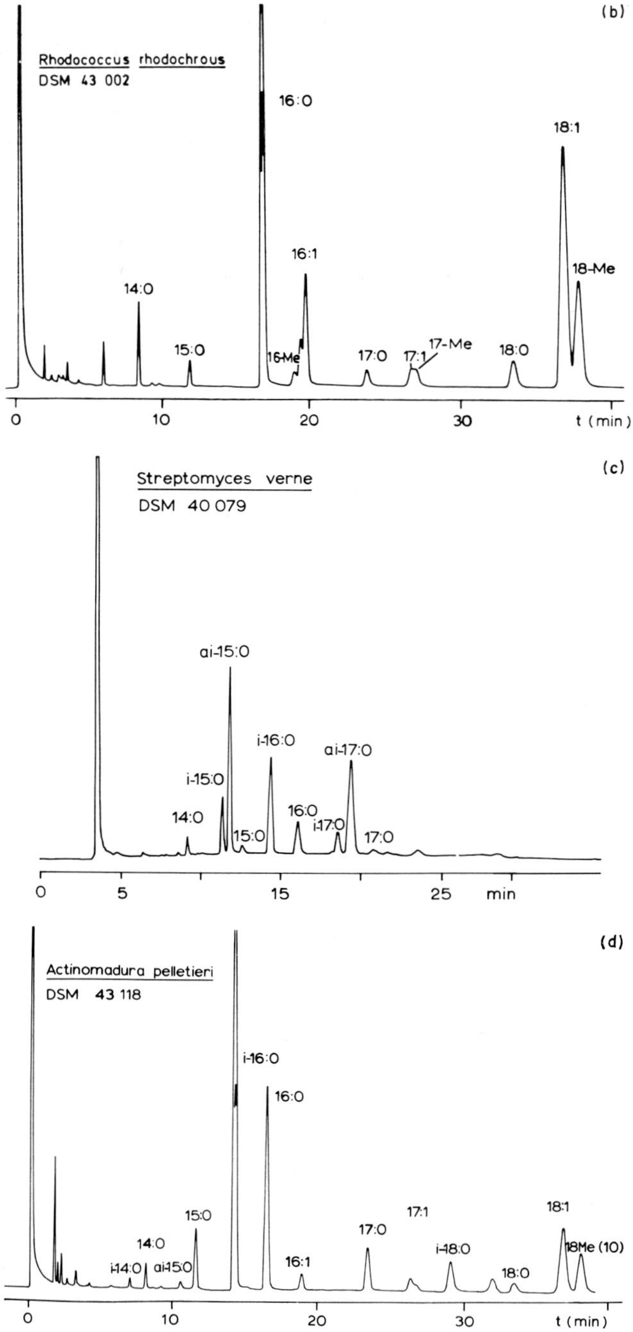

FIG. 2. (*Continued*)

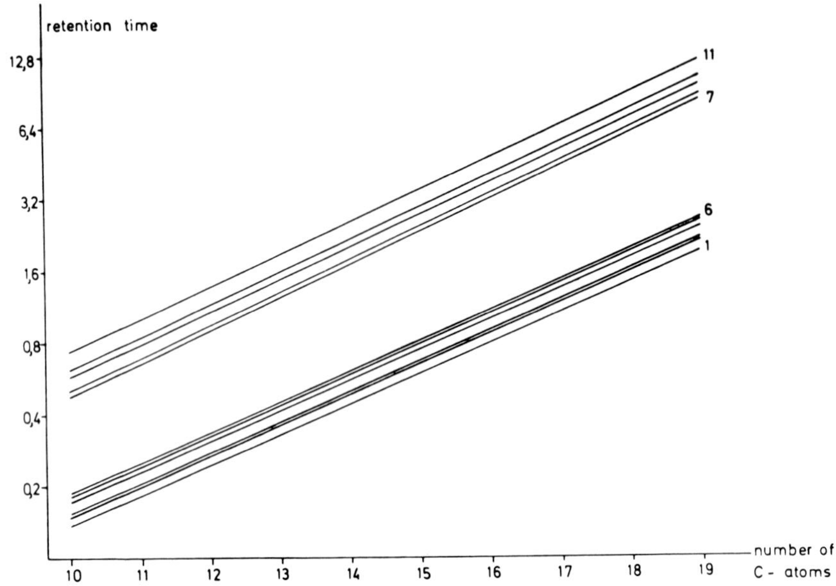

FIG. 3. Relationship between number of carbon atoms and retention times for homologous fatty acid series plotted on semilogarithmic paper. Gas chromatographic separation on WCOT 50 m × 0.24 mm i.d., FFAP, 180°C 2 ml helium/min. Abbreviations: (1) 10-methyl-branched, (2) iso-branched, (3) anteiso-branched, (4) *n*-saturated, (5) *n*-unsaturated or cyclic (ω9), (6) *n*-unsaturated or cyclic (ω7), (7) 2-OH-iso-branched, (8) 2-OH-anteiso-branched, (9) 2-OH-*n*-saturated, (10) 3-OH-iso-branched, and (11) 3-OH-*n*-saturated.

Nocardia asteroides DSM 43 757 for tuberculostearic acid. Unknown fatty acids were identified by using the graph shown in Fig. 3 or by comparing equivalent chain length obtained using the following formula: $ECL = (S_2 - S_1) \log (t_x/t_{s1})/\log (t_{s2}/t_{s1}) + S_1$; where S_1 and S_2 are ECL of standards and t_x, t_{s1} and t_{s2} are retention times of the unknown and two standards (Miwa 1963).

Separation and identification of menaquinones

Actinomycetes show a very high degree of variation in menaquinone composition. Thus their menaquinones differ considerably in the length and degree of saturation of the isoprenylic chain (Collins & Jones 1981). In addition, positional isomers with differences in the position of the isopentyl units in partly hydrogenated isoprenylic side chains, may occur (Fig. 4) (Batrakov *et al.* 1976; Kroppenstedt 1982a; and Yamada *et al.* 1982a,b). Furthermore, the highly saturated menaquinones of actinomycetes are easily destroyed during preparation and storage and some of the altered products are eluted from RP-18 phases at the same time as short-chain menaquinones.

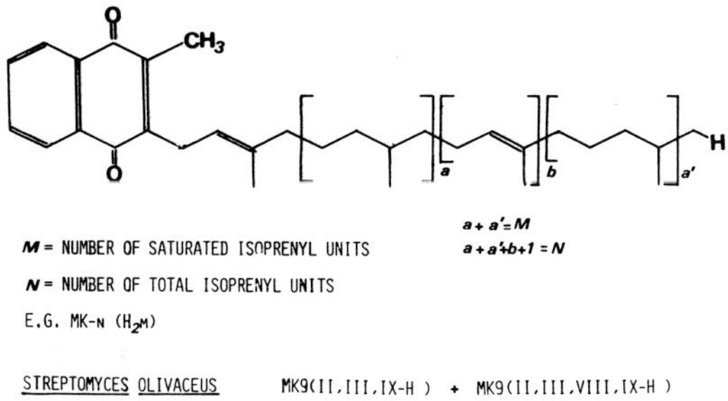

FIG. 4. Structural formula of menaquinones; (2-methyl-3-II, III, IX-hexahydromultiprenyl-1,4-naphthoquinone). According to the IUPAC–IUB (1974) recommendations, menaquinones are referred to by the abbreviation MK-*n* (X,Y,Z . . . -H_m) where: MK, relevance of a compound to the menaquinone series: *n*, the overall number of isoprene and isopentane units; *m*, the doubled number of isopentane units, X, Y, Z . . . ; Roman numerals, the latter in respect to the quinone nucleus.

Therefore, for the separation and identification of highly saturated menaquinones, thin-layer chromatography (reversed phase and argentation) and reversed phase high performance liquid chromatography alone (for methods see Chapter 13) do not achieve the separation and identification of all of these different menaquinones. Again, although the molecular weights of menaquinones may be obtained by mass spectrometry, the point of saturation of the isoprenylic side chain cannot be detected in this way nor can exact quantitative data be obtained (Chapter 13).

As shown for the separation of FAMES by gas-liquid chromatography, the use of stationary phases with different separation behaviours is useful for the identification of such complex menaquinone mixtures. Two stationary phases were developed; separation on a lipophilic reversed phase (RP-18) with acetonitrile–tetrahydrofuran (70:30, v/v) (Kroppenstedt *et al.* 1981) or acetonitrile–isopropanol (75:25, v/v) (R. M. Kroppenstedt unpublished) as mobile phases and a silver-loaded ion exchanger for Ag^+-complexing chromatography with methanol (100%) as mobile phase at elevated temperature (40–60°C) (Kroppenstedt 1982a,c). The elution profile of fully unsaturated menaquinones on octadecylsilane (RP-18) (a) and a silver-loaded ion exchanger (b) is shown in Figs. 5a and 5b. For these menaquinones, both columns show the same retention behaviour; on RP-18 this is due to the lipophilic character of the side chain, which depends on the number of C-atoms and the number of saturated isoprenyl units within the chain, but on the Ag^+ column the separation is due to complex formation between the silver on the ion exchanger and electrons of the double bonds. In addition, the separation of cis–trans isomers on the Ag^+ phase can be seen in Fig. 5b.

The contrasting behaviour of partly saturated menaquinones is shown in

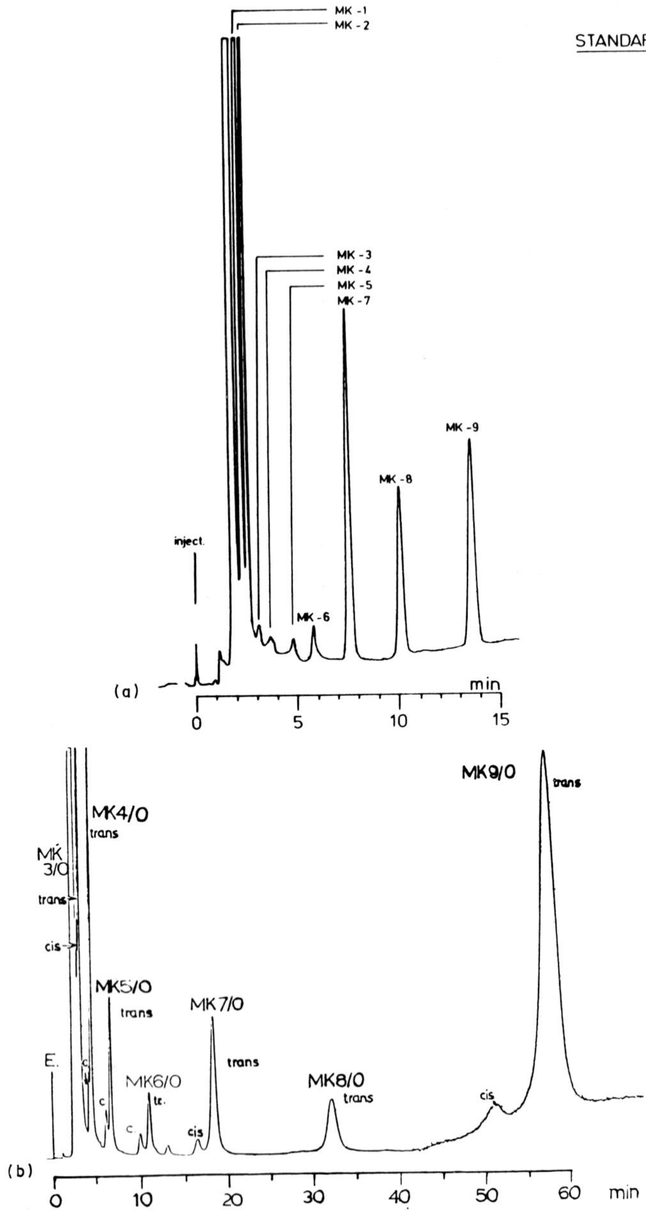

FIG. 5. Separation of menaquinone standard mixtures by HPLC on different stationary phases. (a) Column 125 mm × 4 mm i.d., ODS, particle size 5 μm Hypersil, temperature 30°C, flow rate 0.8 ml/min acetonitrile–tetrahydrofuran (70:30, v/v) (b) Column Ag^+ column Chrompack 250 mm × 4.6 mm i.d., temperature 65°C, flow rate 1.5 ml/min, methanol (100%).

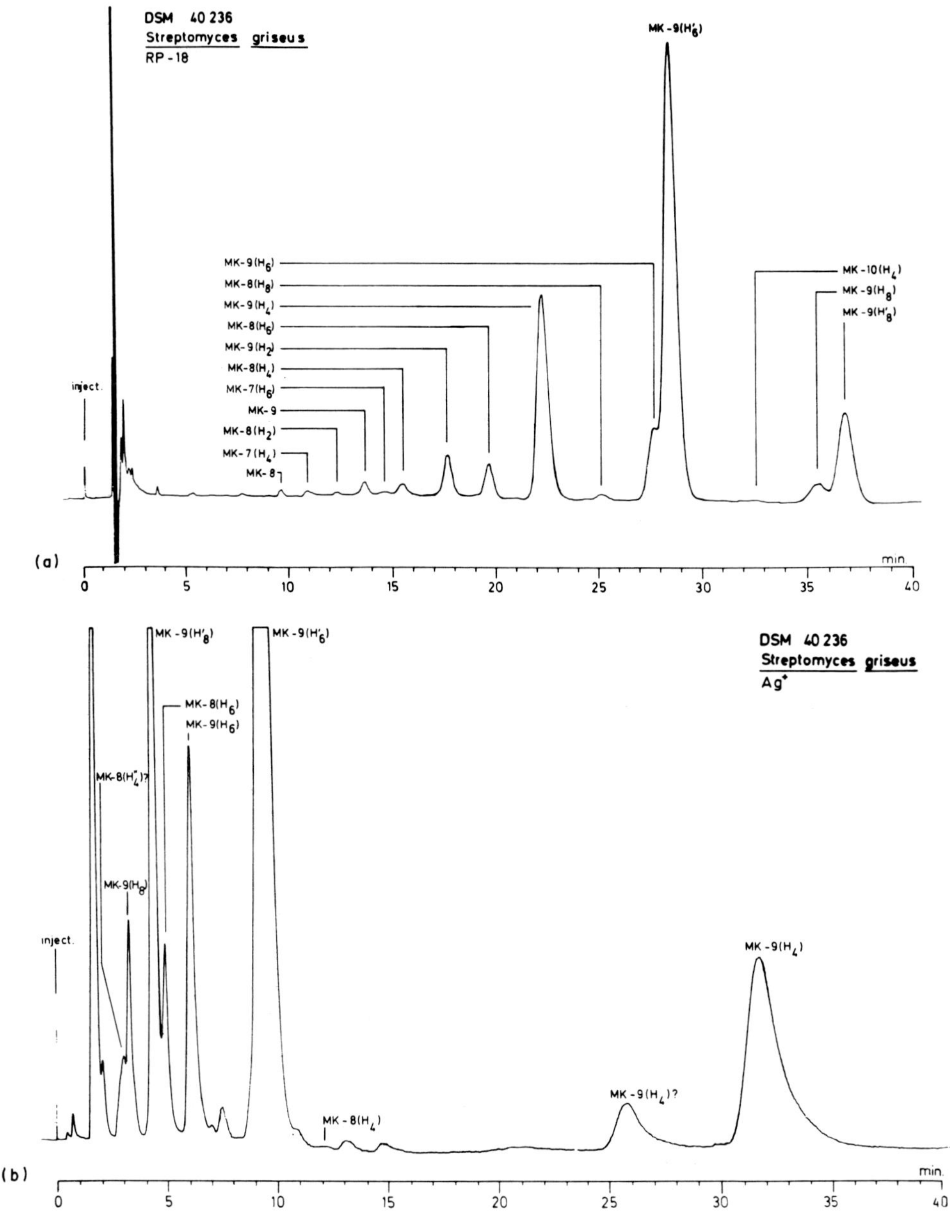

FIG. 6. Separation of bacterial menaquinones by HPLC on different stationary phases. (a) and (c) Column 125 mm × 4 mm i.d., particle size 5 μm Spherisorb ODS-II, flow rate 1.0 ml/min, acetonitrile–isopropanol (75:25, v/v), temperature 40°C; (b) and (d), silver-loaded ion exchanger column 125 mm × 4 mm, Nucleosil SA 5 μm, temperature 45°C, flow rate 1.0 ml/min, methanol (100%).

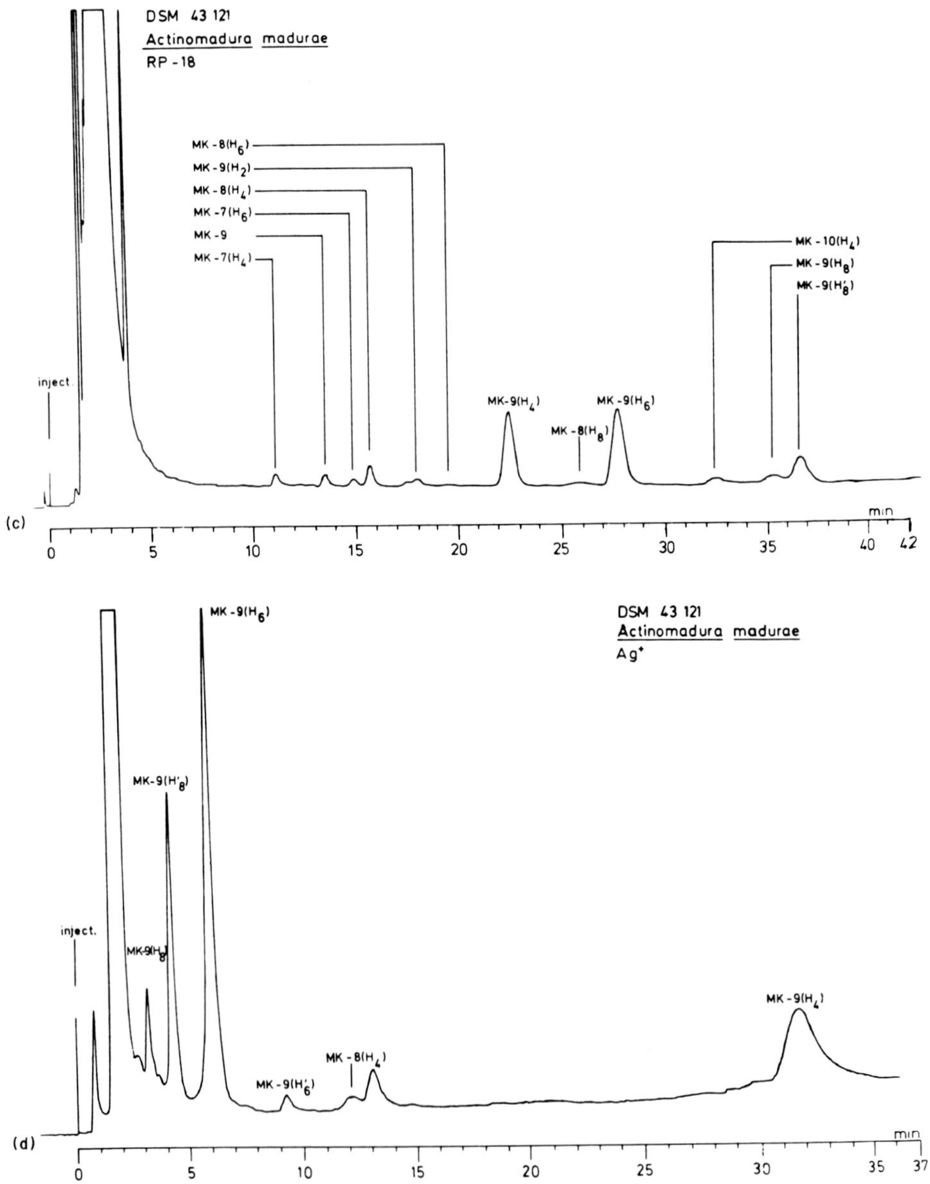

Fig. 6. *(Continued)*

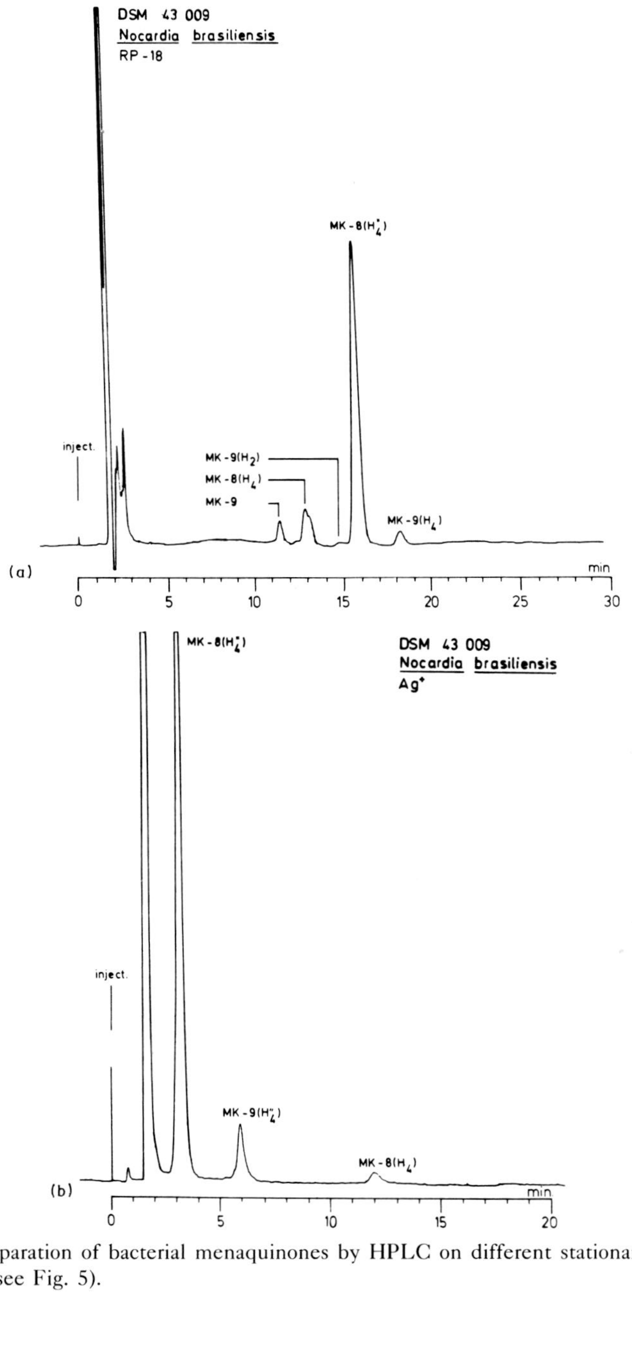

FIG. 7. Separation of bacterial menaquinones by HPLC on different stationary phases (for conditions, see Fig. 5).

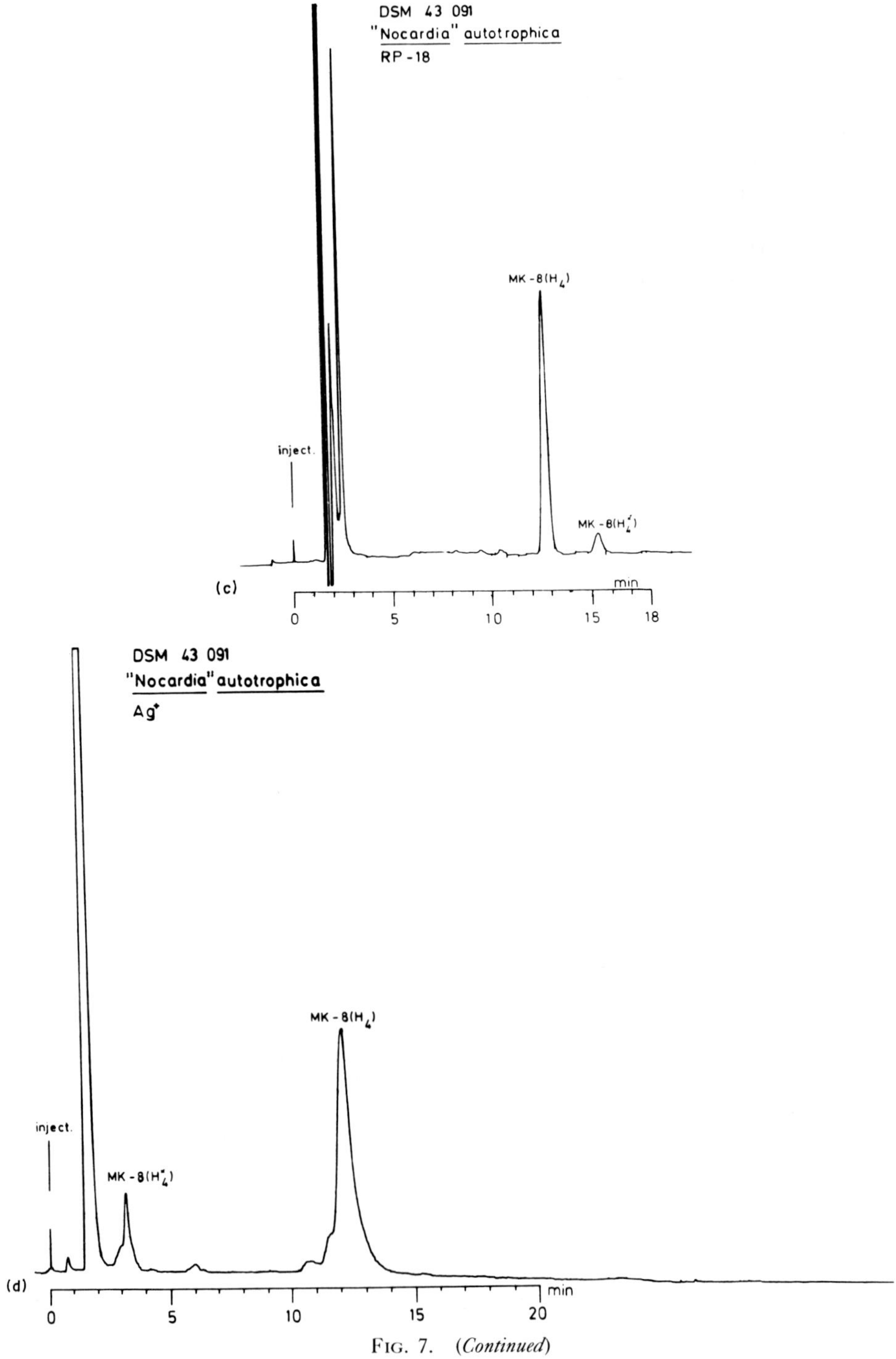

FIG. 7. (*Continued*)

Figs. 6 and 7: on RP-18 these menaquinones are eluted later than their unsaturated homologues (Figs. 6a,c and 7a,c), whereas on the Ag^+ column they appear earlier (Figs. 6b,d and 7b,d). Menaquinones with the same chain length and degree of saturation but with varying positions of their isopentyl units (saturated isoprenyl units) can be distinguished by their different behaviour on the RP-18 and the Ag^+ column (Figs. 6 and 7). MK-9(II,III,IX-H_6) of *Streptomyces griseus* and MK-9(II,III,VIII-H_6) of *Actinomadura madurae* have nearly the same retention time on RP-18 (Figs. 6a and 6c) but can be clearly separated on the Ag^+ column (Figs. 6b and 6d). A similar phenomenon is observed with *Nocardia brasiliensis* and "*Nocardia*" *autotrophica*, which contain MK-8(H_4). Thus the menaquinone of *N. brasiliensis* has a longer retention time on RP-18 than that of "*N.*" *autotrophica* (Figs. 7a and 7c) but a shorter one on the silver column (Figs. 7b and 7d). The chromatographic data indicate that MK-8(H_4) of "*N.*" *autotrophica* is MK-8(II,III-H_4). The structure of the other menaquinone is unknown but it is only found in *Nocardia sensu stricto* (*N. asteroides*, *N. brasiliensis*, *N. brevicatena* and *N. otitidis-caviarum*).

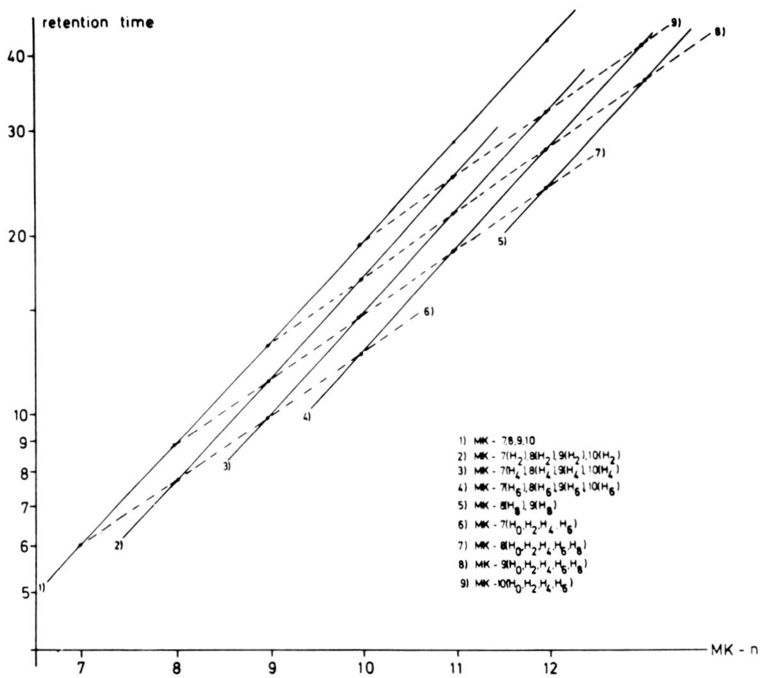

FIG. 8. Relationship between number of isoprene units and retention times (t_r) for homologous menaquinone series plotted on semilogarithmic paper. Menaquinones separated on a reversed phase HPLC column (for conditions, see Fig. 6).

Menaquinones are identified by their retention times. Because standards are commercially unavailable, menaquinone preparations from bacterial strains can be used whose pattern is known from the literature (Collins & Jones 1981). With only a few menaquinone standards the retention times of all other menaquinones can be obtained graphically by plotting the net retention times (time for elution volume substracted from time for void volume) of menaquinone homologues against their number of isoprenylic units (Fig. 8); on semilogarithmic paper straight lines are obtained. It can be seen from Fig. 8 that the five series (1 to 5) of homologous menaquinones differ only in their degree of saturation from a group of five parallel lines; another group of four parallel lines of different slope is formed by four series of homologous menaquinones differing in their number of isoprenyl units (6 to 8). By intra- or extrapolation, the retention data for unknown menaquinones can be obtained by this method. Alternatively, the missing retention data of homologous menaquinones can be calculated by linear regression (Kaiser & Oelrich 1979; Tamaoka *et al.* 1983).

Results and Discussion

The distribution of cellular fatty acids and menaquinones amongst actinomycetes is summarized in Table 1. The relationships based on these lipid data are in good agreement with results derived from the application of other chemotaxonomic techniques (see Table 2) and to a lesser extent with traditional morphological criteria. It is also interesting that some of the distinct fatty-acid and menaquinone patterns are based on specific biosynthetic pathways. In the case of the fatty acids, two pathways are apparent (Table 3). The products of one of these pathways include saturated and unsaturated fatty acids (FA-type 1a), as well as the cyclopropane fatty acids (FA-type 1c), and 10-methyl branched fatty acids (FA-type 1b) derived from the unsaturated compounds. The second pathway yields terminally branched fatty acids (FA-type 2), i.e. iso- and anteiso-branched fatty acids. Members of one group of organisms (FA-type 3) are able to synthesize both types of branched-chain fatty acids: terminally and 10-methyl-branched components. In addition, differences exist in the chain lengths of fatty acids synthesized by, for example, *Actinomadura madurae* 10-Me-C18 (FA-type 3a) and *Actinomadura pusilla* 10-Me-C17 (FA-type 3c) (Fisher *et al.* 1983). Most actinomycetes lack hydroxy fatty acids but some exceptions to this rule are shown in Table 1. In most cases where hydroxy fatty acids have been detected in actinomycetes, it has been shown that the latter were misclassified or that their taxonomic status was questionable (O'Donnell *et al.* 1982; Collins *et al.* 1983).

The value of fatty-acid analyses in the classification of the order *Act-*

Table 1

Distribution of fatty acids and menaquinones among Actinomycetales *and related organisms*

Taxa	Type	Fatty acid pattern						Menaquinone-type (number of isoprenyl units/additional H-atoms) Type	
				Number of carbons					
		Saturated	Unsaturated	Iso-16	Iso-15/17	Anteiso-15/17	10-Methyl 17/18		
Mycolic acid-containing organisms									
Corynebacterium	1a	+++	+++	–	–	–	– –	2a,c	8/2 or 9/2
"Gordona aurantiaca"	1b	++	++	–	–	–	+ +	1b	9/0
Mycobacterium	1b	++	++	–	–	–	– ++	2c	9/2
Nocardia	1b	++	++	–	–	–	– ++	2b	8/4*
Nocardia amarae	1b	++	++	–	–	–	(v) ++	2c	9/2
Rhodococcus rhodochrous	1b	++	++	–	–	–	(v) ++	2a	8/2
Rhodococcus rubropertinctus	1b	++	++	–	–	–	(v) ++	2c	9/2
Dermatophilaceae									
Dermatophilus	1a	++	++	–	–	–	– –	2b	8/4′
Geodermatophilus	2b	+	+	++	++	+	– –	2d	9/4
Monosporic actinomycetes									
Micromonospora	3b	+	+	+	++	++	++(v)	3b	9/4 + 10/4
Saccharomonospora[a]	2a	++	+	+++	+	(+)	– –	3a	8/4, 9/4
Thermoactinomyces candidus	2b	–	(v)	++	+++	+	– –	1a	7/0
Thermoactinomyces "thalpophilus"	2b	–	(v)	++	+++	+	– –	1b	9/0

(*continued*)

TABLE 1 (*Continued*)

Taxa	Fatty acid pattern							Menaquinone-type (number of isoprenyl units/additional H-atoms) Type	
				Number of carbons					
	Type	Saturated	Unsaturated	Iso-16	Iso-15/17	Anteiso-15/17	10-Methyl 17/18		
Thermomonospora curvata	3c	−	−	+++	+	+	+ −	4c	10/4, 10/6
Thermomonospora mesophila	3c	−	−	+++	+	+	+ −	4a	9/2 9/4 9/6
Polysporic actinomycetes:									
Actinomadura flexuosa[b]	3c	+	+	+	++	+++	+ −	4a	9/0, 9/2, 9/4
Actinomadura madurae	3a	+++	++	+	−	−	+ ++	4b	9/4 9/6* 9/8
Actinomadura pusilla	3c	+++	(v)	+++	+	−	++(+)	4a	9/2 9/4 9/6*
Excellospora	3d	+++	+	+++	+++	+	+ ++	4b	9/4 9/6* 9/8
Microbispora	3c	++	−	+++	(v)	(v)	++ +	4a	9/0 9/2 9/4
Micropolyspora angiospora	3c	++	++	−	−	−	(+)++	4a	9/2 9/4 9/6*
Micropolyspora faeni	2d	+	−	+++	++	++	++(+)	3b	9/4, 10/4
Micropolyspora internatus	2d	++	(+)	+++	+	++	− −	4a	9/0 9/2, 9/4
Microtetraspora	3c	++	(v)	+++	(v)	−	++(v)	4a	9/2 9/4 9/6*
"Nocardia" autotrophica	3b	++	++	+++	++	+	++−	2b	8/4
Nocardiopsis	3d	+	++	+++	(+)	+	− +	4c	10/4 10/6
Pseudonocardia	2b	(+)	−	++	+++	+	− −	2b	8/4
Saccharopolyspora	2d	++	−	++	++	++	− −	4a	9/2 9/4
Actinoplanaceae									

Actinoplanes	2c	+	++	++	++	++	–	–	3b	9/4, 10/4
Amorphosporangium	2c	++	–	+	++	++	–	–	2d	9/4
Ampullariella	2d	+	+	+++	++	+	–	–	3b	9/4, 10/4
Dactylosporangium	2d	(v)	(v)	++	++	+	–	–	4b	9/4, 9/6 9/8
Pilimelia	2b	+	+	(v)	+++	(v)	–	–	4a	9/2, 9/4
Planobispora	3c	++	++	(+)	–	–	++	+	4a	9/0 9/2, 9/4
Planomonospora	3a	++	++	(v)	(v)	–	–	++	4a	9/0 9/2 9/4
Spirillospora	3a	++	++	+++	–	–	(+)	++	n.d.	
Streptosporangium	3c	++	+	++	(v)	(v)	++	+	4a	9/0 9/2, 9/4
Streptomycetaceae										
Kitasatoa	2c	+	(v)	+++	+	+++	–	–	4a′	9/6 9/6, 9/8
Microellobosporia	2c	+	(v)	+++	+	+++	–	–	4a′	9/4 9/6, 9/8
Streptomyces	2c	+	(v)	+++	+	+++	–	–	4a′	9/4 9/6, 9/8
Streptoverticillium	2c	+	(v)	+++	+	+++	–	–	4a′	9/4 9/6, 9/8
Actinomycetaceae										
Actinomyces bovis	1a	+++	+++	–	–	–	–	–	4c	10/2, 10/4
Actinomyces israelii	1c[c]	+++	++	–	–	–	–	–	4c	10/2, 10/4
Agromyces	2c	(+)	(+)	(+)	(+)	+++	–	–	1c	11/0, 12/0
Arachnia	2c	+	+	+	+	+++		–	n.d.	
Genera in search of a family										
Intrasporangium	2b	+	+	+++	+++	++	–	–	1	8/0
Nocardioides	3a	+	+	+++	+	(v)	+	++	2b	8/4
Promicromonospora	n.d.								2d	9/4
Sporichthya	3c	+	++	+	–	–	++	+	n.d.	

Note: +, positive; v, variable; ( ), less than 5% and n.d., not determined. * and ′, stereoisomers, 8/4′ = MK-8 (II, III H_4), 8/4* = MK-8 (?, ? H_4), 9/6* = MK-9 (II, III, VIII-H_6), 9/6′ = MK-9 (II, III, IX-H_2).

[a]Contain 2-OH-*iso*-C16 in addition.

[b]2-OH-iso/anteiso fatty acids.

[c]Contain cyclopropane fatty acids.

Table 2

Biochemical classification of Actinomycetales *and related organisms*

Taxa	Phospholipid type[a]	GC (%)	Muramic acid[b]	Sugar pattern[c]	Mycolic acid[d]	Peptidoglycan type[e]
Mycolic acid-containing organisms						
*Corynebacterium**	1	65–70	A/G	A	C	
"Gordona aurantiaca"	2	n.d.	G	A	M	
Mycobacterium	2	62–70	G	A	M	
Nocardia	2	64–69	G	A	N	DL-DAP-direct
Nocardia amarae	2	60–66	G	A	N	
Rhodococcus rhodochrous	2	59–69	G	A	C/N	
Rhodococcus rubropertinctus	2	59–69	G	A	N	
Dermatophilaceae						
Dermatophilus	1	58	n.d.	B	—	DL-DAP-direct
Geodermatophilus	2	74	n.d.	C	—	
Monosporic actinomycetes						
Micromonospora	2	71–73	G	D	—	DL-DAP-direct+Gly
Saccharomonospora	2	67–69	n.d.	A	—	
Thermoactinomyces candidus	2	54	n.d.	C	—	
Thermoactinomyces "thalpophilus"	2	54	n.d.	C	—	DL-DAP-direct
Thermomonospora curvata	2	72	n.d.	C	—	
Thermomonospora mesophila	2	72	n.d.	C	—	
Polysporic actinomycetes						
Actinomadura flexuosa	n.d.		n.d.	C	—	
Actinomadura madurae	1	77	A.	B	—	
Actinomadura pusilla	4		n.d.	B	—	
Excellospora	n.d.		n.d.	B	—	
Microbispora	4	71	A	B	—	
Micropolyspora angiospora	3	68	n.d.	C	—	
Micropolyspora faeni	3		n.d.	A	—	DL-DAP-direct
Micropolyspora internatus	3		n.d.	A	—	
Microtetraspora	4/1		A	B/C	—	
"Nocardia" autotrophica	3		A	A	—	
Nocardiopsis	3		n.d.	C	—	

Pseudonocardia	3	79	n.d.	A	—	
Saccharopolyspora	3		n.d.	A	—	
Actinoplanaceae						
Actinoplanes	2	72	G	D	—	
Amorphosporangium	2	71	G	D	—	
Ampullariella	2	72	G	D	—	DL-DAP-direct+Gly
Dactylosporangium	2	71–73	G	D	—	
Pilimelia	2		n.d.	D	—	
Planobispora	4		n.d.	B	—	
Planomonospora	4	72	n.d.	B	—	
Spirillospora	1/2	70–71	n.d.	B	—	DL-DAP-direct
Streptosporangium	4	69–71	A	B	—	
Streptomycetaceae						
Kitasatoa	2		A	n.c.	—	
Microellobosporia	2	69–70	A	n.c.	—	LL-DAP-Gly
Streptomyces	2	69–74	A	n.c.	—	
Streptoverticillium	2	69–71	A	n.c.	—	
Actinomycetaceae						
Actinomyces bovis	n.d.	58	n.d.	n.c.	—	L-Lys-D-Asp
Actinomyces israellii	n.d.		n.d.			(L-Lys)-D-Glu-Glu-Orn
Agromyces	n.d.	71		n.c.	—	DAB
Arachnia	n.d.	64	n.d.	n.c.	—	LL-DAP-Gly
Genera in search of a family						
Intrasporangium	4	71	n.d.	n.c.	—	LL-DAP-Gly
Nocardioides	1	67	n.d.	n.c.	—	
Promicromonospora	5	73	n.d.	n.c.	—	L-Lys-D-Asp
Sporichthya	n.d.		n.d.	n.c.	—	LL-DAP-Gly

Note: n.d., not determined; —, absent; n.c., not characteristic; *, excluding "*Corynebacterium*" *paurometabolum* (contains long-chain mycolic acid M).
[a]Phospholipid type (Lechevalier *et al.* 1977; D. Bauch & R. M. Kroppenstedt unpublished).
[b]A, *N*-Acetylmuramic acid, G, *N*-glycolylmuramic acid (Uchida & Aida 1977).
[c]Sugar pattern according to Lechevalier *et al.* (1971): A, arabinose + galactose; B, madurose; C, no diagnostic sugars; D, xylose; n.c., sugars not characteristic for these organisms.
[d]Mycolic acids (Minnikin *et al.* 1975, 1980) C, mycolic acids characteristic of corynebacteria; N, mycolic acids characteristic of nocardiae; M, mycolic acids *sensu stricto*; C/N, mycolic acids with intermediate chain lengths.
[e]Peptidoglycan type (Schleifer & Kandler 1972).

TABLE 3

Fatty acid patterns among members of the order Actinomycetales

	Saturated	Unsaturated	Iso- 14/16/18	Iso- 15/17	Anteiso- 15/17	10-Methyl- 17	10-Methyl- 18	Cyclopropane
1. a. *Corynebacterium**	+++	+++	–	–	–	–	–	–
b. *Mycobacterium, Nocardia*	+++	+++	–	–	–	–	++	–
c. *Actinomyces israelii*	+++	+++	–	–	–	–	–	++
2. a. *Saccharomonospora*	++	+	+++	+	(+)	–	–	–
b. *Thermoactinomyces*	(+)	+	++	+++	+	–	–	–
c. *Streptomyces*	+	(v)	+++	+	+++	–	–	–
d. *Actinoplanes*†	+	+	+++	+++	+++	–	–	–
3. a. *Actinomadura madurae*	+++	++	+++	(+)	(+)	(+)	+++	–
b. *Micromonospora*	+	+	+++	+++	++	++	(+)	–
c. *Actinomadura pusilla*	+	+	++	+	+	+++	(+)	–
d. *Nocardiopsis*	+	+	+++	++	+++	(+)	+++	–

Note: (v), variable (less than 5%).
*Excluding *Corynebacterium bovis.*
†Only those Actinoplanes genera containing xylose in the cell wall.

TABLE 4

Menaquinone types in members of the order Actinomycetales

Type			Main menaquinones	Main representatives
Eubacteria (no hydrogenation of the isoprene units)	1	a	MK-7	*Thermoactinomyces candidus*
		b	MK-9	*"Gordona aurantiaca"*
Mycobacterium (only one type of menaquinone occurs, mainly dihydrogenated or tetrahydrogenated quinones with eight or nine isoprene units)	2	a	MK-8(H_2)	*Rhodococcus rhodochrous*
		b	MK-8(H_4)*	*Nocardia*
		c	MK-9(H_2)	*Mycobacterium*
		d	MK-9(H_4)	*Geodermatophilus*
Saccharomonospora (tetrahydrogenated multiprenyl menaquinone)	3	a	MK-8(H_4),MK-9(H_4)	*Saccharomonospora*
		b	MK-9(H_4),MK-10(H_4)	*Actinoplanes*†
Streptomyces (menaquinones with same chain length but different degree of saturation)	4	a	MK-9(H_2),*MK-9(H_4)*,MK-9(H_6)	*Microtetraspora*
		b	MK-9(H_4),*MK-9(H_6)*,MK-9(H_8)'	*Streptomyces*
		c	*MK-10(H_4)*,MK-10(H_6)	*Nocardiopsis*

*May be differentiated into subgroups.
†Overlapping may occur to group two sometimes in *Micromonospora*.

inomycetales may be demonstrated by using a few examples drawn from morphologically similar organisms. Thus true corynebacteria are difficult to distinguish from rhodococci, which also synthesize short-chain mycolic acids; both groups of organisms have many morphological, physiological (Seiler 1983) and chemical properties (Minnikin *et al.* 1978) in common. However, while tuberculostearic acid is invariably found in rhodococci, it is not synthesized by most corynebacteria. Similarly, corynebacteria lack phosphatidylethanolamine (PE) (Table 2, phospholipid type 1) whereas rhodococci always contain this lipid. The separation of *Dermatophilus* and *Geodermatophilus* strains on the basis of their fatty acid profiles (Table 1) is further evidence that these organisms have little in common (Stackebrandt *et al.* 1983a). Further, monosporic thermoactinomycetes and thermomonosporas can readily be distinguished and so can streptomycetes from other polysporic actinomycetes (Table 1).

The biosynthesis of isoprenylic chains of menaquinones also seems to follow defined pathways leading to the formation of special menaquinone patterns in different taxa (Table 4). Menaquinone type 1, which consists of menaquinones without additional hydrogen atoms (eubacteria type) is very rare in actinomycetes having only been found in "*Gordona aurantiaca*" (Goodfellow *et al.* 1978) "*Corynebacterium*" ***paurometabolum*** (Collins 1978; Collins & Jones 1982) and the unusual *Agromyces ramosus* (Collins 1982b; Döpfer *et al.* 1982). The presence of endospores and MK-7 in *Thermoactinomyces* (Cross *et al.* 1968; Collins *et al.* 1982) is consistent with their arrangement to the family *Bacillaceae* (Stackebrandt & Woese 1983). Menaquinone type 2, with only one type of partly saturated menaquinone, is usually found in representatives of the mycolic acid-containing genera (*Corynebacterium, Mycobacterium, Nocardia* and *Rhodococcus*). Type 3 harbours menaquinones of the same degree of hydrogenation but which differ in isoprenylic chain length, whereas type 4 organisms synthesize menaquinones of the same isoprenylic chain length but which have different degree of saturation.

Of special interest is the occurrence of menaquinones with the same isoprenyl chain length and degree of saturation (stereoisomers), which seem to be taxon-specific among actinomycetes, e.g. "*Nocardia*" *autotrophica* MK-8(II,III-H_4) against *Nocardia asteroides* MK-8(?,?-H_4) or *Streptomyces griseus* MK-9(II,III,IX-H_6) against *Actinomadura madurae* MK-9(II,III,VIII-H_6). Detailed qualitative and quantitative fatty-acid and menaquinone data will be published elsewhere.

References

ACKMAN, R. G. 1972 Influence of methyl substituent position on retention times in the GLC of higher monomethylbranched fatty acid esters and hydrocarbons. *Journal of Chromatographic Science* **10**, 243–246.

BALLIO, A. & BARCELLONA, S. 1968 Relations chimiques et immunologiquess ches les *Actinomycetales*. I. Les acides gras de 43 souches d'*Actinomycetales* aerobies. *Annales Institut Pasteur, Paris* **114**, 121–137.

BATRAKOV, S. G., PANOSYAN, A. G., ROSYNOV, B. V., KONOVA, I. V. & BERGELSON, L. D. 1976 Menaquinones of *Actinomyces olivaceus:* on the structures of MK- *9* (H_6), MK- *9* (H_8), MK- *8* (H_6) and MK- *8* (H_8). *Soviet Journal of Bioorganic Chemistry (English Translation)* **2**, 1538–1546.

BRIAN, B. L. & GARDNER, E. W. 1967 Preparation of bacterial fatty acid methyl esters for rapid characterization by gas-liquid chromatography. *Applied Microbiology* **15**, 1499–1500.

BRIAN, B. L. & GARDNER, E. W. 1968 A simple procedure for detecting the presence of cyclopropane fatty acids in bacterial lipids. *Applied Microbiology* **16**, 549–552.

COLLINS, M. D. 1978 Lipids in coryneform taxonomy. Ph.D. Thesis, University of Newcastle upon Tyne.

COLLINS, M. D. 1982a A note on the separation of natural mixtures of bacterial menaquinones using reversed-phase high-performance liquid chromatography. *Journal of Applied Bacteriology* **52**, 457–460.

COLLINS, M. D. 1982b Lipid composition of *Agromyces ramosus* (Gledhill and Casida). *FEMS Microbiology Letters* **14**, 187–189.

COLLINS, M. D. & JONES, D. 1981 Distribution of isoprenoid quinone structural types in bacteria and their taxonomic implications. *Microbiological Reviews* **45**, 316–354.

COLLINS, M. D. & JONES, D. 1982 Lipid composition of *Corynebacterium paurometabolum* (Steinhaus). *FEMS Microbiology Letters* **13**, 13–16.

COLLINS, M. D., PIROUZ, T., GOODFELLOW, M. & MINNIKIN, D. E. 1977 Distribution of menaquinones in actinomycetes and corynebacteria. *Journal of General Microbiology* **100**, 221–230.

COLLINS, M. D., MACKILLOP, G. C. & CROSS, T. 1982 Menaquinone composition of members of the genus *Thermoactinomyces*. *FEMS Microbiology Letters* **13**, 151–153.

COLLINS, M. D., KEDDIE, R. M & KROPPENSTEDT, R. M. 1983 Lipid composition of *Arthrobacter simplex*, *Arthrobacter tumescens* and possibly related taxa. *Systematic and Applied Microbiology* **4**, 18–26.

CROSS, T., WALKER, P. D. & GOULD, G. W. 1968 Thermophilic actinomycetes producing resistant endospores. *Nature, London* **220**, 352–354.

DÖPFER, H., STACKEBRANDT, E. & FIEDLER, F. 1982 Nucleic acid hybridization studies on *Microbacterium, Curtobacterium, Agromyces* and related taxa. *Journal of General Microbiology* **128**, 1697–1708.

DRUCKER, D. B. 1981 *Microbiological Applications of Gas Chromatography*. Cambridge: University Press.

DSM CATALOGUE OF STRAINS 1983 German Collection of Microorganisms, Grisebachstraße 8 D-Göttingen.

FISCHER, A., KROPPENSTEDT, R. M. & STACKEBRANDT, E. 1983 Molecular-genetic and chemotaxonomic studies on *Actinomadura* and *Nocardiopsis*. *Journal of General Microbiology* **129**, 3433–3446.

GOODFELLOW, M. & WAYNE, L. G. 1982 Taxonomy and nomenclature. In *The Biology of Mycobacteria* eds. Ratledge, C. & Stanford, J. L., pp. 471–521. London: Academic Press.

GOODFELLOW, M., ORLEAN, P. A. B., COLLINS, M. D., ALSHAMAONY, L. & MINNIKIN, D. E. 1978 Chemical and numerical taxonomy of strains received as "*Gordona aurantiaca*." *Journal of General Microbiology* **109**, 57–68.

HOFHEINZ, W. & GRISEBACH, H. 1965 Die Fettsäuren von *Streptomyces erythreus* und *Streptomyces halstedii*. *Zeitschrift für Naturforschung* **20B**, 43–53.

IUPAC-IUB 1974 Commission on biochemical nomenclature. Quinones with isoprenoid side-chains. *Archives of Biochemistry and Biophysics* **165**, 1–5.

KAISER, R. E & OELRICH, E. 1979 *Optimierung in der HPLC*. Heidelberg, Basel, New York: Dr. Alfred Hüthig Verlag.

KROPPENSTEDT, R. M. 1977 Untersuchungen zur Chemotaxonomie der Ordnung *Actinomycetales* Buchanan 1917. Ph.D. Thesis, Technische Hochschule Darmstadt, German Federal Republic.

KROPPENSTEDT, R. M. 1982a Separation of bacterial menaquinones by HPLC using reverse phase (RP18) and a silver loaded ion exchanger as stationary phases. *Journal of Liquid Chromatography* **5**, 2359–2367.

KROPPENSTEDT, R. M. 1982b Anwendung von Dünnschicht- und Gas-Chromatographie in der Bakterientaxonomie. *GIT-Fachzeitschrift Laboratorium, Supplement Chromatographie* 1982, pp. 34–40.

KROPPENSTEDT, R. M. 1982c Anwendung chromatographischer HP-Verfahren (HPTLC und HPLC) in der Bakterien-Taxonomie. *GIT Labor-Medizin* **4**, 266–275.

KROPPENSTEDT, R. M. & KUTZNER, H. J. 1976 Biochemical markers in the taxonomy of *Actinomycetales*. *Experientia* **32**, 318.

KROPPENSTEDT, R. M. & KUTZNER, H. J. 1978 Biochemical taxonomy of some problem actinomycetes. *Zentralblatt für Bakteriologie, Parasitenkunde, Infektionskrankheiten und Hygiene. Abteilung 1, Supplement* **6**, 125–133.

KROPPENSTEDT, R. M., KORN-WENDISCH, F., FOWLER, V. J. & STACKEBRANDT, E. 1981 Biochemical and molecular genetic evidence for a transfer of *Actinoplanes armeniacus* into the family *Streptomycetaceae*. *Zentralblatt für Bakteriologie*, **C2**, 254–262.

LECHEVALIER, M. P. & LECHEVALIER, H. A. 1970a Composition of whole-cell hydrolysates as a criterion in the classification of aerobic actinomycetes. In *The Actinomycetales* ed. Prauser, H., pp. 311–316. Jena: VEB-Fischer.

LECHEVALIER, M. P. & LECHEVALIER, H. A. 1970b Chemical composition as a criterion in the classification of aerobic actinomycetes. *International Journal of Systematic Bacteriology* **20** 435–443.

LECHEVALIER, H. A., LECHEVALIER, M. P. & GERBER, N. N. 1971 Chemical composition as a criterion in the classification of actinomycetes. *Advances in Applied Microbiology* **14**, 47–72.

LECHEVALIER, M. P., DE BIEVRE, C. & LECHEVALIER, H. A. 1977 Chemotaxonomy of aerobic actinomycetes: phospholipid composition. *Biochemical Ecology and Systematics* **5**, 249–260.

LEMMER, H. & KROPPENSTEDT, R. M. 1984 Chemotaxonomy and physiology of some actinomycetes isolated from scumming activated sludge. *Systematic and Applied Bacteriology* **5**, 124–135.

MEISEL, C. 1982 Menaquinon-Muster als taxonomisches Merkmal bei *Actinomycetales*. M.S. Thesis Technische Hochschule Darmstadt, German Federal Republic.

MINNIKIN, D. E. & GOODFELLOW, M. 1980 Lipid composition in the classification and identification of acid-fast bacteria. In *Microbial Classification and Identification* eds. Goodfellow, M. & Board, R. G., pp. 189–256. London: Academic Press.

MINNIKIN, D. E., ALSHAMAONY, L. & GOODFELLOW, M. 1975 Differentiation of *Mycobacterium, Nocardia,* and related taxa by thin-layer chromatographic analysis of whole-organism methanolysates. *Journal of General Microbiology* **88**, 200–204.

MINNIKIN, D. E., GOODFELLOW, M. & COLLINS, M. D. 1978 Lipid composition in the classification and identification of coryneform and related taxa. In *Coryneform Bacteria* eds. Bousfield, I. J. & Callely, A. G., pp. 85–160. London: Academic Press.

MINNIKIN, D. E., HUTCHINSON, I. G., CALDICOTT, A. B. & GOODFELLOW, M. 1980 Thin-layer chromatography of methanolysates of mycolic acid-containing bacteria. *Journal of Liquid Chromatography* **133**, 323–329.

MIWA, T. K. 1963 Identification of peaks in gas-liquid chromatography. *Journal of the American Oil Chemists' Society* **40**, 309–313.

O'DONNELL, A. G., GOODFELLOW, M. & MINNIKIN, D. E. 1982 Lipids in the classification of *Nocardioides:* Reclassification of *Arthrobacter simplex* (Jensen Lochhead in the genus *Nocardioides* (Prauser) emend. O'Donnell *et al.* as *Nocardioides simplex* comb. nov. *Archives of Microbiology* **133**, 323–329.

PANOS, C. & HENRIKSON, C. V. 1968 Resolution of positional isomers of bacterial long chain cyclopropane-ring containing fatty acids by capillary column chromatography. *Journal of Gas Chromatography* **6**, 551–553.

SCHEUERBRANDT, G. & BLOCH, K. 1962 Unsaturated fatty acids in microorganisms. *Journal of Biological Chemistry* **237**, 2064–2068.

SCHLEIFER, K. H. & KANDLER, O. 1972 Peptidoglycan types of bacterial cell walls and their taxonomic implications. *Bacteriological Reviews* **36**, 407–477.

SEILER, H. 1983 Identification key for coryneform bacteria derived by numerical taxonomic studies. *Journal of General Microbiology* **129**, 1433–1471.

SKERMAN, V. B. D., McGOWAN, V. & SNEATH, P. H. A. 1980 Approved lists of bacterial names. *International Journal of Systematic Bacteriology* **30**, 225–420.

STACKEBRANDT, E. & WOESE, C. R. 1981 The evolution of prokaryotes. In *Molecular and Cellular Aspects of Microbial Evolution* eds. Carlile, M. T., Collins, M. T., & Moseley, B. E. B., pp. 1–31. Cambridge: University Press.

STACKEBRANDT, E., KROPPENSTEDT, R. M. & FOWLER, V. J. 1983a A phylogenetic analysis of the family *Dermatophilaceae. Journal of General Microbiology* **129**, 1831–1838.

STACKEBRANDT, E., LUDWIG, W., SEEWALD, E. & SCHLEIFER, K. H. 1983b Phylogeny of sporeforming members of the order *Actinomycetales. International Journal of Systematic Bacteriology* **33**, 173–180.

TAMAOKA, J., KATAYAMA-FUJIMURA, Y. & KURAISHI, H. 1983 Analysis of bacterial menaquinone mixtures by high performance liquid chromatography. *Journal of Applied Bacteriology* **54**, 31–36.

UCHIDA, K. & AIDA, K. 1977 Acyl type of bacterial cell wall: its simple identification by colorimetric method. *Journal of General and Applied Microbiology* **23**, 249–260.

YAMADA, Y., INOUYE, G., TAHARA, Y. & KONDO, K. 1976a The menaquinone system in the classification of coryneform and nocardioform bacteria and related organisms. *Journal of General and Applied Microbiology* **22**, 203–214.

YAMADA, Y., INOUYE, G., TAHARA, Y. & KONDO, K. 1976b The menaquinone system in the classification of aerobic Gram-positive cocci in the genera *Micrococcus, Staphylococcus, Planococcus* and *Sporosarcina. Journal of General and Applied Microbiology* **22**, 227–236.

YAMADA, Y., ISHIKAWA, T., TAHARA, Y. & KONDO, K. 1977a The menaquinone system in the classification of the genus *Nocardia. Journal of General and Applied Microbiology* **23**, 207–216.

YAMADA, Y., YAMASHITA, M., TAHARA, Y. & KONDO, K. 1977b The menaquinone system in the classification of the genus *Actinomadura. Journal of General and Applied Microbiology* **23**, 331–335.

YAMADA, Y., AOKI, K. & TAHARA, Y. 1982a The structure of the hexahydrogenated isoprenoid side-chain menaquinone with nine isoprene units isolated from *Actinomadura madurae. Journal of General Microbiology* **28**, 321–429.

YAMADA, Y., HOU, C. F., SASAKI, J., TAHARA, Y. & YOSHIOKA, H. 1982b The structure of the octahydrogenated isoprenoid side-chain menaquinone with nine isoprene units isolated from *Streptomyces albus. Journal of General and Applied Microbiology* **28**, 519–529.

Chemical Composition and Structure of Murein

K. H. Schleifer and P. H. Seidl

Institute of Botany and Microbiology, The Technical University, Munich, Federal Republic of Germany

Most bacteria are surrounded by envelopes that maintain the shape of the cells and preserve them against osmotic disruption. Murein (peptidoglycan) is the main cell-wall polymer of eubacteria and is common to both Gram-negative and Gram-positive bacteria. A few prokaryotic organisms, such as the *Mycoplasmatales* and archaebacteria lack murein (Kandler 1982; Schleifer & Stackebrandt 1983). The latter is a heteropolymer consisting of polysaccharide (glycan) strands cross-linked through short peptides (Fig. 1). The glycan moiety resembles chitin and, as in the case of the latter, it is made up of β (1,4)-glycosidically linked *N*-acetylglucosamine residues. In contrast to chitin, however, each alternate glucosamine residue is substituted by a D-lactic acid ether at its C-3 hydroxyl group. This derivative of glucosamine has been given the trivial name muramic acid. An oligopeptide (stem peptide) containing L- and D-amino acids is linked to the carboxyl group of *N*-acetylmuramic acid. Adjacent stem peptides may be cross-linked directly or via an interpeptide bridge; the resulting macromolecule encompasses the entire bacterial cell.

Chemical Structure of the Murein

Extensive investigation of the chemical structure of the murein of thousands of bacterial strains has demonstrated the existence of almost 100 different murein types (Schleifer & Kandler 1972; Schleifer & Stackebrandt 1983). The glycan moiety of the various mureins is rather uniform and reveals only a few variations such as acetylation or phosphorylation of the muramyl 6-hydroxyl groups or the occasional absence of *N*-acetyl or peptide substituents. In some mycobacteria and nocardiae the *N*-acetyl group of muramic acid is oxidized to give an *N*-glycolyl residue (Lederer *et al.* 1975; Uchida &

CHEMICAL METHODS IN
BACTERIAL SYSTEMATICS

0-12-289675-0

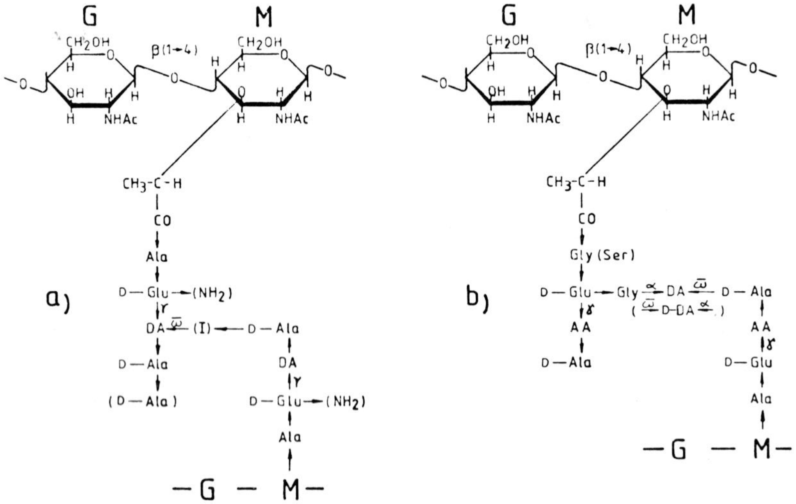

FIG. 1. Fragments of the primary structures of group A (a) and B (b) mureins. Unusual abbreviations: AA, L-amino acid; DA, diamino acid; G, *N*-acetylglucosamine; I, interpeptide bridge; M, *N*-acetylmuramic acid.

Aida 1977), which may be a useful chemotaxonomic marker for the classification of these bacteria.

The stem peptide always consists of alternating L- and D-amino acids. Even *meso*-diaminopimelic acid (m-A_2pm) is bound with its L-asymmetric centre within the stem peptide. However, the amino-acid sequence of the stem peptide shows some variation among the different mureins (Fig. 2). The greatest variation is found in position 3. To date, 11 different amino acids have been detected at this position; *meso*-diaminopimelic acid and L-lysine are the most common ones. Position 4 is almost always occupied by D-alanine, though cells grown in the presence of high amounts of glycine or certain D-amino acids may incorporate these amino acids instead of D-alanine (Hammes *et al.* 1973; Schleifer *et al.* 1976). The carboxyl group of D-alanine in position 4 is often blocked by cross-linkage to an adjacent stem peptide. In noncross-linked stem peptides the D-alanine in position 4 may be split off or substituted by another D-alanine; the latter D-alanine represents a remnant of the murein precursor.

The number of amino acids occurring in a definite murein type is somewhat restricted. Three to six chemically different amino acids can be found in a particular murein. Clean preparations of murein do not contain certain typical constituents of proteins, such as branched amino acids, aromatic amino acids, cysteine, methionine, arginine, histidine and proline.

The stem peptides of murein can be cross-linked in two different ways.

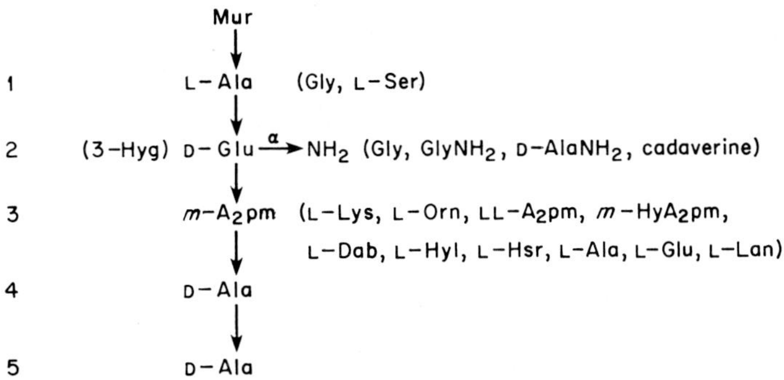

FIG. 2. Variations of the stem peptide (peptide subunit). Amino acids in parentheses may replace the corresponding amino acids or substituents. Unusual abbreviations: A_2pm, 2,6-diaminopimelic acid; Dab, diaminobutyric acid; HyA_2pm, 2,6-diamino-3-hydroxypimelic acid; Hyg, *threo*-β-hydroxyglutamic acid; Hyl, δ-hydroxylysine; Hsr, homoserine; Lan, lanthionine.

Depending on the mode of cross-linking, two main groups of murein, named A and B, have been distinguished by Schleifer & Kandler (1972). The cross-linkage of group A extends, as depicted in Fig. 1a, from the distal amino group of the diamino acid in position 3 of one stem peptide to the carboxyl group of D-alanine in position 4 of an adjacent stem peptide either directly or via an interpeptide bridge. The amino-acid composition and sequence of the interpeptide bridge show considerable variation.

Group B mureins are characterized by a cross-linkage between the α-carboxyl group of D-glutamic acid in position 2 in one stem peptide and the C-terminal D-alanine in position 4 of an adjacent stem peptide (Fig. 1b). There is always a diamino acid present in the interpeptide bridge since two carboxyl groups must be cross-linked. Mureins of group B reveal an interesting correlation between the configuration of the diamino acid present in the interpeptide bridge and its mode of linkage. L-Diamino acids are linked with their distal amino group to D-alanine, whereas D-diamino acids are bound with their α-amino group. Group B mureins are quite rare and are only found in some coryneform bacteria classified in the genera *Agromyces* (Fiedler & Kandler 1973b), *Arthrobacter*, *Brevibacterium*, *Corynebacterium*, *Microbacterium* (Schleifer & Kandler 1972; Döpfer *et al.* 1982), *Aureobacterium* (Collins *et al.* 1983a) and *Curtobacterium* (Yamada & Komagata 1972) and in a few anaerobic organisms such as *Acetobacterium woodii*, *Eubacterium limosum* (Kandler & Schoberth 1979) and *Clostridium barkeri* (O. Kandler, personal communication).

The murein of Gram-negative bacteria is rather uniform. It usually forms a monomolecular layer and is directly cross-linked. In most cases *meso*-diaminopimelic acid (m-A_2pm) is present as the diamino acid in position 3 of

TABLE 1

New murein types of cross-linkage group A published after 1972

Position 1	Position 3	Interpeptide bridge	Abbreviation*	Organism	Reference
Ala	Orn	Direct	A1β	*Spirochaeta stenostrepta*	Schleifer & Joseph 1973
Gly	*m*-A_2pm†	Direct	A1γ′	*Micromonospora, Actinoplanes*	Kawamoto *et al.* 1981
Ala	*m*-Lan	Direct	A1δ	*Fusobacterium nucleatum*	Vasstrand *et al.* 1979
Ala	Lys(Orn)	←Gly_5	A3α	Anaerobic coccus	Weiss 1981
Ala	Lys	←Ser	A3α	"*Bifidobacterium minimum*"	Lauer & Kandler 1983
Ala	Lys	←Ser←Gly_{1-3}	A3α	*Clostridium leptum*	Weiss *et al.* 1981
Ala	Lys	←Ala-Gly-Ala_2	A3α	*Lactobacillus kandleri*	Holzapfel & van Wyk 1982
Gly	Lys	←Gly	A3α′	"*Peptococcus variabilis*"	Schleifer & Nimmermann 1973
Ala	Orn	←Gly	A3β	*Treponema pallidum*	Umemoto *et al.* 1981
Gly	LL-A_2pm	←Gly	A3γ′	*Arachnia propionica*	Weiss *et al.* 1981
Ala	Lys	←D-Ser$\overset{\beta}{\leftarrow}$D-Asp	A4α	*Cellulomonas cartae*	Stackebrandt *et al.* 1978
Ala	Lys	←Thr$\overset{\beta}{\leftarrow}$D-Asp	A4α	*Oerskovia turbata*	Seidl *et al.* 1980
Ala	Lys	←Thr$\overset{\beta}{\leftarrow}$D-Glu	A4α	*Oerskovia turbata*	Seidl *et al.* 1980

Ala	Lys	$\leftarrow$Thr$\overset{\gamma}{\leftarrow}$D-Glu	A4α	*Oerskovia turbata*	Seidl *et al.* 1980
Ala	Lys	$\leftarrow$D-Ser$\leftarrow$D-Glu	A4α	*Bifidobacterium boum*	Lauer & Kandler 1983
Gly	Lys	$\overset{\gamma}{\leftarrow}$D-GluNH$_2$	A4α′	"*Peptococcus aerogenes*"	Schleifer & Nimmermann 1973
Gly	Lys	$\leftarrow$D-Asp	A4α′	Anaerobic cocci	Weiss 1981
Gly	Lys	$\leftarrow$Gly$\leftarrow$D-Asp	A4α′	Anaerobic cocci	Weiss 1981
Ala	Orn	$\overset{\gamma}{\leftarrow}$D-GluNH$_2$	A4β	*Peptococcus indolicus*	Schleifer & Nimmermann 1973
Ala	Orn	$\leftarrow$Ser$\leftarrow$D-Glu	A4β	*Lactobacillus minutus*	N. Weiss personal communication
Gly	Orn	$\leftarrow$D-Asp	A4β′	*Peptostreptococcus micros*	Weiss 1981
Ala	Dab	$\leftarrow$Ala$_2$$\leftarrow$Gly$\leftarrow$Ala$\overset{\beta}{\leftarrow}$Asp	A4δ	*Arthrobacter sp.*	Fiedler & Kandler 1973b
Ala	Lys	$\leftarrow$Lys$\overset{\varepsilon}{\leftarrow}$D-Asp	A5α	*Actinomyces bovis*	N. Weiss, personal communication
Ala	Lys	$\leftarrow$Lys$\leftarrow$D-Glu	A5α	*Arcanobacterium haemolyticum*	N. Weiss, personal communication
Ala	Lys	$\leftarrow$Ala$\leftarrow$Lys$\leftarrow$D-Glu	A5α	*Actinomyces pyogenes*	N. Weiss, personal communication
Ala	Orn	$\leftarrow$Lys$\leftarrow$D-Glu	A5β	*Actinomyces israelii*	Weiss *et al.* 1981

*Abbreviation according to Schleifer & Kandler (1972, 1983).
†Part of *m*-A$_2$pm can be hydroxylated.

TABLE 2

New murein types of cross-linkage group B published after 1972

Position 1	Position 3	Position 2	Interpeptide bridge	Abbreviation	Organism	Reference
Gly	Hsr	D-Glu	$\rightarrow Gly \xrightarrow{\alpha} Dab$	B1β	*Corynebacterium bovis*	Fiedler & Kandler 1973b
Ser	Orn	D-Glu	$\xrightarrow{\delta}$ D-Orn	B2α	*Acetobacterium woodii*	Kandler & Schoberth 1979
Gly	Hsr	D-Hyg	$\rightarrow Gly \xrightarrow{\delta} DHyo$	B2α	*Corynebacterium species*	Schleifer *et al.* 1983

the stem peptide but a few exceptions are known. Thus the murein of spirochetes often contains ornithine instead of *m*-A_2pm and is directly cross-linked as in *Spirochaeta stenostrepta* (Schleifer & Joseph 1973) or contains glycine as the interpeptide bridge as in *Treponema pallidum* (Umemoto *et al.* 1981). Some fusobacteria (*F. gonidiaformans*, *F. necrophorum*, *F. nucleatum*, and *F. russii*) have a murein in which *m*-A_2pm is replaced by its sulphur analogue *m*-lanthionine (Kato *et al.* 1979, 1981; Vasstrand *et al.* 1979, 1982). *F. mortiferum* contained both lanthionine and A_2pm. *Bacteroides assacharolyticus* is another rare Gram-negative organism in which *m*-A_2pm is replaced, in this case by lysine (Vasstrand *et al.* 1982).

Gram-positive bacteria contain a multilayered murein and reveal a great variation in the composition and primary structure of the constituent peptide moiety. Various murein types can be distinguished, depending on the different amino-acid sequence of the stem peptide and/or the interpeptide bridge. Most murein types have been summarized by Schleifer & Kandler (1972), but those more recently described are listed in Tables 1 and 2. The detection of a new kind of interpeptide bridge, subgroup A5, characterized by the presence of a diamino acid and the occurrence of new variations, A1δ and A4δ, are especially noteworthy. Another interesting feature is the replacement of alanine in position 1 by glycine in mureins of cross-linkage group A. Such variations are marked in the abbreviation scheme of Schleifer & Kandler (1972, 1983) by a prime.

Determination of the Chemical Composition of Murein

Cell walls are usually prepared by mechanical disruption of bacterial cells with glass or plastic beads in specially designed cell mills (e.g. Vibrogen cell mill, Bühler, Tübingen, FRG). Other, less frequently applied methods include grinding with aluminum oxide and disruption in a French press or by sonication. All cell-wall preparations must be purified by digestion with proteases (e.g. trypsin) and/or by boiling in 2–4% sodium laurylsulphate. Wall polysaccharides and teichoic acids can be removed, if necessary, with formamide (170°C, 20 min; Schleifer *et al.*, 1968), 10% trichloroacetic acid (4°C, 24–48 h; or 90°C, 10 min; Schleifer, 1975), 0.1 *N* NaOH (18–20°C, 16 h; Endl *et al.*, 1983) or 70% hydrofluoric acid (2°C, 3–6 h; Fiedler *et al.*, 1981). The purified cell-wall preparations are hydrolysed with 4 *N* HCl at 100°C for 16 h. These hydrolysis conditions are usually sufficient to break down the mureins completely to free amino acids and amino sugars. In a few rare cases where threonine or isoasparagine are linked to the distal amino group of lysine or ornithine, stronger hydrolysis conditions (6 *N* HCl, 120°C, 18 h) must be applied.

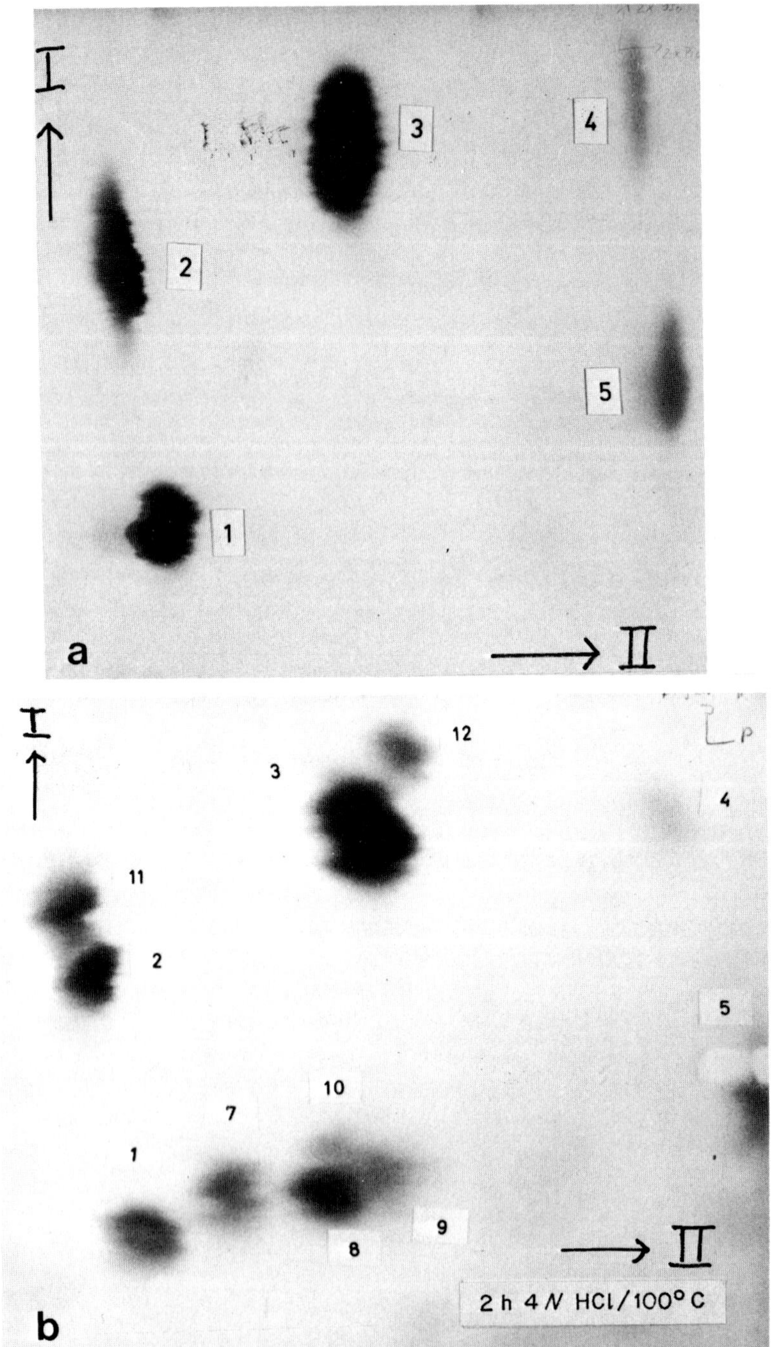

FIG. 3. Two-dimensional chromatograms of total (a, 4 *N* HCl, 100°C, 16 h) and partial acid (b, 4 *N* HCl, 100°C, 2h) hydrolysates of cell walls of *Micrococcus roseus* (murein type: Lys-Ala_3). I, isopropanol–acetic acid–water (75:10:15, by vol.); II, α-picoline–25% NH_4OH–water (70:2:28, by vol.). 1, Lys; 2, Glu; 3, Ala; 4, muramic acid; 5, glucosamine; 7, Lys-D-Ala; 8, N^6-Ala-Lys; 9, N^6-Ala_2-Lys; 10, N^6-Ala-Lys-D-Ala; 11, Ala-D-Glu; 12, L-Ala_2 and D-Ala-L-Ala.

Two-dimensional descending paper or ascending thin-layer cellulose chromatography are usually sufficient to determine qualitative amino-acid and amino-sugar composition. The most suitable solvent systems are isopropanol–acetic acid–water (75:10:15, v/v) for the first and α-picoline (2-methylpyridine)–25% NH_4OH–water (70:2:28, v/v) for the second direction. A hydrolysate of 1–2 mg of a purified cell-wall preparation is sufficient for the qualitative determination of the amino-acid and amino-sugar composition by paper chromatography (Fig. 3). An unequivocal determination of the diamino acids requires an additional one-dimensional paper chromatography run, using the solvent system methanol–pyridine–water–10 *N* HCl (32:4:7:1, v/v) or methanol–pyridine–water–formic acid (80:10:19:1, v/v).

Quantitative determination of the amino compounds is performed with an amino-acid analyzer (Schleifer & Kandler 1972). The configuration of amino acids can be determined enzymatically, using D-amino oxidase (Larsen *et al.* 1971) or L-amino acid decarboxylases (lysine, glutamic acid, ornithine; Kandler & König 1978). Another possibility for determining the configuration of most amino acids is to measure the optical rotatory dispersion of their 2,4-dinitrophenyl (DNP) derivatives (Bricas *et al.* 1967; Kandler *et al.* 1968a) and to compare them with those of authentic L- and D-DNP-amino acids.

Determination of Murein Type

The primary structure of the mureins of *Escherichia coli* (Weidel & Pelzer 1964) and *Staphylococcus aureus* (Ghuysen & Strominger 1963a,b) was determined by analyzing fragments (muropeptides) obtained from the degradation of bacterial walls with wall-lytic enzymes. Once the basic structure of the murein was known, it was possible to introduce a rather simple chemical method to elucidate the primary structure of the peptide moieties (Schleifer & Kandler 1967, 1972). The main procedures of this method are as follows:

(a) The amino acid composition of purified cell walls must be determined quantitatively (see above). In cases where an interpeptide bridge is present, it is also often necessary to determine the configuration of the amino acids.

(b) The N- and C-terminal amino acids of intact cell walls can be established by dinitrophenylation (Takebe 1965) and hydrazinolysis (Hammes *et al.* 1973), respectively. In particular, the determination of the N-terminal amino acid is important since it gives an indication of the N-terminus of the interpeptide bridge. A certain pecentage of interpeptide bridges are not cross-linked and reveal an unsubstituted amino group. Dinitrophenylated amino acids can be identified by one-dimensional thin-layer chromatography on silica gel, using the following solvent systems: benzene–

acetic acid (8:2, v/v), chloroform–methanol–acetic acid (95:5:1, v/v) or chloroform–methanol–acetic acid–water (65:25:13:8, v/v).

(c) The isolation and identification of oligopeptides from partial acid hydrolysates (4 *N* HCl at 100°C for 15–60 min) of purified cell walls is the crucial step in the determination of the murein type. The partial acid hydrolysates are separated by two-dimensional paper chromatography, using the same solvent systems as for the qualitative amino-acid determination (Fig. 3b). A detailed description of the method, including R_{Ala} values of various peptides, pictures of two-dimensional paper chromatograms and an elution profile of a partial acid hydrolysate on an amino-acid analyzer, has been published by Schleifer & Kandler (1972).

(d) In most cases, these three steps are sufficient to establish the murein type. There are, however, a few examples of mureins with a rather complicated structure where additional information is needed to determine their primary structure unequivocally. Such information can be gained from the isolation of nucleotide-activated murein precursors or by the isolation and identification of muropeptides (for further detail, see Hammes *et al.* 1973).

Rapid Methods for the Determination of Amino Acid Composition or Sequence of Murein

Simple and rapid procedures are the prerequisite for using the chemical composition or the primary structure of murein as chemotaxonomic criteria in the routine classification of Gram-positive bacteria.

Amino-acid composition of murein

In cases where the chemical structure of the murein is quite distinct, simple, qualitative amino-acid analysis of the cell wall is often sufficient to provide useful chemotaxonomic information. There are various examples such as the separation of *Streptococcus faecalis* from other enterococci (Kandler *et al.* 1968b; Kilpper-Bälz *et al.* 1982), distinguishing *Propionibacterium freudenreichii* and "*P. shermanii*" from other propionibacteria (Schleifer & Kandler 1972), separation of staphylococci from micrococci (Schleifer & Kloos 1976), and differentiation of oerskoviae (Seidl *et al.* 1980), certain micrococci (Schleifer *et al.* 1982) and bifidobacteria (Lauer & Kandler 1983).

A modification of the method of Park & Hancock (1960), as proposed by Schleifer & Kandler (1972), is usually sufficient for a rapid preparation of cell-wall material. In short, an overnight culture (30–50 ml) is harvested, resuspended in 10% (w/v) trichloroacetic acid (E_{650} = 1.0 at a 1:10 dilution) and boiled for 20 min. The suspension is then centrifuged, washed and incubated in trypsin–phosphate buffer (2 mg of trypsin/10 ml of 0.1 *M* phosphate buffer, pH 7.9) at 37°C on a shaker for about 2 h. Following centrifugation and washing with distilled water, a sample of the pellet is hydrolysed with 6 *N* HCl at 105°C for 6 h to determine the amino-acid composition of the cell wall by two-dimensional paper chromatography using the solvent systems isopropanol and α-picoline (see above). It is advisable to prepare two chromatograms with different concentrations of the hydrolysate in order to distinguish minor contaminating amino acids from the murein amino acids.

Chemical determination of murein type

Simple qualitative, or even quantitative, amino-acid determination of the cell wall is in many cases not sufficient to characterize unequivocally the murein type. There are some instances where the amino-acid composition of the wall is identical but the murein type is quite different, e.g. *Bifidobacterium breve* and *Micrococcus luteus*, *Arthrobacter aurescens* and *Arthrobacter citreus*, *Lactobacillus viridescens* and *Leuconostoc cremoris*, and *Clostridium leptum* and various coagulase-negative staphylococci. Moreover, the allocation of mureins to the cross-linkage groups A and B is in most cases only possible after establishing the amino-acid sequence of the murein or at least after the detection of characteristic oligopeptides.

Known murein types can be easily recognized by applying the rapid screening method proposed by Schleifer & Kandler (1972). Cell walls are prepared by the trichloroacetic acid method as described above. A sample of the wall preparation is partially acid hydrolysed (4 *N* HCl, 100°C, 45–60 min) and subjected to two-dimensional paper chromatography (Schleicher and Schüll 2043b Mgl 29 by 30 cm), using the solvent systems isopropanol and α-picoline. Each direction is run twice for about 16–18 h. The type of murein can be recognized in most cases from the characteristic fingerprint of its partial acid hydrolysate. In some cases, however, it is also necessary to determine the quantitative amino-acid composition of the walls. For a correct interpretation of fingerprints, it is advisable to compare them with those of known murein types. Since fingerprints can sometimes be slightly different, depending on the conditions used for chromatography or because of minor variations in the solvent systems, it is helpful to run in parallel a two-dimensional chromatogram of a partial acid hydrolysate of cell walls of a known murein type to standardize the system.

Serological detection of certain murein types

Many murein types differ from each other in the amino-acid composition or sequence of their interpeptide bridges. Since not all mureins studied so far are completely cross-linked, a part of the interpeptide bridges always show unsubstituted amino groups. It is known that proteins substituted with oligopeptides containing free amino groups yield antisera that show specificity to the N-terminal part of the peptide determinants (Schechter *et al.* 1971).

Oligopeptides, pentaglycyl-ϵ-amino-*n*-hexanoic acid and trialanyl- ϵ-amino-*n*-hexanoic acid, with structural similarity to the interpeptide bridges of staphylococci or certain micrococci, respectively, can be covalently linked to human serum albumin (Seidl & Schleifer 1978a). Antisera to these synthetic peptidyl–protein immunogens react predominantly with the N-terminal parts of the oligopeptide substituents. Antisera to (pentaglycyl-ϵ-Ahx)–albumin react with all staphylococcal mureins but not with other mureins lacking oligoglycine interpeptide bridges. On the other hand, antisera to (trialanyl-ϵ-Ahx)–albumin react with mureins containing oligoalanine interpeptide bridges (e.g. certain micrococci, streptococci) but not with mureins of staphylococci (Seidl & Schleifer 1978a). These specific antisera can be used for the serological separation of staphylococci from micrococci (Seidl & Schleifer 1978b).

Trichloroacetic acid (10%, 100°C, 30 min) extracted cells of staphylococci can be specifically agglutinated with antisera to (pentaglycyl-ϵ-Ahx)–albumin. Other bacteria lacking oligoglycine interpeptide bridges in their murein do not agglutinate. Antisera to (trialanyl -ϵ-Ahx)–albumin agglutinate trichloroacetic acid-extracted cells of bacteria containing oligoalanine interpeptide bridges in their murein. The test is highly specific and very sensitive; 10^6–10^7 cells/ml can be detected. It may be used for the serological detection of bacteria that differ in the interpeptide bridge composition or sequence of their mureins.

Chemical Structure of Murein and Its Use as a Taxonomic Marker

Schleifer & Kandler (1972) discussed the taxonomic relevance of murein structure in detail. They drew the conclusion that murein structure, at least among Gram-positive eubacteria, was a valuable taxonomic marker since it fulfills all of the prerequisites of a taxonomically valuable macromolecule.

(a) It shows a widespread distribution since it is a typical constituent of almost all eubacteria.
(b) The biosynthesis of the murein requires at least 20 enzymes. The involvement of so many genes guarantees a certain stability and means that

convergent formation of identical variations is rather unlikely. No single-step mutations are known that lead to an altered murein structure (Schleifer *et al.* 1976).

(c) Considerable mutational changes in the chemical structure of the murein of Gram-positive eubacteria were achieved during the course of evolution. This diversification of the chemical structure is obviously compatible with survival since many different murein types are still conserved, especially amongst Gram-positive bacteria.

(d) Intensive studies on the effect of endo- and exogenous factors on the primary structure of mureins have shown that the structure and composition of the latter are quite stable (Schleifer *et al.* 1976). However, the amino-acid composition of the growth medium can influence the composition and structure of murein; this has been shown in intensive studies of staphylococci (Schleifer *et al.* 1969, 1976). The interpeptide bridges of the mureins of staphylococci can vary to a certain extent, depending on the concentration of alanine, serine and glycine in the growth medium. Under unusual conditions, two genetically different strains can even appear to be phenotypically similar, e.g. a strain of *S. aureus* grown in a serine-enriched medium shows a similar murein type to that of *S. epidermidis* grown in a glycine-enriched and serine-deficient medium. The genetic differences become obvious, however, when the same strains are cultivated in a balanced medium containing sufficient glycine.

Phenotypic variations in mureins seem to be an exception rather than a rule, since studies with other eubacteria have shown that the amino-acid composition of their cell walls did not change significantly when they were grown in various media. Moreover, even with the staphylococci the phenotype variations were only found with staphylococci when different strains were grown under unusual conditions. Indeed, the murein types of staphylococci are of taxonomic value, provided that cells are grown in a balanced medium with sufficient glycine (Schleifer 1973, 1981; Schleifer & Kocur 1973; Schleifer & Kloos 1976).

False identifications of mureins have been repeatedly reported in the literature. The majority of these mistakes are based on incorrect or insufficient amino-acid determinations. Examples of false identifications have been summarized in recent reviews by Kandler & Schleifer (1980) and Schleifer & Stackebrandt (1983).

Studies on murein structure have had considerable impact on the taxonomy of many Gram-positive eubacteria, whereas with most Gram-negative bacteria this macromolecule is chemically uniform and therefore unsuited as a chemotaxonomic marker. There are only a few examples of Gram-negative bacteria, such as the spirochaetes and fusobacteria, that can be distinguished

TABLE 3

*Murein structure and its use as a chemotaxonomic marker**

Groups of organisms	Taxonomic implication	References
Actinomycetales	Murein types and phylogenetic relationship	Stackebrandt & Schleifer 1984
Arthrobacter sensu stricto	Definition on the basis of murein structure	Collins & Kroppenstedt 1983
Bacillus pasteurii	Separation from other bacilli	Ranftl & Kandler 1973
Bacteroidaceae	Occurrence of different mureins	Vasstrand *et al.* 1982
Bifidobacterium	Species differentiation	Lauer & Kandler 1983
Cellulomonas	Species differentiation	Fiedler & Kandler 1973a
Clostridium	Differentiation of some species	Weiss *et al.* 1981
Corynebacterium ilicis	Transfer to *Arthrobacter*	Collins *et al.* 1981
Corynebacterium pyogenes, C. haemolyticum	Exclusion from *Corynebacterium*	Collins *et al.* 1982a,b; Collins & Jones 1982
Coryneforms with murein of group B	Transfer to *Curtobacterium, Microbacterium, Aureobacterium*	Döpfer *et al.* 1982; Collins *et al.* 1983a,b
Eubacterium limosum	Relationship to *Clostridium barkeri* and *Acetobacterium woodii*	Weiss *et al.* 1981
Gemella	Separation from *Neisseria*	Stackebrandt *et al.* 1982
Micrococcaceae	Separation of staphylococci and micrococci	Schleifer & Kloos 1976
Micrococcus	Species differentiation	Schleifer *et al.* 1981
"*Micrococcus candidus,*" "*M. caseolyticus*"	Transfer to *Staphylococcus*	Schleifer *et al.* 1982
"*M. conglomeratus*"	Relationship with *Arthrobacter* species	Schleifer & Lang 1980
Oerskovia	Species differentiation	Seidl *et al.* 1980
Peptococcus, Peptostreptococcus	Separation from other cocci and species differentiation	Schleifer & Nimmermann 1973; Weiss 1981; Huss 1983
Peptococcus saccharolyticus	Transfer to *Staphylococcus*	Kilpper-Bälz & Schleifer 1981
P. indolicus	Relationship to *P. aerogenes*	Huss *et al.* 1982
Prochloron	Occurrence of murein	Stackebrandt & Kandler 1982
Staphylococcus	Species differentiation	Schleifer & Kocur 1973; Kloos & Schleifer 1981
Stomatococcus	Separation from other cocci	Stackebrandt *et al.* 1983

*Examples not included in the review of Schleifer and Kandler (1972).

on the basis of murein structure. The value of the murein structure as a chemotaxonomic marker is outlined in Table 3. The determination of murein structure is of value in the differentiation of genera, species or subspecies, and in the recognition of misclassified strains. One of the main strengths of such studies is that the results correlate well with other useful chemotaxonomic markers from lipid (Minnikin & Goodfellow 1980) and DNA base analyses (Schleifer 1973). The taxonomic information derived from murein structure is, generally speaking, comparable with that from DNA base composition studies.

Occurrence of the same murein type in organisms does not necessarily mean that they are closely related, e.g. *Escherichia coli* and *Bacillus megaterium* have the same murein type. On the other hand, closely related eubacteria usually possess the same murein type, and eubacteria with different murein types are not normally closely related. There are very few exceptions to this rule. However, quite recently it was shown that some anaerobic cocci (Hare group I and III) with two murein types [Lys-(Orn)-Lys-D-Glu and Lys-(Orn)-Gly_5, respectively] were nevertheless closely related on the basis of their DNA–DNA homology values (Huss 1983).

Last, but not least, the presence of murein is one of the key characteristics for separating eubacteria from archaebacteria (Woese *et al.* 1978; Kandler 1982). The presence or absence of the typical murein constituents, muramic acid and D-amino acids in cell-wall or even whole-cell hydrolysates allows a clear distinction to be drawn between eubacteria and archaebacteria, with the exception of mycoplasmas. The cell envelopes of archaebacteria are structurally and chemically much more diverse than those of eubacteria (Kandler 1982).

References

BRICAS, E., GHUYSEN, J. M. & DEZÉLÉE, PH. 1967 The cell wall peptidoglycan of *Bacillus megaterium* KM. I. Studies on the stereochemistry of α-, α'-diaminopimelic acid. *Biochemistry* **6**, 2598–2607.

COLLINS, M. D. & JONES, D. 1982 Reclassification of *Corynebacterium pyogenes* (Glage) in the genus *Actinomyces*, as *Actinomyces pyogenes* comb.nov. *Journal of General Microbiology* **128**, 901–903.

COLLINS, M. D. & KROPPENSTEDT, R. M. 1983 Lipid composition as guide to the classification of some coryneform bacteria—containing an A4α type peptidoglycan (Schleifer and Kandler). *Systematic and Applied Microbiology* **4**, 95–104.

COLLINS, M. D., JONES, D. & KROPPENSTEDT, R. M. 1981 Reclassification of *Corynebacterium ilicis* (Mandel, Guba and Litsky) in the genus *Arthrobacter*, as *Arthrobacter ilicis* comb.nov. *Zentralblatt für Bakteriologie, Mikrobiologie und Hygiene. I. Abteilung Originale* **C2**, 318–323.

COLLINS, M. D., JONES, D., KROPPENSTEDT, R. M. & SCHLEIFER, K. H. 1982a Chemical studies as a guide to the classification of *Corynebacterium pyogenes* and "*Corynebacterium haemolyticum*". *Journal of General Microbiology* **128,** 335–341.

COLLINS, M. D., JONES, D. & SCHOFIELD, G. M. 1982b Reclassification of "*Corynebacterium haemolyticum*" (MacLean, Liebow and Rosenberg) in the genus *Arcanobacterium* gen.nov. as *Arcanobacterium haemolyticum* nom.rev., comb.nov. *Journal of General Microbiology* **128,** 1279–1281.

COLLINS, M. D., JONES, D., KEDDIE, R. M., KROPPENSTEDT, R. M. & SCHLEIFER, K. H. 1983a Classification of some coryneform bacteria in a new genus *Aureobacterium*. *Systematic and Applied Microbiology* **4,** 236–252.

COLLINS, M. D., JONES, D. & KROPPENSTEDT, R. M. 1983b Reclassification of *Brevibacterium imperiale* (Steinhaus) and "*Corynebacterium laevaniformis*" (Dias and Bhat) in a redefined genus *Microbacterium* (Orla-Jensen) as *Microbacterium imperiale* comb.nov. and *Microbacterium laevaniformis* nom.rev., comb.nov. *Systematic and Applied Microbiology* **4,** 65–78.

DÖPFER, H., STACKEBRANDT, E. & FIEDLER, F. 1982 Nucleic acid hybridization studies on *Microbacterium, Curtobacterium, Agromyces* and related taxa. *Journal of General Microbiology* **128,** 1697–1708.

ENDL, J., SEIDL, P. H., FIEDLER, F. & SCHLEIFER, K. H. 1983 Chemical composition and structure of cell wall teichoic acids of staphylococci. *Archives of Microbiology* **135,** 215–223.

FIEDLER, F. & KANDLER, O. 1973a Die Mureintypen in der Gattung *Cellulomonas* Bergey *et al.* *Archiv Für Mikrobiologie* **89,** 41–50.

FIEDLER, F. & KANDLER, O. 1973b Die Aminosäuresequenz von 2,4-Diaminobuttersäure enthaltenden Mureinen bei verschiedenen coryneformen Bakterien und *Agromyces ramosus*. *Archiv für Mikrobiologie* **89,** 51–66.

FIEDLER, F., SCHÄFFLER, M. J. & STACKEBRANDT, E. 1981 Biochemical and nucleic acid hybridization studies on *Brevibacterium linens* and related strains. *Archives of Microbiology* **129,** 85–93.

GHUYSEN, J. M. & STROMINGER, J. L. 1963a Structure of the cell wall of *Staphylococcus aureus*, strain Copenhagen. I. Preparation of fragments by enzymatic hydrolysis. *Biochemistry* **2,** 1110–1119.

GHUYSEN, J. M. & STROMINGER, J. L. 1963b Structure of the cell wall of *Staphylococcus aureus*, strain Copenhagen. II. Separation and structure of disaccharides. *Biochemistry* **2,** 1119–1125.

HAMMES, W., SCHLEIFER, K. H. & KANDLER, O. 1973 Mode of action of glycine on the biosynthesis of peptidoglycan. *Journal of Bacteriology* **116,** 1029–1053.

HOLZAPFEL, W. H. & VAN WYK, E. P. 1982 *Lactobacillus kandleri* sp. nov., a new species of the subgenus *Betabacterium*, with glycine in the peptidoglycan. *Zentralblatt für Bakteriologie, Mikrobiologie und Hygiene. I. Abteilung Originale* **C3,** 495–502.

HUSS, V. A. R. 1983 Untersuchungen zur spektrophotometrischen Bestimmung von Desoxyribonukleinsäure-Homologien am Beispiel der Gattungen *Peptococcus* und *Peptostreptococcus*. Ph.D. Dissertation, Technische Universität München.

HUSS, V. A. R., SCHLEIFER, K. H., LINDAL, E., SCHWAN, O. & SMYTH, C. J. 1982 Peptidoglycan type, base composition of DNA, and DNA-DNA homology of *Peptococcus indolicus* and *Peptococcus asaccharolyticus*. *FEMS Microbiology Letters* **15,** 285–289.

KANDLER, O. 1982 Cell wall structures and their phylogenetic applications. *Zentralblatt für Bakteriologie, Mikrobiologie und Hygiene. I. Abteilung Originale* **C3,** 149–160.

KANDLER, O. & KÖNIG, H. 1978 Chemical composition of the peptidoglycan-free cell walls of methanogenic bacteria. *Archives of Microbiology* **118,** 141–152.

KANDLER, O. & SCHOBERTH, S. 1979 Murein structure of *Acetobacterium woodii*. *Archives of Microbiology* **120,** 181–183.

KANDLER, O. & SCHLEIFER, K. H. 1980 Systematics of bacteria. *Progress in Botany* **42,** 234–252.

KANDLER, O., KOCH, D. & SCHLEIFER, K. H. 1968a Die Aminosäuresequenz eines glycinhaltigen Mureins einiger Stämme von *Lactobacillus bifidus*. *Archiv für Mikrobiologie* **61**, 181–186.

KANDLER, O., SCHLEIFER, K. H. & DANDL, R. 1968b Differentiation of *Streptococcus faecalis* Andrews and Horder and *Streptococcus faecium* Orla-Jensen based on the amino acid composition of their murein. *Journal of Bacteriology* **96**, 1935–1939.

KATO, K., UMEMOTO, T., SAGAWA, H. & KOTANI, S. 1979 Lanthionine as an essential constituent of cell wall peptidoglycan of *Fusobacterium nucleatum*. *Current Microbiology* **3**, 147–152.

KATO, K., UMEMOTO, T., FUKUHARA, H., SAGAWA, H. & KOTANI, S. 1981 Variation in dibasic amino acid in the cell wall peptidoglycan of bacteria of genus *Fusobacterium*. *FEMS Microbiology Letters* **10**, 81–85.

KAWAMOTO, I., OKA, T. & NARA, T. 1981 Cell wall composition of *Micromonospora olivoasterospora* and related organisms. *Journal of Bacteriology* **146**, 527–534.

KILPPER-BÄLZ, R. & SCHLEIFER, K. H. 1981 Transfer of *Peptococcus saccharolyticus* Foubert and Douglas to the genus *Staphylococcus: Staphylococcus saccharolyticus* (Foubert and Douglas) comb.nov. *Zentralblatt für Bakteriologie, Mikrobiologie und Hygiene. I. Abteilung Originale* **C2**, 324–331.

KILPPER-BÄLZ, R., FISCHER, G. & SCHLEIFER, K. H. 1982 Nucleic acid hybridization of group N and group D streptococci. *Current Microbiology* **7**, 245–250.

KLOOS, W. E. & SCHLEIFER, K. H. 1981 The genus *Staphylococcus*. In *The Prokaryotes* eds. Starr, M. P., Stolp, H., Trüper, H. G., Balows, A. & Schlegel, H. G., pp. 1548–1569. Berlin: Springer-Verlag.

LARSEN, D. H., SNETSINGER, D. C. & WAIBEL, P. E. 1971 Procedure for determination of D-amino acids. *Analytical Biochemistry* **39**, 395–401.

LAUER, E. & KANDLER, O. 1983 DNA-DNA homology, murein types and enzyme patterns in the type strains of the genus *Bifidobacterium*. *Systematic and Applied Microbiology* **4**, 42–64.

LEDERER, E., ADAM, A., CIORBARU, R., PETIT, J.-F. & WIETZERBIN, J. 1975 Cell walls of mycobacteria and related organisms; chemistry and immunostimulant properties. *Molecular and Cellular Biochemistry* **7**, 87–104.

MINNIKIN, D. E. & GOODFELLOW, M. 1980 Lipid composition in the classification and identification of acid-fast bacteria. In *Microbiological Classification and Identification* eds. Goodfellow, M. & Board, R. G., pp. 189–256. London: Academic Press.

PARK, J. T. & HANCOCK, R. 1960 A fractionation procedure for studies of the synthesis of cell wall mucopeptide and other polymers in cells of *Staphylococcus aureus*. *Journal of General Microbiology* **22**, 249–258.

RANFTL, H. & KANDLER, O. 1973 D-Aspartyl-L-Alanin als Interpeptidbrüke im Murein von *Bacillus pasteurii* Migula, *Zeitschrift für Naturforschung* **28C**, 4–8.

SCHECHTER, I., CLERICI, E. & ZAZEPITZKI, E. 1971 Distinct antigenic specificities of alanine peptide determinants attached to protein carriers *via* terminal amino or carboxyl groups. *European Journal of Biochemistry* **18**, 561–572.

SCHLEIFER, K. H. 1973 Chemical composition of staphylococcal cell walls. In *Staphylococci and Staphylococcal Infections: Recent Progress* ed. Jeljaszewicz, J., pp. 13–23. Basel: S. Karger.

SCHLEIFER, K. H. 1975 Chemical structure of the peptidoglycan, its modifiability and relation to the biological activity. *Zeitschrift für Immunitätsforschung* **149**, 157–164.

SCHLEIFER, K. H. 1981 Die Klassifikation von *Staphylococcus* und *Micrococcus*. Ein Beispiel für die moderne Bakteriensystematik. *Forum Mikrobiologie* **4**, 272–278.

SCHLEIFER, K. H. & JOSEPH, R. 1973 A directly cross-linked L-ornithine containing peptidoglycan in cell walls of *Spirochaeta stenostrepta*. *FEBS Letters* **36**, 83–86.

SCHLEIFER, K. H. & KANDLER, O. 1967 Zur chemischen Zusammensetzung der Zellwand der

Streptokokken. I. Die Aminosäuresequenz des Mureins von *Str. thermophilus* und *Str. faecalis*. *Archiv für Mikrobiologie* **57**, 335–364.

Schleifer, K. H. & Kandler, O. 1972 The peptidoglycan types of bacterial cell walls and their taxonomic implications. *Bacteriological Reviews* **36**, 407–477.

Schleifer, K. H. & Kandler, O. 1983 Primary structures of murein and pseudomurein. In *The Target of Penicillin* eds. Hakenbeck, R., Höltje, J. V. & Labischinski, H., pp. 11–17. Berlin: Walter de Gruyter.

Schleifer, K. H. & Kloos, W. E. 1976 Separation of staphylococci from micrococci. *Zentralblatt für Bakteriologie, Mikrobiologie und Hygiene. I. Abteilung, Supplement* **5**, 3–9.

Schleifer, K. H. & Kocur, M. 1973 Classification of staphylococci based on chemical and biochemical properties. *Archiv für Mikrobiologie* **93**, 65–85.

Schleifer, K. H. & Lang, K. 1980 Close relationship among strains of *Micrococcus conglomeratus* and *Arthrobacter species*. *FEMS Microbiology Letters* **9**, 223–226.

Schleifer, K. H. & Nimmermann, E. 1973 Peptidoglycan types of strains of the genus *Peptococcus*. *Archiv für Mikrobiologie* **93**, 245–258.

Schleifer, K. H. & Stackebrandt, E. 1983 Molecular systematics of prokaryotes. *Annual Review of Microbiology* **37**, 143–187.

Schleifer, K. H., Ried, M. & Kandler, O. 1968 Die Aminosäuresequenz des Mureins von *Staphylococcus epidermidis* (Winslow and Winslow) Evans, Stamm 66. *Archiv für Mikrobiologie* **62**, 198–208.

Schleifer, K. H., Huss, L. & Kandler, O. 1969 Die Beeinflussung der Aminosäuresequenz des serinhaltigen Mureins von *Staphylococcus epidermidis* Stamm 24 durch die Nährbodenzusammensetzung. *Archiv für Mikrobiologie* **68**, 387–404.

Schleifer, K. H., Hammes, W. P. & Kandler, O. 1976 Effect of endogenous and exogenous factors on the primary structures of bacterial peptidoglycan. *Advances in Microbial Physiology* **13**, 245–292.

Schleifer, K. H., Kloos, W. E. & Kocur, M. 1981 The genus *Micrococcus*. *In The Prokaryotes* eds. Starr, M. P., Stolp, H., Trüper, H. G., Balows, A. & Schlegel, H. T., pp. 1539–1547. Berlin: Springer-Verlag.

Schleifer, K. H., Kilpper-Bälz, R., Fischer, U., Faller, A. & Endl, J. 1982 Identification of '*Micrococcus candidus*' ATCC 14852 as a strain of *Staphylococcus epidermidis* and of '*Micrococcus caseolyticus*' ATCC 13548 and *Micrococcus varians* ATCC 29750 as members of a new species, *Staphylococcus caseolyticus*. *International Journal of Systematic Bacteriology* **32**, 15–20.

Schleifer, K. H., Hayn, I., Seidl, P. H. & Firl, J. 1983 Threo-β-hydroxyornithine: a natural constituent of the peptidoglycan of *Corynebacterium species* Co 112. *Archives of Microbiology* **34**, 243–246.

Seidl, P. H. & Schleifer, K. H. 1978a Specific antibodies to the N-termini of the interpeptide bridges of peptidoglycan. *Archives of Microbiology* **118**, 185–192.

Seidl, P. H. & Schleifer, K. H. 1978b A rapid test for the serological separation of staphylococci from micrococci. *Applied and Environmental Microbiology* **35**, 479–482.

Seidl, P. H., Faller, A. H., Loider, R. & Schleifer, K. H. 1980 Peptidoglycan types and cytochrome patterns of strains of *Oerskovia turbata* and *O. xanthineolytica*. *Archives of Microbiology* **127**, 173–178.

Stackebrandt, E. & Kandler, O. 1982 The murein type of *Prochloron*. *Zentralblatt für Bakteriologie, Mikrobiologie und Hygiene. I. Abteilung Originale* **C3**, 354–357.

Stackebrandt, E. & Schleifer, K. H. 1984 Studies on the phylogeny of the order *Actinomycetales*. In *Actinomycete Biology* eds. Bojalil, L. F. & Ortiz-Ortiz, L. New York: Academic Press.

Stackebrandt, E., Fiedler, F. & Kandler, O. 1978 Peptidoglycantyp und Zusammensetzung der Zellwandpolysaccharide von *Cellulomonas cartalyticum* und einigen coryneformen Organismen. *Archives of Microbiology* **117,** 115–118.

STACKEBRANDT, E., WITTEK, B., SEEWALDT, E. & SCHLEIFER, K. H. 1982 Physiological, biochemical and phylogenetic studies on *Gemella haemolysans*. *FEMS Microbiology Letters* **13,** 361–365.

STACKEBRANDT, E., SCHEUERLEIN, C. & SCHLEIFER, K. H. 1983 Phylogenetic and biochemical studies on *Stomatococcus mucilaginosus*. *Systematic and Applied Microbiology* **4,** 207–217.

TAKEBE, I. 1965 Extent of cross-linkage in the murein sacculus of *Escherichia coli* B cell wall. *Biochimica et Biophysica Acta* **101,** 124–126.

UCHIDA, K. & AIDA, K. 1977 Acyl type of bacterial cell wall: Its simple identification by colorimetric method. *Journal of General and Applied Microbiology* **23,** 249–260.

UMEMOTO, T., OTA, T., SAGAWA, H., KATO, K., TAKADA, H., TSUJIMOTO, M., KAWASAKI, A., OGAWA, T., HARADA, K. & KOTANI, S. 1981 Chemical and biological properties of a peptidoglycan isolated from *Treponema pallidum* Vazan. *Infection and Immunity* **31,** 767–774.

VASSTRAND, E. N., HOFSTAD, T., ENDRESEN, C. & JENSEN, H. B. 1979 Demonstration of lanthionine as a natural constituent of the peptidoglycan of *Fusobacterium nucleatum* Fev1. *Infection and Immunity* **25,** 775–780.

VASSTRAND, E. N., JENSEN, H. B., MIRON, T. & HOFSTAD, T. 1982 Composition of peptidoglycan in *Bacteriodaceae:* determination and distribution of lanthionine. *Infection and Immunity* **36,** 114–122.

WEIDEL, W. & PELZER, H. 1964 Bagshaped macromolecules: a new outlook on bacterial cell walls. *Advances in Enzymology* **26,** 193–232.

WEISS, N. 1981 Cell wall structure of anaerobic cocci. *Revue de l'Institut Pasteur de Lyon* **14,** 53–59.

WEISS, N., SCHLEIFER, K. H. & KANDLER, O. 1981 The peptidoglycan types of gram-positive anaerobic bacteria and their taxonomic implications. *Revue de l'Institut Pasteur de Lyon* **14,** 3–12.

WOESE, C. R., MAGRUM, L. J. & FOX, G. E. 1978 Archaebacteria. *Journal of Molecular Evolution* **11,** 245–252.

YAMADA, K. & KOMAGATA, K. 1972 Taxonomic studies on coryneform bacteria. V. Classification of coryneform bacteria. *Journal of General and Applied Microbiology* **18,** 417–431.

Simple Rapid Methods of Cell Wall Analysis as an Aid in the Identification of Aerobic Coryneform Bacteria

I. J. BOUSFIELD

National Collection of Industrial & Marine Bacteria Ltd., Torry Research Station, Aberdeen, UK

R. M. KEDDIE

Department of Microbiology, University of Reading, Reading, UK

T. R. DANDO

National Collection of Industrial & Marine Bacteria Ltd., Torry Research Station, Aberdeen, UK

S. SHAW

Watney Mann and Truman Brewers Ltd., London, UK

Following morphological examination, one of the most valuable next steps in the identification of coryneform bacteria is to examine the simple cell-wall composition, using one of the so-called "rapid" methods. These methods allow the components most useful in the differentiation of genera to be determined, i.e. the peptidoglycan diamino acid and certain diagnostic sugars (see e.g. Keddie & Bousfield 1980; Chapter 10, this volume).

For components that occur mainly or entirely in the cell wall, such as the isomers of diaminopimelic acid (DAP) and arabinose, chromatographic analyses of acid hydrolysates of whole cells will suffice. However, for the determination of other diamino acids and sugars, acid hydrolysates of purified cell-wall material must be examined. Sufficiently pure wall material may be prepared by the alkali treatment of whole cells, a rapid method that gives results comparable with those obtained by traditional methods of cell-wall analysis (Keddie & Cure 1977).

CHEMICAL METHODS IN
BACTERIAL SYSTEMATICS

0-12-289675-0

Having determined the simple cell-wall composition, further identification may be achieved using a combination of chemotaxonomic and conventional characters. Examples of these characters are given in Tables 1 and 2 together with the determinatively useful cell-wall components in the various aerobic coryneform genera. More detailed information may be had from the reviews by Keddie & Cure (1978), Minnikin *et al.* (1978), Keddie & Bousfield (1980) and Keddie & Jones (1981).

Preparation of Hydrolysates

Growth of cells

Cells are grown in a suitable liquid or solid medium such as TSX broth (Keddie *et al.* 1966) or nutrient broth (Oxoid CM1) or agar (Oxoid CM3) (Bousfield *et al.* 1983). An incubation temperature of 25°C is suitable for most saprophytic aerobic coryneform bacteria. The alkali method usually requires about 50 ml of shaken liquid culture or the growth from two or three agar slopes (Universal bottles). Less material is required for whole-cell analysis: about 10 ml of liquid culture or the growth from one agar slope. (If solid media are used, however, fragments of agar may inadvertently be harvested and hydrolysed with the cells, possibly resulting in the contamination of hydrolysates with galactose. Therefore, if cell-wall galactose is sought, hydrolysates should be prepared from cells grown in liquid media). After harvesting, cells are washed twice with distilled water and either freeze-dried or stored frozen at −18°C if not used immediately. If freeze-drying is used, bacteria should be killed before harvesting by overnight incubation with 5% (v/v) formalin, so that potentially infective dry aerosols are not produced during subsequent handling. In any event, known or suspected pathogens should always be killed with formalin before harvesting.

Hydrolysis of whole cells

The method of hydrolysis of whole cells is based on those of Becker *et al.* (1964) and Murray & Proctor (1965).

About 50 mg wet weight of cells are suspended in 95% (v/v) ethanol and centrifuged. The supernatant is discarded and the pellet is air dried in an oven at 55°C. The dried material is resuspended in 1 ml of 6 *M* HCl and transferred to a glass Bijou bottle (about 6 ml), which is sealed tightly with a polypropylene cap. If freeze-dried material is used, the alcohol/air drying step is omitted and about 1–5 mg of cells are suspended directly in the acid. The bottle is heated in an oven at 105°C for 18 h and allowed to cool. The hydrolysate is then removed using a plastic syringe and filtered through a

TABLE 1

Some conventional and chemotaxonomic features of coryneform genera that contain meso-*diaminopimelic acid in the cell-wall peptidoglycan*[a]

	Mycelium produced	Rod–coccus[b] cycle	Oxygen requirements	Mycolic acids present	Arabinose in wall	% G + C	Major[c] menaquinone
Brevibacterium	—	+	Aerobic	—	—	60–64	MK-8(H_2)
Caseobacter	—	—	Aerobic	+	+	60–67	MK-9(H_2)
Corynebacterium	—	—	Facultative[d]	+	+	51–61[e]	MK-8(H_2) or MK-9(H_2)
Rhodococcus[f]	±	±	Aerobic	+	+	59–69	MK-8(H_2) or MK-9(H_2)

[a]Data from Keddie & Bousfield (1980); Keddie & Jones (1981); Collins & Jones (1981); Pitcher (1983).
[b]Similar to that in *Arthrobacter globiformis*.
[c]MK-8, MK-9, etc. indicate the number of isoprene units in the menaquinone; (H_2), (H_4), etc. indicate the number of double bonds hydrogenated.
[d]A few aerobic.
[e]*Corynebacterium kutscheri* has lower values; *Corynebacterium bovis* has higher values (Pitcher 1983).
[f]Usually considered a member of the nocardioform bacteria but included here for comparison with *Corynebacterium*.

TABLE 2

Some conventional and chemotaxonomic features of coryneform genera that contain diamino acids other than meso-*diaminopimelic acid in the cell wall peptidoglycan*[a]

	Cell-wall diamino acid	Peptido-glycan[b] group	Glycine in wall	Galactose in wall	Mycelium produced	Rod–coccus[c] cycle	Oxygen requirements	Acid from[d] glucose	Cellulo-lysis	% G + C	Major[e] menaquinone(s)
Arthrobacter[f,g]	Lysine	A	—	+	—	+	Aerobic	—	—	59–66[g]	MK-9 (H_2) or MK-8 and/or MK-9
Microbacterium[h]	Lysine	B	+	+	—	—	Equivocal	+	—	69–75	MK-11, MK-12
Renibacterium	Lysine	ND[i]	+	—	—	—	Aerobic	—	—	53–54	MK-9
Oerskovia[j]	Lysine	A	—	±	+	—	Facultative	+	±	71–75	MK-9(H_4)
Aureobacterium	Ornithine	B	+	(+)	—	—	Aerobic	W	—	65–75	MK-11, MK-12, MK-13
Cellulomonas[k]	Ornithine	A	—	—	—	—	Most facultative	+	+	71–76	MK-9(H_4)
Curtobacterium[l]	Ornithine	B	+	(+)[m]	—	—	Aerobic	W[n]	—	67–75	MK-9
"*Arthrobacter*"[q] *simplex* & "*Arthrobacter*" *tumescens*	LL-DAP	A	+	+	—	+	Aerobic	—	—	70–74	MK-8(H_4)
"DAB-containing"[p] plant pathogenic species[o]	DAB[p]	A	+	ND	—	—	Aerobic	(+)	—	67–78	MK-9, MK-10

[a]Data from Schleifer & Kandler (1972); Dye & Kemp (1977); Keddie & Cure (1977, 1978); Keddie & Bousfield (1980); Seidl *et al.* (1980); Keddie & Jones (1981); Collins & Jones (1981); Lechevalier & Lechevalier (1981); Collins (1982); Döpfer *et al.* (1982); Collins *et al.* (1983a,b,c).

[b]Group A: cross-linkage between positions 3 and 4 of two peptide subunits; Group B cross-linkage between positions 2 and 4 of two peptide subunits (Schleifer & Kandler 1972).

[c]Similar to that in *Arthrobacter globiformis*.

[d]In peptone-based media.

[e]MK-8, MK-9, etc. indicate the number of isoprene units in the menaquinone; (H_2), (H_4), etc. indicate the number of double bonds hydrogenated.

[f]As defined by Keddie and Jones (1981).

[g]*Arthrobacter atrocyaneus* has a value of ~70 mol % (see Keddie & Jones 1981).

[h]As defined by Collins *et al.* (1983b).

[i]Data not available or incomplete.

[j]A nocardioform genus, which Stackebrandt *et al.* (1982) have proposed should be combined with *Cellulomonas*.

[k]As defined by Keddie & Jones (1981).

[l]Excluding *Curtobacterium saperdae* and *Curtobacterium testaceum*, which Collins *et al.* (1983a) propose should be transferred to the new genus *Aureobacterium*.

[m]Most strains positive.

[n]Weak reaction.

[o] "*Corynebacterium*" *michiganese*, *C. insidiosum*, *C. iranicum*, *C. nebraskense*, *C. rathayi*, *C. sepedonicum* and *C. tritici*.

[p]DAB, diaminobutyric acid.

[q]It has been proposed by O'Donnell *et al.* (1982) that "*Arthrobacter*" *simplex*, but not "*Arthrobacter*" *tumescens*, be transferred to the genus *Nocardioides*. On the other hand, Suzuki & Komagata (1983) have proposed the new genus *Pimelobacter* to accommodate both "*Arthrobacter*" *simplex* and "*Arthrobacter tumescens* together with a new species *Pimelobacter jensenii*.

Whatman No. 1 filter-paper disc cut to fit a Swinnex 25-mm filter holder (Millipore UK Ltd., London, England UK) into a clean Bijou bottle. The filter paper is then washed through with a few drops of distilled water. Hydrolysates are evaporated to dryness over a boiling-water bath and the residue is taken up in 1 ml of distilled water. After evaporation to dryness again, the residue is taken up in 0.04 ml of distilled water. The hydrolysate may be stored at −18°C until required.

"Alkali" methods

A method of obtaining cell walls by digesting other cellular constituents with alkali was described by Boone & Pine in 1968. Two modifications of this method, both based on an earlier modification by I. J. Bousfield (unpublished), have since been described (Keddie & Cure 1977; Bousfield *et al.* 1983). Both methods are given here and for convenience are referred to as the "Reading" method and the "Aberdeen" method, respectively. The Reading method is closer to Bousfield's original modification; the Aberdeen method is quicker.

Reading method

Preparation of cell-wall material by alkali treatment

About 200 mg wet weight of cells are dispersed in 5% (w/v) KOH with a Rota-mixer (Hook and Tucker Instruments Ltd., Surrey, England UK) to give a suspension of turbidity about equal to that of Brown's opacity tube No. 2 (Wellcome Reagents Ltd.). Experience has shown that material for subsequent sugar analysis is usually required in greater quantity but needs shorter alkali treatment than that for amino-acid analysis. Accordingly, the suspension is divided between two glass Bijou bottles, one-third for amino acid analysis and two-thirds for sugar analysis. The bottles are sealed tightly with polypropylene caps and then heated at 100°C in an oven having a glass door to allow observation. Treatment is continued until the suspensions first turn yellow-brown and then just begin to clear. At this point, the portion for sugar analysis is removed from the oven and cooled immediately. Gram stains may be prepared to check that few whole cells remain. Treatment of the portion for amino acid analysis is allowed to continue until some clearing has occurred; this gives cleaner preparations. Treatment times are usually in the range 15–40 min (rarely $<$ 10 min or $>$ 60 min) for the sugar portion and 2–30 min longer for the amino-acid portion. It must be emphasized that considerable care is required in judging treatment times in order to obtain clean cell-wall preparations and thus reproducible results. Figures 1 and 2 show cells before and after alkali treatment for 40 min. An equal volume of distilled water is added to the cooled suspensions, which are then cen-

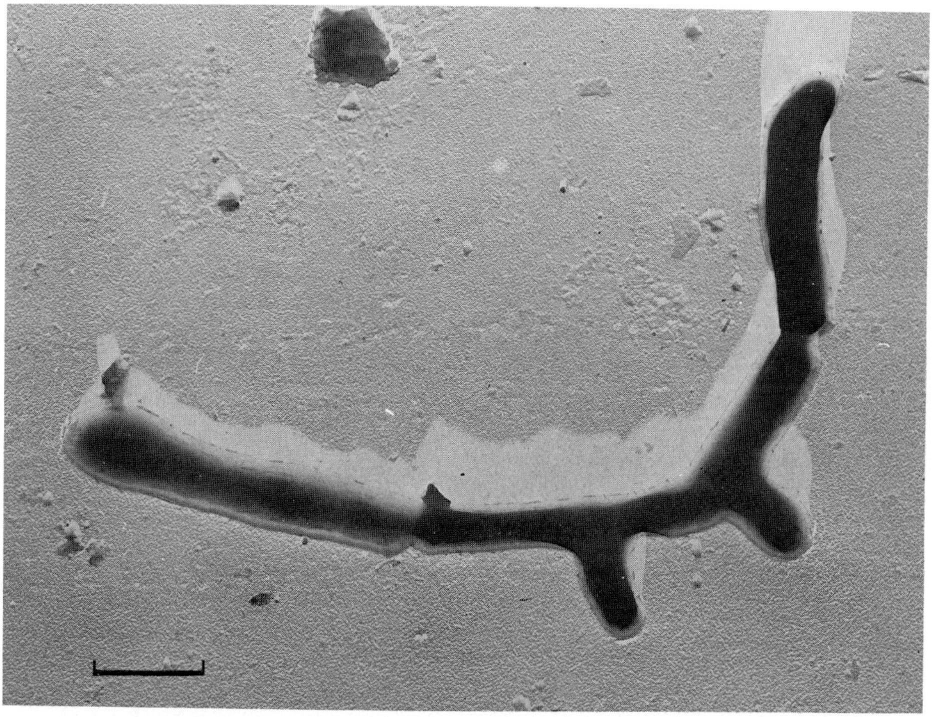

FIG. 1. Whole cells of *Arthrobacter globiformis* NCIB 9792. Bar marker represents 1 μm.

trifuged at 38,000 *g* for 5 min. The supernatants are discarded, and the deposits are washed twice in 10-ml amounts of distilled water.

Hydrolysis of cell-wall material

The wall deposit for sugar analysis is hydrolysed in about 0.5 ml of 2 *M* HCl at 100°C for 2 h and evaporated to dryness *in vacuo* over silica gel in a desiccator containing a small dish of solid NaOH. The deposit for amino-acid analysis is hydrolysed in about 0.5 ml of 6 *M* HCl at 105°C for 18 h and evaporated to dryness over a boiling-water bath. Both hydrolysates are resuspended in about 0.5 ml of distilled water and stored at −18°C if not required immediately.

Aberdeen method

Preparation of cell-wall material by alkali treatment

About 5–10 mg of freeze-dried weight or 200 mg wet weight of cells are suspended in 2 ml of 600 m*M* KOH in a glass Bijou bottle tightly closed with

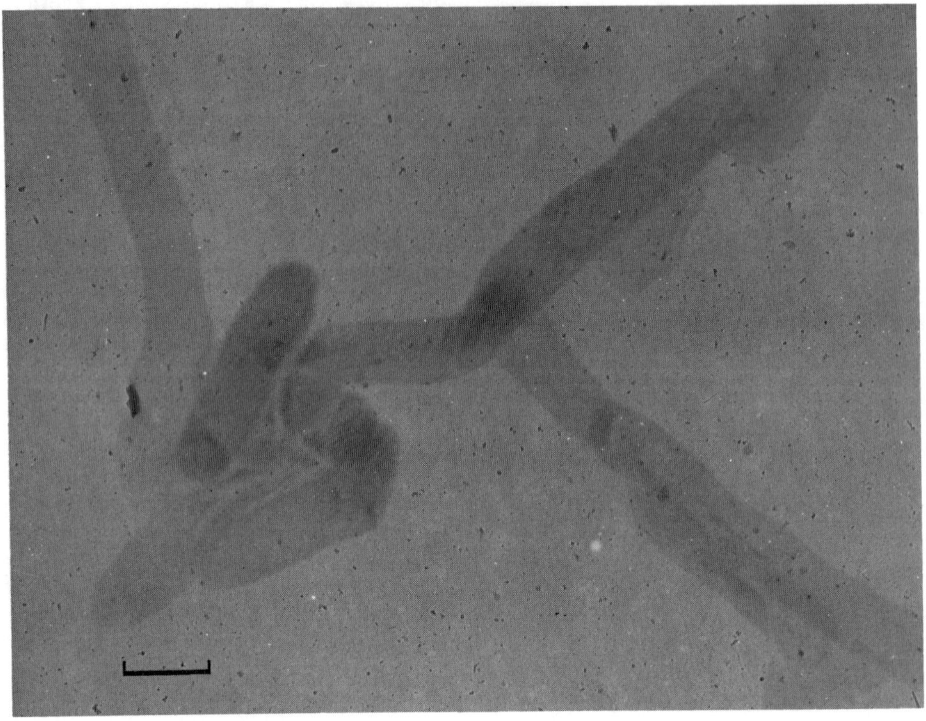

FIG. 2. Wall material of *Arthrobacter globiformis* NCIB 9792 obtained by boiling cells for 40 min with 5% (w/v, aqueous) KOH. Bar marker represents 1 μm.

a polypropylene screw cap. The suspension is autoclaved at 1 atm (101.3 kPa) for 5 min in a domestic pressure cooker, which is then cooled immediately by plunging into cold water. If the cap on the Bijou bottle has not been screwed down tightly, the sudden drop in pressure will cause the contents to boil over violently and most will be lost. After cooling, some clearing should have occurred but the suspension should still be turbid. If it has cleared completely, which happens only rarely, a fresh suspension should be prepared and heated in a boiling-water bath or oven at 100°C and cooled after some clearing has taken place, but while still turbid. Figure 3 shows cells after 5 min autoclaving with alkali. The cooled suspension is diluted to 10 ml with 2 *M* HCl (which prevents any further clearing) and centrifuged at 3200 *g* for 20 min. The deposit is then washed twice in 10-ml amounts of distilled water.

Hydrolysis of cell-wall material

After washing, the cell-wall deposit is resuspended in 0.5 ml of 10 *M* HCl. Experience has shown that water left in the deposit after decanting the

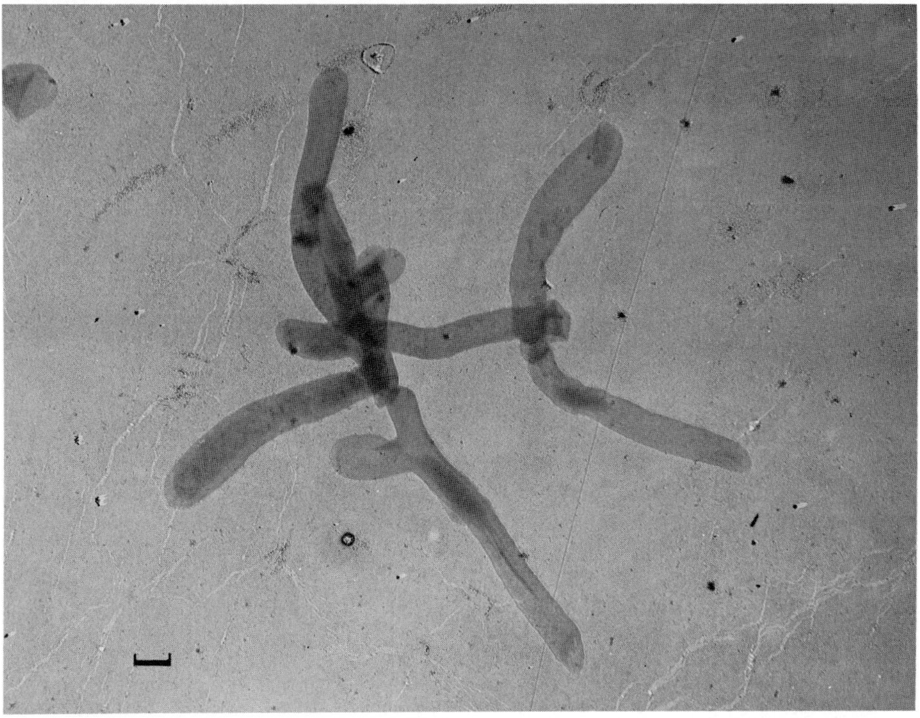

FIG. 3. Wall material of *Arthrobacter globiformis* NCIB 9792 obtained by autoclaving cells for 5 min at 1 atm in 600 m*M* KOH. Bar marker represents 1 μm.

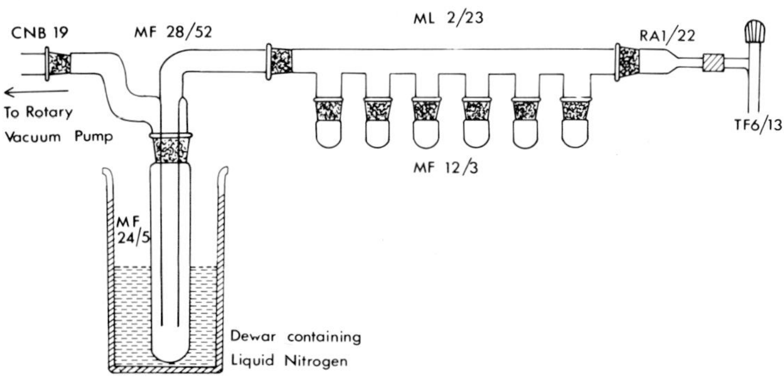

FIG. 4. Glass apparatus for freeze-drying cell wall hydrolysates. All glass parts are standard Quickfit items and catalogue numbers are shown. Operating procedure: (a) Ensure all joints are greased with high-vacuum silicone grease. (b) Open the stopcock fully and switch on the pump. (c) Close the stopcock slowly until the hydrolysates start to bubble gently. Do not allow violent bubbling or the hydrolysates will be blown into the manifold. (d) When all bubbling has ceased, carefully close the stopcock fully. Frost should quickly form on the outside of the flasks. Lack of frost indicates a poor vacuum connection somewhere in the system, which will greatly increase the drying time. (e) When the frost has disappeared and the flasks are no longer cold to the touch, the hydrolysates are dry.

washings will dilute this acid to about 6 *M*. The suspension is divided equally between two clean glass Bijou bottles fitted with polypropylene screw caps. The portion to be used for sugar analysis is diluted with twice its own volume of distilled water to reduce the molarity of the acid to about 2 *M* and is then hydrolysed at 100°C for 2.5 h. The portion for amino-acid analysis is diluted with twice its own volume of 6 *M* HCl and is hydrolysed at 100°C preferably overnight, but for at least 3.5 h. The longer the time of hydrolysis, the more concentrated will be the cell-wall amino acids in the final hydrolysate. Hydrolysates are then freeze-dried in a glass apparatus, water and HCl being trapped in a liquid-nitrogen-cooled glass trap. Such an apparatus is shown diagrammatically in Fig. 4. Alternatively, hydrolysates may be dried as described in the Reading method. Residues are taken up in about 0.05 ml of 10% (v/v) isopropanol and are stored at −18°C if not required immediately.

Chromatography

Hydrolysates may be examined by paper or thin-layer chromatography (TLC). The paper chromatographic methods described below have been used by Keddie and Shaw to detect most of the amino acids and all the sugars present in cell wall hydrolysates. The TLC methods, used by Bousfield and Dando, are rapid analyses aimed at detecting only the cell wall diamino acid and arabinose. If full analyses are required, however, standard TLC methods for the determination of amino acids and sugars can be used (see, e.g. Randerath 1963).

Paper chromatography

Preparation of chromatograms

The hydrolysates are not obtained from weighed amounts of wall material, so the amounts required for chromatography differ between samples. The hydrolysates are, therefore, applied to chromatograms in about 2.5-μl amounts using a 5-μl micropipette, and the spots are dried in a stream of hot air between applications until light straw-coloured spots are obtained. Usually the amounts required are in the range 2.5–20 μl both for amino acids and sugars.

Detection of amino acids

The method of detection of amino acids is applicable only to purified cell-wall material. It is not suitable for whole-cell hydrolysates, which contain cytoplasmic as well as cell-wall amino acids. Two-dimensional ascending chromatograms on Whatman No. 1 paper (200 × 200 mm) are developed in a BTL multisheet chromatographic tank (Baird & Tatlock Ltd., Romford, Essex, England UK) using the solvent systems described by Cummins & Johnson (1971). The first solvent system, *n*-butanol–acetic acid–water (60:15:25 v/v) is run for 20 h at about 20°C. After drying, the chromatogram is run in the second dimension for 16 h at about 20°C, using as solvent phenol–water (4:1, w/v) with 1 ml of NH_4OH (specific gravity 0.880) per 200 ml of solvent. After drying, the chromatogram is dipped in 0.2% (w/v) ninhydrin in acetone and heated at 100°C for 2 min to locate the amino-acid spots.

To facilitate identification, a reference mixture of amino acids should be run with each batch of chromatograms to provide a map. The mixture is prepared by dissolving each of the following at a concentration of 50 m*M* in 10% (v/v) isopropanol: L-alanine, L-aspartic acid, DAP, diaminobutyric acid (DAB), L-glutamic acid, glucosamine, glycine, L-lysine, L-ornithine and L-serine. The mixture is applied to chromatograms in 3-μl amounts.

In the solvent systems used, lysine and ornithine should just separate. However, even with a reference map there may be doubt about which amino acid is present in a particular cell-wall hydrolysate. Therefore, to obtain an unambiguous identification, the hydrolysate should be co-chromatographed with lysine if ornithine is suspected and with ornithine if lysine is thought to be present. Also, although the presence of DAP can be determined with these solvent systems, the *meso* and LL isomers are not separated. To identify which isomer is present, hydrolysates containing DAP must be chromatographed as described below.

Detection of DAP isomers

The method for detecting DAP isomers is applicable both to whole-cell hydrolysates and to hydrolysates of purified cell-wall material. One-dimensional descending chromatograms on Whatman No. 1 paper (160 × 520 mm) are developed in the solvent system methanol–pyridine–10 *M* HCl–water (80:10:2.5:17.5, v/v; Rhuland *et al.* 1955) for 17.5 h at about 20°C. After drying, the amino-acid spots are located as described above. A reference spot containing 3 μl of 0.05 *M* commercially available *meso*-DAP in 10% (v/v) isopropanol should be run on each chromatogram. Since commercially available *meso*-DAP contains small amounts of LL-DAP, it serves as a reference for both isomers.

Detection of sugars

The method for detecting sugars is applicable to both whole-cell and cell-wall hydrolysates. One-dimensional chromatograms on Whatman No. 1 paper (160 × 520 mm) are developed for about 36 h at about 20°C in the upper phase of the biphasic solvent system *n*-butanol–water–pyridine–toluene (5:3:3:4, v/v; Lechevalier, 1968). However, 6-deoxytalose runs off the paper at around 36 h and therefore duplicate chromatograms must be run for 30 h to detect this sugar. After drying, sugar spots are located by dipping chromatograms in 200 m*M* *p*-anisidine + 100 m*M* phthalic acid in 96% (v/v) ethanol (Block *et al.* 1955) and heating at 100°C for 2 min. Hexoses give brown spots and pentoses pink spots.

To facilitate identification, 12 μl of a reference mixture of sugars should be run on each chromatogram. The mixture is prepared by dissolving each of the following at a concentration of 10 m*M* in 10% (v/v) isopropanol: arabinose, fructose, galactose, glucosamine, glucose, mannose, rhamnose, ribose and xylose. The sugar 6-deoxytalose runs further than any of these (i.e. just beyond rhamnose), and positive identification may be obtained by running a reference hydrolysate of a strain that contains this sugar (Keddie & Cure 1977).

Thin layer chromatography

If hydrolysates are obtained from dried, weighed material, spots containing 10–15 μl should yield satisfactory chromatograms. If unweighed, wet material is used, the volume of hydrolysate applied to chromatograms should be gauged as described above for paper chromatography.

Detection of lysine, ornithine, and DAB

The method for detecting lysine, ornithine, and DAB is applicable only to purified cell-wall material. One-dimensional chromatograms on glass plates precoated with 0.25-mm silica gel 60 (no. 5721; Merck) and activated at 100°C for 1 h are developed by continuous elution for 5–6 h at about 20°C in the lower phase of the biphasic solvent system chloroform–methanol–ammonia (specific gravity 0.88)–water (8:8:1:3 v/v; Randerath 1963). Sometimes this solvent mixture does not separate into two phases, in which case it should be discarded and a fresh batch prepared. Adding the chloroform first and the methanol last usually results in biphasis. After drying, chromatograms are sprayed with 0.3% (w/v) ninhydrin in 96% (v/v) *n*-butanol and heated at 100°C for a few minutes to locate the amino-acid spots (Block *et al.* 1955) (Fig. 5). A reference spot containing 2 μl each of lysine, ornithine, and DAB [all 1% (w/v) in 10% (v/v) isopropanol] should be run on each chromatogram.

Detection of DAP isomers

The method for detecting DAP isomers is applicable both to whole-cell hydrolysates and to hydrolysates of purified cell-wall material. One-dimen-

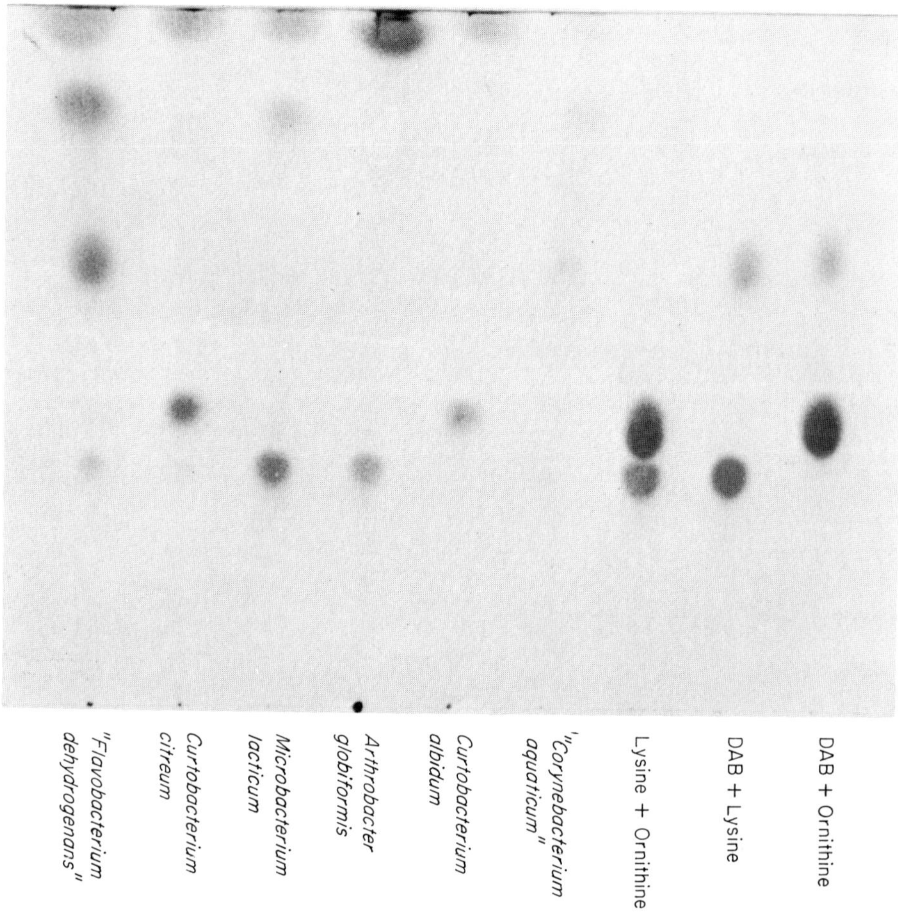

FIG. 5. Detection of lysine, ornithine and DAB, using TLC.

sional chromatograms on aluminum plates precoated with 0.10-mm cellulose (no. 5552; Merck) and activated at 100°C for 1 h are developed for 3–4 h in the solvent system of Rhuland *et al.* (1955) described above. After drying, chromatograms are sprayed with ninhydrin in *n*-butanol and heated (see above) to locate the amino-acid spots (Fig. 6). A reference spot containing 2 μl of 1% (w/v) commercially available DAP should be run on each chromatogram.

Detection of arabinose

The method of detecting arabinose is applicable both to whole-cell hydrolysates and to hydrolysates of purified cell-wall material. One-dimensional

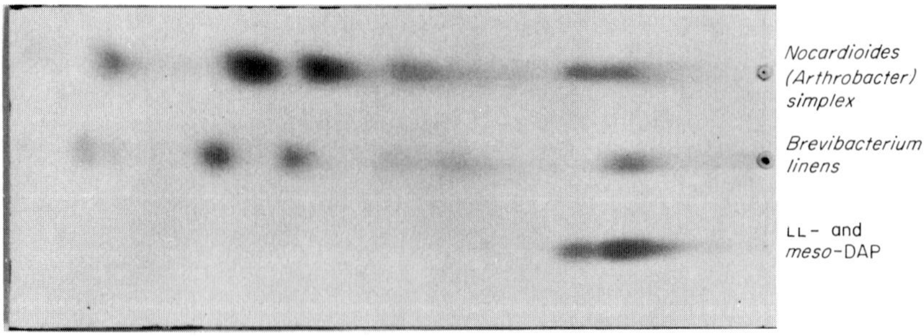

FIG. 6. Detection of DAP isomers, using TLC.

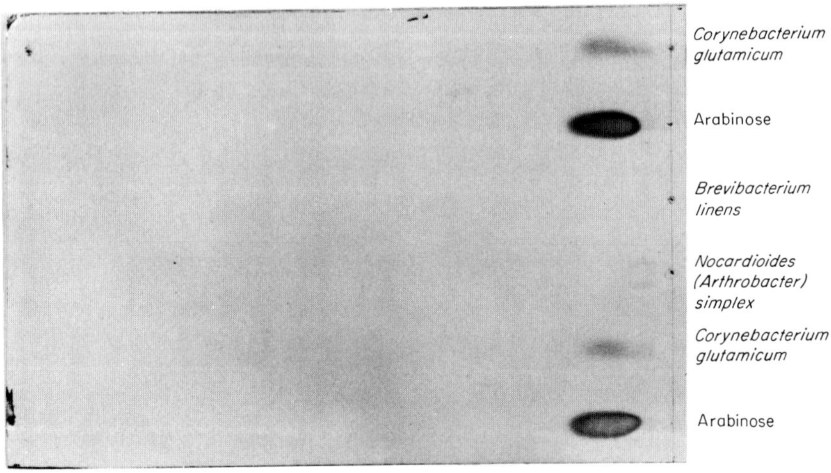

FIG. 7. Detection of arabinose, using TLC.

chromatograms on activated cellulose plates (see above) are developed twice in the solvent system ethyl acetate–pyridine–water (8:2:1, v/v; Block *et al.* 1955). Plates are dried between runs, each of which takes about 45 min. Spots are located by spraying with the *p*-anisidine–phthalic acid reagent (freshly prepared) described above (Fig. 7). A reference spot containing 2 μl of 1% (w/v) arabinose in 10% (v/v) isopropanol should be run on each chromatogram.

References

BECKER, B., LECHEVALIER, M. P., GORDON, R. E. & LECHEVALIER, H. A. 1964 Rapid differentiation between *Nocardia* and *Streptomyces* by paper chromatography of whole-cell hydrolysates. *Applied Microbiology* **12**, 421–423.

BLOCK, R. J., DURRUM, G. & ZWEIG, G. 1955 *A Manual of Paper Chromatography and Paper Electrophoresis*. New York: Academic Press.

BOONE, C. J. & PINE, L. 1968 Rapid method for characterisation of actinomyces by cell wall composition. *Applied Microbiology* **16**, 279–284.

BOUSFIELD, I. J., SMITH, G. L., DANDO, T. R. & HOBBS, G. 1983 Numerical analysis of total fatty acid profiles in the identification of coryneform, nocardioform and some other bacteria. *Journal of General Microbiology* **129**, 375–394.

COLLINS, M. D. 1982 Lipid composition of *Renibacterium salmoninarum* (Sanders and Fryer). *FEMS Microbiology Letters* **13**, 295–297.

COLLINS, M. D. & JONES, D. 1981 Distribution of isoprenoid quinone structural types in bacteria and their taxonomic implications. *Microbiological Reviews* **45**, 316–354.

COLLINS, M. D., JONES, D., KEDDIE, R. M., KROPPENSTEDT, R. M. & SCHLEIFER, K. H. 1983a Classification of some coryneform bacteria in a new genus *Aureobacterium*. *Systematic and Applied Microbiology* **4**, 236–252.

COLLINS, M. D., JONES, D. & KROPPENSTEDT, R. M. 1983b Reclassification of *Brevibacterium imperiale* (Steinhaus) and "*Corynebacterium laevaniformans*" (Dias and Bhat) in a redefined genus *Microbacterium* (Orla-Jensen), as *Microbacterium imperiale* comb. nov. and *Microbacterium laevaniformans* nom. rev.; comb. nov. *Systematic and Applied Microbiology* **4**, 65–78.

COLLINS, M. D., KEDDIE, R. M. & KROPPENSTEDT, R. M. 1983c Lipid composition of *Arthrobacter simplex*, *Arthrobacter tumescens* and possibly related taxa. *Systematic and Applied Microbiology* **4**, 18–26.

CUMMINS, C. S. & JOHNSON, J. L. 1971 Taxonomy of the clostridia: wall composition and DNA homologies in *Clostridium butyricum* and other butyric acid producing clostridia. *Journal of General Microbiology* **67**, 33–46.

DÖPFER, H. E., STACKEBRANDT, E. & FIEDLER, F. 1982 Nucleic acid hybridization studies on *Microbacterium*, *Curtobacterium*, *Agromyces* and related taxa. *Journal of General Microbiology* **128**, 1697–1708.

DYE, D. W. & KEMP, W. J. 1977 A taxonomic study of plant pathogenic *Corynebacterium* species. *New Zealand Journal of Agricultural Research* **20**, 563–582.

KEDDIE, R. M. & BOUSFIELD, I. J. 1980 Cell wall composition in the classification and identification of coryneform bacteria. In *Microbiological Classification and Identification* eds. Goodfellow, M. & Board, R. G., pp. 167–187. London: Academic Press.

KEDDIE, R. M. & CURE, G. L. 1977 The cell wall composition and distribution of free mycolic acids in named strains of coryneform bacteria and in isolates from various natural sources. *Journal of Applied Bacteriology* **42**, 229–252.

KEDDIE, R. M. & CURE, G. L. 1978 Cell wall composition of coryneform bacteria. In *Coryneform Bacteria* eds. Bousfield, I. J. & Callely, A. G., pp. 47–83. London: Academic Press.

KEDDIE, R. M. & JONES, D. 1981 Saprophytic, aerobic coryneform bacteria. In *The Prokaryotes* eds. Starr, M. P., Stolp, H., Trüper, H. G., Balows, M. & Schlegel, H. G., pp. 1838–1878. Berlin, Heidelberg, New York: Springer-Verlag.

KEDDIE, R. M., LEASK, B. G. S. & GRAINGER, J. M. 1966 A comparison of coryneform bacteria from soil and herbage: cell wall composition and nutrition. *Journal of Applied Bacteriology* **29**, 17–43.

LECHEVALIER, M. P. 1968 Identification of aerobic actinomycetes of clinical importance. *Journal of Laboratory and Clinical Medicine* **71**, 934–944.

LECHEVALIER, H. A. & LECHEVALIER, M. P. 1982 Actinomycete genera "in search of a family". *Oerskovia*. In *The Prokaryotes* eds. Starr, M. P., Stolp, H., Trüper, H. G., Balows, M. & Schlegel, H. G., pp. 2121–2122. Berlin, Heidelberg, New York: Springer-Verlag.

MINNIKIN, D. E., GOODFELLOW, M. & COLLINS, M. D. 1978 Lipid composition in the classification and identification of coryneform bacteria and related taxa. In *Coryneform Bacteria* eds. Bousfield, I. J. & Callely, A. G., pp. 85–160. London: Academic Press.

MURRAY, I. G. & PROCTOR, A. G. J. 1965 Paper chromatography as an aid to the identification of *Nocardia* species. *Journal of General Microbiology* **41**, 163–167.

PITCHER, D. G. 1983 Deoxyribonucleic acid base composition of *Corynebacterium diphtheriae* and other corynebacteria with cell wall type IV. *FEMS Microbiology Letters* **16**, 291–295.

O'DONNELL, A. G., GOODFELLOW, M. & MINNIKIN, D. E. 1982 Lipids in the classification of *Nocardioides:* reclassification of *Arthrobacter simplex* (Jensen) Lochhead in the genus *Nocardioides* (Prauser) emend. O'Donnell *et al.* as *Nocardioides simplex* comb. nov. *Archives of Microbiology* **133**, 323–329.

RANDERATH, K. 1963 *Thin Layer Chromatography* (English Translation). London: Academic Press.

RHULAND, L. E., WORK, E., DENMAN, R. F. & HOARE, D. S. 1955 The behaviour of the isomers of αε-diaminopimelic acid on paper chromatograms. *Journal of the American Chemical Society* **77**, 4844–4846.

SANDERS, J. E. & FRYER, J. L. 1980 *Renibacterium salmoninarum* gen. nov., sp. nov., the causative agent of bacterial kidney disease in salmonid fishes. *International Journal of Systematic Bacteriology* **30**, 496–502.

SCHLEIFER, K. H. & KANDLER, O. 1972 Peptidoglycan types of bacterial cell walls and their taxonomic implications. *Bacteriological Reviews* **36**, 407–477.

SEIDL, P. H., FALLER, A. H., LOIDER, R. & SCHLEIFER, K. H. 1980 Peptidoglycan types and cytochrome patterns of strains of *Oerskovia turbata* and *Oerskovia xanthineolytica*. *Archives of Microbiology* **127**, 173–178.

STACKEBRANDT, E., SEILER, H. & SCHLEIFER, K. H. 1982 Union of the genera *Cellulomonas* Bergey *et al.* and *Oerskovia* Prauser *et al.* in a redefined genus *Cellulomonas*. *Zentralblatt für Bakteriologie, Mikrobiologie und Hygiene C, Originale 1 Abteilung* **3**, 401–409.

SUZUKI, K. & KOMAGATA, K. 1983 *Pimelobacter* gen. nov. a new genus of coryneform bacteria with LL-diaminopimelic acid in the cell wall. *Journal of General and Applied Microbiology* **29**, 59–71.

Systematic Analysis of Complex Mycobacterial Lipids

G. DOBSON, D. E. MINNIKIN, S. M. MINNIKIN,
J. H. PARLETT, AND M. GOODFELLOW

Departments of Microbiology and Organic Chemistry, The University of Newcastle upon Tyne, Newcastle upon Tyne, UK

M. RIDELL

Department of Medical Microbiology, University of Gothenburg, Gothenburg, Sweden

M. MAGNUSSON

Tuberculin Department, Statens Seruminstitut, Copenhagen, Denmark

Mycobacterial cell envelopes contain high proportions of lipids whose diverse structures have obvious chemosystematic potential (Asselineau 1966; Asselineau & Asselineau 1978a; Minnikin & Goodfellow 1980; Minnikin 1982). Systematic studies have been hampered by the wide range of different structural types encountered, ranging in polarity from very non-polar waxes to highly polar glycophospholipids. Consequently it is very difficult to devise systems for the comprehensive comparative analysis of the full range of mycobacterial lipids.

Lipids can be broadly divided into "free" lipids, which can be released simply by soaking in appropriate organic solvents, and "bound" lipids, whose release requires the use of chemical agents capable of cleaving covalent bonds. In pioneering studies, Anderson and co-workers (1941) used a system of sequential solvent extraction to fractionate mycobacterial lipids into groups for further study; the scheme used (Asselineau 1966) is summarised in Fig. 1. In the absence of modern analytical and chromatographic techniques, this procedure provided a quite reliable means for the reproducible preliminary separation of mycobacterial lipids and allowed the great

CHEMICAL METHODS IN
BACTERIAL SYSTEMATICS

0-12-289675-0

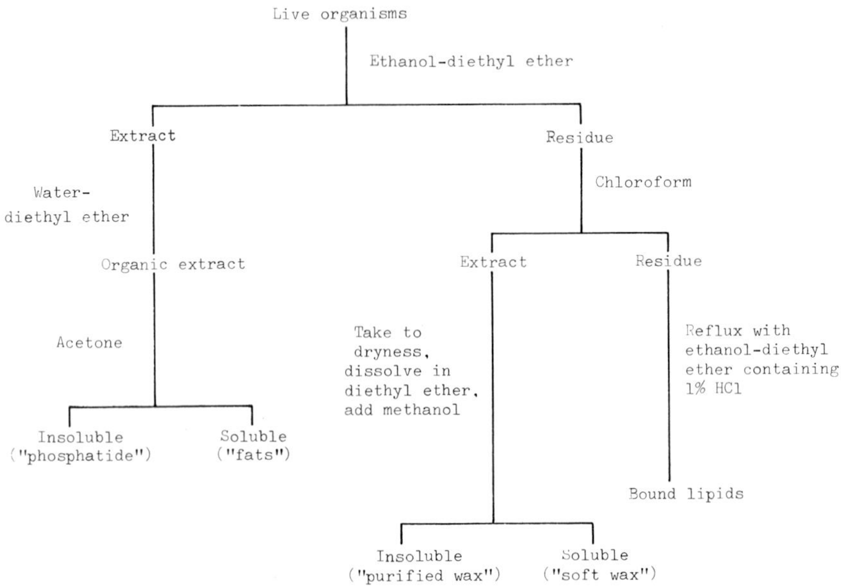

FIG. 1. Extraction of mycobacterial lipids by differential solvent extraction.

diversity of these lipids to be recognized (Asselineau 1966). In early studies, different classes of long-chain compounds were obtained by mild or vigorous alkaline hydrolysis (Chanley & Polgar 1950) and the use of column chromatography gave relatively homogeneous lipid fractions (Asselineau 1966).

The introduction of the new spectroscopic and chromatographic techniques since the early 1960s has revolutionized the approaches to the analysis of these relatively intractable natural products. Thin-layer, gas and high-performance liquid chromatographic techniques now allow the preparation of reproducible lipid profiles and the rapid small-scale isolation of individual components. Mass spectrometry and nuclear magnetic resonance spectroscopy facilitate the precise determination of molecular weights and the characterization of functional groups. In spite of these advances, however, problems have remained in the development of rapid small-scale procedures for the comparative analysis of the full polarity range of mycobacterial lipids. This article describes extraction procedures and chromatographic systems that, in principle, allow profiles of all the known mycobacterial diagnostic lipids to be recorded in a semiquantitative manner.

Extraction of Free Lipids

The extraction procedure employed is essentially that developed for the isolation of isoprenoid quinones and polar lipids from single small samples of

BIOMASS (50 mg, in Corning 99449-13 8.5 ml P.T.F.E. capped tube).

- Methanol—0.3% aqueous NaCl (100:10 v/v) (2 ml) and petroleum ether (1 ml).
- Mix on tube rotator (15 min), transfer upper layer to small vial.
- Add petroleum ether (1 ml), mix (15 min).
- Dry down combined upper layers under N_2 (<37°C). → NON-POLAR LIPIDS

RESIDUE

- Heat in boiling water bath (5 min).
- Cool in water bath at 37°C (5 min).
- Chloroform-methanol-0.3% aqueous NaCl (90 : 100 : 30 v/v) (2.3 ml).
- Mix (60 min), centrifuge, transfer supernatant to 8.5 ml tube.
- Chloroform-methanol-0.3% aqueous NaCl (50 : 100 : 40 v/v) (0.75 ml).
- Mix (30 min), centrifuge, combine supernatants.
- Chloroform-methanol-0.3% aqueous NaCl (50 : 100 : 40 v/v) (0.75 ml).
- Mix (30 min), centrifuge, combine supernatants. → DEFATTED CELLS
- Add chloroform (1.3 ml) and 0.3% aqueous NaCl (1.3 ml) to supernatants.
- Mix (5 min), centrifuge, discard upper layer.
- Dry down lower layer under N_2 (<37°C).

POLAR LIPIDS

FIG. 2. Extraction of mycobacterial free lipids.

bacterial biomass (Minnikin *et al.* 1984a; see Chapter 7, this volume). The method (Fig. 2) involves treatment of organisms with a biphasic mixture of methanolic saline and petroleum ether (bp 60–80°C) followed by extraction of the cells and residual aqueous phase with monophasic chloroform–methanol–saline according to the well-proven Bligh & Dyer (1959) procedure. The petroleum-ether extract from the first stage is composed of readily extractable nonpolar lipids and addition of chloroform and saline to the second stage extract gives the remaining more polar lipids.

Analysis of Free Non-Polar Lipids

A total of five specially developed two-dimensional thin-layer chromatographic (TLC) systems are necessary to cover the polarity range of both non-polar and polar mycobacterial lipids (Fig. 3). The least polar system is exemplified in Fig. 4, and it is used (Minnikin *et al.* 1983a) to detect the presence of triacylglycerols (**I**), menaquinones (**II**) and the waxes composed of mycocerosic acids (**III**) diesterified to members of the phthiocerol family (**IV–VI**). Further structural information on the composition of triacylglycerols can be obtained by degradation to component fatty acids (Walker *et al.* 1970), but such information remains to be exploited systemat-

I 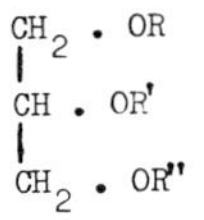

R, R,'' R''' = long chain acyl

II 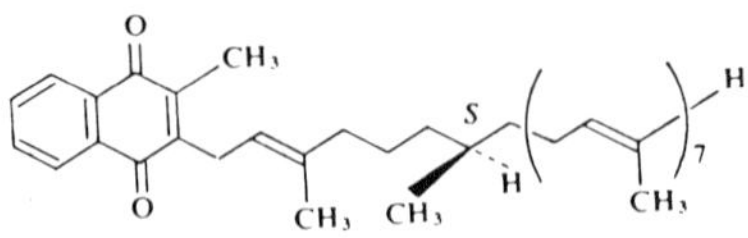

III 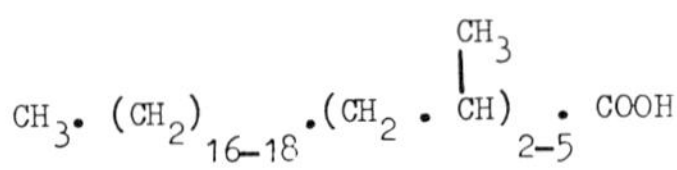

IV
V
VI
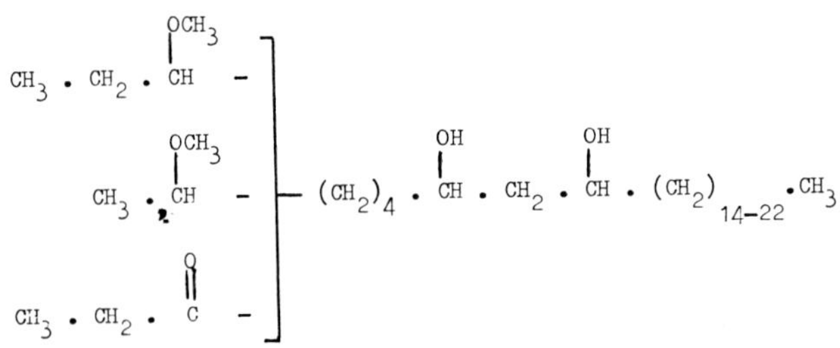

VII 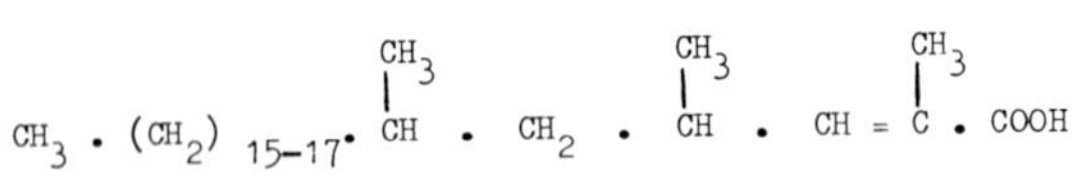

ically. Menaquinones can be isolated and analysed by mass spectrometry (Collins *et al.* 1977), and reverse-phase and silver-ion TLC and high-performance liquid chromatography (see Chapters 9 and 13). Mycobacteria examined so far contain dihydrogenated menaquinones with nine isoprene units (Minnikin & Goodfellow 1980; Minnikin 1982; Collins *et al.* 1984).

Mycocerosates (**III**) of the phthiocerol family (**IV–VI**) are proving to be powerful characters for the classification and identification of a limited number of species of *Mycobacterium*. These waxes may be efficiently degraded by reductive fission and the resulting alcohols converted to *tert*-

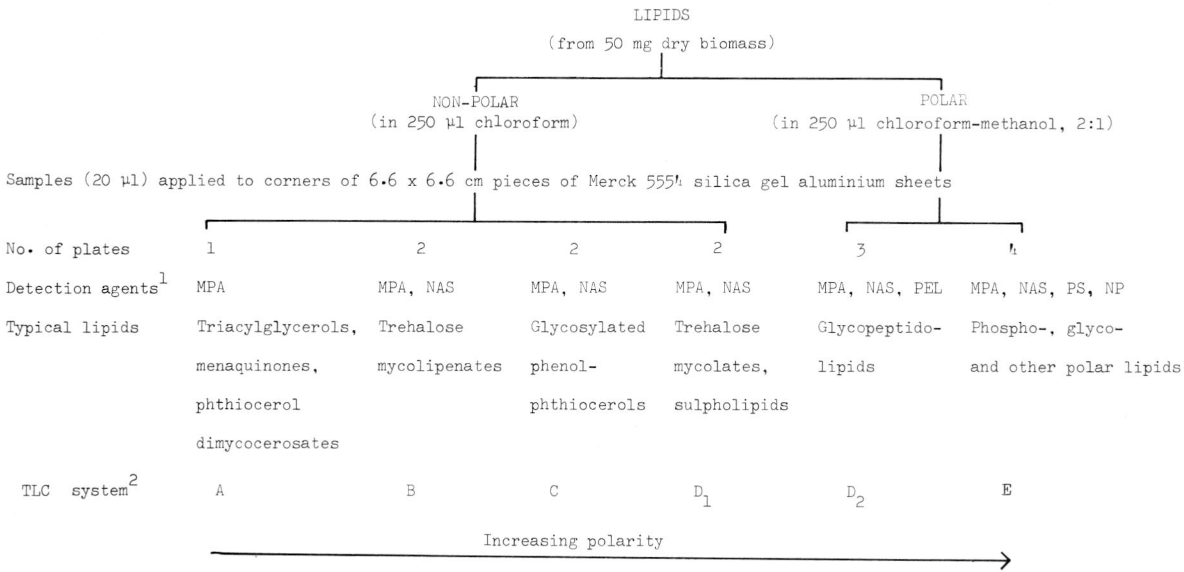

FIG. 3. Scheme for TLC analysis of full polarity range of mycobacterial lipids. [1]Detection spray reagents: MPA, 5% ethanolic molybdophosphoric acid followed by heating at 180°C for 15 min for all lipids; NAS, α-naphthol-sulphuric acid (Jacin & Mishkin 1965) followed by heating at 120°C for sugars; PEL, for peptidolipids the plate is exposed to chlorine gas in a tank for 20 min and blown-in air for 10 min and then sprayed with an aqueous solution of starch (1%) and potassium iodide (1%) (Stahl 1969); PS, periodate–Schiff spray for α-glycols according to Shaw (1968); NP, 0.2% ninhydrin in water-saturated 1-butanol for amino groups (Consden & Gordon 1948), with heating at 110°C, followed by lipid phosphate reagent of Dittmer & Lester (1964). [2]Details of developing solvent systems can be found as follows: A, Fig. 4; B, Fig. 5; C, Fig. 6; D_1 and D_2, Figs. 7 and 8; E, Fig. 10.

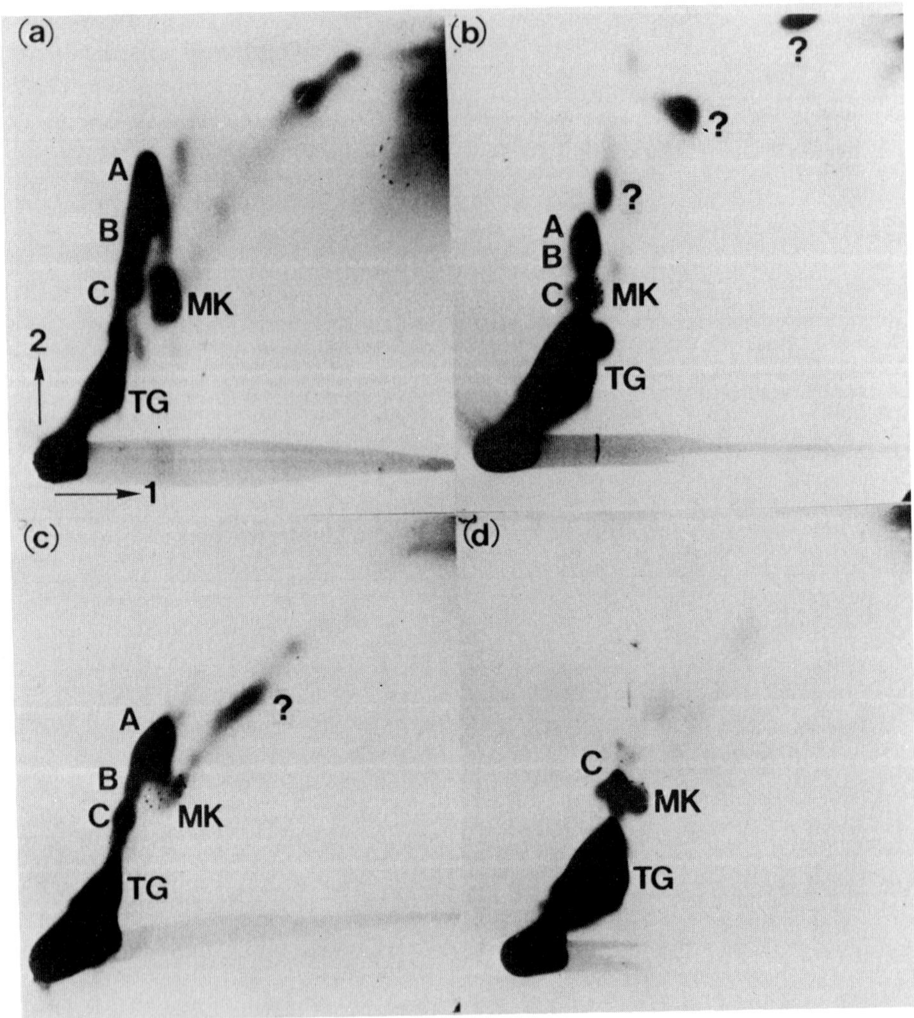

FIG. 4. Two-dimensional TLC of least polar class of lipids in mycobacterial non-polar lipid fractions (Fig. 3, TLC system A). First direction, petroleum-ether (bp 60–80°C): ethyl acetate (98:2, v/v, thrice); second direction, petroleum-ether (bp 60–80°C): acetone (98:2, v/v, once). Organisms: (a) *Mycobacterium tuberculosis* MNC 3, (b) *Mycobacterium bovis* MNC 27, (c) *Mycobacterium marinum* MNC 170, (d) *Mycobacterium kansasii* ATCC 12478; MNC numbers refer to collection of M. Magnusson (see Minnikin *et al.* 1984c). Cultivation: (a)–(c) Sauton's medium (Minnikin *et al.* 1984c), (d) 37°C on 7H 11 medium. Abbreviations: A–C, dimycocerosates (**III**) of the phthiocerol family (**IV–VI**); MK, menaquinone; TG, triacylglycerol; ?; unknown. Detection: molybdophosphoric acid (MPA, Fig. 3) for all lipids; MK also detected by 254-nm ultraviolet light.

TABLE 1

Distribution of members of the mycocerosic acid and phthiocerol families in mycobacteria

Mycocerosic acid (III)	Taxon	Phthiocerol family (IV–VI)
C_{27}–C_{32}	*M. bovis*[a]	C_{34}, C_{36}
	M. microti[a]	
	M. tuberculosis[b]	
	M. kansasii[a,c]	C_{27}
C_{30}–C_{34}	*M. leprae*[d]	C_{30}, C_{32}
C_{27}–C_{30}	*M. marinum*[a]	C_{28}, C_{30}

[a]G. Dobson, D. E. Minnikin, P. Draper, M. Magnusson and M. Ridell, unpublished results.
[b]Minnikin *et al.* 1983a; Mallet *et al.* 1984.
[c]Dimycocerosates of phthiodiolone A (**VI**) only detected in *M. kansasii*, others have dimycocerosates of all members (**IV–VI**) of the phthiocerol family.
[d]Draper *et al.* 1983.

butyldimethylsilyl (TBDMS) ethers. Simple gas chromatographic analysis (Draper *et al.* 1983) provides information on the general size of the mycocerosates and phthiocerols but combined gas chromatography–mass spectrometry is a sensitive method for the detailed analysis of these long-chain compounds (Mallet *et al.* 1984). The distribution of the various structural types of mycocerosates and phthiocerols is summarised in Table 1; dimycocerosates of the phthiocerol family have also been detected by TLC in representatives of *Mycobacterium ulcerans* (D. E. Minnikin, F. Portaels, J. H. Parlett, unpublished results).

The second least polar TLC system (Fig. 5) is used for the identification of glycolipids found only so far in petroleum-ether extracts (Figs. 2 and 3) of *M. tuberculosis*. These glycolipids are composed of trehalose esterified with undetermined numbers of mycolipenic (**VII**) or hydroxymycolipenic (**VIII**) acids (D. E. Minnikin and G. Dobson, unpublished results). In addition to substantial amounts of unesterified fatty acids, a number of other unidentified lipids are detected, using this TLC system.

A slightly more polar TLC system (Fig. 6) enables glycosides of acylated phenolphthiocerols to be identified in petroleum-ether extracts (Figs. 2 and 3) of a limited number of mycobacteria. These lipids are structurally related to the dimycocerosates of phthiocerol, having mycocerosic acids (**III**) diesterified to a phenolphthiocerol (**IX**) backbone. Only four structural types of these glycolipids have been characterized. The simplest types, found in extracts of *Mycobacterium bovis* and *Mycobacterium marinum*, have 2-*O*-methyl-D-rhamnose and 3-*O*-methyl-L-rhamnose, respectively, glycosidically linked to the phenol part of the phenolphthiocerol (**IX**) (Minnikin 1982).

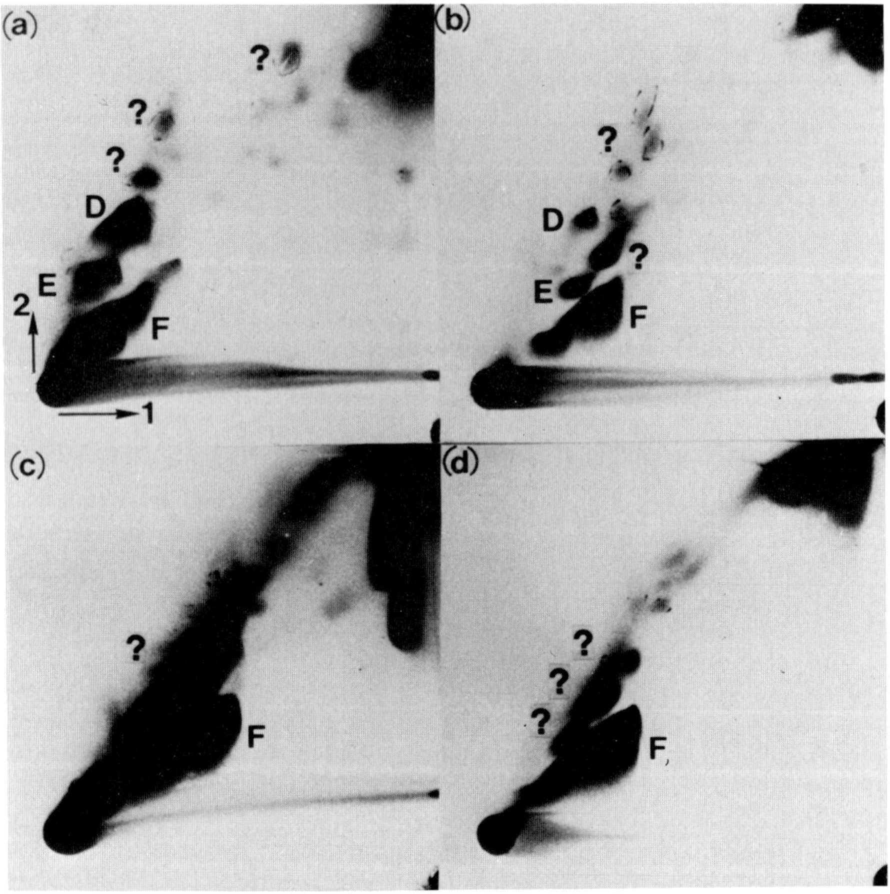

FIG. 5. Two-dimensional TLC of second least polar class of lipids in mycobacterial non-polar lipid fractions (Fig. 3, TLC system B). First direction, petroleum-ether (bp 60–80°C)–acetone (92:8, v/v, thrice); second direction, toluene–acetone (95:5, v/v, once). Organisms: (a) *Mycobacterium tuberculosis* C (Central Veterinary Laboratory, Weybridge, UK.; see Minnikin *et al.* 1980), (b) *Mycobacterium tuberculosis* MNC 3, (c) *Mycobacterium bovis* MNC 27, (d) *Mycobacterium kansasii* ATCC 12478; for source and cultivation of (b)–(d) see Fig. 4. Abbreviations: D and E, mycolipenates (**VII, VIII**) of trehalose; F, free unesterified fatty acids; ?, unknown. Detection: molybdophosphoric acid (MPA, Fig. 3) for all lipids; D and E also detected by α-naphthol-sulphuric acid (NAS, Fig. 3).

Trisaccharides, composed of 2-*O*-methyl-fucose, 2-*O*-methyl-rhamnose and 2,4-di-*O*-methyl-rhamnose and 3-*O*-methyl-rhamnose, 2,3-di-*O*-methyl-rhamnose, and 3,6-di-*O*-methyl-glucose are glycosidically linked to phenolphthiocerol (**IX**) in the glycolipids from *Mycobacterium kansasii* and *Mycobacterium leprae*, respectively (Hunter & Brennan 1981; Hunter *et al.* 1982).

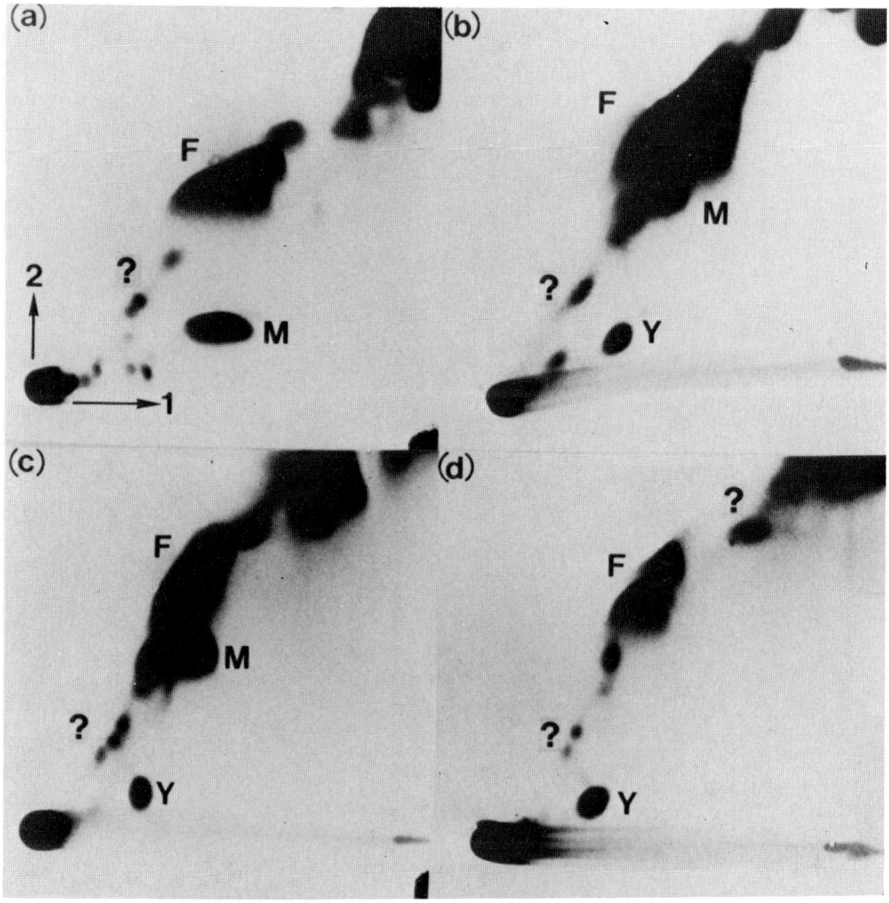

FIG. 6. Two-dimensional TLC of second most polar class of lipids in mycobacterial non-polar lipid fractions (Fig. 3, TLC system C). First direction, chloroform–methanol (96:4, v/v); second direction, toluene–acetone (80:20, v/v). Organisms: (a) *Mycobacterium kansasii* ATCC 12478, (b) *Mycobacterium marinum* MNC 170, (c) *Mycobacterium bovis* MNC 27; (d) *Mycobacterium tuberculosis* 3; see Fig. 4 for source and cultivation. Abbreviations: M, glycosides of dimycocerosates (**III**) of phenolphthiocerol (**IX**); F, free fatty acids; ? and Y, unknowns. Detection: molybdophosphoric acid (MPA, Fig. 3) for all lipids; components labeled M were the only ones giving a positive reaction with α-naphthol-sulphuric acid (NAS, Fig. 3); unknown lipid Y gave no reaction with any of the specific spray reagents described in Fig. 3.

As can be seen in Fig. 6, the glycolipids from *M. bovis* and *M. marinum* are co-chromatographed very closely together, as would be expected from the very small structural difference between these compounds. These lipids can, however, be clearly distinguished by spraying the chromatograms with α-naphthol-sulphuric acid. The glycolipid from *M. bovis* is initially stained a

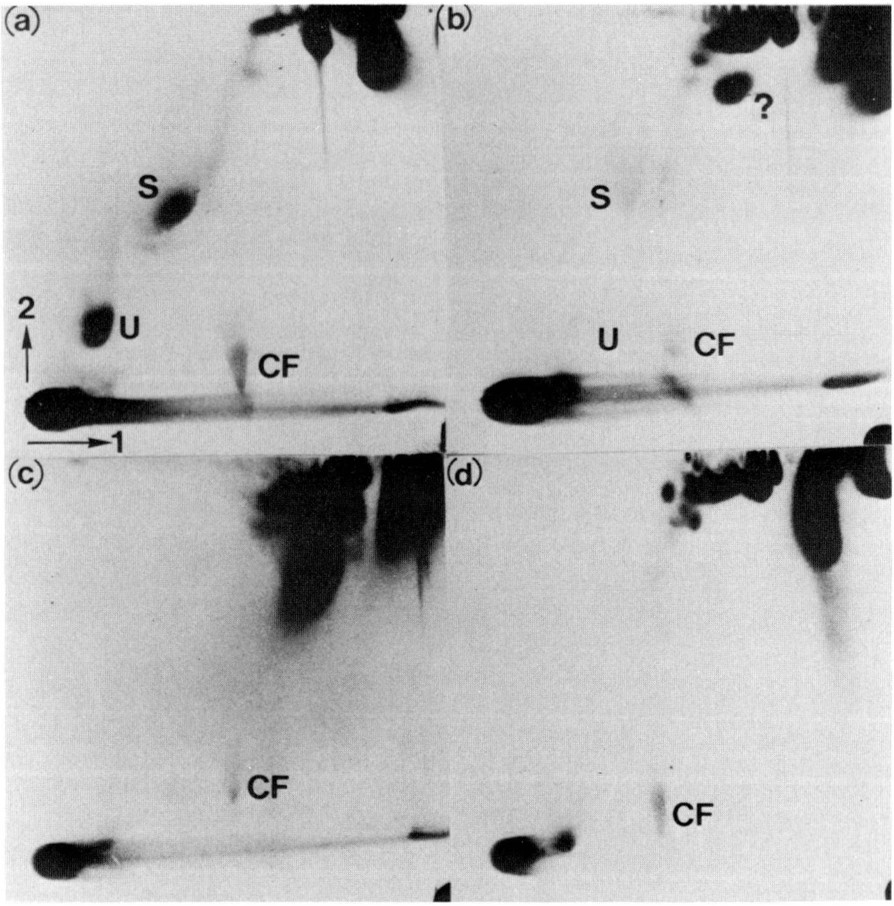

FIG. 7. Two-dimensional TLC of most polar class of lipids in mycobacterial non-polar lipid fractions (Fig. 3, TLC system D). First direction, chloroform–methanol–water (100:14:0.8, v/v); second direction, chloroform–acetone–methanol–water (50:60:2.5:3, v/v). Organisms: (a) *Mycobacterium tuberculosis* H37 Rv (Sauton's medium), (b) *Mycobacterium tuberculosis* MNC 3, (c) *Mycobacterium bovis* MNC 27, (d) *Mycobacterium kansasii* ATCC 12478; see Fig. 4 for source and cultivation of (b)–(d). Abbreviations: CF, trehalose dimycolates; S, sulphoglycolipids; U, uncharacterized glycolipid (see Fig. 8) ?, unknown. Detection: molybdophosphoric acid (MPA, Fig. 3) for all lipids; components CF, S and U also react positively with α-naphthol-sulphuric acid (NAS, Fig. 3).

brown colour, which changes at room temperature during a period of up to 16 h to orange; for the glycolipid from *M. marinum*, the opposite colour change from initial orange to final brown takes place. It has also been shown that *Mycobacterium microti* produces a lipid that is co-chromatographed with and stained in the same way as the glycolipid from *M. bovis* (G. Dobson, D. E.

VIII

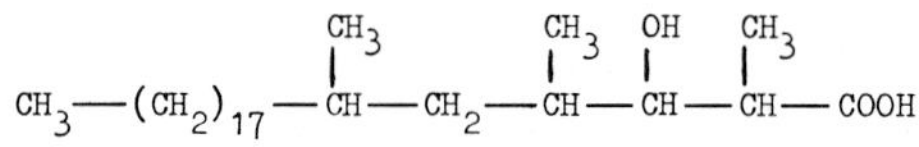

IX

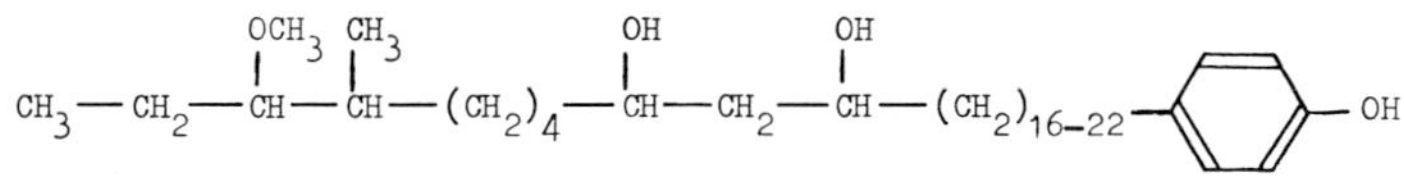

X

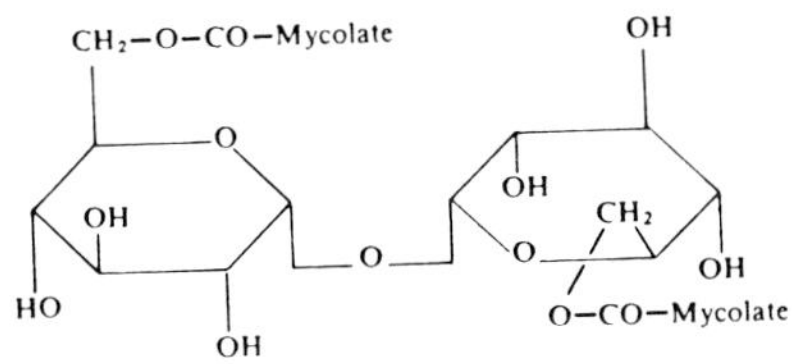

XI

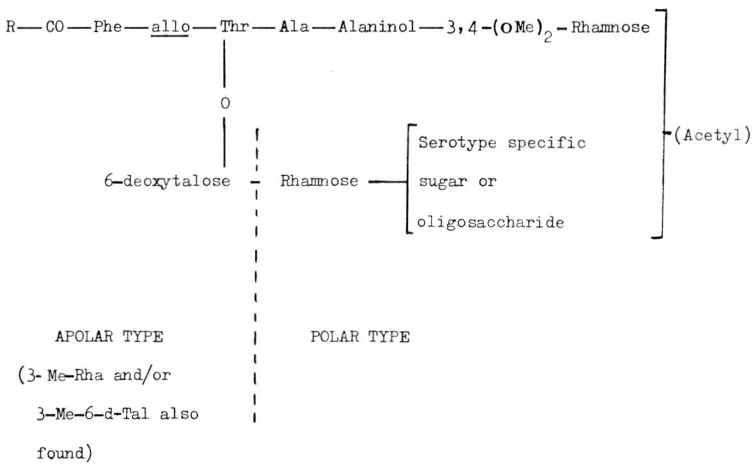

Minnikin and P. Draper, unpublished results). Glycolipids of this type were absent from *M. tuberculosis* (Fig. 6). Minor variants of these glycolipids, esterified with mycolic acids, have been described by Gastambide-Odier (1973a,b) but with the present analytical systems they were not detected. Large amounts of free fatty acids are detected in this TLC system (Fig. 6).

The most polar components of petroleum ether extracts of mycobacteria (Figs. 2 and 3) are analysed using the system shown in Fig. 7, the most characteristic components being trehalose dimycolates (**X**) and sulphoglycolipids. The former lipids (**X**) are the so-called "cord-factors" and they have been found in most mycobacteria examined so far (Asselineau & Asselineau 1978b; Minnikin & Goodfellow 1980; Minnikin 1982). Sulpholipids have only been characterized in detail from *M. tuberculosis* (Goren & Brennan 1979; Minnikin 1982), though it has been suggested that they are much more widespread (Prabhudesai *et al.* 1981).

The same TLC system is also used to analyse the least polar constituents of the chloroform–methanol–water extract (Figs. 2 and 3) from mycobacteria (Fig. 8). The most characteristic lipids found in this fraction are the glycopeptidolipids (**XI**) whose structures have been mainly studied from representatives of the *Mycobacterium avium–Mycobacterium intracellulare–Mycobacterium scrofulaceum* complex (Goren & Brennan 1979; Brennan 1981; Brennan *et al.* 1981a,b). These glycolipids are the so-called "C-mycosides" (Asselineau & Asselineau 1978a), and they occur as families of both non-polar and polar types (**XI**) (Brennan 1981). The sugar components of the non-polar lipids are limited to 3-*O*-methyl- or 3,4-di-*O*-methyl-rhamnose on the alaninol and 6-deoxytalose or 3-*O*-methyl-6-deoxytalose linked to *allo*-threonine (**XI**). The polar variants also have alaninol-linked 3,4-di-*O*-methyl-rhamnose but characteristic oligosaccharides are linked via rhamnose to the 6-deoxytalose on the *allo*-threonine, these lipids being the serotype-specific Schaefer antigens (Brennan 1981; Brennan *et al.* 1981a,b; 1982).

These glycopeptidolipids also have acetyl substituents so that the TLC analysis of the native extract reveals the presence of complex mixtures as exemplified for *M. avium* in Fig. 8. The TLC patterns may be simplified, without significant loss of diagnostic value, by brief alkali treatment of the lipid extracts to remove the acetyl substituents (Brennan *et al.* 1978, 1982). For the *M. avium–M. intracellulare–M. scrofulaceum* complex it was shown that the same six non-polar components were identified in all strains by single-dimensional TLC (Brennan *et al.* 1978). The deacetylated polar glycopeptidolipids are restricted to usually one or two characteristic components and single-dimensional TLC systems have been selected for their analysis (Brennan *et al.* 1978, 1982). The analysis of these deacetylated glycopeptidolipids with one- and two-dimensional TLC systems is shown in Fig. 9. It has been shown (Brennan *et al.* 1978, 1982) that practically all of the serotypes in the *M. avium–M. intracellulare–M. scrofulaceum* complex may be identified by TLC analysis of these deacetylated glycopeptidolipids. It is also valuable, however, to be able to detect the presence or absence of glycopeptidolipids by two-dimensional TLC of the intact lipid extracts (Fig. 8).

An unknown lipid (Fig. 8) having chromatographic behaviour similar to that of the nondeacetylated glycopeptidolipids is found in extracts of H37Rv (Fig. 8) and some other strains of *M. tuberculosis*. This lipid has the staining properties of a glycopeptidolipid but since it also absorbs strongly under shortwave (254 nm) ultraviolet light, its structure must be different, in-

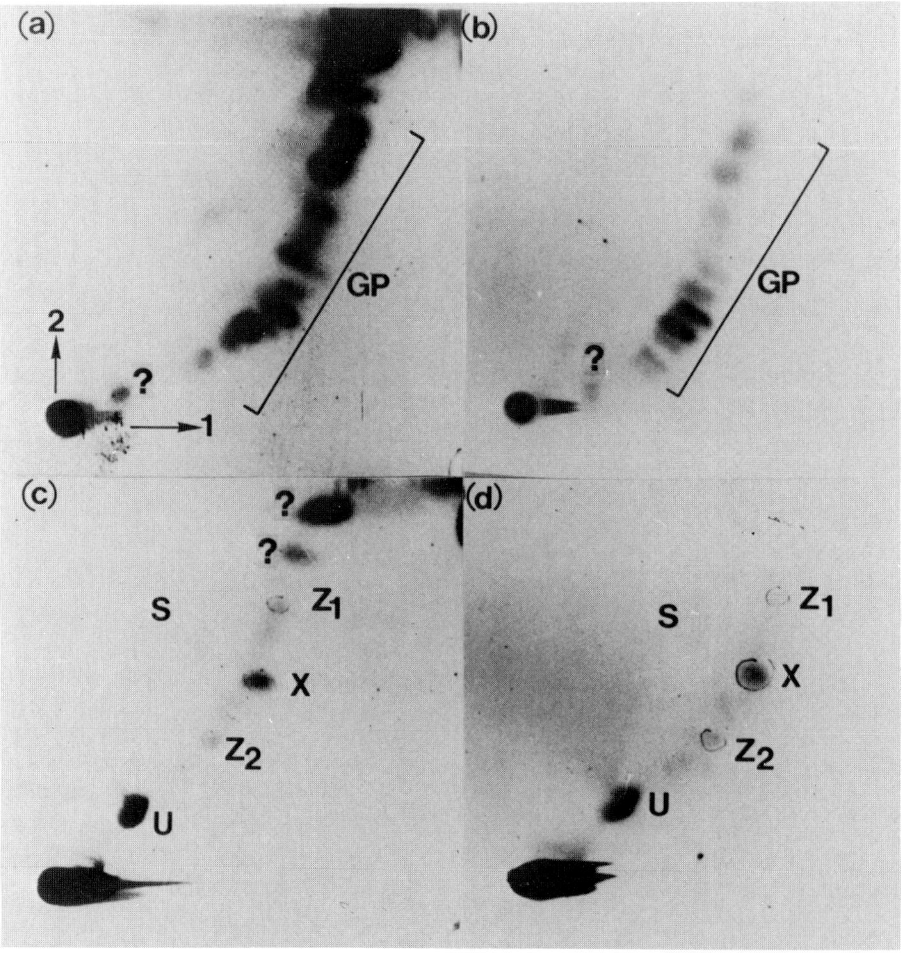

FIG. 8. Two-dimensional TLC of least polar class of lipids in mycobacterial polar lipid fractions (Fig. 3, TLC system D_2). Developing solvents as for TLC system D_1 (Figs. 3 and 7). Organisms: (a) and (b) *Mycobacterium avium* 1403 (Central Veterinary Laboratory, Weybridge, UK.; see Minnikin *et al.* 1980); (c) and (d) *Mycobacterium tuberculosis* H37Rv (see Fig. 7). Abbreviations: GP, glycopeptidolipids; U, uncharacterized glycolipid (see Fig. 7); S, traces of sulphoglycolipids; X, ultraviolet-absorbing possible glycopeptidolipid; Z_1 and Z_2, ultraviolet-absorbing minor glycolipids; ?, unknown. Detection: (a) and (c) molybdophosphoric acid (MPA, Fig. 3) for all lipids; (b) and (d) α-naphthol-sulphuric acid (NAS, Fig. 3) for sugars; using 254-nm ultraviolet light, glycolipids GP absorbed weakly and X and Z absorbed strongly.

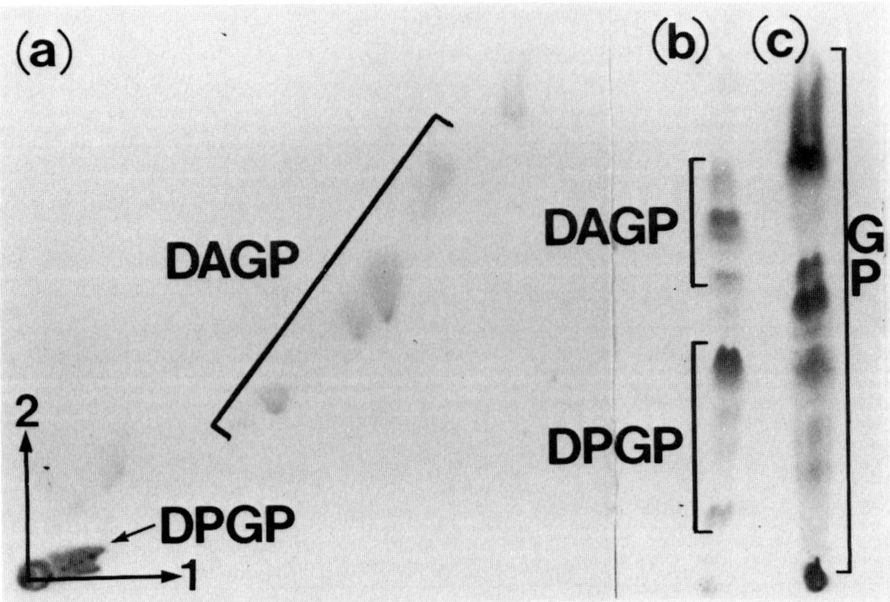

FIG. 9. TLC of glycopeptidolipids from *Mycobacterium intracellulare* ATCC 13950 cultivated on 7H 11 medium at 37°C. Patterns (a) and (b) are for lipids deacetylated according to Brennan *et al.* (1978), and (c) is the intact lipid extract. Developing solvents: (a) chloroform–methanol–water (90:12:0.4, v/v) in the first direction, followed by chloroform–acetone–methanol–water (50:60:2.5:3, v/v); (b) and (c) chloroform–methanol–water (60:16:2, v/v). Abbreviations: DAGP, deacetylated non-polar glycopeptidolipids; DPGP, deacetylated polar glycopeptidolipids; GP, glycopeptidolipids; ?, unknown. Detection: spraying with 0.1% orcinol in 40% H_2SO_4 followed by heating at 120°C for 15 min.

as much as the glycopeptidolipids absorb very weakly. As indicated in Fig. 8, small amounts of sulpholipids are also detected in the chloroform–aqueous methanol extracts of *M. tuberculosis* H37Rv in addition to their presence in the petroleum ether fraction (Fig. 7). The glycopeptidolipids from *M. avium* and related organisms are usually confined to the more polar extract; however, when very large amounts are present, they may also be detected in petroleum-ether extracts. Another unknown glycolipid (U, Figs. 7 and 8) is found in both non-polar and polar lipid extracts of *M. tuberculosis* H37Rv. Preliminary studies (D. E. Minnikin and G. Dobson, unpublished results) indicate that this lipid is a more polar version of the trehalose mycolipenates detected by the second least polar TLC system (Fig. 5).

The most polar TLC system (Fig. 10) is designed to analyse both the conventional polar lipids and some of the more characteristic types found only in mycobacteria and related actinomycetes. The developing solvents used are based on those introduced by Pangborn and McKinney (1966). The

TABLE 2

Structures of the principal mycobacterial polar lipids

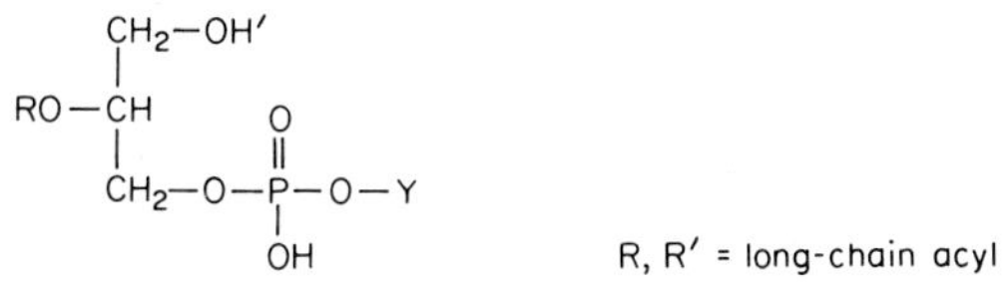

Y = ethanolamine, phosphatidylethanolamine (PE); Y = glycerol, phosphatidylglycerol (PG); Y = PG, diphosphatidylglycerol (DPG); Y = inositol, phosphatidylinositol (PI)

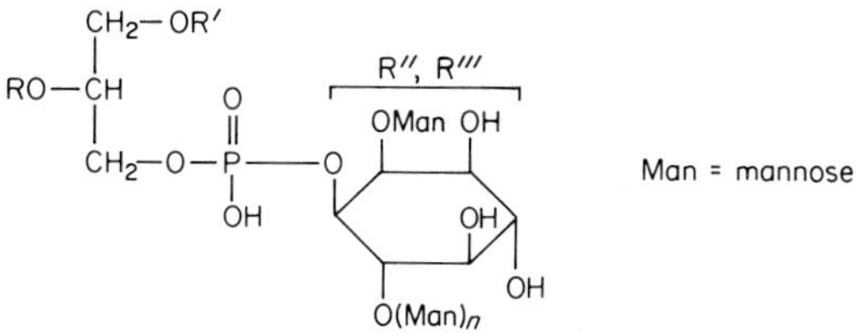

Phosphatidylinositol mannosides:
R, R′, R″ = acyl; R‴ absent; *n* = 1: monoacylphosphatidylinositol dimannoside (MPIDM)
R, R′, R″, R‴ = acyl; *n* = 1: diacylphosphatidylinositol dimannoside (DPIDM)
R, R′, R″ = acyl; R‴ absent; *n* = 4: monoacylphosphatidylinositol pentamannoside (MPIPM)
R, R′, R″, R‴ = acyl; *n* = 4: diacylphosphatidylinositol pentamannoside (DPIPM)

structures of some of the polar lipids are given in Table 2. Reliable identification of the various polar lipid types is provided by using the combination of spray reagents shown (Figs. 3 and 10). In particular, phosphatidylinositol (Table 2) gives a characteristic final brown colour with the periodate–Schiff spray (Shaw 1968) but the more polar phosphatidylinositol penta- or hexamannosides (Table 2) do not show very positively with the phospholipid spray.

The type of pattern of polar lipids recorded for *M. bovis* (Fig. 10) is typical of that present in many mycobacteria (Minnikin and Goodfellow 1980; Minnikin 1982). Certain species, however, contain additional very characteristic polar glycolipids (Brennan *et al.* 1982; Jenkins 1981) and, for example, those present in *M. kansasii* can be readily detected using the TLC system shown in Fig. 10.

Analysis of Whole-Organism Long-Chain Components

Long-chain compounds are found as components of the free lipids considered in the previous sections but mycobacteria also contain substantial amounts of covalently bound fatty acids (Minnikin 1982). The bound acids

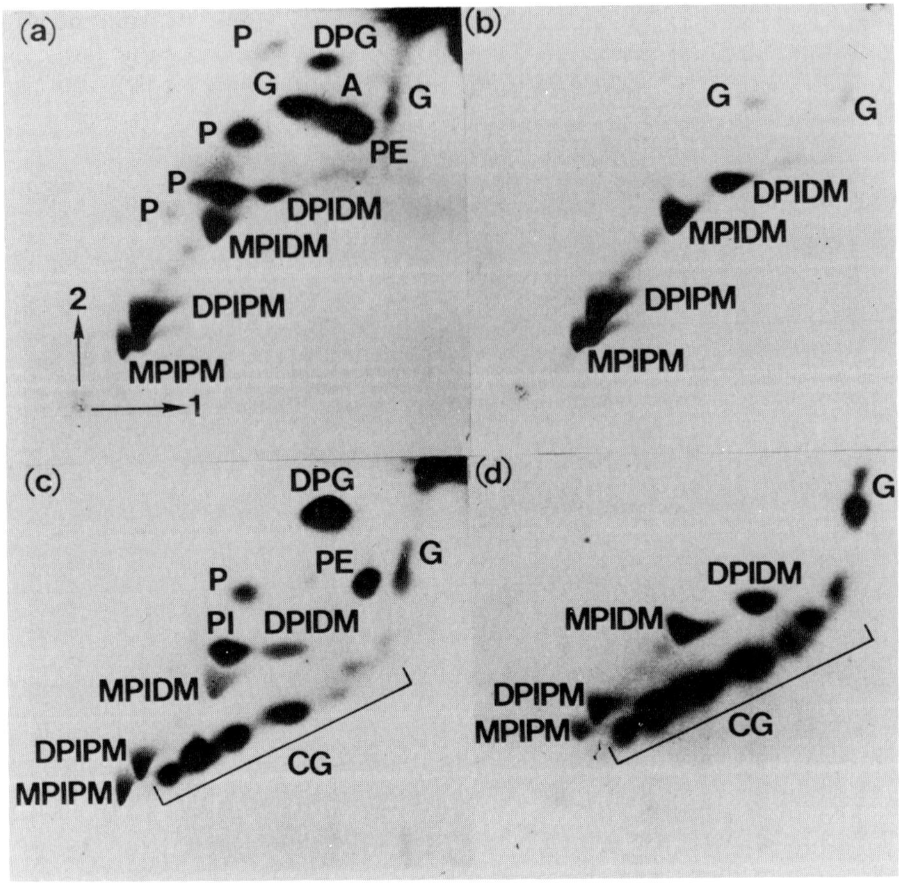

FIG. 10. Two-dimensional TLC of most polar class of lipids in mycobacterial polar lipid fractions (Fig. 3, TLC system E). First direction, chloroform–methanol–water (60:30:6, v/v); second direction, chloroform–acetic acid–methanol–water (40:25:3:6, v/v). Organisms: (a) and (b) *Mycobacterium bovis* MNC 27; (c) and (d) *Mycobacterium kansasii* ATCC 12478 (see Fig. 4 for source and cultivation). Abbreviations: DPG, diphosphatidylglycerol; PE, phosphatidylethanolamine; P, uncharacterized phospholipid; G, uncharacterized glycolipid; PI, phosphatidylinositol; DPIDM and MPIDM, di- and mono-acylphosphatidylinositol dimannosides; DPIPM and MPIPM, di- and mono-acylphosphatidylinositol pentamannosides; CG, characteristic polar glycolipids; A. aminolipid. Detection: (a) and (c), molybdophosphoric acid (MPA, Fig. 3) for all lipids; (b) and (d), α-naphthol-sulphuric acid (NAS, Fig. 3) for sugars. Lipids were further characterized by use of periodate–Schiff sprays and ninhydrin followed by lipid phosphate reagent (see Fig. 3).

may be analysed using the defatted cells remaining after the extraction of free lipids but most studies of long-chain compounds have been performed on whole organisms. In this section, a small-scale procedure for the analysis of patterns of whole-organism long-chain compounds will be described.

The principal bound fatty acids in mycobacteria are the mycolic acids, which are high-molecular-weight, 2-alkyl-branched, 3-hydroxy fatty acids (Asselineau 1966). Mycolic acids are considered to be bound to a peptidoglycan-linked arabinogalactan (Lederer *et al.* 1975) and this lipid matrix may form the foundation layer of a characteristic outer membrane permeability barrier in mycobacteria (Minnikin 1982). Mycolic acids without oxygen functions in addition to the 3-hydroxy acid system are termed "α-mycolic acids"; representative structures are shown in Table 3. A selection of oxygenated mycolic acids having methoxy, keto, epoxy and ester groups is found in extracts of most mycobacteria and typical structures are given in Table 3. Hydrolytic procedures usually result in wax-ester mycolates being degraded to ω-carboxymycolates and long-chain alcohols, homologous with 2-eicosanol (Minnikin *et al.* 1980).

Mycolic acids were isolated in early studies by alkaline hydrolysis followed by esterification of the free acids with diazomethane (Asselineau 1966). In order to produce a rapid small-scale method suitable for the systematic analysis of many strains, an acid methanolysis procedure was developed (Minnikin *et al.* 1975, 1980). Two-dimensional thin-layer chromatography of acid methanolysates (Minnikin *et al.* 1980) has enabled broad patterns of mycobacterial mycolates to be established (Minnikin & Goodfellow 1980; Minnikin 1982). In particular, very characteristic polar mycolic acids were discovered in *M. fortuitum*, *M. smegmatis* (Minnikin *et al.* 1980), *M. farcinogenes*, and *M. senegalense* (Ridell *et al.* 1982) by use of acid methanolysis. These polar mycolates were derived from epoxymycolates (Table 3) (Daffé *et al.* 1981a; Minnikin *et al.* 1982a). The acid methanolysis procedure also revealed the presence of a very characteristic simple pattern of α- and α′- mycolic acids in *M. chelonae* (Minnikin *et al.* 1982b).

A disadvantage of the acid methanolysis procedure is the possible degradation of cyclopropane rings by addition of the elements of methanol (Minnikin 1972). Alkaline hydrolysis may also result in the production of isomers formed by racemization of the C-2 chiral carbon atom (Minnikin & Polgar 1966; Etémadi 1967a). To overcome this latter problem, an alkaline methanolysis procedure, involving digestion with tetramethylammonium hydroxide followed by conversion of tetramethylammonium salts to methyl esters by treatment with iodomethane in dimethylformamide, has been developed for mycolic acid analysis (Minnikin *et al.* 1982a, 1984b,c). This method gives clear patterns of mycolic acid methyl esters, but the procedure is rather slow and involved when compared with the acid methanolysis technique (Min-

TABLE 3

Essential structures of representative mycolic acids[*]

Mycolic acid type	Structure
α-Mycolate	$CH_3(CH_2)_{15-19}X(CH_2)_{14-16}Y(CH_2)_{11-17}CH(OH)-CH(COOH)(CH_2)_{17-23}CH_3$
	X = CH=CH (*cis*) or CH—CH bridged by CH_2 (*cis*); Y = X or $CH(CH_3)CH$=CH (*trans*)
α′-Mycolate	$CH_3(CH_2)_{17}CH$=$CH(CH_2)_{17}CH(OH)-CH(COOH)(CH_2)_{17-23}CH_3$ (CH=CH *cis*)
Methoxymycolate	$CH_3(CH_2)_{15,17}CH(CH_3)-CH(OCH_3)$— } $(CH_2)_{14,16}Y(CH_2)_{16-19}CH(OH)-CH(COOH)(CH_2)_{17-23}CH_3$
Ketomycolate	$CH_3(CH_2)_{15,17}CH(CH_3)-C(=O)$— } $(CH_2)_{14,16}Y(CH_2)_{16-19}CH(OH)-CH(COOH)(CH_2)_{17-23}CH_3$
Wax ester mycolate	$CH_3(CH_2)_{15,17}CH(CH_3)-O-C(=O)$— } $(CH_2)_{14,16}Y(CH_2)_{16-19}CH(OH)-CH(COOH)(CH_2)_{17-23}CH_3$
Epoxymycolate	$CH_3(CH_2)_{15,17}CH(CH_3)-CH-CH$ (epoxide O bridging CH—CH, *trans*) } $(CH_2)_{14,16}Y(CH_2)_{16-19}CH(OH)-CH(COOH)(CH_2)_{17-23}CH_3$
	Y = CH=CH (*cis*), CH—CH bridged by CH_2 (*cis*), $CH(CH_3)CH$=CH (*trans*) or $CH(CH_3)-CH-CH$ bridged by CH_2 (*trans*)

[*]For further structural details see Minnikin (1982); see also Minnikin *et al.* (1982a,b).

BIOMASS (0.50 mg, in Corning 8.5 ml P.T.F.E. capped tube).

Methanol (1 ml), 30% aqueous KOH (1 ml), toluene (0.1 ml).
Heat in oven at 75°C overnight. Cool.
Acidify to pH 1 with 3.6% HCl (36% diluted ten times).
Extract with diethyl ether (3 × 1 ml).
Evaporate to dryness under N_2 (<37°C) in 8.5 ml tube.

↓

FATTY ACIDS

Dichloromethane (1 ml), catalyst solution[1] (1 ml), iodomethane (25 ml).
Mix (30 min), centrifuge at low speed (5 min).
Discard upper layer, dry down lower layer under N_2 (<37°C).

↓

FATTY ACID METHYL ESTERS AND OTHER LONG-CHAIN COMPOUNDS

[1]NaOH (0.8 g) and tetrabutylammonium hydrogen sulphate (3.39 g) in water (100 ml): according to Gyllenhaal & Ehrsson (1975)

FIG. 11. Preparation of methyl esters of mycobacterial long-chain compounds.

nikin *et al* 1980). A more convenient procedure involves alkaline hydrolysis followed by esterification of free acids with iodomethane, using phase-transfer catalysis as shown in Fig. 11.

In order to demonstrate the value of patterns of mycolic esters and related long-chain compounds, extracts were prepared from the representative mycobacteria *M. avium, M. chelonae, M. fortuitum* and *M. tuberculosis* by both alkaline and acidic procedures. The alkaline hydrolysis and esterification method was that shown in Fig. 11, and acid methanolysis was performed according to Minnikin *et al.* (1980) with the modification that the mixture was kept at 50 rather than 75°C overnight. Methanolysates of both extracts from the test strains were examined by two-dimensional TLC and the results are shown in Figs. 12 and 13. The patterns obtained for acid and alkaline methanolysates of *M. avium, M. chelonae* and *M. tuberculosis* (Figs. 12 and 13) were essentially comparable but the epoxymycolates present in alkaline digests of *M. fortuitum* (Fig. 13) were converted to characteristic more polar mycolates in acid methanolysates (Fig. 12). The results reported for *M. fortuitum* (Figs. 12 and 13) correspond closely to patterns recorded previously (Minnikin *et al.* 1982a) for a strain of *M. senegalense* using acid methanolysis (Minnikin *et al.* 1980) and the alkaline methanolysis technique using tetramethylammonium hydroxide (Minnikin *et al.* 1984b). If the analysis of mycolic acid composition is limited to TLC, it is useful, therefore, to use acid methanolysis to detect the presence of epoxymycolates since these esters are co-chromatographed very closely with ketomycolates (Fig. 13).

It is also difficult, using TLC alone, to distinguish between α′- and methoxymycolates (Figs. 12 and 13); to achieve this distinction reliably, mycolic

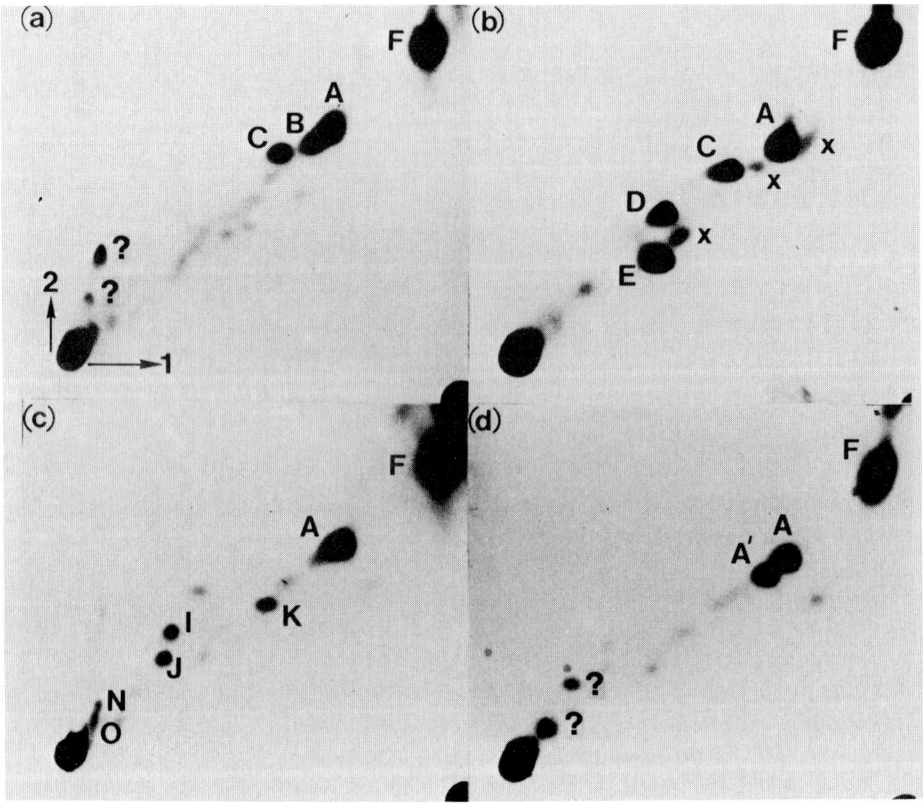

FIG. 12. Two-dimensional TLC of mycobacterial acid methanolysates. The procedure was according to Minnikin *et al.* (1980) but at 50°C rather than 75°C. First direction, petroleum ether (bp 60–80°C)–acetone (95:5, v/v, thrice); second direction, toluene–acetone (97:3, v/v, once). Organisms: (a) *Mycobacterium tuberculosis* C, (b) *Mycobacterium avium* D4 (Central Veterinary Laboratory, Weybridge, U.K.; see Minnikin *et al.* 1980), (c) *Mycobacterium fortuitum* M233 (see Minnikin *et al.* 1984d), (d) *Mycobacterium chelonae* M352 (see Minnikin *et al.* 1982b); (c) and (d) cultivated according to Minnikin *et al.* 1982b. Abbreviations: A, α-mycolate; A′, α′-mycolate; B, methoxymycolate; C, ketomycolate; D, ω-carboxymycolate; E, 2-eicosanol and homologous alcohols; F, nonhydroxylated fatty-acid methyl esters; K, unidentified mycolate; I and J, hydroxymethoxymycolates (derived from epoxymycolates); X, minor components possibly arising by acid-catalysed addition of the elements of methanol to unsaturations; ?, unknown. Detection: molybdophosphoric acid (MPA, Fig. 3).

acid methyl esters may be converted to *tert*-butyldimethylsilyl (TBDMS) ethers and analysed by TLC. The total mycolate fraction was isolated by preparative two-dimensional TLC using the solvent systems shown in Figs. 12 and 13 and 10 × 10-cm pieces of Merck 5735 plastic-backed silica gel sheets. Separated components were located by spraying with 0.01% eth-

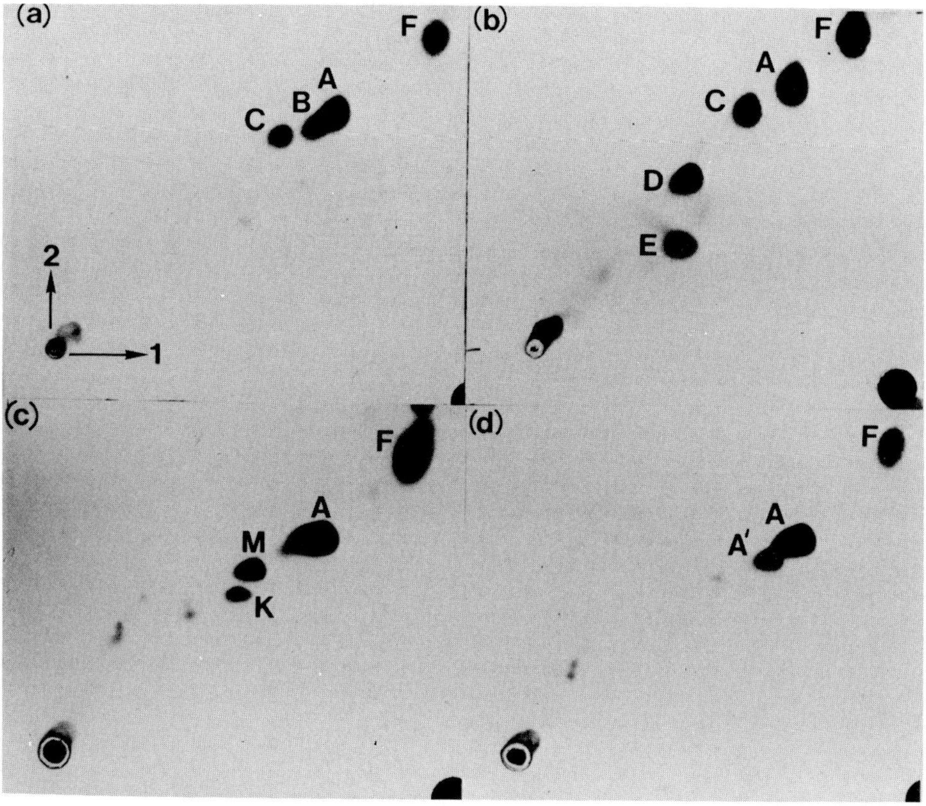

FIG. 13. Two-dimensional TLC of mycobacterial alkaline methanolysates. The procedure used was as in Fig. 11: the developing solvents, organisms, and detection systems were as in Fig. 12. Abbreviations: M, epoxymycolate; others as Fig. 12.

anolic Rhodamine 6G, and the total mycolates and long-chain alcohols were extracted from the sheets by shaking with diethyl ether (2 × 1 ml). Long-chain compounds were treated at 75°C overnight with toluene (0.15 ml) and a solution (0.15 ml) of *tert*-butyldimethylsilyl chloride (0.15 g) and imidazole (0.17 g) in dimethylformamide (1 ml). After cooling, the reaction mixtures were extracted thrice with petroleum ether (bp 60–80°C) (0.3 ml) and the extracts were passed through 1-cm columns of neutral aluminium oxide, prewashed with diethyl ether. The columns were washed twice with diethyl ether (0.5 ml) and the combined extracts were evaporated to dryness (Minnikin and Patel 1979).

The TBDMS ethers of mycolates and other long-chain components were analysed by TLC as shown in the examples in Fig. 14. It is clear that α′- and methoxymycolates can be readily separated by TLC of these derivatives

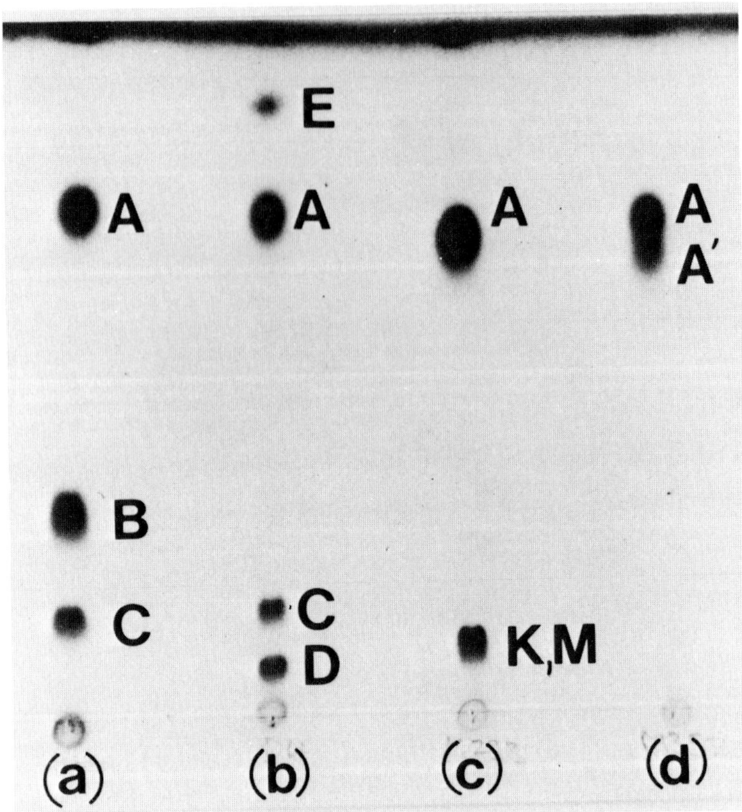

FIG. 14. TLC of *tert*-butyldimethylsilyl (TBDMS) ethers of components of methanolysates of mycobacteria. Organisms (a)–(d) and detection systems were as in Figs. 12 and 13. A single development with petroleum ether (bp 60–80°C)–toluene (70:30, v/v) was used. Abbreviations: A–E, K, M refer to TBDMS ethers of the long-chain components shown in Fig. 13.

(Fig. 14). It should also be noted that the TBDMS ethers of epoxymycolates and "mycolate-K" from *M. fortuitum* (Figs. 12 and 13) chromatographed together but were distinguishable from the derivatives of keto- or ω-carboxy-mycolates.

By use of the analytical systems outlined above, the mycolic-acid patterns of hundreds of representatives of most mycobacterial species have been determined as summarised in Table 4. In addition to determining the pattern of the main types of mycolic acids, further systematic information may be obtained by analysing the individual mycolate types in more detail. Mass spectrometry of mycolic acid methyl esters (Etémadi 1967b; Minnikin & Goodfellow 1980; Asselineau *et al.* 1981; Minnikin *et al.* 1982b) or their TBDMS ethers (Minnikin *et al.* 1982a) provides a range of molecular weights

TABLE 4
Mycobacterial mycolic acid TLC patterns

Mycolic acid pattern[1]	Species
α only	*M. fallax*,[2–4] *M. triviale*[2–4]
α, α′	"*M. borstelense*",[5] *M. chelonae*[5]
α, α′, methoxy	*M. agri*[4,6]
α, α′, keto	*M. simiae*[7]
α, α′, methoxy, keto	*M. thermoresistible*[7]
α, keto	*M. bovis* BCG,[8] *M. leprae*[9]
α, methoxy, keto	*M. asiaticum*,[7] *M. bovis*,[7] *M. gastri*,[7] *M. gordonae*,[7,10] *M. kansasii*,[7] *M. marinum*,[7] *M, microti*,[11] *M. szulgai*,[3,10] *M. ulcerans*[3,4]
α, α′, epoxy[12]	"*M. chitae*",[6] *M. farcinogenes*,[13,14] *M. fortuitum*,[14–16] "*M. giae*",[7] "*M. peregrinum*",[6,16] *M. senegalense*,[13,14] *M. smegmatis*[14–16]
α, keto, wax ester	*M. aichiense*,[6] *M. aurum*,[6,17] *M. avium*,[3,7,15] *M. flavescens*,[7,10] *M. gadium*,[6] "*M. gallinarum*",[6] *M. intracellulare*,[3,7,15] *M. lepraemurium*,[18] *M. nonchromogenicum*,[3,7] *M. neoaurum*,[6] "*M. novum*",[7] *M. paratuberculosis*,[7,15] *M. phlei*,[15] *M. rhodesiae*,[6] *M. scrofulaceum*,[3,7] *M. terrae*,[3,7] *M. tokaiense*,[6] *M. xenopi*[3,7]
α, α′, keto, wax ester	*M. chubuense*,[6] *M. duvalii*,[6] *M. gilvum*,[6] *M. parafortuitum*,[6] *M. vaccae*[19]
α, keto, methoxy, wax ester	*M. komossense*[6]

[1]See Table 3 for representative mycolic acid structures.
[2]Lévy-Frébault *et al.* 1983.
[3]D. E. Minnikin, S. M. Minnikin, M. Goodfellow, G. Dobson, J. H. Parlett, and F. Portaels, unpublished results.
[4]Minor amounts of more polar components were also detected in certain strains.
[5]Minnikin *et al.* 1982b.
[6]D. E. Minnikin, S. M. Minnikin, M. Goodfellow, and J. H. Parlett, unpublished results.
[7]Minnikin *et al.* 1984c.
[8]Minnikin *et al.* 1983b; one strain also had a methoxymycolate.
[9]Draper *et al.* 1982 and references cited therein.
[10]Daffé *et al.* 1981b; in *M. szulgai* the presence of wax ester mycolates was indicated.
[11]Davidson *et al.* 1982.
[12]Precise distribution of α′ mycolates requires further study.
[13]Ridell *et al.* 1982.
[14]Minnikin *et al.* 1982a.
[15]Minnikin *et al.* 1980.
[16]Minnikin *et al.* 1984d.
[17]In certain strains, α′-mycolates also detected.
[18]D. E. Minnikin, G. Dobson, P. Draper, F. Portaels, and M. Goodfellow, unpublished results.
[19]D. E. Minnikin, G. Dobson, S. M. Minnikin, and M. Goodfellow, unpublished results.

and the size of the chain in the 2-position. Gas chromatography of methyl esters of mycolic acids leads to pyrolytic breakdown to give straight-chain methyl esters derived from the chain in the 2-position (Etémadi 1967b; Lechevalier *et al.* 1971). Esters of C_{26} acids are released on pyrolysis of the mycolates from *M. bovis* and *M. tuberculosis* and closely related mycobacteria but other species give varying proportions of C_{20}, C_{22}, or C_{24} esters (Lechevalier *et al* 1971; Minnikin & Goodfellow 1980). The overall size of mycolic acids may also be determined by reverse-phase high-performance liquid chromatography of derivatives such as phenacyl esters (Qureshi *et al.* 1978; Steck *et al.* 1978) or pentafluorobenzoates of methyl esters (Draper *et al.* 1982) and also by high-temperature gas chromatography of trimethylsilyl ethers of methyl esters (Toriyama *et al.* 1980). Detailed chemical analysis of the various structural types also provides further data of systematic value (Minnikin 1982).

Degradation of mycobacteria to give methyl esters of long-chain compounds also produces derivatives of non-hydroxylated fatty acids (Figs. 12 and 13). Gas chromatography of fatty-acid methyl esters has been developed for the identification of mycobacterial species (Tisdall *et al.* 1979; Julák *et al.* 1980; Larsson *et al.* 1981). Certain species contain unusual fatty acids; *Mycobacterium gordonae* and *M. kansasii,* for example, produce acids having a methyl branch in the 2-position (Julák *et al.* 1980). *Mycobacterium gordonae* is the only species in the genus so far examined that lacks tuberculostearic (10-methyloctadecanoic) acid (Tisdall *et al.* 1979; Julák *et al.* 1980). As noted by Guerrant *et al.* (1981), if total methanolysates are analysed, then the resulting profiles may or may not contain products of mycolate pyrolysis, depending on the injection-port temperature of the gas chromatograph. By use of selected ion recording gas chromatography–mass spectrometry, very small amounts of components such as tuberculostearic and mycocerosic (**III**) acids can be detected from biological samples such as sputum, allowing rapid, sensitive and early diagnosis of mycobacterial disease (Larsson *et al.* 1981). The same sensitive technique has been applied to the analysis of derivatives of both the mycocerosate (**III**) and phthiocerol (**IV–VI**) components of the waxes analysed, as described above, by the least polar free lipid TLC system (Fig. 4) (Mallet *et al.* 1984).

Conclusions

Analyses of the characteristic lipids of mycobacteria are providing data that are of value in classification and identification (Minnikin and Goodfellow 1980; Goodfellow & Wayne 1982; Minnikin 1982). The data remain incomplete; the procedures given in this chapter will, however, allow lipid

TABLE 5

Representative mycobacterial lipid chemotypes

Mycolic acid pattern*	Representative species		
A	*M. fallax*		
	M. triviale		
A, A′	*M. chelonae*		
A, A′, B	*M. agri*		
A, A′, C	*M. simiae*		
A, A′, B, C	*M. thermoresistible*		
A, C	*M. leprae*		
	M. bovis BCG	Glycosyl phenolphthiocerols	Phthiocerol dimycocerosates
A, B, C	*M. bovis*		
	M. kansasii		
	M. marinum	Sulpholipids, trehalose mycolipenates	
	M. tuberculosis		
	M. gordonae		
A, A′, M	*M. farcinogenes*	No tuberculostearate	
	M. fortuitum		
	M. senegalense	Glycopeptidolipids	
	M. smegmatis		
A, C, D	*M. avium*		
	M. intracellulare		
	M. scrofulaceum		
	M. lepraemurium		
A, A′, C, D	*M. vaccae*		
A, B, C, D	*M. komossense*		

*For abbreviations of mycolic acid types, see Figs. 12 and 13.

profiles to be gathered in a much more systematic and reliable manner. Indeed, the chromatographic systems are designed so that the whole range of lipid polarities are covered, allowing the detection of novel as well as known lipids. Lipid analysis provides a way of assessing the interrelatedness between mycobacterial species as shown for some of the best-studied taxa in Table 5.

Traditional classifications of mycobacteria have been largely based on methods developed for clinical identification. Chemotaxonomy offers an alternative viewpoint on mycobacterial systematics and thereby contributes much to an understanding of the integrity and diversity of the genus *Mycobacterium*. It has also been repeatedly pointed out that lipid analysis is of prime importance in separating mycobacteria from other mycolic acid-containing taxa (Minnikin *et al.* 1978; Minnikin & Goodfellow 1980; Goodfellow & Wayne 1982; Goodfellow & Minnikin 1984).

Thanks are principally due to the British Leprosy Relief Association (LEPRA) for a studentship (for G. Dobson). The work was also supported by grants from the Science and Engineering Research Council (GRA 88651), Medical Research Council (G974/522/S and G8216538SB) and the IMMLEP Steering Committee of the UNDP/World Bank/WHO Special Programme for Research and Training in Tropical Diseases (T16/181/L4/29). The present manuscript was prepared by Doreen Moor and Margaret Herron.

References

ANDERSON, R. J. 1941 Structural peculiarities of acid-fast bacterial lipids. *Chemical Reviews* **29**, 225–243.

ASSELINEAU, J. 1966 *The Bacterial Lipids*. Paris: Hermann.

ASSELINEAU, C. & ASSELINEAU, J. 1978a Lipides spécifiques des mycobactéries. *Annales de Microbiologie Institut Pasteur* **129A**, 49–69.

ASSELINEAU, C. & ASSELINEAU, J. 1978b Trehalose-containing glycolipids. *Progress in the Chemistry of Fats and Other Lipids* **16**, 59–99.

ASSELINEAU, C., CLAVEL, S., CLÉMENT, F., DAFFÉ, M., DAVID, H., LANÉELLE, M. A. & PROMÉ, J. C. 1981 Constituants lipidiques de *Mycobacterium leprae* isolé de tatou infecté expérimentalement. *Annales de Microbiologie Institut Pasteur* **132A**, 19–30.

BLIGH, E. G. & DYER, W. J. 1959 A rapid method of total lipid extraction and purification. *Canadian Journal of Biochemistry and Physiology* **37**, 911–917.

BRENNAN, P. J. 1981 Structures of the typing antigens of atypical mycobacteria: A brief review of present knowledge. *Reviews of Infectious Diseases* **3**, 905–913.

BRENNAN, P. J., SOUHRADA, M., ULLOM, B., MCCLATCHY, J. K. & GOREN, M. B. 1978 Identification of atypical mycobacteria by thin-layer chromatography of their surface antigens. *Journal of Clinical Microbiology* **8**, 374–379.

BRENNAN, P. J., ASPINALL, G. O. & NAM SHIN, J. E. 1981a Structure of the specific oligosaccharides from the glycopeptidolipid antigens of serovars in the *Mycobacterium avium-Mycobacterium intracellulare-Mycobacterium scrofulaceum* complex. *Journal of Biological Chemistry* **256**, 6817–6822.

BRENNAN, P. J., MAYER, H., ASPINALL, G. O. & NAM SHIN, J. E. 1981b Structures of the glycopeptidolipid antigens from serovars in the *Mycobacterium avium/Mycobacterium intracellulare/Mycobacterium scrofulaceum* serocomplex. *European Journal of Biochemistry* **115**, 7–15.

BRENNAN, P. J., HEIFETS, M. & ULLOM, B. P. 1982 Thin-layer chromatography of lipid antigens as a means of identifying nontuberculous mycobacteria. *Journal of Clinical Microbiology* **15**, 447–455.

CHANLEY, J. D. & POLGAR, N. 1950 Dextrorotatory acids of tubercle bacilli lipids. *Nature, London* **166**, 693–695.

COLLINS, M. D., PIROUZ, T., GOODFELLOW, M. & MINNIKIN, D. E. 1977 Distribution of menaquinones in actinomycetes and corynebacteria. *Journal of General Microbiology* **100**, 221–230.

COLLINS, M. D., GOODFELLOW, M., MINNIKIN, D. E. & ALDERSON, G. 1984 Menaquinone composition of mycolic acid-containing actinomycetes and some sporoactinomycetes. *Journal of Applied Bacteriology*. (In press.)

CONSDEN, R. & GORDON, A. H. 1948 Effect of salt on partition chromatograms. *Nature, London* **162**, 180–181.

Daffé, M., Lanéelle, M. A. Puzo, G. & Asselineau, C. 1981a Acide mycolique epoxydique: Un nouveau type d'acide mycolique. *Tetrahedron Letters* **22,** 4515–4516.

Daffé, M., Lanéelle, M. A., Promé, D. & Asselineau, C. 1981b Étude des lipides de *Mycobacterium gordonae* comparativement à ceux de *M. leprae* et de quelques mycobactéries scotochromogenes. *Annales de Microbiologie Institut Pasteur* **132B,** 3–12.

Davidson, L. A., Draper, P. & Minnikin, D. E. 1982 Studies on the mycolic acids from the walls of *Mycobacterium microti*. *Journal of General Microbiology* **128,** 823–828.

Dittmer, J. C. F. & Lester, R. L. 1964 A simple, specific spray for the detection of phospholipids on thin-layer chromatograms. *Journal of Lipid Research* **5,** 126–127.

Draper, P., Dobson, G., Minnikin, D. E. & Minnikin, S. M. 1982 The mycolic acids of *Mycobacterium leprae* harvested from experimentally infected nine-banded armadillos. *Annales de Microbiologie Institut Pasteur* **133B** 39–47.

Draper, P., Payne, S. N., Dobson, G. & Minnikin, D. E. 1983 Isolation of a characteristic phthiocerol dimycocerosate from *Mycobacterium leprae*. *Journal of General Microbiology* **129,** 859–863.

Etémadi, A. H. 1967a Isomerisation de mycolates de methyle en milieu alcalin. *Chemistry and Physics of Lipids* **1,** 165–175.

Etémadi, A. H. 1967b The use of pyrolysis gas chromatography and mass spectroscopy in the study of the structure of mycolic acids. *Journal of Gas Chromatography* **5,** 447–456.

Gastambide-Odier, M. 1973a Variantes de mycosides caractérisées par des résidus glycosidiques substitués par des chaînes acyles. I. Spectres de masse des mycosides G′ & A′ peracetyles. *Organic Mass Spectrometry* **7,** 845–860.

Gastambide-Odier, M. 1973b Variantes de mycosides caractérisées par des résidus glycosidiques substitués par des chaînes acyles. Nature mycolique des mycosides G′. *European Journal of Biochemistry* **33,** 81–86.

Goodfellow, M. & Minnikin, D. E. 1984 Circumscription of the genus. In *The Mycobacteria: A Sourcebook* eds. Kubica, G. P. & Wayne, L. G., Vol. 1, New York: Marcel Dekker. (In press.)

Goodfellow, M. & Wayne, L. G. 1982 Taxonomy and nomenclature. In *The Biology of the Mycobacteria* eds. Ratledge, C. & Stanford, J. L., Vol. 1, pp. 471–521. London: Academic Press.

Goren, M. B. & Brennan, P. J. 1979 Mycobacterial lipids: Chemistry and biologic activities. In *Tuberculosis* ed. Youmans, G. P., pp. 63–193. Philadelphia: W. B. Saunders.

Guerrant, G. O., Lambert, M. A. & Moss, C. W. 1981 Gas chromatographic analysis of mycolic acid cleavage products in mycobacteria. *Journal of Clinical Microbiology* **13,** 899–907.

Gyllenhaal, O. & Ehrsson, H. 1975 Determination of sulphonamides by electron-capture gas chromatography. *Journal of Chromatography* **107,** 327–333.

Hunter, S. W. & Brennan, P. J. 1981 A novel phenolic glycolipid from *Mycobacterium leprae* possibly involved in immunogenicity and pathogenicity. *Journal of Bacteriology* **147,** 728–735.

Hunter, S. W., Fujiwara, T. & Brennan, P. J. 1982 Structure and antigenicity of the major specific glycolipid antigen of *Mycobacterium leprae*. *Journal of Biological Chemistry* **257,** 15072–15078.

Jacin, H. & Mishkin, A. R. 1965 Separation of carbohydrates on borate-impregnated silica gel G. plates. *Journal of Chromatography* **18,** 170–173.

Jenkins, P. A. 1981 Lipid analysis for the identification of mycobacteria: an appraisal. *Reviews of Infectious Diseases* **3,** 862–865.

Julák, J., Tureček, F. & Miková, Z. 1980 Identification of characteristic branched-chain fatty acids of *Mycobacterium kansasii* and *Mycobacterium gordonae* by gas chromatography-mass spectrometry. *Journal of Chromatography* **190,** 183–187.

Larsson, L. Mårdh, P. A., Odham, C. & Westerdahl, G. 1981 Use of selected ion monitoring for detection of C_{32} mycocerosic acid in mycobacteria and in five-day-old cultures of sputum specimens from patients with pulmonary tuberculosis. *Acta Pathologica et Microbiologica Scandinavica, Section B* **89**. 245–251.

LECHEVALIER, M. P., HORAN, A. C. & LECHEVALIER, H. A. 1971 Lipid composition in the classification of nocardiae and mycobacteria. *Journal of Bacteriology* **105**, 313–318.

LEDERER, E., ADAM, A., CIORBARU, R., PETIT, J.-F. & WIETZERBIN, J. 1975 Cell walls of mycobacteria and related organisms; chemistry and immunostimulant properties. *Molecular and Cellular Biochemistry* **7**, 87–104.

LÉVY-FRÉBAULT, V., RAFIDINARIVO, E., PROMÉ, J. C., GRANDRY, J., BOISVERT, H. & DAVID, H. L. 1983 *Mycobacterium fallax* sp. nov., *International Journal of Systematic Bacteriology* **33**, 336–343.

MALLET, A. I., MINNIKIN, D. E. & DOBSON, G. 1984 Gas chromatography-mass spectrometry of *tert*-butyldimethylsilyl ethers of phthiocerols and mycocerosic alcohols from *Mycobacterium tuberculosis*. *Biomedical Mass Spectrometry*. **11**, 79–86.

MINNIKIN, D. E. 1972 Ring location in cyclopropane fatty acid esters by boron trifluoride-catalysed methoxylation followed by mass spectroscopy. *Lipids* **7**, 398–403.

MINNIKIN, D. E. 1982 Lipids: Complex lipids, their chemistry, biosynthesis and roles. In *The Biology of the Mycobacteria* eds. Ratledge, C. & Stanford, J. L., Vol. 1, pp. 95–184. London: Academic Press.

MINNIKIN, D. E. & GOODFELLOW, M. 1980 Lipid composition in the classification and identification of acid-fast bacteria. In *Microbiological Classification and Identification* eds. Goodfellow, M. & Board, R. G., pp. 189–256. London: Academic Press.

MINNIKIN, D. E. & PATEL, D. V. 1979 Butyldimethylsilyl ethers of iodine-catalysed solvolysis products of long-chain epoxides. *Chemistry and Physics of Lipids* **23**, 173–178.

MINNIKIN, D. E., & POLGAR, N. 1966 Stereochemical studies on the mycolic acids. *Chemical Communications* pp. 648–649.

MINNIKIN, D. E., ALSHAMAONY, L. & GOODFELLOW, M. 1975 Differentiation of *Mycobacterium, Nocardia* and related taxa by thin-layer chromatographic analysis of whole-cell methanolysates. *Journal of General Microbiology* **88**, 200–204.

MINNIKIN, D. E., GOODFELLOW, M. & COLLINS, M. D. 1978 Lipid composition in the classification and identification of coryneform and related taxa. In *Coryneform Bacteria* eds. Bousfield, I. J. & Callely, A. G., pp. 85–160. London: Academic Press.

MINNIKIN, D. E., HUTCHINSON, I. G., CALDICOTT, A. B. & GOODFELLOW, M. 1980 Thin-layer chromatography of methanolysates of mycolic acid-containing bacteria. *Journal of Chromatography* **188**, 221–233.

MINNIKIN, D. E., MINNIKIN, S. M. & GOODFELLOW, M. 1982a The oxygenated mycolic acids of *Mycobacterium fortuitum, M. farcinogenes* and *M. senegalense*. *Biochimica et Biophysica Acta* **712**, 616–620.

MINNIKIN, D. E., MINNIKIN, S. M., GOODFELLOW, M. & STANFORD, J. L. 1982b The mycolic acids of *Mycobacterium chelonei*. *Journal of General Microbiology* **128**, 817–822.

MINNIKIN, D. E., DOBSON, G. & HUTCHINSON, I. G. 1983a Characterization of phthiocerol dimycocerosates from *Mycobacterium tuberculosis*. *Biochimica et Biophysica Acta* **753**, 445–449.

MINNIKIN, D. E., MINNIKIN, S. M., DOBSON, G., GOODFELLOW, M., PORTAELS, F., VAN DEN BREEN, L. & SESARDIC, D. 1983b Mycolic acid patterns of four vaccine strains of *Mycobacterium bovis* BCG. *Journal of General Microbiology* **129**, 889–891.

MINNIKIN, D. E., O'DONNELL, A. G., GOODFELLOW, M., ALDERSON, G., ATHALYE, M., SCHAAL, A. & PARLETT, J. H. 1984a An integrated procedure for the extraction of bacterial isoprenoid quinones and polar lipids. *Journal of Microbiological Methods* **2**, 233–241.

MINNIKIN, D. E., MINNIKIN, S. M., O'DONNELL, A. G. & GOODFELLOW, M. 1984b Extraction of mycobacterial mycolic acids and other long-chain compounds by an alkaline methanolysis procedure. *Journal of Microbiological Methods* **2**, 243–249.

Minnikin, D. E., Minnikin, S. M., Parlett, J. H., Goodfellow, M. & Magnusson, M. 1984c Mycolic acid patterns of some species of *Mycobacterium*. *Archives of Microbiology*. (In press.)

Minnikin, D. E., Minnikin, S. M., Hutchinson, I. G., Goodfellow, M. & Grange, J. M. 1984d Mycolic acid patterns of representative strains of *Mycobacterium fortuitum*, *Mycobacterium peregrinum* and *Mycobacterium smegmatis*. *Journal of General Microbiology* **130**, 363–367.

Pangborn, M. C. & McKinney, J. A. 1966 Purification of serologically active phosphoinositides of *Mycobacterium tuberculosis*. *Journal of Lipid Research* **7**, 627–633.

Prabhudesai, A. V., Kaur, S. & Khuller, G. K. 1981 Sulpholipids of *Nocardia* species—a preliminary report. *Indian Journal of Medical Research* **73**, 181–183.

Qureshi, N., Takayama, K., Jordi, H. C. & Schnoes, H. K. 1978 Characterisation of the purified components of a new homologous series of α-mycolic acids from *Mycobacterium tuberculosis* H37 Ra. *Journal of Biological Chemistry* **253**, 5411–5417.

Ridell, M., Goodfellow, M., Minnikin, D. E., Minnikin, S. M. & Hutchinson, S. G. 1982 Classification of *Mycobacterium farcinogenes* and *Mycobacterium senegalense* by immunodiffusion and thin-layer chromatography of long-chain components. *Journal of General Microbiology* **128**, 1299–1307.

Shaw, N. 1968 The detection of lipids on thin-layer chromatograms with the periodate-Schiff reagent. *Biochimica et Biophysica Acta* **164**, 435–436.

Stahl, E. 1969 *Thin-layer Chromatography. A Laboratory Handbook*. Berlin: Springer.

Steck, P. A., Schwartz, B. A., Rosendahl, M. S. & Gray, G. R. 1978 Mycolic acids. A reinvestigation. *Journal of Biological Chemistry* **253**, 5625–5629.

Tisdall, P. A., Roberts, G. D. & Anhalt, J. P. 1979 Identification of clinical isolates of mycobacteria with gas liquid chromatography alone. *Journal of Clinical Microbiology* **10**, 506–514.

Toriyama, S., Yano, I., Masui, M., Kusunose, E., Kusunose, M. & Akimori, N. 1980 Regulation of cell wall mycolic acid biosynthesis in acid-fast bacteria. I. Temperature-induced changes in mycolic acid molecular species and related compounds in *Mycobacterium phlei*. *Journal of Biochemistry* **88**, 211–221.

Walker, R. W., Barakat, H. & Jung, J. G. C. 1970 The positional distribution of fatty acids in the phospholipids and triglycerides of *Mycobacterium smegmatis* and *M. bovis* BCG. *Lipids* **5**, 684–691.

Isoprenoid Quinone Analyses in Bacterial Classification and Identification

M. D. COLLINS

Department of Microbiology, National Institute for Research in Dairying, Shinfield, Reading, UK

Lipid analyses are increasingly providing criteria of value in microbial classification and identification. A class of terpenoid lipids that have considerable potential as taxonomic markers are the isoprenoid or respiratory quinones. Although the physiological importance of isoprenoid quinones has been known for a long time, it is only during the last few years that their importance in microbial taxonomy has been fully realized. Jeffries and colleagues were the first to demonstrate the taxonomic value of isoprenoid quinones during an examination of the menaquinone composition of some aerobic Gram-positive cocci (Jeffries *et al.* 1967a,b,c). A few years later Yamada and associates (1968, 1969) showed that ubiquinones were similarly of use in the classification of acetic acid bacteria. Since these early investigations, a large number of systematic studies, most notably those of Yamada and colleagues (e.g. Yamada *et al.* 1976a,b, 1977a,b,c, 1982a,b,c) and Collins, Minnikin and associates (e.g. Collins *et al.* 1977, 1979a,b, 1980a,b, 1981, 1982; Collins & Jones 1979a,b, 1980; Shah & Collins 1980) have further established the value of isoprenoid quinones in the classification and identification of many taxa (see Collins and Jones 1981a, for review).

The two major types of bacterial respiratory quinones are menaquinones (formerly vitamin K_2) and ubiquinones (or coenzyme Q). Both menaquinones (2-methyl-3-polyprenyl-1,4-naphthoquinone) and ubiquinones (2,3-dimethoxy-5-methyl-6-polyprenyl-1,4-benzoquinone) form large classes of molecules in which the length of the polyprenyl side chains varies from 1 to 14 isoprene units (Fig. 1). In addition to variations in the length of the side chain, varying degrees of saturation or hydrogenation of the C-3 multiprenyl side chain of menaquinones have been reported (Collins & Jones 1981a).

CHEMICAL METHODS IN
BACTERIAL SYSTEMATICS

0-12-289675-0

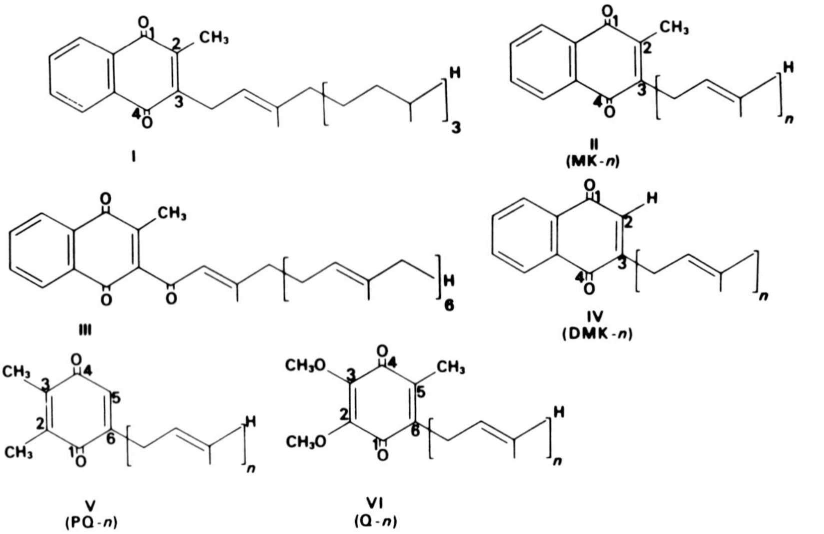

FIG. 1. Structures of phylloquinone (**I**), menaquinone (**II**), chlorobiumquinone (**III**), demethylmenaquinone (**IV**), plastoquinone (**V**) and ubiquinone (**VI**).

The value of menaquinones and ubiquinones as taxonomic criteria basically lies in this variation in the length and degree of saturation of the side chains. In addition to these variations, other modifications are occasionally found in menaquinones and ubiquinones. Thus demethylmenaquinones, which lack the ring methyl substituent on C-2 (Fig. 1, compound **IV**), have been isolated from many bacteria. To date, demethylmenaquinones with polyprenyl side chains varying in length from one to nine isoprene units have been described (see Collins & Jones 1981a). An unusual modification of the multiprenyl side chain of menaquinones has been observed in chlorobiumquinone (Fig. 1, compound **III**), which is produced by certain green photosynthetic bacteria. Chlorobiumquinone (1′-oxomenaquinone-7) is the only example of a bacterial isoprenoid quinone containing a side-chain carbonyl group. In rhodoquinone, a purple quinone isolated from *Rhodospirillum rubrum* by Glover & Threlfall (1962), the methoxyl group in position 3 of the ubiquinone is replaced by an amino group (Fig. 2, compound **VII**). The extremely thermophilic, acidophilic archaebacteria *Sulfolobus* possess a very unusual quinone, designated caldariellaquinone {6-(3,7,11,15,19,23-hexamethyltetracosyl)-5-methylthiobenzo-[b]-thiopen-4,7-quinone} (DeRosa *et al.* 1977; Collins & Langworthy 1983). Until recently, caldariellaquinone (Fig. 2, compound **IX**) was the only known sulphur-containing respiratory quinone. Ishii *et al.* (1983), however, have recently described a novel 2-

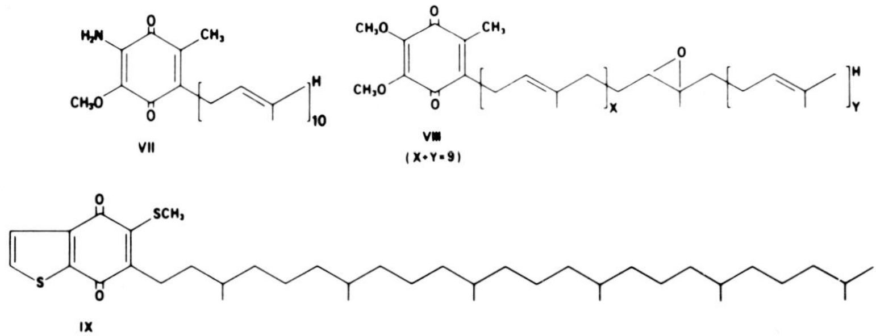

FIG. 2. Structures of rhodoquinone (**VII**), epoxyubiquinone (**VIII**) and caldariellaquinone (**IX**).

methylthio-1,4-naphthoquinone, designated "methionaquinone" from a thermophilic hydrogen bacterium.

Extraction and Purification

Respiratory quinones are quite rapidly photooxidized in the presence of oxygen and strong light and are also particularly susceptible to alkaline conditions. Thus it is preferable to conduct extraction and subsequent purification procedures fairly rapidly, avoiding extremes of pH and strong light. It is not necessary, however, to work in a nitrogen atmosphere or in very dim light. Isoprenoid quinones are soluble in the usual lipid solvents such as petroleum ether, diethyl ether, acetone and chloroform; extraction procedures adopted utilize either one of the above solvents or a mixture of two. Extraction procedures normally yield a mixture of lipids, plus a small amount of nonlipid material, from which the quinones can be readily isolated by simple chromatographic procedures.

Chloroform–methanol extraction procedure

Dry bacterial cells (~50 mg) are extracted with a small volume (~25–50 ml) of chloroform–methanol (2:1, v/v) for ~1–2 h. Some taxa produce very low levels of quinones (e.g., *Streptococcus faecalis*, *Bacteroides* spp.) and larger quantities of cells ~100–200 mg may be required. The cell–solvent mixture is filtered (Whatman filter paper) to remove cell debris, collected in a flask and evaporated to dryness under reduced pressure. The lipid extract is resuspended in a small volume of chloroform–methanol and applied as a narrow band (~3–5 cm long) to a Kieselgel $60F_{254}$ thin layer chromatography

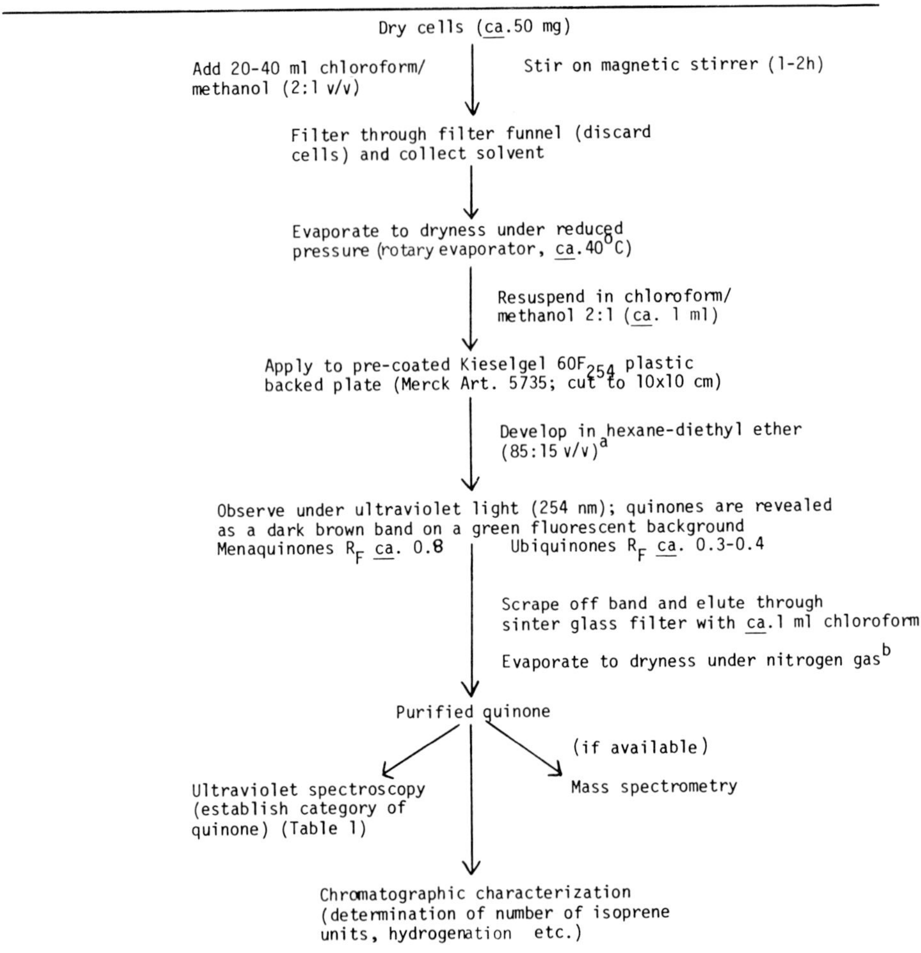

* Based on the procedure of Collins *et al.* (1977)

[a] For ubiquinones hexane-diethyl ether (80:20, v/v) is preferred.

[b] Use oxygen-free nitrogen.

FIG. 3. Extraction and purification of isoprenoid quinones.

(TLC) plate (Merck Art. 5735; plastic-backed 10 × 10 cm). The plate is developed in hexane–diethyl ether (85:15, v/v), and purified quinones are revealed by brief irradiation with ultraviolet (UV) light at 254 nm. The quinones are scraped from the plate and eluted through a sintered-glass filter with a suitable solvent such as chloroform. The purified quinones are finally evaporated to dryness under a stream of nitrogen gas (see Fig. 3). This procedure gives excellent quantitative recovery of quinones. It is worth noting, however, that the presence of water during extraction or subsequent purification steps can cause a reduction in the yield of quinones. Purified quinones awaiting analysis should therefore preferably be kept dry and refrigerated.

Chromatographic Analysis

Reverse-phase, partition, thin-layer chromatography

The composition of natural mixtures of ubiquinones and menaquinones can be investigated using reverse-phase, partition thin-layer chromatography. This procedure facilitates the separation of ubiquinone and menaquinone components according to their overall physical properties, which depend mainly on the length of their polyprenyl side chains, and in the case of menaquinones, on the degree of hydrogenation of the side chain. In reverse-phase, partition chromatography a stationary, nonpolar, nonvolatile hydrocarbon, such as liquid paraffin or hexadecane, and polar developing mixtures (e.g. acetone–water or alcohol–water mixtures) are normally used. In the past, TLC plates were impregnated with the nonpolar stationary phase by dipping activated plates into a solution of the hydrocarbon (e.g. 5% paraffin) in petroleum ether and allowing the volatile solvent to evaporate (e.g. Dunphy & Brodie 1971). Impregnation of the silica gel with the hydrocarbon phase is, however, often difficult, and the method is further hampered by lack of reproducibility and loading problems. Recently, ready-made high-performance, reverse-phase plates (Merck HPTLC RP18F_{254}), in which the nonpolar phase (C_{18}) is chemically bonded to the silica gel, have been employed in ubiquinone and menaquinone analyses (Collins *et al.* 1980b; Collins & Jones 1981b). It is worth noting that the use of ready-made, reverse-phase plates avoids the tedious process of achieving satisfactory impregnation with hydrocarbons. Furthermore, the resolution, reproducibility and developing times (~15 min) are superior to those of conventional plates. The higher loading capacity of the Merck plates also facilitates preparative work.

Reverse-phase TLC of ubiquinones

Ubiquinone samples (in acetone) are spotted (~1 cm from bottom) onto Merck HPTLC RP18F_{254} (Art. 13724) reverse-phase, thin-layer plates (10 × 10 cm) and developed in acetone–acetonitrile (80:20, v/v) (Collins & Jones 1981b). Separated components are observed under UV light (254 nm);

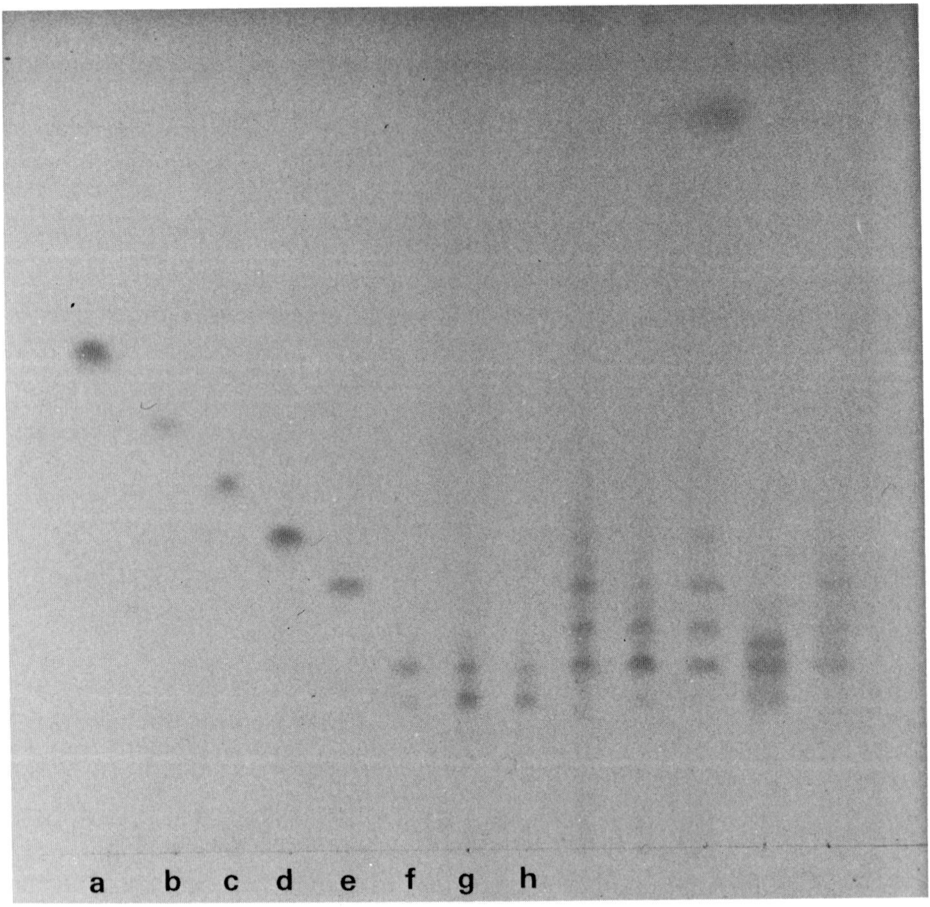

FIG. 4. Reverse-phase, partition TLC of ubiquinones: a, Q-6; b, Q-7; c, Q-8; d, Q-9; e, Q-10; f, Q-12 and Q13 (from *Legionella pneumophila*); g, Q-12 and Q-13 (from *Legionella micdadei*).

quinones appear as dark spots on a blue fluorescent background (see Fig. 4). This system allows clear separation of Q-3 through to Q-14. Since bacterial ubiquinones vary only in the length of the C-6 polyprenyl side chain, this method is sufficient for the qualitative determination of ubiquinone mixtures (see Collins & Jones 1981b). Standard ubiquinones (Q-6 through to Q-10) may be obtained from Sigma Chemical Company.

Reverse-phase TLC of menaquinones

Samples (in acetone) are spotted onto Merck HPTLC RP18F$_{254}$ (Art. 13724) reverse-phase, thin-layer plates (10 × 10 cm) as before and a polar developing mixture of acetone–water (99:1, v/v) (Collins *et al.* 1980b) is used. The method allows clear separation of unsaturated menaquinones (MK-1 through to MK-14). The chromatographic system also facilitates the separation of menaquinone components with the same number of isoprene units but differing degrees of saturation [e.g. MK-9, MK-9(H_2), MK-9(H_4), MK-9(H_6), MK-9(H_8)]. It is worth noting that if the menaquinones under investigation are of the simple unsaturated types, for example, like those found in Gram-negative taxa such as *Bacteroides*, then this method is sufficient for their characterization. However, if partially saturated menaquinones are involved, for example, like those found in most coryneform and actinomycete taxa, then the method by itself is insufficient for their characterization.

Argentation (Ag^+)-modified TLC of menaquinones

Since natural mixtures of bacterial menaquinones often vary both in the degree of saturation and polyprenyl chain length, reverse-phase, partition TLC by itself cannot be used to define the composition of an unknown mixture. This method is, therefore, normally performed in conjunction with either argentation-modified TLC or mass spectrometry.

The ability of silver ions (Ag^+) to form coordination complexes with olefinic bonds is well established. Thus by employing silica gel impregnated with Ag^+ ions it is possible to achieve separation of menaquinone components according to the number of double bonds they contain (e.g., Dunphy & Brodie 1971; Dunphy *et al.* 1971). Layers (20 × 20 cm) may be prepared from a slurry of silica gel G containing $AgNO_3$ (i.e. slurry prepared from a 5% aqueous $AgNO_3$ solution). Plates are allowed to dry in air and are activated by heating at 100°C for ~30 min (Dunphy *et al.* 1971). For rhodamine 6G-impregnated silver layers (to aid visualization) 1–5 ml of 2% rhodamine 6G solution is mixed with aqueous silver nitrate before addition of the adsorbent. The most commonly used solvent systems for Ag^+ plates are acetone–pentane or 2-butanone–hexane mixtures (10–20%, v/v), depending on the number of isoprene units and degree of hydrogenation (see Dunphy *et al.* 1971). Although Ag^+-impregnated plates facilitate the separation of menaquinones according to their chain length and degree of saturation, this system is often not as reproducible as reverse-phase partition TLC.

An alternative to preparing TLC plates is to use ready-made plates and impregnate them with Ag^+ ions. Pre-coated silica gel plates (e.g. Camlab Polygram Sil G/UV_{254} or Merck Kieselgel $60F_{254}$), for example, can be impregnated with Ag^+ by immersing them in a solution of ~2–5% $AgNO_3$ in aqueous acetone (dissolve $AgNO_3$ in 10 ml of H_2O and add 90 ml of acetone). The plates are carefully removed, allowed to dry at room temperature (~10 min) and then heated in an oven at 80°C for ~15 min. The Ag^+ plates should then be stored in a dark, moisture-free container until

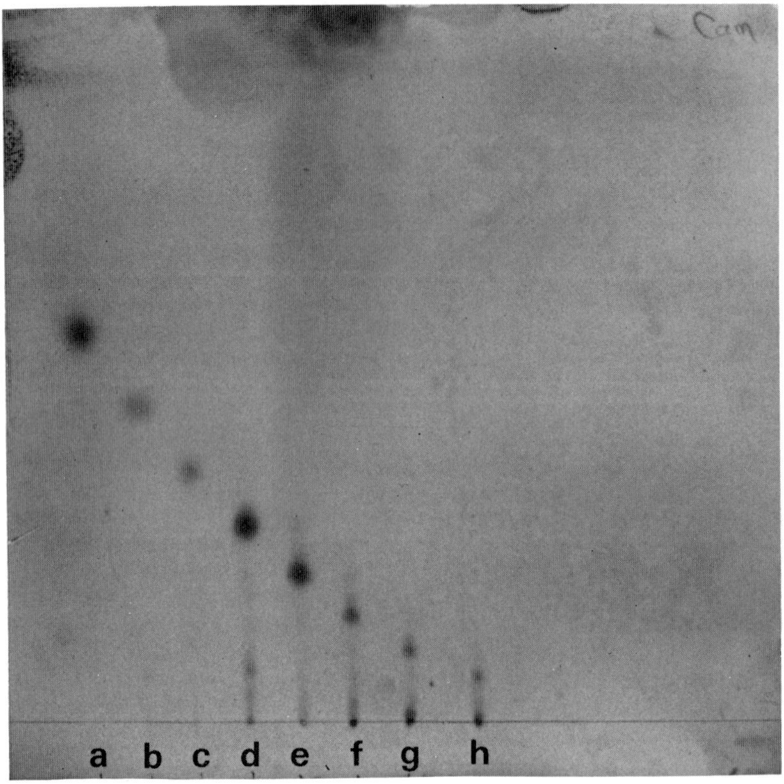

FIG. 5. Ag^+-impregnated TLC (Polygram Sil G/UV_{254} plastic sheets) (10 × 10 cm; 0.25 mm thick) of menaquinones (a–h, MK-1–MK-8). Solvent, hexane–acetone (85:15, v/v)

required. The activated plates can be developed in one of several nonpolar solvent mixtures (e.g. hexane–acetone, 85:15 v/v) and menaquinones are revealed as blue spots on a yellow background by spraying with 10% dodecamolybdophosphoric acid in ethanol and heating at 140°C for 15 min.

The separation of unsaturated menaquinones (MK-1 through to MK-8) using Ag^+-impregnated precoated layers is shown in Fig. 5.

Reverse-phase partition high-performance liquid chromatography

The composition of natural mixtures of bacterial ubiquinones and menaquinones can also be investigated by reverse-phase partition high-performance liquid chromatography (e.g. Collins & Jones 1981b; Collins *et al.* 1982; Kroppenstedt 1982; Collins & Gilbart 1983; Tamaoka *et al.* 1983). The use of high-performance liquid chromatography (HPLC) gives greater resolving power, is more sensitive and quicker than thin-layer chromatographic procedures; the technique also has the additional advantage of providing accurate quantitative data.

Reverse-phase partition HPLC of quinones is normally performed on octadecylsilica (ODS) supports with polar eluents. As with reverse-phase, partition TLC, separation by this method is determined by the lipophilic character of the quinones, which depends mainly on the length of the isoprenyl side chain and degree of saturation.

Reverse-phase HPLC of ubiquinones

Ubiquinones (injection solvent 1-chlorobutane) may be separated isocratically, using a spherisorb ODS (5 μm) (Laboratory Data Control) column (250 × 4.6 mm i.d.). Ubiquinones are monitored with a UV detector at 270 nm and eluted with methanol–1-chlorobutane (100:5, v/v) at 1.5 ml/min. Although this system provides excellent resolution of the most commonly encountered ubiquinones, the level of the nonpolar phase may be increased (e.g. methanol–1-chlorobutane, 100:10 v/v) to facilitate the elution of exceptionally long ubiquinone prenologues, for example, Q-12, Q-13, or Q-14 (Collins & Gilbart 1983); this not only speeds up elution times of the very large prenologues but also prevents "spreading" of the peaks (see Fig. 6).

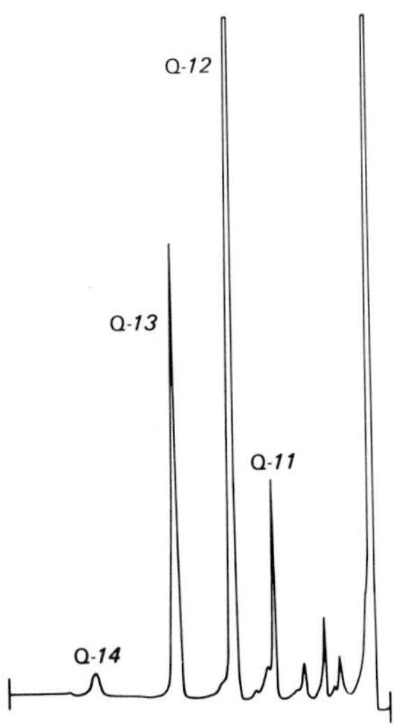

FIG. 6. High-performance liquid chromatogram of ubiquinones from *Legionella pneumophila*. Spherisorb ODS (5 μm) column, using methanol–1-chlorobutane (100:10, v/v) as mobile phase (1.5 ml/min).

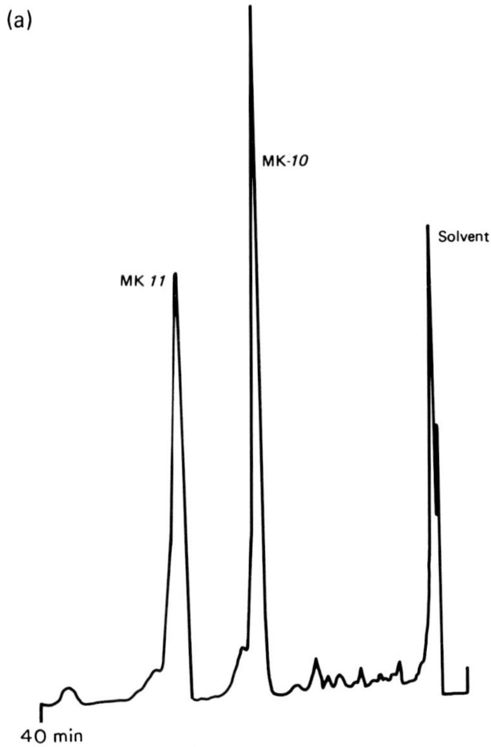

FIG. 7. Examples of high-performance liquid chromatograms of (a) unsaturated and (b) hydrogenated menaquinone mixtures. Spherisorb ODS (5 μm) column using methanol–1-chlorobutane (100:10, v/v) as eluant (1.5 ml/min).

Reverse-phase HPLC of menaquinones

Menaquinones (injection solvent 1-chlorobutane) can be separated on a spherisorb ODS (5 μm) (Laboratory Data Control) column (250 × 4.6 mm i.d.) using methanol–1-chlorobutane (100:5 or 100:10, v/v) as mobile phase (1.5 ml/min) (Collins *et al.* 1982). Alternatively, methanol–isopropyl ether (3:1, v/v) at 1 ml/min may be used as eluent (Tamaoko *et al.* 1983). Menaquinones may be monitored at 270 nm and quantitation achieved with computer integration.

Although the above chromatographic systems give good separation of the majority of menaquinone components, the relative proportions of the solvents of the mobile phases may be altered either to speed up or improve particular separations. For instance, the methanol–1-chlorobutane (100:10, v/v) system (Collins *et al.* 1982) gives excellent separation of long, unsaturated menaquinones, MK-7 through to MK-14, and most complex mixtures of partially

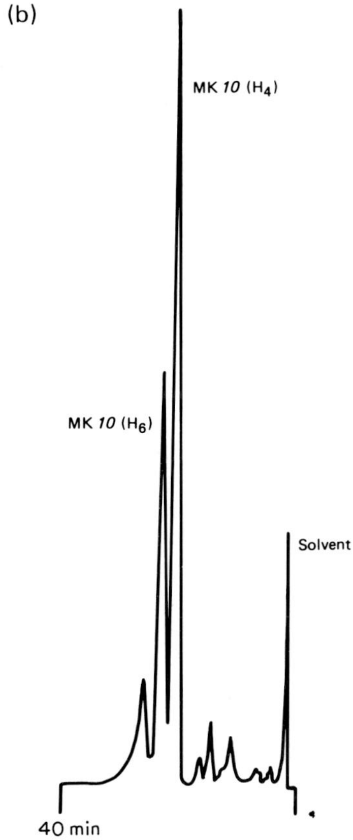

FIG. 7. (*Continued*)

hydrogenated menaquinones (see Fig. 7). Improved separation of short-chain, unsaturated (e.g. MK-1 to MK-6) or small, hydrogenated [MK-6(H_2), MK-5(H_2)] menaquinones can, however, be achieved by reducing the proportion of nonpolar solvent in the mobile phase (e.g. methanol–1-chlorobutane, 100:1, v/v) (see Collins & Langworthy 1983). Similarly with some of the larger unsaturated (e.g. MK-12, MK-13, MK-14, MK-15) and partially hydrogenated [e.g. MK-11(H_6), MK-11(H_8), MK-12(H_6), MK-12(H_8)] menaquinones, the solvent system (methanol–isopropyl ether, 3:1, v/v) of Tomaoka *et al.* (1983) can result in unacceptably long retention times and "spreading" of peaks. Shorter elution times and narrower peaks may be obtained by increasing the proportion of nonpolar solvent (isopropyl ether). As mentioned earlier, some bacteria such as certain actinomycetes contain very complex mixtures of partially saturated menaquinones (see Collins & Jones 1981a; Fig. 8). Although the majority of these partially saturated menaquinones can be readily

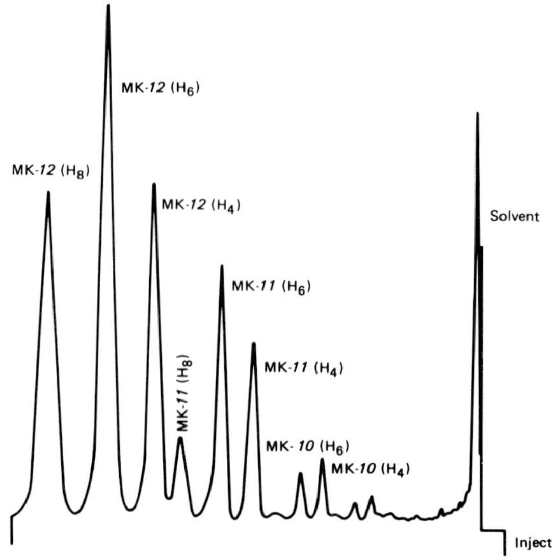

FIG. 8. Example of high-performance liquid chromatogram of a very complex mixture of hydrogenated menaquinones from *M. echinospora* subsp. *pallida*. Spherisorb ODS (5 μm) column using methanol–1-chlorobutane (100:10, v/v) as eluant (1.5 ml/min).

identified using either of the reverse-phase partition HPLC systems, for the unequivocal assignment of some components this technique should be used in conjunction with either silver-ion HPLC or mass spectrometry.

A cautionary note concerning the quantitative HPLC determination of respiratory quinones is necessary. As menaquinones and ubiquinones degrade quite rapidly, it is preferable to perform analyses as soon as possible after completion of sample purification, i.e. the same day or as soon as possible thereafter. This practice is essential if accurate and reliable quantitative fingerprints are to be obtained. The presence of water in the sample (during purification or storage) or exposure to strong light can cause not only a reduction in quinone yield but may also result in the formation of artefacts (particularly near solvent peak). Thus during the final stages of sample purification and storage it is essential to keep the quinone free from water. For long-term storage, quinone samples should be kept dry (preferably freeze-dried or sealed under N_2) and refrigerated. It is worth noting that partially saturated menaquinones are particularly susceptible to degradation.

Silver-ion HPLC of menaquinones

A stainless-steel column (250 × 4.6 mm i.d.) custom-packed (Chrompack Nederland, Middleburg, Netherlands) with silver-loaded Nucleosil^{R} 10SA

(Macherey, Nagel & Co., Düren, FRG) is used. The column is maintained at 65°C (by use of a precision thermostat), and menaquinones are eluted using methanol as a mobile phase (1.5 ml/min, 40 atm pressure) (Kroppenstedt 1982). Menaquinones are monitored at 269 nm.

As mentioned earlier, separation by reverse-phase partition chromatography depends on the lipophilic character of the quinones. In contrast, separation by the silver-ion exchange chromatographic system is determined by the ability of silver ions to form complexes with olefinic bonds (hence separation is primarily according to the number of double bonds in the menaquinone). Thus menaquinones with fewer olefinic bonds are eluted before those with a larger number. Similarly, as expected from Ag^+ TLC data, menaquinones with the same number of double bonds but longer isoprenoid chain length are eluted earlier. Thus, for example, MK-9(H_4) is eluted before MK-8(H_2) even though both contain seven double bonds. Unlike reverse-phase partition chromatography, the silver-ion chromatographic system also facilitates the separation of cis and trans isomers. *cis*-Menaquinones have shorter retention times than *trans*-menaquinones (Kroppenstedt 1982). Because of the strong nature of the olefinic–Ag^+ bonds, the above system can only be used for menaquinones with nine or fewer double bonds.

Physicochemical Analysis

Ultraviolet spectroscopy

Ultraviolet spectroscopy provides a simple method for determining the class or category to which an unknown isoprenoid quinone belongs. Details of the UV absorption characteristics of some of the major types of bacterial quinones are shown in Table 1. Ubiquinones have a λ_{max} at ~270–275 nm (Fig. 9). The replacement of the methoxyl group in position 3 of ubiquinones by an amino group, as in rhodoquinone (2-methoxy-3-amino-5-methyl-6-nonaprenyl-1,4-benzoquinone), causes a marked change in the spectrum with three absorption maxima (251, 280, and 320 nm) and one point of inflection (500 nm) produced (Glover & Threlfall 1962).

Menaquinones and phylloquinones display qualitatively identical UV spectra, with five absorption maxima (λ_{max}) at 242, 248, 260, 269 and 326 nm and one point of inflection at 238 nm (Fig. 10). Absorption bands at 242, 248 and 238 (inflection) nm are due to benzenoid contributions, while bands at 260 and 269 nm are due to quinone contributions. The replacement of the methyl group in position 2 of menaquinones by a hydrogen atom, as in demethylmenaquinones, causes a shift in the quinone absorption contribution of about 6 nm to shorter wavelengths (λ_{max} 254 and 263 nm), whereas the benzenoid contributions (λ_{max} 242, 248 and 238 nm) remain unaltered.

TABLE 1

Ultraviolet absorption characteristics of menaquinones, ubiquinones, and related compounds

Quinone	Solvent	λ_{max} (nm)*
Menaquinone	Isooctane	242, **248,** 260, 269, 326, 238(inf)
Phylloquinone	Isooctane	242, **248,** 260, 269, 326, 238(inf)
Demethylmenaquinone	Isooctane	242, **248,** 254, 263, 326, 238(inf)
Chlorobiumquinone	Ethanol	**254,** 265(inf)
Ubiquinone	Ethanol	**275,** 405
Rhodoquinone	Cyclohexane	251, **280,** 320, 500(inf)
	Ethanol	253, **283,** 320, 500(inf)
Plastoquinone	Isooctane	**254,** 262
Caldariellaquinone	Methanol†	241, 283, 333
	Isooctane‡	237, 272, 278, 322
Methionaquinone**	Ethanol	238, 262, 310

*Each major peak is in boldface; inf, point of inflection.
†Data from DeRosa *et al.* (1977).
‡Data from Collins & Langworthy (1983).
**2-methylthio-3-VI,VII-tetrahydroheptaprenyl-1,4-naphthoquinone (Ishii *et al.* 1983).

Ultraviolet spectrophotometry thus provides a simple method for distinguishing menaquinones from demethylmenaquinones. Chlorobiumquinone (1′-oxo-MK-7) exhibits absorption characteristics quite distinct from those of menaquinones (or demethylated derivatives), showing λ_{max} at 254 nm and a point of inflection at 265 nm (Powls *et al.* 1968).

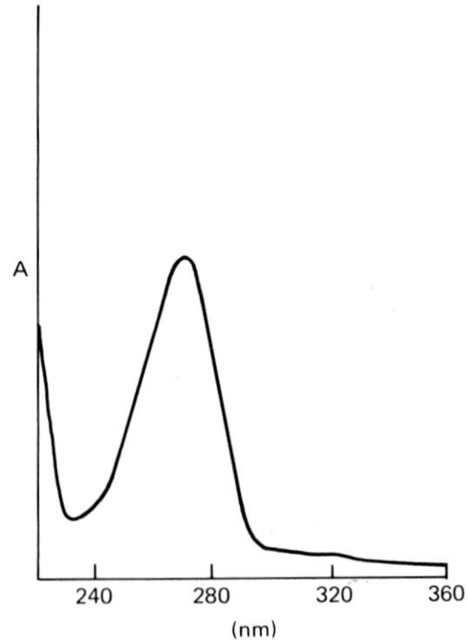

FIG. 9. Ultraviolet spectrum of Q-10 (isooctane).

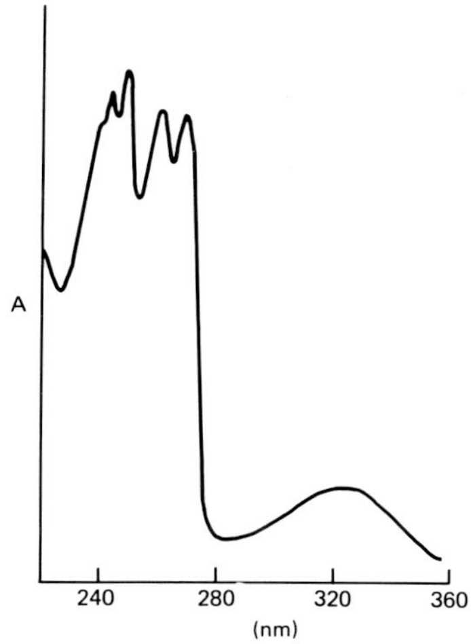

FIG. 10. Ultraviolet spectrum of MK-7 (isooctane).

The sulphur-containing quinone "caldariellaquinone," from *Sulfolobus*, exhibits a very characteristic UV spectrum (λ_{max} 237, 272, 278 and 322 nm; Fig. 11), which is quite distinct from those of any other described bacterial quinone (Table 1).

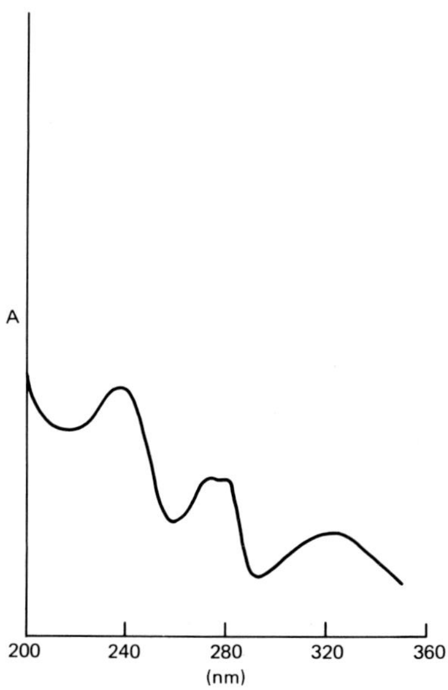

FIG. 11. Ultraviolet spectrum of caldariellaquinone (isooctane)

Mass spectrometry

The most precise and sensitive method for determining respiratory quinone structure is mass spectrometry. When subjected to mass spectrometry, isoprenoid quinones produce characteristic fragmentation patterns that provide structural information on, for instance, the nature of the ring system and on the length and degree of saturation of the multiprenyl side chain. Ubiquinones produce base peaks at *m/e* 235, with a second intense peak at *m/e* 197. These peaks correspond to the pyrilium and benzelium ions, respectively, and are produced from the benzoquinone portion of the molecule. The base peaks in the mass spectra of menaquinones (including phylloquinone) occur at *m/e* 225, with a second intense peak at *m/e* 187, and are similarly derived from the naphthoquinone nucleus (Fig. 12, compound **XII**). Demethylmenaquinones have corresponding nuclear fragments at *m/e* 211 and 173. Strong peaks corresponding to molecular ions (M^+) are observed in the high-mass region of the spectra of ubiquinones, menaquinones and related compounds, thereby facilitating accurate molecular-weight determination. Peaks at $M^+ - 15$ correspond to the loss of a methyl group from the molecular ions. The pattern of fragmentation of the multiprenyl substituent of ubiquinones and menaquinones is characteristic of the cracking pattern exhibited by polyisoprenoid chains in general. Fragmentation of the side chain under electron impact solely involves diallylic bonds, following the rule $(M - 69) - (68 \times N)$, where N is the number of isoprene units in the side chain minus one. Using the mass spectrum of MK-6 as an illustration (Fig. 13), fragmentation of the side chain involves the loss of a terminal isoprenyl unit $(M - 69)^+$ followed by four successive losses of 68 mass units.

The unusual sulphur-containing terpenoid "caldariellaquinone" produces a very characteristic mass spectrum. The base peak normally occurs at *m/e* 225 and is derived from the quinone nucleus (Fig. 14); peaks of comparable intensity occur at *m/e* 192 and 193 formed by the elimination of SH and S, respectively, from the above ion. Strong peaks attributable to molecular ions

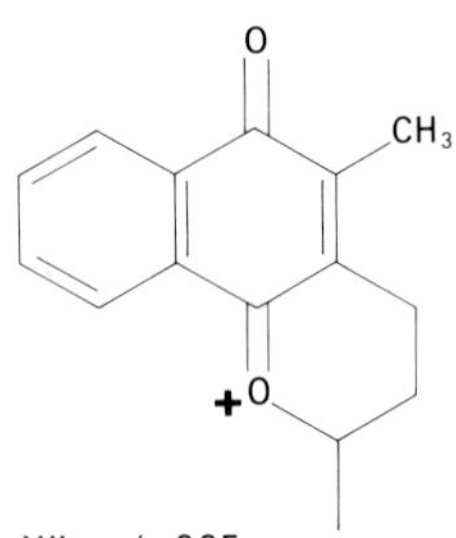

XII *m/e* 225

FIG. 12. Mass spectral fragments at *m/e* 225 derived from the naphthoquinone nucleus of menaquinones.

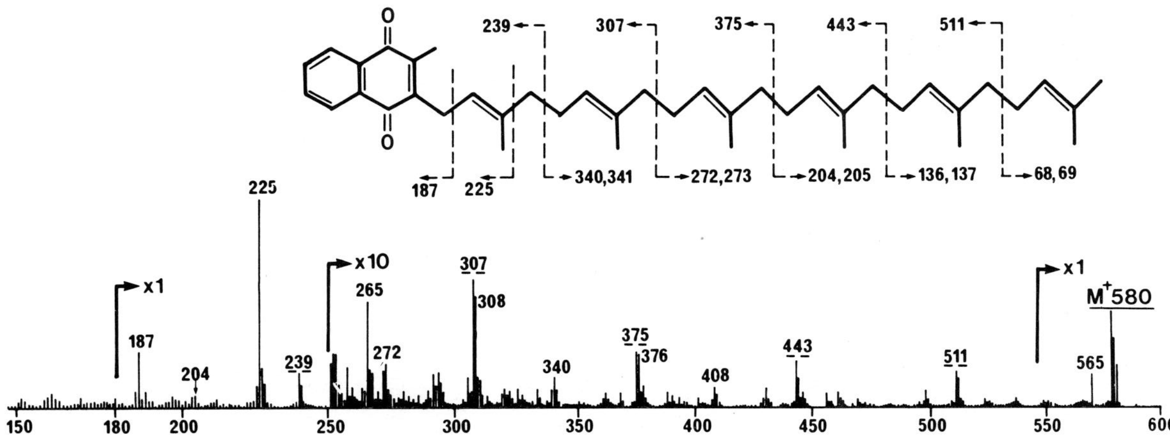

FIG. 13. Mass spectrum of MK-6 from *Capnocytophaga ochracea*.

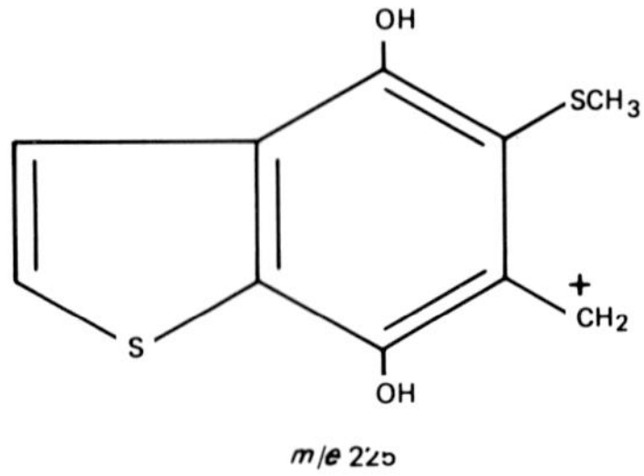

FIG. 14. Mass spectral fragments at *m/e* 225 from caldariellaquinone.

(M^+) are present in the high-mass region, although peaks of greater intensity may be observed at $M^+ - 32$, formed by the elimination of S from the molecular ion (DeRosa *et al.* 1977; Collins & Langworthy 1983). The characteristic side-chain fragmentation pattern of menaquinones (and related compounds) is, however, not present in caldariellaquinone (owing to saturation).

As noted earlier, certain bacteria (e.g. some coryneform and actinomycete taxa) are rich in hydrogenated menaquinones. The detailed structures of only a few of these hydrogenated menaquinones have been established inasmuch as it is not normally possible to determine the points of saturation directly by mass spectrometry. Menaquinones can, however, be converted to chromenyl acetates and subjected to ozonolysis. Reduction of the ozonides (by dimethyl sulphide) to the corresponding aldehydes, followed by mass spectrometry, can provide information on the exact points of saturation of the side chain (e.g. Yamada *et al.* 1982a,b).

A cautionary note on the analysis of isoprenoid quinone mixtures by mass spectrometry alone is necessary. Although mass spectrometry provides accurate molecular weights and structural information, it does not give reliable quantitative information. The intensity of peaks (corresponding to molecular ions) may vary considerably and does not always accurately reflect the relative proportions of prenologues in a mixture. This problem becomes particularly acute when either large isoprenologues (e.g. MK-10, MK-11, MK-12, MK-13) and/or complex mixtures of hydrogenated quinones [e.g. MK-10(H_4), MK-10(H_6), MK-10(H_8), MK-11(H_6), MK-11(H_8)] are present (see Collins & Gilbart 1983; Collins *et al.* 1983; Collins & Kroppenstedt 1983). Mass spectrometry should, therefore, be used in conjunction with either thin-layer or high-performance liquid chromatographic techniques.

Conclusions

The distribution of isoprenoid quinones in the prokaryotes, and their possible value as taxonomic markers, has been extensively reviewed (Collins &

Jones 1981a). There is every indication from the available information that these lipids can provide valuable additional material for elucidating the taxonomic relationships among many bacterial groups. It is worth emphasizing that until recently the majority of quinone studies generated only qualitative–semiquantitative information (relying on the intensity of spots on thin-layer chromatograms or peaks in mass spectra). Recent studies have demonstrated that high-performance liquid chromatography can provide taxonomists with a new, rapid and sensitive tool for obtaining detailed quantitative quinone fingerprints (Collins & Jones 1981b; Collins *et al.* 1982; Kroppenstedt 1982; Collins & Gilbart 1983; Tamaoka *et al.* 1983). Indeed, the relative ease with which isoprenoid quinones can now be isolated and characterized make these compounds ideal chemotaxonomic markers.

References

COLLINS, M. D. & GILBART, J. 1983 New members of the coenzyme Q series from *Legionellaceae*. *FEMS Microbiology Letters* **16**, 251–255.

COLLINS, M. D. & JONES, D. 1979a Isoprenoid quinone composition as a guide to the classification of *Sporolactobacillus* and possibly related bacteria. *Journal of Applied Bacteriology* **47**, 293–297.

COLLINS, M. D. & JONES, D. 1979b The distribution of isoprenoid quinones in streptococci of serological groups D and N. *Journal of General Microbiology* **114**, 27–33.

COLLINS, M. D. & JONES, D. 1980 Lipids in the classification and identification of coryneform bacteria containing peptidoglycans based on 2,4-diaminobutyric acid (DAB). *Journal of Applied Bacteriology* **48**, 459–470.

COLLINS, M. D. & JONES, D. 1981a Distribution of isoprenoid quinone structural types in bacteria and their taxonomic implications. *Microbiological Reviews* **45**, 316–354.

COLLINS, M. D. & JONES, D. 1981b A note on the separation of natural mixtures of bacterial ubiquinones using reverse-phase partition thin-layer chromatography and high performance liquid chromatography. *Journal of Applied Bacteriology* **51**, 129–134.

COLLINS, M. D. & KROPPENSTEDT, R. M. 1983 Lipid composition as a guide to the classification of some coryneform bacteria containing an A4α type peptidoglycan (Schleifer and Kandler). *Systematic and Applied Microbiology* **4**, 95–104.

COLLINS, M. D. & LANGWORTHY, T. A. 1983 Respiratory quinone composition of some acidophilic bacteria. *Systematic and Applied Microbiology* **4**, 295–304.

COLLINS, M. D., PIROUZ, T., GOODFELLOW, M. & MINNIKIN, D. E. 1977 Distribution of menaquinones in actinomycetes and corynebacteria. *Journal of General Microbiology* **100**, 221–230.

COLLINS, M. D., GOODFELLOW, M. & MINNIKIN, D. E. 1979a Isoprenoid quinones in the classification of coryneform and related bacteria. *Journal of General Microbiology* **110**, 127–136.

COLLINS, M. D., JONES, D., GOODFELLOW, M. & MINNIKIN, D. E. 1979b Isoprenoid quinone composition as a guide to the classification of *Listeria, Brocothrix, Erysipelothrix* and *Caryophanon*. *Journal of General Microbiology* **111**, 453–457.

COLLINS, M. D., GOODFELLOW, M. & MINNIKIN, D. E. 1980a Fatty acid, isoprenoid quinone and polar lipid composition in the classification of the genus *Curtobacterium* and possibly related taxa. *Journal of General Microbiology* **118**, 29–37.

COLLINS, M. D., SHAH, H. N. & MINNIKIN, D. E. 1980b A note on the separation of natural mixtures of bacterial menaquinones using reverse-phase partition thin-layer chromatography. *Journal of Applied Bacteriology* **48,** 277–282.

COLLINS, M. D., ROSS, H. N. M., TINDALL, B. J. & GRANT, W. D. 1981 Distribution of isoprenoid quinones in halophilic bacteria. *Journal of Applied Bacteriology* **50,** 559–565.

COLLINS, M. D., MCCARTHY, A. J. & CROSS, T. 1982 New highly saturated members of the vitamin K_2 series from *Thermomonospora. Zentralblatt für Bakteriologie, Mikrobiologie und Hygiene I. Abteilung, Originale* **C3,** 358–363.

COLLINS, M. D., JONES, D., KROPPENSTEDT, R. M. 1983 Reclassification of *Brevibacterium imperiale* (Steinhaus) and "*Corynebacterium laevaniformans*" (Dias and Bhat) in a redefined genus *Microbacterium* (Orla-Jensen), as *Microbacterium imperiale* comb. nov. and *Microbacterium laevaniformans* nom. rev., comb. nov. *Systematic and Applied Microbiology* **4,** 65–78.

DE ROSA, M., DE ROSA, S., GAMBACORTA, A., MINALE, L., THOMSON, R. H. & WORTHINGTON, R. D. 1977 Caldariellaquinone, a unique benzo-b-thiopen-4, 7 quinone from *Caldariella acidophila,* an extremely thermophilic and acidophilic bacterium. *Journal of the Chemical Society, Perkin Transactions* **1,** 653–657.

DUNPHY, P. J. & BRODIE, A. F. 1971 The structure and function of quinones in respiratory metabolism. *Methods in Enzymology* eds. McCormick, D. B. & Wright, L. D., Vol. 18, pp. 407–461.

DUNPHY, P. J., PHILLIPS, P. G. & BRODIE, A. F. 1971 Separation and identification of menaquinones in microorganisms. *Journal of Lipid Research* **12,** 442–449.

GLOVER, J. & THRELFALL, D. R. 1962 A new quinone (rhodoquinone) related to ubiquinone in the photosynthetic bacterium *Rhodospirillum rubrum. Biochemical Journal* **85,** 14P–15P.

ISHII, M., KAWASUMI, T., IGARASHI, Y., KODAMA, T. & MINODA, Y. 1983 2-Methylthio-1,4-naphthoquinone, a new quinone from an extremely thermophilic hydrogen bacterium. *Agricultural and Biological Chemistry* **47,** 167–169.

JEFFRIES, L., HARRIS, M. & PRICE, S. A. 1967a Atypical menaquinone pattern in a strain of *Staphylococcus aureus. Nature, London* **216,** 808–809.

JEFFRIES, L., CAWTHORNE, M. A., HARRIS, M., DIPLOCK, A. T., GREEN, J. & PRICE, S. A. 1967b Distribution of menaquinones in aerobic *Micrococcaceae. Nature, London* **215,** 257–259.

JEFFRIES, L., CAWTHORNE, M. A., HARRIS, M., DIPLOCK, A. T., GREEN, J. & PRICE, S. A. 1967/65c Menaquinone patterns in *Micrococcaceae;* their possible taxonomic significance. *Spisy Přirodovědecké Fakulty Universita J. E. Purkyne Brne* **K40,** 230–235.

KROPPENSTEDT, R. M. 1982 Separation of bacterial menaquinones by hplc using reverse-phase (RP18) and a silver loaded ion exchanger as stationary phases. *Journal of Liquid Chromatography* **5,** 2539–2569.

POWLS, R., REDFEARN, E. R. & TRIPPETT, S. 1968 The structure of chlorobiumquinone. *Biochemical Biophysical Research Communications* **33,** 408–411.

SHAH, H. N. & COLLINS, M. D. 1980 Fatty acid and isoprenoid quinone composition in the classification of *Bacteroides melaninogenicus* and related taxa. *Journal of Applied Bacteriology* **48,** 75–87.

TAMAOKA, J., KATAYAMA-FUJIMURA, Y. & KURAISHI, H. 1983 Analysis of bacterial menaquinone mixtures by high performance liquid chromatography. *Journal of Applied Bacteriology* **54,** 31–36.

YAMADA, Y., AIDA, K. & UEMURA, T. 1968 Distribution of ubiquinone-10 and -9 in acetic acid bacteria and its relation to the classification of genera *Gluconobacter* and *Acetobacter,* especially the so-called intermediate strains. *Agricultural and Biological Chemistry* **32,** 786–788.

YAMADA, Y., AIDA, K. & UEMURA, T. 1969 Enzymatic studies on the oxidation of sugar and sugar alcohol. V. Ubiquinone of acetic acid bacteria and its relation to classification of genera *Gluconobacter* and *Acetobacter,* especially of the so-called intermediate strains. *Journal of General and Applied Microbiology* **15,** 181–196.

YAMADA, Y., INOUYE, G., TAHARA, Y. & KONDO, K. 1976a The menaquinone system in the classification of coryneform and nocardioform bacteria and related organisms. *Journal of General and Applied Microbiology* **22**, 203–214.

YAMADA, Y., INOUYE, G., TAHARA, Y. & KONDO, K. 1976b The menaquinone system in the classification of aerobic gram-positive cocci in the genera *Micrococcus*, *Staphylococcus*, *Planococcus* and *Sporosarcina*. *Journal of General and Applied Microbiology* **22**, 227–236.

YAMADA, Y., TAKINAMI, H., TAHARA, Y. & KONDO, K. 1977a The menaquinone system in the classification of radiation-resistant micrococci. *Journal of General and Applied Microbiology* **23**, 105–108.

YAMADA, Y., ISHIKAWA, T., TAHARA, Y. & KONDO, K. 1977b The menaquinone system in the classification of the genus *Nocardia*. *Journal of General and Applied Microbiology* **23**, 207–216.

YAMADA, Y., YAMASHITA, M., TAHARA, Y. & KONDO, K. 1977c The menaquinone system in the classification of the genus *Actinomadura*. *Journal of General and Applied Microbiology* **23**, 331–335.

YAMADA, Y., AOKI, K. & TAHARA, Y. 1982a The structure of the hexahydrogenated isoprenoid side-chain menaquinone with nine isoprene units isolated from *Actinomadura madurae*. *Journal of General and Applied Microbiology* **28**, 321–329.

YAMADA, Y., HOU, C. F., SASAKI, J., TAHARA, Y. & YOSHIOKA, H. 1982b The structure of the octahydrogenated isoprenoid side-chain menaquinone with nine isoprene units isolated from *Streptomyces albus*. *Journal of General and Applied Microbiology* **28**, 519–529.

YAMADA, Y., TAKINAMI-NAKAMURA, H., TAHARA, Y., OYAIZU, H. & KOMAGATA, K. 1982c The ubiquinone systems in the strains of *Pseudomonas* species. *Journal of General and Applied Microbiology* **28**, 7–12.

Lipids in Archaebacterial Taxonomy

H. N. M. Ross and W. D. Grant

Department of Microbiology, University of Leicester, Leicester, UK

J. E. Harris

AFRC Food Research Institute, Norwich, UK

Since the original definition of archaebacteria by Woese and colleagues (Woese & Fox 1977; Woese *et al.* 1978), a number of archaebacterial traits have been catalogued and the detection of some of them has been used as a rapid screening procedure for possible new archaebacterial isolates (Ross *et al.* 1981; Kessel & Klink 1981). The thermophilic isolates to date have been clearly phenotypically distinct but the assignment of a new methanogenic or halophilic isolate to genus or species requires the application of a number of laborious procedures since these isolates are often phenotypically rather similar. An analysis of the methanogenic phenotype has shown that oligonucleotide cataloguing and nucleic acid hybridization are often essential in determining taxa at genus and species levels (Balch *et al.* 1979; Tu *et al.* 1982). Similarly, within the halophile and haloalkaliphilic phenotype Tindall *et al.* (1984) have used 16S rRNA–DNA hybridization to distinguish taxa amongst these phenotypically almost indistinguishable archaebacteria.

Chemotaxonomic procedures such as cell-wall (Schleifer & Kandler 1972), menaquinone (Collins & Jones 1981) and polar lipid analyses (Collins *et al.* 1980) have proved useful in defining eubacterial taxa and are now widely applied. One might expect that the application of certain of these relatively rapid techniques to archaebacteria might prove of equal value. Cell-wall composition has already been shown to be of some taxonomic value within methanogens (Balch *et al.* 1979), although the other archaebacteria have not been examined with any precise taxonomic purpose in mind. A recent analysis of menaquinone composition in halophiles (Collins *et al.* 1981) indicated that menaquinones were of little taxonomic value within the archaebacterial halophile phenotype, although archaebacterial halophiles could clearly be

CHEMICAL METHODS IN
BACTERIAL SYSTEMATICS

0-12-289675-0

distinguished from their eubacterial counterparts. Methanogenic bacteria have not yet been examined for isoprenoid quinones in any systematic way.

Despite only a small number of studies (Kushwaha *et al.* 1976, 1981; Evans *et al.* 1980; Tindall *et al.* 1984), polar lipids appear to have considerable taxonomic potential within the archaebacterial kingdom. All archaebacteria have complex lipids based on characteristic glycerol isoprenyl ethers, and methanogens, thermophiles and halophiles, including haloalkaliphiles, can be readily distinguished from each other by the structure of these core diether and tetraether lipids (De Rosa *et al.* 1982). Complex lipids have only been examined in a small number of halophilic isolates (Kushwaha *et al.* 1976, 1982; Evans *et al.* 1980; Tindall *et al.* 1983), and one methanogenic isolate (Kushwaha *et al.* 1981). However, the taxonomic value of polar lipid patterns was pointed to by Tindall *et al.* (1984) who, in a study of certain halophiles and haloalkaliphiles, were able to show that polar lipid patterns reflected major 16S rRNA–DNA hybridization homology groups.

The techniques of Ross *et al.* (1981) and Tindall *et al.* (1984) are of general applicability to any archaebacterial isolate. This chapter summarizes these techniques and presents an extension of the work on halophiles described by Tindall *et al.* (1984), showing a clear correlation between polar lipid composition and 16S rRNA–DNA hybridization homology groups. Some preliminary studies of methanogenic bacteria are also presented, which indicate that the analysis of polar lipid patterns may be a useful, rapid procedure for identifying new isolates at genus and species levels.

The Methodology of Lipid Extraction and Analysis

Polar lipids

Polar lipids are extracted by the procedure of Bligh and Dyer (1959) as modified by Collins *et al.* (1980). Freeze-dried cells (100–150 mg) are stirred for 1 min in 2.75 ml of aqueous methanol–NaCl (10 ml, 0.3%, w/v, aqueous NaCl in 100 ml CH_3OH), and then heated to 100°C for 5 min. After cooling, 0.75 ml of aqueous NaCl (0.3%, w/v) is added followed by 1.25 ml of chloroform. The mixture is then stirred for 2–3 h at room temperature and cell debris is centrifuged at top speed for 2 min in an MSE bench centrifuge. The supernatant is carefully removed with a pipette and retained and 2 ml of chloroform–methanol–aqueous NaCl (50:100:40, v/v) is added to the remaining debris, which is stirred for a further 15 min. The mixture is centrifuged as before and the second supernatant is removed and pooled with the first. To the pooled supernatants, 0.75 ml of aqueous NaCl and 1.75 ml of chloroform are added and, after shaking, the layers are separated by bench centrifugation for 3–4 min. The upper aqueous layer is removed and discarded and the lower layer containing the polar lipids is evaporated by a stream of oxygen-free nitrogen.

The dried preparation of polar lipids is redissolved in a small amount of chloroform–methanol (2:1, v/v). Material is spotted on one corner of a thin-layer silica gel plate (Merck HPTLC F524), which is developed in two dimensions after Collins *et al.* (1980). The first solvent is chloroform–methanol–water (65:25:4, v/v) and the second is chloroform–methanol–glacial acetic acid–water (80:12:15:4, v/v). The plates are allowed to dry thoroughly in air after each development.

To determine the types of lipids present on the chromatogram, a variety of differential stains are used.

(a) Total lipids; a 10% (w/v) solution of dodecamolybdophosphoric acid in absolute ethanol is sprayed liberally over the chromatograms, which are then heated at 150–160°C for 10–15 min. All lipids appear as black spots on a yellow background (Ross *et al.* 1981).
(b) Phospholipids; the Zinzadze reagent of Dittmer & Lester (1964) is sprayed lightly over the chromatograms. Phosphate-containing lipids appear as light blue spots on a white background within 5 min at room temperature.
(c) Aminolipids; chromatograms are liberally sprayed with a 0.2% (w/v) solution of ninhydrin in butanol saturated with water and then heated at 105°C for 10 min. Lipids containing free amino groups appear as purple-mauve spots on a white background.
(d) Glycolipids; chromatograms are sprayed with the 1-naphthol reagent of Jacin & Mishkin (1965) and heated at 120°C for 10 min. Glycolipids appear as purple spots on an orange-brown background.
(e) Vicinal glycols; chromatograms are sprayed with a 1% (w/v) solution (in water) of sodium periodate until saturated. They are left at room temperature for 5 min. The oxidation reaction is then stopped by exposure to sulphur dioxide gas until the thin-layer chromatography (TLC) plates are decolourized. The plates are sprayed lightly with aqueous Schiff reagent and finally exposed again to sulphur dioxide. Vicinal glycols show immediately as purple-blue spots. Glycolipids appear as purple-blue spots after 12 h at room temperature (Shaw 1968).

Whole-cell acid methanolysis and analysis of core diether moieties

Freeze-dried cells (100–200 mg) are mixed with 3.5 ml of toluene, 3.5 ml of methanol and 0.1 ml of concentrated sulphuric acid in Pyrex reaction tubes fitted with teflon-lined caps. The tubes are heated at 50°C overnight (16 h)

(Minnikin *et al.* 1975). After cooling, the preparation is extracted with 1.5 ml of hexane. The hexane extract is neutralised with a small amount of ammonium carbonate and then transferred to a clean vial and stored at 4°C.

Products of acid methanolysis are analysed by TLC on silica gel plates developed in petroleum ether (bp 60–80°C)–diethyl ether (85:15, v/v) (Minnikin *et al.* 1975). Lipid components are visualised using dodecamolybdophosphoric acid as described earlier.

Glycerol diether moieties can be further resolved by using a double development system (Ross *et al.* 1981). The chromatogram is first developed in petroleum ether (60–80°C)–acetone (95:5, v/v), allowed to dry, and then redeveloped in the same direction with toluene–acetone (97:3, v/v). Lipid components are visualised as before.

Halophilic Archaebacteria

Two-dimensional TLC polar lipid patterns of a large number of strains were divided into eight distinct groups, examples of which are shown in Fig. 1. The freeze-dried starting material for extraction was obtained from cells grown to stationary phase in the medium of Payne *et al.* (1960) for halophiles and that of Tindall *et al.* (1984) for haloalkaliphiles. Identification of individual lipids was based on data generated by Kates (1978) and Kushwaha *et al.* (1982). Whilst the lipid patterns of strains within any one group were essentially similar, minor differences were often detected between strains (Figs. 1 and 2).

The detection of core diether lipids, using the procedures described by Ross *et al.* (1981), has been used as a rapid test for archaebacteria. The complete reliance placed in this test must be questioned following the discovery of diether lipids in an organism of doubtful archaebacterial status (Langworthy *et al.* 1983). However, the procedures are still of value in detecting the presence and nature of core lipids in archaebacteria. Figure 2 illustrates the separation of different core lipids including the universal C_{20}–C_{20} diether originally described by Kates (1972), the C_{40}–C_{40} tetraether common to most methanogens, and the C_{20}–C_{25} diether recently shown to be a major core diether in certain haloalkaliphiles (De Rosa *et al.* 1982).

FIG. 1. Halophile polar lipid groups: (a) *Halobacterium salinarum* CCM 2084; (b) *Halobacterium volcanii* NCMB 2012; (c) *Halococcus morrhuae* NCMB 787; (d) *Natronobacterium gregoryi* NCMB 2198; (e) *Halobacterium vallismortis* ATCC 29715; (f) *Halobacterium saccharovorum* NCMB 2081; (g) *Halobacterium halobium* NCMB 777; (h) *Halobacterium cutirubrum* NCMB 763. Abbreviations: PGP, phosphatidylglycerophoshate; PG, phosphatidylglycerol; PGS, phosphatidylglycerosulphate; GLS, glycolipid sulphate; DGD, glycodiosyl diether; S-DGD, sulphated glycodiosyl diether; TGD-2, triglycosyl diether; 1, probably S-DGD, 2, unidentified glycolipid.

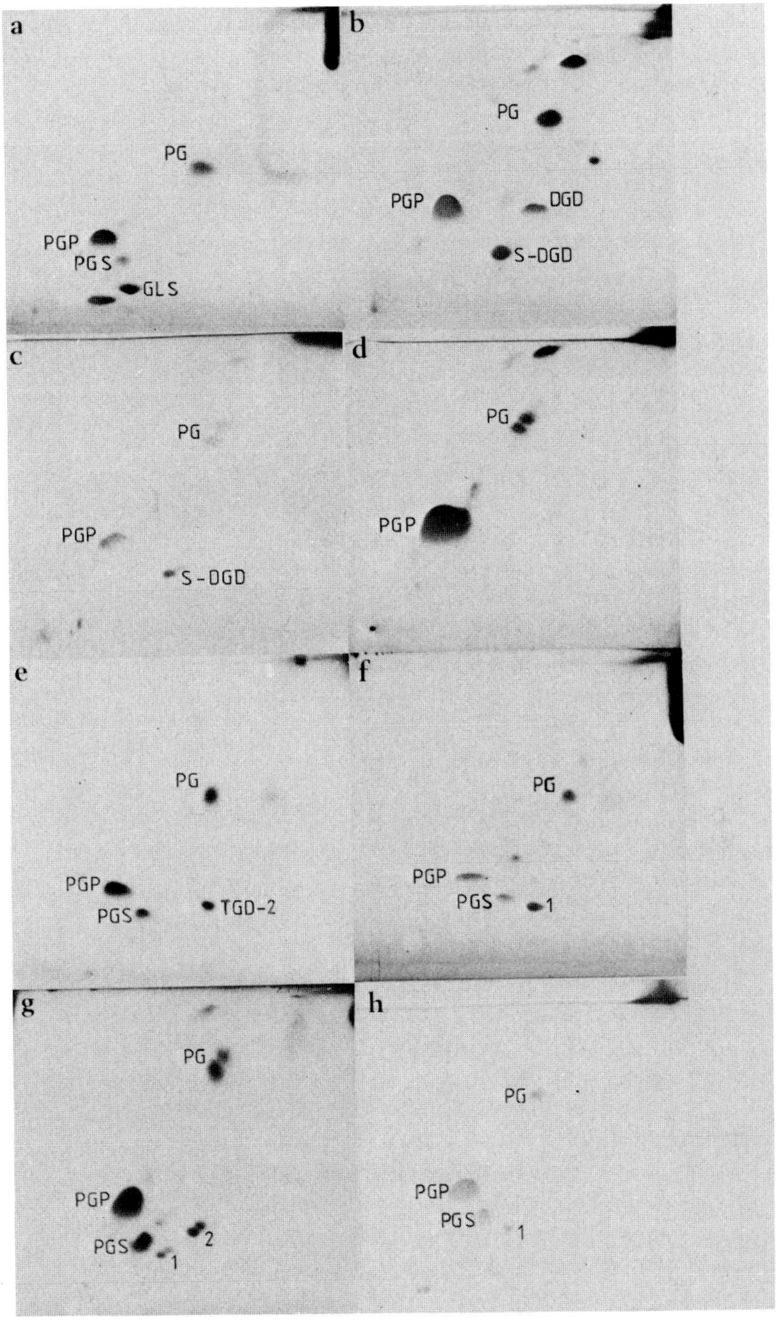
a
PG
PGP
PGS
GLS
b
PG
PGP
DGD
S-DGD
c
PG
PGP
S-DGD
d
PG
PGP
e
PG
PGP
PGS
TGD-2
f
PG
PGP
PGS
1
g
PG
PGP
PGS
2
1
h
PG
PGP
PGS
1

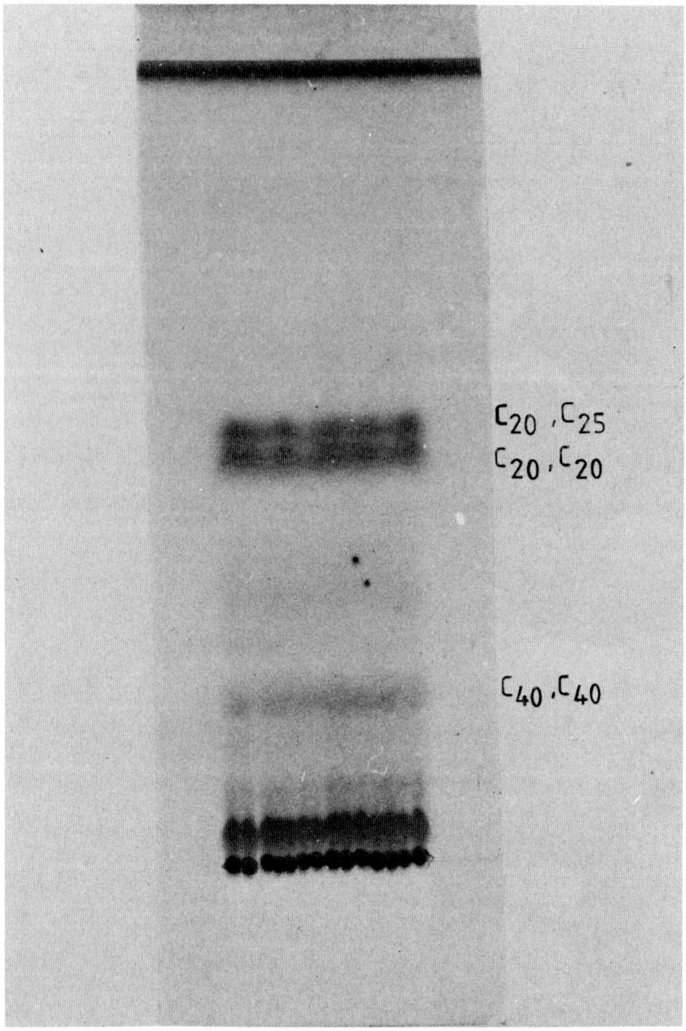

FIG. 2. The resolution of archaebacterial core lipids by thin-layer chromatography. The C_{20}–C_{20} and C_{20}–C_{25} diether core lipids were purified from acid methanolysates of *Natronobacterium magadii* NCMB 2190, and the C_{40}–C_{40} tetraether core lipid from acid methanolysates of *Methanospirillum hungatii* DSM 864.

Table 1 summarizes the major polar lipid groups found among halophilic archaebacteria. It can also be seen that the core diether content of all strains within any one group is very consistent. All strains have the C_{20}–C_{20} diether, whereas strains in three groups additionally possess the C_{20}–C_{25} diether. De Rosa *et al.* (1983) have recently shown that at least one other strain in the

TABLE 1

*Polar lipid composition in archaebacterial halophiles**

Polar lipid groups	PGP	PG	PGS	GLS	TGD-1	TGD-2	S-DGD	DGD	Core lipids
Halobacterium cutirubrum CCM 2088									
Halobacterium halobium CCM 2090	+++	++	++	++	—	Tr	—	—	$C_{20}C_{20}$
Halobacterium salinarium CCM 2084									
Halobacterium volcanii NCMB 2012	+++	+++	—	—	—	—	++	++	$C_{20}C_{20}$
Halobacterium sp. CCM 3361	+++	++	—	—	—	—	++	+	
Halobacterium halobium NCMB 777	+++	+++	+	?	—	?	—	—	$C_{20}C_{20}$,
Halobacterium salinarium NCMB 787	+++	+++	Tr	?	—	?	—	—	$C_{20}C_{25}$
Halobacterium marismortui (M. Ginzburg)	+++	+++	++	—	++	—	—	—	$C_{20}C_{20}$
Halobacterium vallismortis ATCC 29715									
Halobacterium saccharovorum NCMB 2081									
Halobacterium sodomense (G. E. Fox)	+++	++	++	—	—	—	++	—	$C_{20}C_{20}$
Halobacterium trapanicum NRC 34021									
Halobacterium cutirubrum NCMB 763	+++	++	+	+	—	+	++	—	$C_{20}C_{20}$
Halococcus morrhuae NCMB 787	++	+++	—	Tr	—	—	+	+	$C_{20}C_{20}$,
Halococcus morrhuae NCMB 749	++	+++	—	Tr	—	—	+	+	$C_{20}C_{25}$
Natronobacterium gregoryi NCMB 2189									
Natronobacterium magadii NCMB 2190	++	++	—	—	—	—	—	—	$C_{20}C_{20}$,
Natronobacterium pharaonis DSM 2160									$C_{20}C_{25}$
Natronococcus occultus NCMB 2192									

*+++, denotes major component; ++, denotes more than 50% of major component; +, denotes less than 50% of major component; Tr, denotes trace amounts. See legend to Fig. 1 for key to abbreviations.

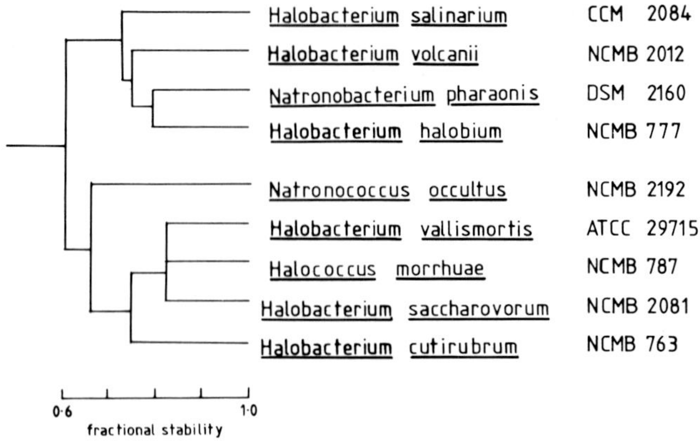

FIG. 3. Relationships within extreme halophiles based on fractional stability of 16S rRNA–DNA hybrids.

haloalkaliphilic group has an additional C_{25}–C_{25} diether, which can also be resolved using the procedures of Ross *et al.* (1981).

The polar lipid groupings show a close correlation with homology groups generated by 16S rRNA–DNA hybridization studies on the same strains (Tindall *et al.* 1984; H. N. M. Ross & W. D. Grant, in preparation). Figure 3 shows a dendrogram generated from fractional stability studies of 16S rRNA–DNA hybrids as calculated by the method of Tu *et al.* (1982). The polar lipid groups correspond closely with rRNA–DNA groups that define taxa at what is probably the generic level. The only exception to the general rule is to be found amongst the haloalkaliphiles where the coccoid strain *Natronococcus occultus* shares a common lipid pattern with the rod-shaped haloalkaliphilic *Natronobacterium* spp. and yet appears unrelated on the basis of rRNA–DNA homology—indeed the isolate seems little related to any of the other archaebacterial halophiles (Fig. 3). It is possible that the alkaline environment imposes particular restraints on lipid structure unrelated to phylogeny—for example, haloalkaliphiles are the only halophiles to lack glycolipids.

Methanogenic Archaebacteria

Freeze-dried material from a number of strains grown to stationary phase has also been subjected to two-dimensional TLC analysis of polar lipids. *Methanococcus* and *Methanogenium* spp. were grown to stationary phase at 37 and 25°C, respectively, in medium 3 of Balch *et al.* (1979) supplemented with 0.001 g $litre^{-1}$ resazurin. However, the trace mineral solution additionally contained 0.01 g $litre^{-1}$ NiCl and 0.01 g $litre^{-1}$ Na_2SeO_3. All other meth-

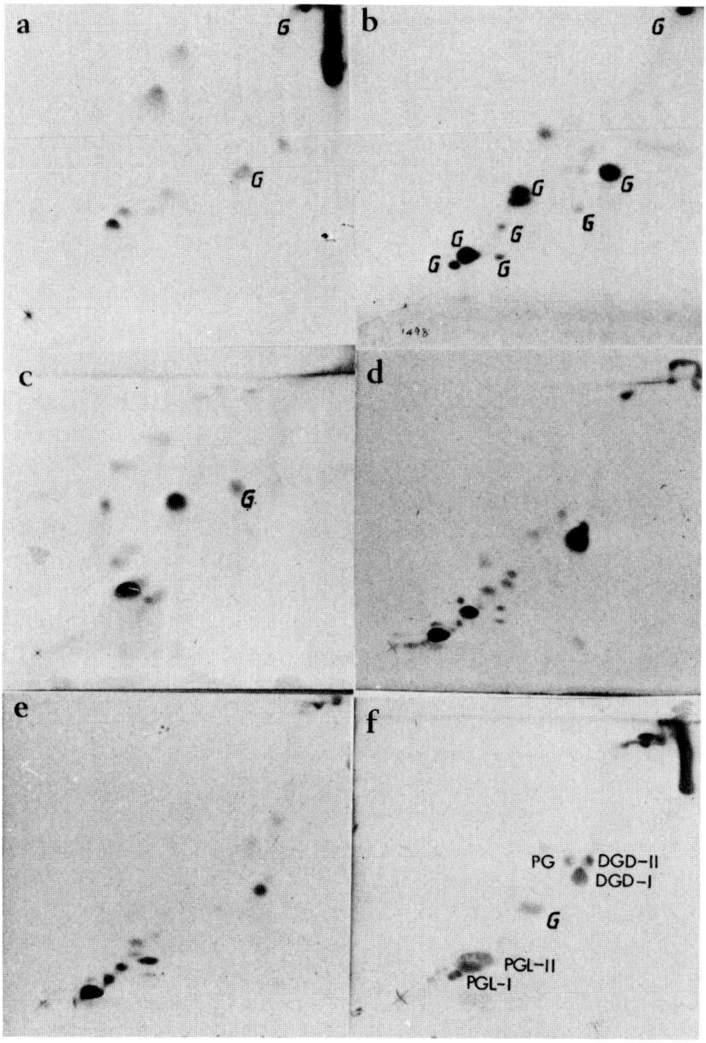

FIG. 4. Polar lipid patterns of species from different methanogen genera: (a) *Methanococcus vannielii* DSM 1224; (b) *Methanogenium marisnigri* DSM 1498; (c) *Methanosarcina barkeri* DSM 800; (d) *Methanobrevibacter ruminantium* DSM 1093;* (e) *Methanobacterium thermoautotrophicum* DSM 1053;* (f) *Methanospirillum hungatii* DSM 864. Abbreviations: PGL I, PGL II, bidiphytanylphosphoglycolipids; DGD I, DGD II, diglycosyldiphytanyl lipids (Kushwaha *et al.* 1981); PG phosphatidyl glycerol; G, unidentified glycolipids. *All lipid spots stain positively for glycolipids.

anogens were grown to stationary phase at 37°C except *Methanobacterium thermoautotrophicum* which was grown at 65°C, in M3 medium of the following composition (g $litre^{-1}$); yeast extract, 5.0; Na formate, 2.0; Na acetate, 2.0; NaCl, 0.9; $(NH_4)_2SO_4$, 0.9; K_2HPO_4, 0.9; KH_2PO_4, 0.45; $CaCl_2 \cdot 6H_2O$, 0.2; $MgSO_4 \cdot 7H_2O$, 0.2; $FeSO_4 \cdot 7H_2O$, 0.01; 10 ml $litre^{-1}$ trace element solution as above. All media were boiled out and cooled under 80% N_2–20% CO_2, and Na_2CO_3 (2 g $litre^{-1}$) was added to the M3 medium. The pH of each medium was adjusted to 6.8 before reduction by addition of cysteine hydrochloride (0.4 g $litre^{-1}$) and $Na_2S \cdot 9H_2O$ (0.4 g $litre^{-1}$). After inoculation, the dead-space gas was replaced with 80% H_2–20% CO_2 at 2 atm pressure and replenished daily, except in the case of *Methanosarcina* spp., where the growth substrate was CH_3OH added prior to inoculation at 1% (w/v).

Cultures were grown in 400-ml amounts in 1l Duran bottles modified to incorporate a glass tube fitted with a butyl-rubber septum similar to that described by Balch *et al.* (1979); incubation was initially stationary for 24 h followed by shaking at 150 rpm.

Some of the total polar lipid patterns of strains from six genera are shown in Fig. 4. It can be seen that it is possible to clearly distinguish examples from the different genera defined by Balch *et al.* (1979). Individual species within a genus can also be distinguished although they share the same overall patterns. These preliminary results suggest that different species within a genus also share the same core diether pattern (Langworthy *et al.* 1983). The lipids have still to be identified although the majority in most strains are glycolipids (*Methanosarcina* spp. and *Methanococcus* spp. are exceptions to the rule). Kushwaha *et al.* (1981) have analysed *Methanospirillum hungatii* in detail and also found a majority of glycolipids. Although Kushwaha *et al.* (1981) used different solvent systems, it is possible to deduce the likely identity of the major lipid spots exhibited by *Methanospirillum hungatii* in this analysis; this is tentatively shown in Fig. 4f. However, this is of little help in identifying major lipids in other strains that have few major lipids in common. It is clear that methanogens have extremely complex polar lipid compositions—in part because some are based on C_{20}–C_{20} diethers and others on C_{40}–C_{40} tetraethers. This complexity is likely to be of considerable taxonomic value in the future.

In conclusion, it can be said that polar lipid analysis can be a useful, rapid and simple procedure in classifying new isolates of methanogenic and halophilic archaebacteria.

References

BALCH, W. E., FOX, G. E., MAGRUM, L. J., WOESE, C. R. & WOLFE, R. S. 1979 The methanogens: re-evaluation of a unique biological group. *Microbiological Reviews* **43**, 260–296.

BLIGH, E. G. & DYER, W. J. 1959 A rapid method of total lipid extraction and purification. *Canadian Journal of Biochemistry and Physiology* **37**, 911–917.

COLLINS, M. D. & JONES, D. 1981 Distribution of isoprenoid quinone structural types in bacteria and their taxonomic implications. *Microbiological Reviews* **45**, 316–354.

COLLINS, M. D., GOODFELLOW, M. & MINNIKIN, D. E. 1980 Fatty acid, isoprenoid quinone and polar lipid composition in the classification of *Curtobacterium* and related taxa. *Journal of General Microbiology* **118**, 29–37.

COLLINS, M. D., ROSS, H. N. M., TINDALL, B. J. & GRANT, W. D. 1981 Distribution of isoprenoid quinones in halophilic bacteria. *Journal of Applied Bacteriology* **50**, 559–565.

DE ROSA, M., GAMBACORTA, A., NICOLAUS, B., ROSS, H. N. M., GRANT, W. D. & BU'LOCK, J. D. 1982 An asymmetric archaebacterial diether lipid from alkaliphilic halophiles. *Journal of General Microbiology* **128**, 343–348.

DE ROSA, M., GAMBACORTA, A., NICOLAUS, B. & GRANT, W. D. 1983 A $C_{25}C_{25}$ diether core lipid from archaebacterial haloalkaliphiles. *Journal of General Microbiology* **129**, 2333–2337.

DITTMER, J. C. & LESTER, R. L. 1964 A simple, specific spray for the detection of phospholipids on thin-layer chromatograms. *Journal of Lipid Research* **5**, 126–127.

EVANS, R. W., KUSHWAHA, S. C. & KATES, M. 1980 The lipids of *Halobacterium marismortui*, an extremely halophilic bacterium of the Dead Sea. *Biochimica et Biophysica Acta* **619**, 533–544.

JACIN, H. & MISKHIN, A. R. 1965 Separation of carbohydrates on borate impregnated silica gel plates. *Journal of Chromatography* **18**, 170–176.

KATES, M. 1972 Ether-linked lipids in extremely halophilic bacteria. In *Ether Lipids, Chemistry and Biology* ed. Snyder, F., pp. 352–398. Academic Press, New York.

KATES, M. 1978 The phytanyl ether-linked polar lipids and isoprenoid neutral lipids of extremely halophilic bacteria. *Progress in the Chemistry of Fats and Other Lipids* **15**, 301–342.

KESSEL, M. & KLINK, F. 1981 Two elongation factors from the extremely halophilic archaebacterium *Halobacterium cutirubrum*. Assay systems and purification at high salt concentrations. *European Journal of Biochemistry* **114**, 481–486.

KUSHWAHA, S. C., KATES, M. & STOECKENIUS, W. 1976 Comparison of purple membrane from *Halobacterium cutirubrum* and *Halobacterium halobium*. *Biochimica et Biophysica Acta* **426**, 703–710.

KUSHWAHA, S. C., KATES, M., SPROTT, G. D. & SMITH, I. C. P. 1981 Novel polar lipids from the methanogen *Methanospirillum hungatei* GP1. *Biochimica et Biophysica Acta* **664**, 156–173.

KUSHWAHA, S. C., KATES, M., JUEZ, G., RODRIGUEZ-VALERA, F. & KUSHNER, D. J. 1982 Polar lipids of an extremely halophilic strain (R4) isolate from salt pans in Spain. *Biochimica et Biophysica Acta* **711**, 19–25.

LANGWORTHY, T. A., HOLZER, G., ZEIKUS, J. G. & TORNABENE, T. G. 1983 *Iso*- and *anteiso*-branched glycerol diethers of the thermophilic anaerobe *Thermodesulfobacterium commune*. *Systematic and Applied Microbiology* **4**, 1–17.

MINNIKIN, D. E., ALSHAMAONY, L. & GOODFELLOW, M. 1975 Differentiation of *Mycobacterium*, *Nocardia* and related taxa by thin layer chromatographic analysis of whole organism methanolysates. *Journal of General Microbiology* **88**, 200–204.

PAYNE, J. I., SEHGAL, S. N. & GIBBONS, N. E. 1960 Immersion refractometry of some halophilic bacteria. *Canadian Journal of Microbiology* **6**, 9–15.

ROSS, H. N. M., COLLINS, M. D., TINDALL, B. J. & GRANT, W. D. 1981 A rapid procedure for the detection of archaebacterial lipids in halophilic bacteria. *Journal of General Microbiology* **123**, 75–80.

SCHLEIFER, K. H. & KANDLER, O. 1972 Peptidoglycan types of bacterial cell walls and their taxonomic implications. *Bacteriological Reviews* **36**, 407–477.

SHAW, N. 1968 The detection of lipids on chromatograms with the periodate-Schiff reagents. *Biochimica et Biophysica Acta* **164**, 435–436.

Tindall, B. J., Mills, A. A. & Grant, W. D. 1980 An alkalophilic red halophilic bacterium with low magnesium requirement from a Kenyan soda lake. *Journal of General Microbiology* **116,** 257–260.

Tindall, B. J., Ross, H. N. M. & Grant, W. D. 1984 *Natronobacterium* gen. nov. and *Natronococcus* gen. nov., two genera of haloalkaliphilic archaebacteria. *Systematic and Applied Microbiology* **5,** 41–57.

Tu, J., Prangishvilli, D., Huber, H., Wildgruber, G., Zillig, W. & Stetter, K. O. 1982 Taxonomic relations between archaebacteria including 6 novel genera examined by cross-hybridization of DNAs and 16S RNAs. *Journal of Molecular Evolution* **18,** 109–114.

Woese, C. R. & Fox, G. E. 1977 Phylogenetic structure of the prokaryote domain. The primary kingdoms. *Proceedings of the National Academy of Sciences of the United States of America* **74,** 5088–5090.

Woese, C. R., Magrum, L. J. & Fox, G. E. 1978 Archaebacteria. *Journal of Molecular Evolution* **11,** 245–252.

Chromatographic Analysis of Volatile End Products of Gram-Positive Bacteria

D. B. DRUCKER AND A. TAVAKKOL[1]

Department of Bacteriology and Virology, University of Manchester, Manchester, UK

End products are formed as the result of distinct processes. Neutral and acidic products tend to result from fermentative carbohydrate metabolism during which electrons are accepted by organic compounds, e.g. pyruvate. Reduction of pyruvate leads to formation of various neutral and acidic end products. Various fermentation types are known and are characteristic of specific taxonomic groups. The lactic, ethanolic, mixed acid, butanediol, butyric, butylic, acetone and propionic fermentations are discussed elsewhere (Drucker 1981a). Amines may for convenience be regarded as basic end products. In reality, amines are intermediates which are further metabolised by most organisms. They are formed in several ways. The most important source of amines is decarboxylation of amino acids to form an amine and to liberate CO_2:

$$RCHNH_2COOH \xrightarrow{\text{decarboxylase}} RCH_2NH_2 + CO_2$$

The amino acid decarboxylase used may be constitutive or inducible (Morris & Pardee 1965). Amines such as spermine or spermidine are produced as essential components of polyribosomes. Amines can also be derived from reductive amination of alkanals (aldehydes):

$$NADH + H^+ + RCHO + NH_3 \rightarrow RCH_2NH_2 + NAD^+ + H_2O$$

Whereas neutral and acidic fermentation products are formed as characteristic end products chiefly of anaerobes, amines can also be produced by

[1]Present address: Basic Medical Sciences, School of Medicine, Memorial University of Newfoundland, St. John's, Newfoundland, Canada A1B 3V6.

0-12-289675-0

TABLE 1

Properties of acidic and neutral end products

Class	End product			pK	bp (°C)	Solubility in 100 parts		
	Trivial name	Systematic name	Formula			Water	Alcohol	Ether
Monocarboxylic acids	Formic	Methanoic	HCOOH	3.75	100.8	∞	∞	···
	Acetic	Ethanoic	CH_3COOH	4.75	118.1	∞	∞	∞
	Propionic	Propanoic	CH_3CH_2COOH	4.88	141.1	∞	∞	∞;∞Chl
	Isobutyric	2-Methylpropanoic	$(CH_3)_2CH\ COOH$	4.81	154.5	$20^{20°}$	∞;∞Chl	∞
	n-Butyric	Butanoic	$CH_3(CH_2)_2COOH$	4.82	163.5	∞	∞	∞
	Isovaleric	3-Methylbutanoic	$(CH_3)_2CHCH_2COOH$	—	176	$4.2^{20°}$	∞;∞Chl	∞
	n-Valeric	Pentanoic	$CH_3(CH_2)_3COOH$	4.81	185.4	$3.3^{16°}$	∞	∞
	Isocaproic	4-Methylpentanoic	$(CH_3)_2CH(CH_2)_2COOH$	—	—	v.sl.s.	s.	s.
	n-Caproic	Hexanoic	$CH_3(CH_2)_4COOH$	—	205	$1{,}10^{20°}$	s.	s.
Alkanols (alcohols)	Ethyl alcohol	Ethanol	CH_3CH_2OH	—	78.4	∞	∞Chl	∞
	Isopropyl alcohol	2-Propanol	$(CH_3)_2CHOH$	—	82.4	∞	∞	∞
	n-Propyl alcohol	Propanol	$CH_3(CH_2)_2OH$	—	97.2	∞	∞	∞
	tert-Butyl alcohol	2-Methyl-2-propanol	$(CH_3)_3C\ OH$	—	82.6	∞	∞	∞
	Isobutyl alcohol	2-Methyl-1-propanol	$(CH_3)_2CH\ CH_2OH$	—	108.1	$10^{15°}$	∞	∞
	n-Butyl alcohol	1-Butanol	$CH_3(CH_2)_3OH$	—	118	$9^{15°}$	∞	∞
	Isoamyl alcohol	3-Methyl-1-butanol	$(CH_3)_2CH(CH_2)_2OH$	—	132	$2^{14°}$	∞	∞
Diol	Butylene glycol	2,3-Butanediol	$CH_3CHOHCHOHCH_3$	—	181.7*	···	···	···
Aldehydes (alkanals)	Acetaldehyde	Ethanal	CH_3CHO	—	20.2	∞	∞	∞
	Butyraldehyde	Butanal	$CH_3(CH_2)_2CHO$	—	—	4	∞	∞
Ketone (alkanone)	Acetone	Propanone	CH_3COCH_3	—	56.2	∞	∞	∞
Acyloin	Acetoin	3-Hydroxybutanone	$CH_3COCHOHCH_3$	—	148	∞	v.s.	sl.s.; i.lg
Diketone	Diacetyl	Butane-2.3-dione	$CH_3COCOCH_3$	—	87.8	$25^{15°}$	∞	∞

Note: Neutral products have no p*K* value; *at 740 mm Hg Data from Lange (1952). Abbreviations: s., soluble; sl., slight or slightly; ∞, miscible, i.e., soluble in all proportions; v., very; h., hot; i., insoluble; >, greater than; <, less than; ±, about or near; ···, not known; sl., slightly; lg., ligroin; Chl, Chloroform.

TABLE 2

*Properties of basic end products (amines)**

Amine			pK	bp (°C)	Related amino acid	Solubility in 100 parts		
Trivial name	Systematic name	Formula				Water	Alcohol	Ether
Methylamine	Aminomethane	CH_3NH_2	10.62	-7.6[b]	Glycine	$959^{25°}$[a]	s.	...
Ethylamine	Aminoethane	$CH_3CH_2NH_2$	10.63	16.6	Alanine	∞	∞	∞
n-Propylamine	Aminopropane	$CH_3(CH_2)_2NH_2$	10.58	49–50[c]	α-Aminobutyric acid	∞	∞	∞
Isobutylamine	2-Methylaminopropane	$(CH_3)_2CHCH_2NH_2$	10.61[a]	68–69	Valine	∞	∞	∞
Isopropanolamine	1-Amino-2-propanol	$CH_3CHOHCH_2NH_2$	—	160–161	Threonine	v.s.	s.	i.
Isoamylamine	3-Methylaminobutane	$(CH_3)_2CH(CH_2)_2NH_2$	10.63[a]	95	Leucine	∞	∞	∞
2-Methylbutylamine	2-Methylaminobutane	$CH_3CH_2CH(CH_3)CH_2NH_2$	—	91–92	Isoleucine	...	...	...
Histamine	β-Imidazolyl-4-ethylamine	 $CH_2CH_2NH_2$	—	209–210[d]	Histidine	v.s.,s.h.	s.	i.
Cadaverine	1,5-Diaminopentane	$NH_2(CH_2)_5NH_2$	—	178–180	Lysine	v.s.	v.s.	sl.s.
Tyramine	2-(4-Hydroxyphenyl)ethylamine	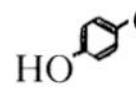 $CH_2CH_2NH_2$	—	175–181[e]	Tyrosine	<10h	10±h	sl.s.h
Putrescine	1,4-Diaminobutane	$NH_2(CH_2)_4NH_2$	—	158–160	Ornithine	v.s.	...	...
β-Phenylethylamine	2-Phenylethylamine	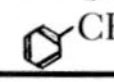 $CH_2CH_2NH_2$	—	198	Phenylalanine	s.	∞	∞

*Abbreviations, see footnote to Table 1.

[a]Data for *n* isomers of the amines. [b]At 758 mm Hg. [c]At 761 mm Hg. [d]At 18 mm Hg. [e]At 8 mm Hg.

facultative organisms. In order to appreciate the problems of analysing fermentation products, it is helpful to consider the p*K*, solubility and volatility of each product. Table 1 lists data for acids and neutral compounds. Table 2 provides similar data for amines. Metabolism of amino acids can also lead to formation of acids by deamination:

$$H^+ + NADH + (CH_3)_2CHCHNH_2COOH \rightarrow (CH_3)_2CHCH_2COOH + NH_3 + NAD^+$$

Standardisation of Growth Conditions

Several studies have revealed discrepancies between fermentation analysis results for the same organism in different laboratories. Lambert and Armfield (1979) found that their data for peptostreptococci differed from the results published by the V.P.I. Anaerobe Laboratory. In a controlled experiment (Bohannon *et al.* 1978), three laboratories each analysed fermentation products of four clostridial species. For *Clostridium perfringens*, the laboratories found acetic acid with smaller amounts of butyric, methylmalonic and malonic acids; acetic and butyric with smaller amounts of propionic, lactic and succinic acids; and lastly, acetic and butyric with less propionic acid. Apart from the need to standardise analysis and extraction conditions, growth conditions must be vigorously controlled. In the case of *Clostridium difficile*, 48 h incubation is needed for production of "typical" profiles. Additional glucose suppresses formation of isobutyric and isocaproic acids, while alcohol production is stimulated by the presence of calcium carbonate (Turton *et al.* 1982).

In the case of glucose concentration, the greatest effect on fermentation product profiles occurs between 0.0 and 0.1% (w/v) glucose with *Clostridium bifermentans*, *Clostridium sporogenes* and *Peptostreptococcus anaerobius* (Turton *et al.* 1983a).

In a study of the effect of eight growth media on fermentation profiles of 10 anaerobic bacteria (Turton *et al.* 1983a), it was found that "differences in end product profile were at least as great between different media used to grow a single anaerobic species, as between different anaerobes grown in a single type of medium." Even the "same" medium from different manufacturers will yield different results. For research purposes, all studies should ideally be performed using a medium of a single manufacturer's batch. There are no published data on the effect of growth conditions on amine production. Studies in other laboratories indicate that amines may be further metabolised upon prolonged incubation of cultures (Gale 1942). Glucose appears to stimulate production of amines; presumably, glucose fermentation lowers pH to the optimum for activity of amino-acid decarboxylases (Eggerth 1939). Certain media already contain fermentation products of interest, e.g. acetic and lactic acids. Any medium used for fermentation product analysis must be tested by analysis of an uninoculated control. The medium used ought to be one that supports good growth of the organism under investigation. Each laboratory ought to compare its findings with a library of data for known organisms analysed in the same way in the laboratory. The most widely used medium is probably peptone–yeast extract–glucose (PYG) (Holdeman *et al.* 1977), which is also described in detail by Drucker (1981a).

Preparation of Samples for GC Analysis

Preparation of samples varies widely according to laboratory and class of end products. Volatile fermentation products can be analysed in four fundamentally different ways: head-space gas analysis, direct injection of aqueous samples, extraction into organic solvent and formation of less polar chemical derivatives. A much simpler technique is the use of thin-layer chromatography (TLC), which is summarized below.

Thin-Layer Chromatographic Analysis

Any of the standard TLC texts will include separations for acids. The advantages of TLC over gas liquid chromatography (GLC) are cheapness, simplicity and the ability to separate simultaneously volatile and nonvolatile acids. The disadvantages are the lack of a convenient permanent record, relative insensitivity, longer analysis time and quantitation problems. One method specifically developed for a very wide range of bacterial end products is the separation recommended by Drucker & Melville (1966), which uses continuous TLC overnight in order to separate neutral compounds, volatile, and nonvolatile acids in a single analysis.

Head-Space Gas Analysis

This has not been widely used, but it offers the advantages of simplicity and rapidity. Products are analysed by sampling head-space gas and injecting the gas sample directly onto a gas chromatography column. A selection of techniques is reviewed elsewhere (Drucker 1981a). Acids and alcohols have been analysed successfully on columns of 5% free fatty acid phase Carbowax 20M treated with nitroterephthalic acid (FFAP) on Chromosorb G (Larsson *et al.* 1978a). Amines have been analysed by injection onto the porous polymers Chromosorb 103 or Porapak Q or onto 10% Carbowax 20M plus 2% potassium hydroxide on Chromosorb W (Larsson *et al.* 1978b).

Direct Injection of Aqueous Samples

Most stationary phases used in gas chromatography will not tolerate injection of aqueous samples that can react with column packings. Porous polymers do not react with water and can withstand injection of aqueous samples. Usually, cultures are centrifuged to remove cells and the pH of the supernatant is adjusted prior to analysis. In the case of acids, the addition of a mineral acid, e.g. hydrochloric acid, to $<$ pH 2 ensures that virtually all of the organic acids present are at a pH 3 units below their p*K* values (Table 1). At such a pH, $\log([A^-]/[HA]) = 3$, or 99.9% of acid is present as the unionised (more volatile) form, since $pH = pKa + \log([A^-]/[HA])$.

In the case of bases, the opposite argument applies. The pH is raised 3 units above the p*K* value for the amine (Table 2) prior to analysis.

Chromosorb 101 is a suitable porous polymer for analysis of acids and neutral compounds. The use of this column for fermentation product analysis was pioneered by Carlsson (1973) who subsequently found that prior extraction of acids by ion-exchange chromatography was unnecessary (personal communication). Chromosorb 101 offers three major advantages. It can be used to detect a comprehensive range of products and derivatization procedures are not necessary even for formic acid, which does not normally give a response with a flame ionization detector. Finally, extremely small volumes can be analysed, e.g. 2 μl of liquid from an agar plate.

Chromosorb 103 is a suitable porous polymer for analysis of amines. Recently, a technique has been developed that permits direct analysis of up to 13 amines in aqueous solution (Tavakkol & Drucker 1983a). This technique offers the advantage of simplicity but suffers the disadvantages of failure to detect a few higher boiling amines, e.g. indole, and of the possible detection of other classes of compounds in the crude sample.

One possible disadvantage of direct injection techniques is that, by their very nature, it is not possible to concentrate samples. Minor products may not therefore be detected.

Extraction into Organic Solvents

There is a possibility of failure to extract all the solute of interest if organic solvents are used to extract solutes. Such failures, which reflect a solute's partition ratio between organic phase and water, could produce misleading results. To minimise the problem, modification of pH is used to convert acids (or bases) to their un-ionised forms (see above). The un-ionised forms and neutral products become less soluble in water if the ionic strength of the solution is raised by addition of sodium chloride (salting out). For acids and neutral compounds, diethyl ether is the best solvent. Ether sometimes contains detectable impurities which should be sought before use. Ether extracts may be injected onto either porous polymer columns or onto packings of stationary phases coated onto supports. Chloroform is less suitable as a solvent. It attacks porous polymers, frequently contains ethanol, and is a less efficient extraction solvent for many fermentation products in aqueous solution. For injection onto stationary phases, ether extracts need to be dried.

A suitable procedure for acids and alcohols is as follows: 9 ml of culture supernatant is acidified with 1 ml of 2 *M* sulphuric acid after addition of 2 g of sodium chloride; it is then extracted by shaking vigorously with 10 ml of analytical grade diethyl ether. Any emulsion that forms can be "broken" by centrifugation. The ether layer can be removed and dried over anhydrous sodium sulphate. For analysis of acids only, a concentration step can be introduced by removal of acids from ether by 0.5 ml of 0.5 *M* sodium hydroxide. After addition of 0.5 ml of *M* sulphuric acid and 0.2 g of sodium chloride, concentrated acids can be extracted into 1 ml of diethyl ether. If amines are extracted into solvents, they can be derivatised prior to analysis (see above).

A suitable procedure for extraction of amines is as follows: Culture supernatant fluids are first acidified (pH 1–2) with concentrated H_2SO_4 and then saturated with sodium chloride. The clear supernatant fluid obtained after centrifugation of cultures (3000 g for 10 min) is then extracted with ether to remove nonbasic extractable metabolites. The ether layer is removed and discarded. Next, the residual aqueous phase is made alkaline (pH 13–14) with 8 *M* sodium hydroxide and amines are thrice extracted with equal volumes of chloroform. The chloroform extract containing amines is then treated with 3 ml of 5 *N* HCl to form salts of the amines, and the aqueous phase is retrieved and evaporated at 90–95°C. The method has the following advantages over other available techniques for extraction of amines.

(a) The amines are greatly concentrated when amine salts are formed from free bases.
(b) More volatile amines that might otherwise be lost because of evaporation of solvent are effectively preserved.
(c) Interfering substances such as lipids and other less volatile compounds are removed from the samples.
(d) The volatility of amine salts is less than that of their parent compounds; thus, they can be stored in the laboratory for months if immediate analysis is not convenient.
(e) The salts can either be derivatised as their trifluoroacetamides or converted to their free bases by addition of a normal solution of sodium hydroxide, allowing their analysis as free bases.

Formation of Less Polar Derivatives

If it is intended to analyze nonvolatile acids, less polar (more volatile) derivatives must be formed. Usually, methyl esters are made. Esterification methods are described elsewhere (Drucker 1981b). When methyl esters are prepared, volatile acids will also usually be methylated. This can be advantageous if formic acid is to be analysed; methyl formate elicits a greater flame ionisation detector (FID) response than free formic acid can on Chromosorb 101. Amines have been analysed as their heptafluorobutyryl derivatives (Kellogg *et al.* 1978), which are very suitable for electron capture detector (ECD) instruments, as trifluoroacetamides (Brooks & Moore 1969) or as Schiff bases (Tavakkol & Drucker 1983b). In our laboratory, amines are derivatised as trifluoroacetamides and as Schiff bases. Identification is less tentative if two different derivatisation procedures and analyses yield the same tentative identifications.

Trifluoroacetamides are formed as follows: Amines are dissolved in dichloromethane and then treated with a 10-fold excess of a mixture of trifluoroacetic anhydride–pyridine (14:1, v/v) either at 60°C for 30 min or at room temperature for 18 h. The excess reagents and by-products are removed by washing, first with 1 *N* HCl for 10 min and second with 0.1 *N* NaOH for the same period or until the sample is rendered colourless. The cloudy organic phase is then dried over sodium sulphate prior to GC analysis. The general reaction is:

$$RNH_2 + (CF_3CO)_2O \rightarrow RNHCOCF_3 + CF_3COOH$$

Amine + Trifluoroacetic anhydride → Trifluoroacetamide + Trifluoroacetic acid

Schiff bases are formed as follows. Amines are dissolved in amyl alcohol and a 20-fold molar excess of benzaldehyde is then added. The reaction mixture is allowed to react at room temperature for 18 h or at 60°C for 2 h. Excess reagent is removed by treating the sample with 3% hydrogen peroxide plus 0.1 *M* sodium hydroxide to form sodium benzoate. The organic layer is then separated and dried over anhydrous sodium sulphate. The general reaction is:

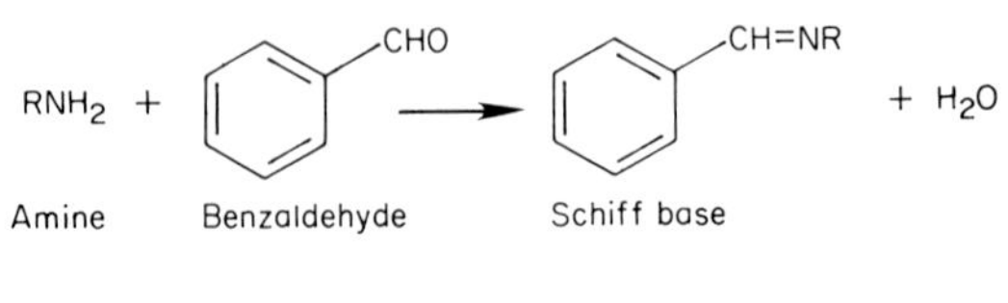

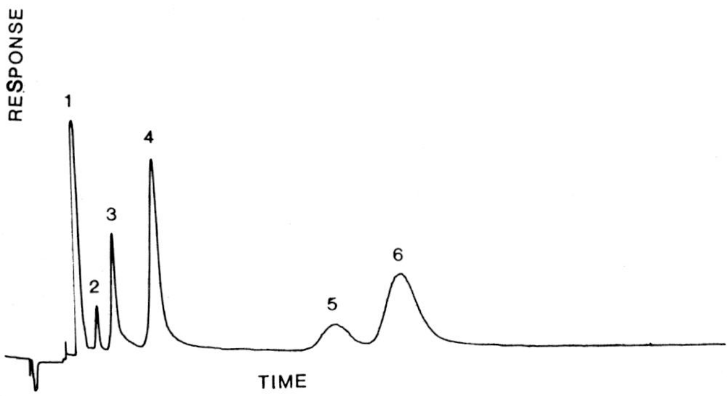

FIG. 1. Gas chromatographic separation of a mixture of acidic and neutral end products analysed isothermally (170°C) on a column of 1.52 m × 0.4 cm i.d. packed with Chromosorb 101 porous polymer. Peaks: 1,water; 2,ethanol; 3,isopropanol; 4,acetic acid; 5,lactic acid; and 6,succinic acid. (From *J. Chromatogr. Sci.* **8**, 489–490, by permission.)

Conditions for Gas Chromatographic Analysis

Suitable columns for analysis of ether extracts of acids and neutral products are Chromosorb 101 (Fig. 1), the coated packings Carbowax 20 M terminated with terephthalic acid (TPA), 6% (w/w) on Gas Chromosorb Q (Thurton *et al.* 1983a) or "DAP," which is diethylene glycol adipate (DEGA), 10%

TABLE 3

Retention data for neutral fermentation products

Product*	Retention relative to ethanol		
Acetaldehyde	0.914	—	—
Acetone	0.967	0.84	1.23
tert-Butyl alcohol	0.940	—	—
Ethyl alcohol	1.000	1.00	1.00
Isopropyl alcohol	—	1.00	1.29
Diacetyl	1.005	1.26	1.89
tert-Amyl alcohol	1.052	—	—
n-Propyl alcohol	1.076	1.47	1.55
Isobutyl alcohol	—	1.47	1.94
n-Butyl alcohol	1.212	2.63	2.63
Isoamyl alcohol	1.330	—	—
Acetoin	1.639	—	—
Column	SP1000 capillary 130°C	DAP 100°C	Chromosorb 101 195°C

*See Table 1 for systematic names.

TABLE 4

Retention data for volatile acidic fermentation products

Acid*	Retention relative to acetic acid		
Formic	—	—	0.76
Acetic	1.00	1.000	1.00
Propionic	1.34	1.29	1.65
Isobutyric	1.49	1.59	2.36
n-Butyric	1.88	2.08	2.78
Isovaleric	2.13	2.44	4.12
n-Valeric	2.94	3.40	4.91
Isocaproic	3.90	4.57	7.59
n-Caproic	4.69	5.49	8.80
Column	SP1000 capillary 130°C	DAP 130°C	Chromosorb 101 195°C

*See Table 1 for systematic names.

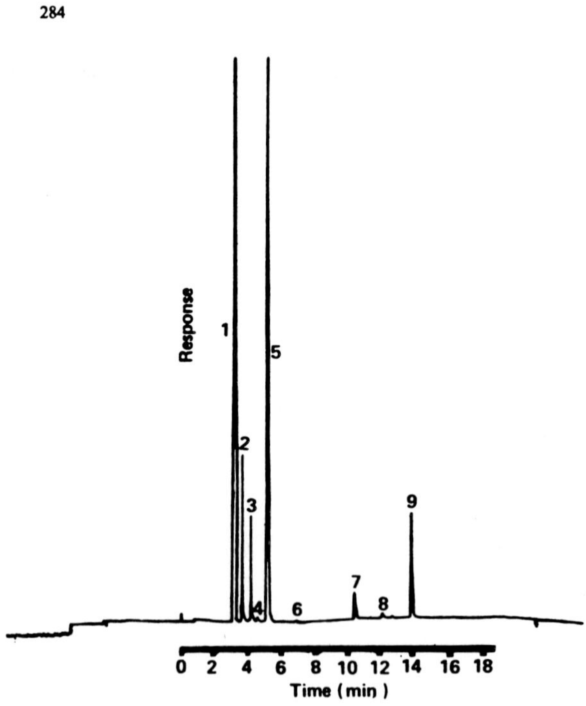

FIG. 2. Gas chromatographic separation of ether extract of supernatant fluid of *Clostridium tetani* analysed on a 30 m × 0.5 mm i.d. SP 1000 WCOT column, programed from 110 to 170°C at 10°C/min after a 7-min hold. Peaks: 1, solvent; 2, ethanol; 3, 1-propanol; 4, 2-methyl-1-propanol; 5, 1-butanol; 6, 1-pentanol; 7, acetic acid; 8, 2-methylpropanoic acid; and 9, *n*-butyric acid. (From *Journal of Chromatography* **208**, 279–285, by permission.)

with phosphoric acid, 2% on celite (Drucker 1970). Retention data with analysis temperatures are shown in Tables 3 and 4 for neutral and acidic products, respectively. Recently, a powerful capillary column separation has been described (Drucker 1981b) for volatile acids and neutral products (Fig. 2) as well as for the methyl esters of nonvolatile acids. Table 5 displays retention data for amines analysed as their trifluoroacetamides and as their Schiff bases.

It is difficult to recommend specific analysis temperatures. Retention times are affected not only by solute volatility, which is temperature dependent, but by flow rates, length of column and other factors. Temperatures are provided as guidelines in the tables. When lower-boiling (Tables 1 and 2) solutes are of interest, the use of lower temperatures will improve resolution of fast-eluting peaks. If higher-boiling solutes are of interest, it is wise to use

TABLE 5

Retention data for basic end products

Amine	Retention relative to *n*-butylamine*	
Derivative	Schiff bases	Trifluoroacetamides
Methylamine	0.37	—
Ethylamine	0.51	0.53
Isopropylamine	0.57	0.28
n-Propylamine	0.75	0.63
Triethylamine	—	—
sec-Butylamine	0.80	0.59
Isobutylamine	0.87	0.75
n-Butylamine	1.0	1.0
2-Methylbutylamine	—	—
Isoamylamine	1.11	1.09
n-Amylamine	1.23	1.21
Pyrrolidine	—	1.33
n-Hexylamine	1.43	1.34
di-*n*-Butylamine	—	1.55
Ethanolamine	—	—
Isopropanolamine	1.29	—
1,3-Diaminopropane	2.72	1.78
β-Phenylethylamine	2.08	1.99
Putrescine	2.88	2.18
Cadaverine	3.0	2.49
Tyramine	—	2.57
Indole	—	2.95
Spermidine	—	2.98
Spermine	—	3.64
n-Propanolamine	1.54	—

*All data based on temperature-programed analysis on 2.74 m by 2 mm i.d. columns of OV-17/Chromosorb W AW DMCS.

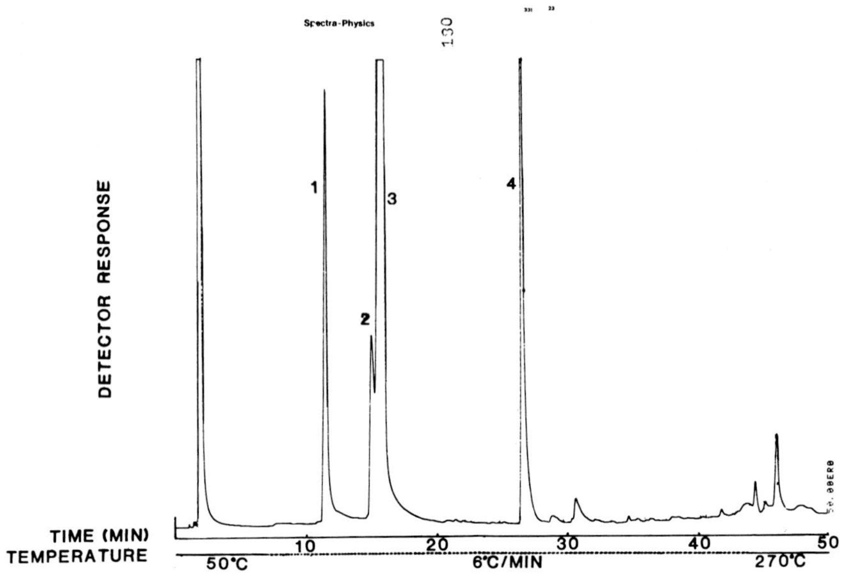

FIG. 3. Gas chromatographic separation of amines of *Clostridium tetani* converted to their trifluoroacetamides and analysed on a 2.73 m × 0.2 cm i.d. column of OV-17 on Chromosorb W AW DMCS, programed from 50°C at 6°C/min after a 7-min hold. Peaks: 1, *n*-butylamine; 2, 2-methylbutylamine; 3, isoamylamine; and 4, β-phenylethylamine.

higher temperatures. Where combinations of low- and high-boiling solutes are to be analysed, use of temperature programing may be advantageous (Thurton *et al.* 1983b). Fermentation patterns for acids and alcohols of Gram-positive anaerobic bacteria are listed in the VPI Anaerobe Laboratory Manual (Holdeman *et al.* 1977). Detailed amine profiles have not yet been published. The amines of *Clostridium tetani* are shown in Fig. 3.

Confirmation of Peak Identities

Gas chromatography is a separation, not an identification technique. However, on the basis of retention data compared to results for standards it is possible to make tentative peak identifications. If absolute identification of peaks is required, the following procedures may be followed.

(a) Co-chromatography. A sample can be "spiked" with the authentic compound and reanalysed to seek a relative increase in size of the peak of interest.
(b) Reanalysis on other columns. If a sample is reanalysed on a column of differing polarity, the retention data for solutes will alter markedly. A

peak that fails to appear where predicted on the second column must be misidentified. An alternative, though less satisfactory, procedure is to reanalyse samples on the same column at a lower, or higher temperature.

(c) On-line mass spectrometry. This is feasible only in large organisations with a chemical analysis section that has a mass spectrometer. Chemical ionization mass spectrometry of peaks, as they are eluted from a gas

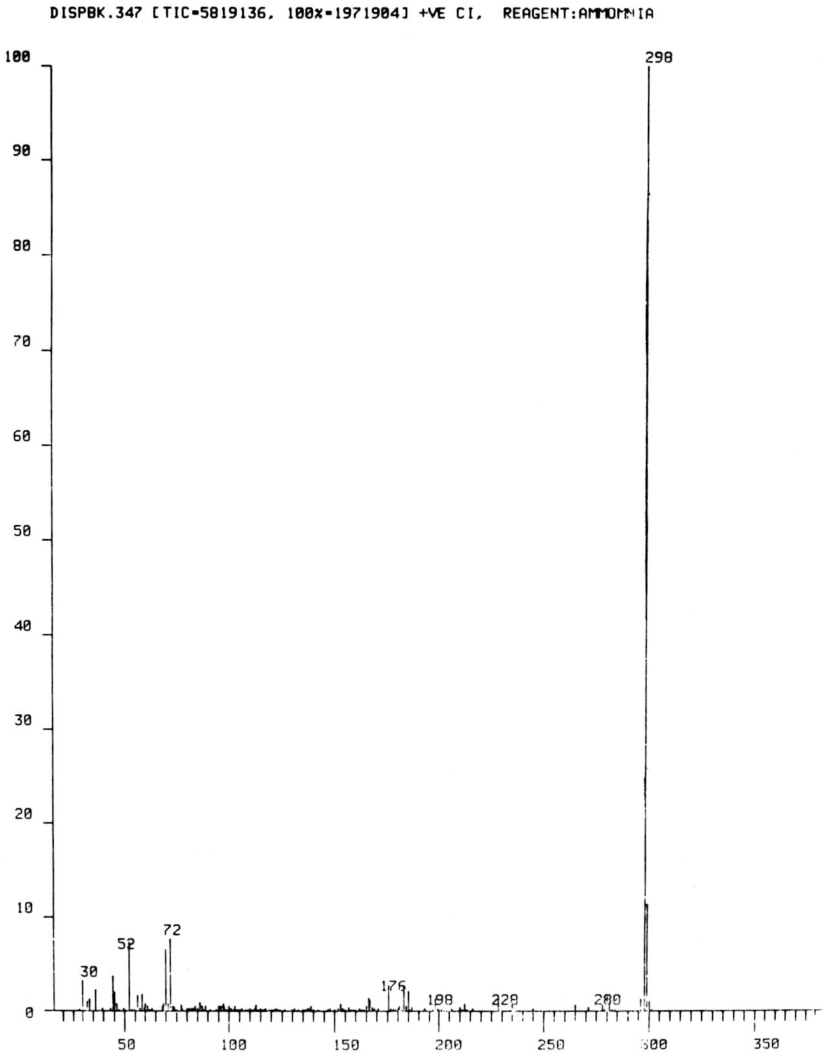

FIG. 4. Chemical ionization mass spectrum of putrescine trifluoroacetamide. Base peak at *m/e* 298 corresponds to molecular weight of TFA-putrescine.

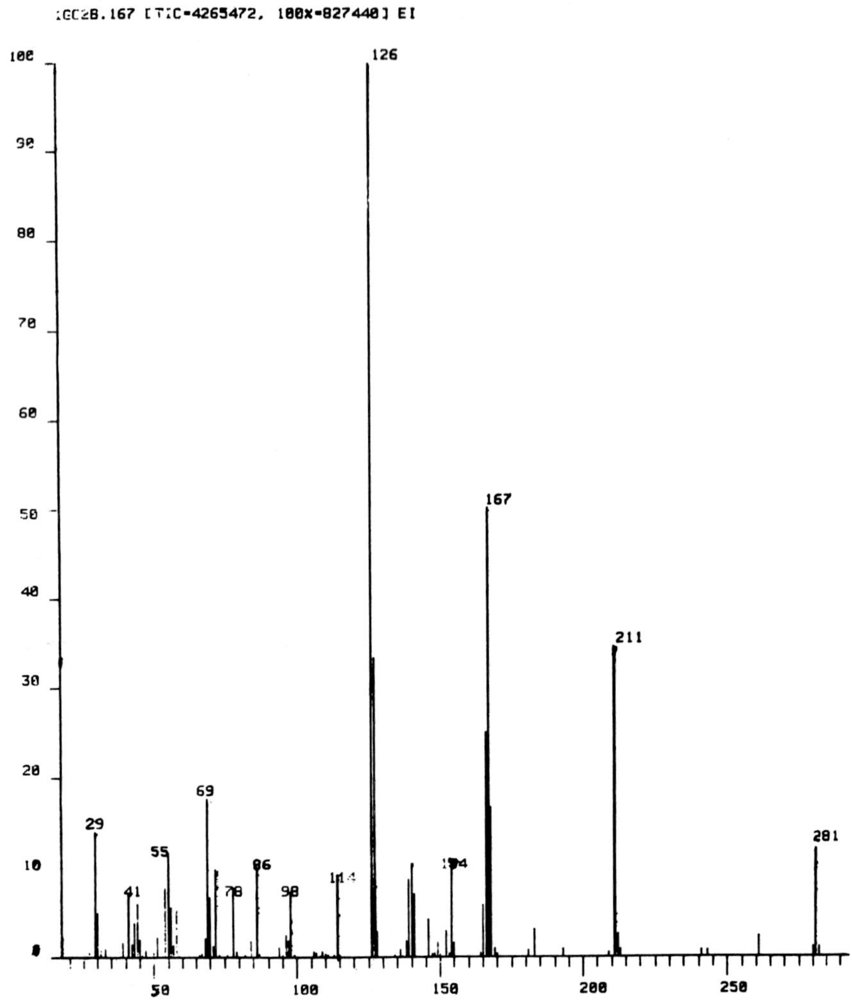

FIG. 5. Electron ionization mass spectrum of putrescine trifluoroacetamide.

chromatography column, provides data on molecular weights (Fig. 4). Electron-ionisation mass spectrometry tends to fragment molecules so that a strong molecular ion may be absent; on the other hand, it will produce a characteristic "fingerprint" of fragment ions (Fig. 5).

Quantitative Analysis

Usually, quantitative analyses are not needed. Quantitative analysis requires admixture with a known amount of an internal standard prior to analysis. For each solute of interest, a calibration graph is usually prepared using a known range of concentrations of solute and a standard quantity of internal standard. The same quantity of internal standard is added to the sample and the relative peak area of solute:standard is measured. The amount of solute can then be determined. Full details are described elsewhere (Drucker 1981a).

References

BOHANNON, T. E., MANIUS, G., MAMARIL, F. & LI WEN, L.-F. 1978 Quantitative methods for the gas chromatographic characterization of acidic fermentation by-products of anaerobic bacteria. *Journal of Chromatographic Science* **16**, 28–35.

BROOKS, J. B. & MOORE, W. E. C. 1969 Gas chromatographic analysis of amines and other compounds produced by several species of *Clostridium*. *Canadian Journal of Microbiology* **15**, 1433–1437.

CARLSSON, J. 1973 Simplified gas chromatographic procedure for identification of bacterial metabolic products. *Applied Microbiology* **25**, 287–289.

DRUCKER, D. B. 1970 Analysis of lactic acid and volatile fatty acids extracted from bacterial fermentations. *Journal of Chromatographic Science* **8**, 489–490.

DRUCKER, D. B. 1981a *Microbiological Applications of Gas Chromatography*. Cambridge: University Press.

DRUCKER, D. B. 1981b Wall-coated open-tubular capillary column system for gas chromatographic analysis of neutral and acidic fermentation endproducts. *Journal of Chromatography* **208**, 279–285.

DRUCKER, D. B. & MELVILLE, T. H. 1966 A high resolution system for chromatographic analysis of end-products of bacterial fermentations. *Journal of Applied Bacteriology* **29**, 536–539.

EGGERTH, A. H. 1939 The production of histamine in bacterial cultures. *Journal of Bacteriology* **37**, 205–222.

GALE, E. F. 1942 The oxidation of amines by bacteria. *Biochemical Journal* **36**, 64–75.

HOLDEMAN, L. V., CATO, E. P. & MOORE, W. E. C. 1977 *Anaerobe Laboratory Manual*, 4th Edn. Blacksburg: V.P.I., Anaerobe Laboratory.

KELLOGG, D. S., BROOKS, J. B. & SHEPHERD, M. E. 1978 In *Immunobiology of Neisseria gonorrhoeae* ed. Brooks, G. F., pp. 360–361. Washington: American Society for Microbiology.

LAMBERT, M. A. & ARMFIELD, A. Y. 1979 Differentiation of *Peptococcus* and *Peptostreptococcus* by gas-liquid chromatography of cellular fatty acids and metabolic products. *Journal of Clinical Microbiology* **10**, 464–476.

LANGE, N. A. 1952 *Lange's Handbook of Chemistry*. New York: McGraw Hill.

LARSSON, L., MARDH, P.-A. & ODHAM, G. 1978a Detection of alcohols and volatile fatty acids by head-space gas chromatography in identification of anaerobic bacteria. *Journal of Clinical Microbiology* **7**, 23–27.

LARSSON, L., MARDH, P.-A. & ODHAM, G. 1978b Analysis of amines and other bacterial products by head-space gas chromatography. *Acta Pathologica et Microbiologica Scandinavica, Section B: Microbiology* **86B**, 207–213.

MORRIS, D. R. & PARDEE, A. B. 1965 A biosynthetic ornithine decarboxylase in *Escherichia coli*. *Biochemical Biophysical Research Communications* **20** 697–702.

TAVAKKOL, A. & DRUCKER, D. B. 1983a Gas chromatographic analysis of bacterial amines as their free bases. *Journal of Chromatography* **274**, 37–44.

TAVAKKOL, A. & DRUCKER, D. B. 1983b Qualitative gas chromatographic analysis of bacterial amines as their Schiff bases and other derivatives. *Chromatographia* **17**, 318–321.

TURTON, L. J., DRUCKER, D. B. & GANGULI, L. A. 1982 Effect of incubation time and calcium carbonate and glucose in the growth medium, upon the fermentation end-product profile of *Clostridium difficile*. *Microbios* **35**, 7–16.

TURTON, L. J., DRUCKER, D. B. & GANGULI, L. A. 1983a Effect of glucose concentration in the growth medium upon neutral and acidic fermentation end-products of *Clostridium bifermentans*, *Clostridium sporogenes* and *Peptostreptococcus anaerobius*. *Journal of Medical Microbiology* **16**, 61–67.

TURTON, L. J., DRUCKER, D. B., HILLIER, V. F. & GANGULI, L. A. 1983b Effects of eight growth media upon fermentation profiles of ten anaerobic bacteria. *Journal of Applied Bacteriology* **54**, 295–304.

Detection of Acidic End Products of Metabolism of Anaerobic Gram-Negative Bacteria

H. N. SHAH, R. A. NASH AND J. M. HARDIE

Oral Microbiology, London Hospital Medical College, London, UK

D. A. WEETMAN, D. A. GEDDES AND T. W. MACFARLANE

University of Glasgow Dental Hospital and School, Glasgow, UK

A complex spectrum of metabolic end products is usually formed when fermentative bacteria metabolise carbohydrates or nitrogenous substrates. Although some anaerobic bacteria possess an electron-transport chain, organic molecules such as pyruvate or fumarate, rather than oxygen, act as electron acceptors for metabolism and energy conservation (Rizza *et al.* 1968; Macy *et al.* 1975; Morris 1975; Kroger 1977; Mountfort & Robinson 1977; Thauer *et al.* 1977; Reddy & Peck 1978; Stouthamer 1980; Shah & Williams 1981). The implications of cytochrome patterns (Reddy & Bryant 1977; Chapter 19), the electron carrier protohaem (Shah *et al.* 1979; Shah & Collins 1983) and menaquinone composition (Shah & Collins 1980; Shah & Collins 1983) to the classification and identification of Gram-negative anaerobic rods have recently been reported. However, it is the acid end products of these metabolic processes that have been most widely used in both research and routine diagnostic laboratories.

Acid end product analysis is now regarded as one of the essential biochemical tests for the characterization of anaerobic bacteria. The types of acids produced by a particular species are reasonably constant under a given set of conditions but end products do vary, both qualitatively and quantitatively, in response to variations in cultural conditions; the latter include availability of substrate, growth rate, pH and gaseous atmosphere (see, for example, Yamada & Carlsson 1976). Thus any attempt to use the acid end products of bacteria as diagnostic features requires that the organisms are grown under standardised conditions.

CHEMICAL METHODS IN
BACTERIAL SYSTEMATICS

0-12-289675-0

Acid end products derived from fermentative metabolism have long been known (Wolf & Telfer 1917), but the original techniques were very laborious and time consuming and were consequently rarely used in bacterial systematics. However, the introduction of gas–liquid chromatography (GLC) as a simple and rapid method of identifying acid end products soon led to the widespread use of the short-chain fatty acids as valuable and reliable criteria for definitive identification of many types of bacteria, particularly anaerobes (Sutter *et al.* 1972; Dowell & Hawkins 1974; Holdeman *et al.* 1977; Deacon *et al.* 1978; Hammann & Werner 1980.)

Other methods available for detecting acid end products have not hitherto had the same universal acceptance as GLC. Advances in high-performance liquid chromatography (HPLC) will undoubtably play a major role in future analysis of metabolic end products but, to date, this technique has had limited application (Turkelson & Richards 1978; Ehrlich *et al.* 1981). High-pressure liquid chromatography has several advantages over GLC, and minimum sample preparation is required (see below). The most sensitive method of analysis, however, is by isotachophoresis where nanomole levels of acid end products are detected. As with HPLC, complex extraction or derivatisation of the acids is not required and the technique is rapid and reliable. None of the methods outlined distinguish between the D(−) and L(+) isomers of lactic acid, and for this purpose enzymatic methods are required. It is important to detect the isomers of lactic acid for the definitive identification of many lactic acid bacteria (Sharpe 1979), but the taxonomic potential of these data for the Gram-negative anaerobic bacteria is only just becoming apparent.

One of the most modern techniques applied in the analysis of metabolic end products is proton "spin-echo" nuclear magnetic resonance spectroscopy (R. A. Iles & H. N. Shah, unpublished data). With this technique all metabolic end products can be detected at concentrations of 0.5 m*M* upwards and alcohols, ketones, esters, and basic organic compounds can be examined simultaneously. Although analysis is rapid (5–10 min) and samples require no pretreatment, the value of this technique is likely to be in *in vivo* studies such as in the detection of normal and abnormal organic metabolites in plasma and urine in patients (Iles *et al.* 1983) and not as an analytical tool for culture supernatants. Indeed, a comprehensive review of the technique and its applications to the examination of metabolites in living tissue has been published (Iles *et al.* 1982).

In this chapter, an assessment is made of the potential of some of the currently available techniques for the detection of acid end products of metabolism using both standard mixtures of acids and culture supernatants from a wide range of Gram-negative anaerobic rods.

Gas–Liquid Chromatography

Gas–liquid chromatography (GLC) has many applications in microbiology (Drucker 1981). It has been used for the detection of neutral and basic fermentation products (Garibaldi & Bayne 1970) and for the detection of microorganisms in various materials including food, clinical specimens and samples from outer space. The technique has also been adapted for analysis of structural components of microorganisms, such as lipids, proteins, nucleic acids, polysaccharides, teichoic acids and peptidoglycans. In general, the sample is injected through a rubber septum into the carrier-gas stream and is then carried through the column which contains a column packing. The type of column packing used depends on the nature of the sample. The column is mounted in an accurately temperature-controlled oven as the speed and efficiency of separation are temperature dependent. A detector at the column outlet responds to the emergence of the separated compounds and produces an electrical signal which may be amplified and displayed on a recorder. The technique employed in these laboratories separates the acidic components in broth culture supernatants mainly on the basis of the number of carbon molecules within each component. The greater the number of carbon molecules, the slower the component flows through the column.

In the present study, two methods of analysis were performed, one for volatile (method 1) and the other for nonvolatile fatty acids (method 2).

Method 1

Cell-free broth culture supernatant (1.5 ml) was mixed with 1.5 ml of the ion-exchange resin DUOLITE C225 in the H^+ form (B. D. H. Ltd., Chadwell Heath, UK). Conversion of the resin to the H^+ form was facilitated by adding resin (500 g) to 1 litre of 1 *M* HCl and stirring overnight. The fluid was then filtered off and fresh molar HCl was added for a further 18 h. The resin was then washed free of HCl with distilled water. The mixture was shaken and then allowed to stand and settle for approximately 60 min. A volume of 100 μl of the internal standard solution (20 mmol of aqueous 2-propanol) was added to 600 μl of the resin-treated fluid, and 1 μl of this resin-free mixture was injected into the column of a PYE model 104 instrument (PYE-UNICAM, Cambridge, UK). The conditions for the analysis are as listed below:

Dual glass columns	1.8 × 4 mm i.d.
Column packing	Chromosorb 101
Injection port temperature	250°C

Oven temperature, isothermal run	185°C
Detector temperature	250°C
Flame ionization detectors (FID)	
Carrier gas (Nitrogen)	Flow rate 50 ml/min
Hydrogen (for FID)	Flow rate 50 ml/min
Air (for FID)	Flow rate 500 ml/min

Integration of peaks was made using a Shimadzu C-RIB recording integrator (Dyson Instruments, Hetton, UK). The integrator was calibrated for absolute retention times and relative response factors (to the internal standards) using a single calibration curve. Ten consecutive injections of a carefully prepared acid mixture were injected into the column and the resulting responses were used for calibration (Table 1).

Calibration was checked for accuracy by injecting known concentrations of acid, both singly and in combination. For the volatile acids, reproducibility was better than ± 4%, whereas for lactic and succinic acids a figure of ± 11% was obtained. A typical chromatogram obtained following the injection of the acid standards is shown in Fig. 1a.

Method 2

The nonvolatile fatty acids (NVFA) were methylated according to a modification of the procedure described below (Bulletin 748F, Supelco UK, Ltd.) (Holdeman *et al.* 1977). Benzoic acid was added to the sample as an internal standard prior to extraction (Mayhew & Gorbach 1977).

TABLE 1

Relative times and response factors for components of aqueous acid standards using Chromosorb 101 as column packing

	Concentration (m*M*)	Relative response factor	Absolute retention time (min)
Ethanol	—	—	0.8
Internal standard	20	1	0.97
Acetic acid	5	0.81	1.57
Propionic acid	5	0.33	2.46
Isobutyric acid	5	0.21	3.42
n-Butyric acid	5	0.21	4.0
Isovaleric acid	5	0.18	5.68
n-Valeric acid	5	0.19	6.96
Lactic (D and L) acid	45	2.07	8.88
Hexanoic acid	10	0.08	12.0
Succinic acid	20	1.10	23.9

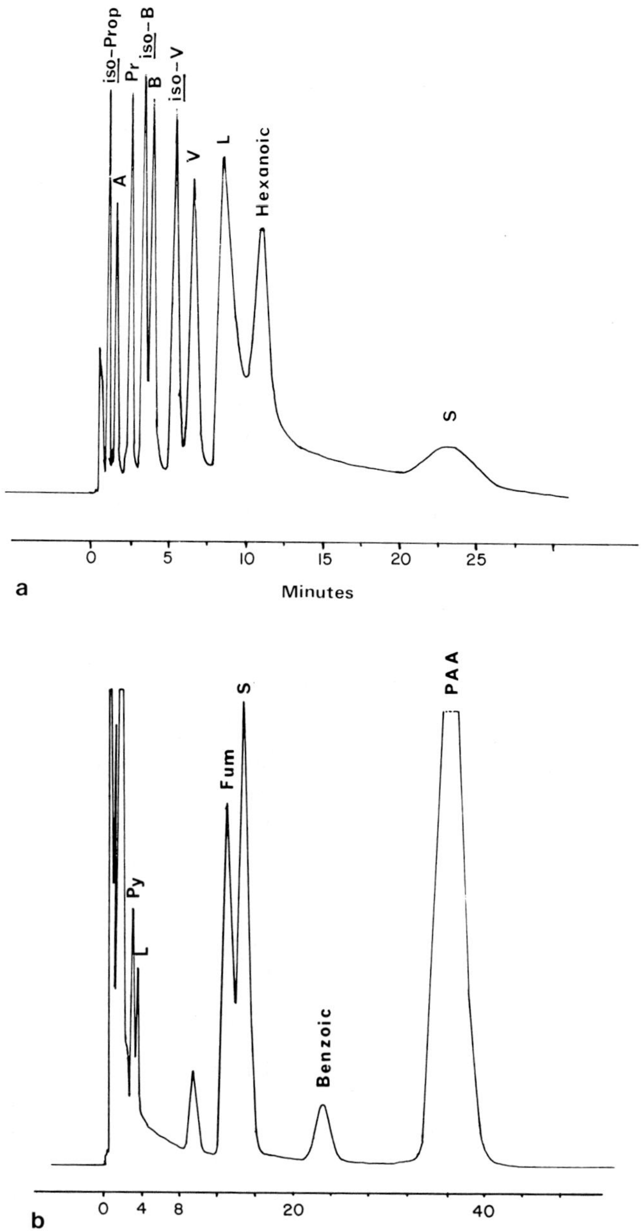

FIG. 1. (a) Chromatograph of volatile fatty acid standard solution (Table 1); 1 μl of duolite C225 treated sample injected. (b) Chromatograph of nonvolatile fatty acid standard solution (Table 2); 2 μl of methylated sample injected. Abbreviations: A, acetic; B, *n*-butyric; iso-B, isobutyric; F, formic; Fum, fumaric; L, lactic; PAA, phenylacetic; iso-P, isopropanol; Pr, propionic; Py, pyruvic; S, succinic; iso-V, isovaleric; V, valeric; UV, ultraviolet; RI, refractive index.

TABLE 2

Relative times and response factors for components of methylated acid standards using Chromosorb 101 as column packing

	Concentration (m*M*)	Relative response factor	Absolute retention time (min)
Internal standard	10	1	24.44
Pyruvic acid	10	2.33	3.16
Lactic acid	10	3.39	3.56
Fumaric acid	10	0.28	14.18
Succinic acid	10	0.20	15.8
Phenylacetic acid	10	0.07	38.77

To 1.0 ml of culture supernatant 0.5 ml of aqueous benzoic acid (10 mmol), 2.0 ml of methanol, and 0.5 ml of H_2SO_4 (50%) were added. The contents were then mixed and left at room temperature overnight or held at 55°C for 30 min. After incubation, 1.0 ml of distilled water and 0.5 ml of chloroform were added to the mixture and the contents were mixed and centrifuged (1000 *g*). The chloroform layer was withdrawn, and 2 μl of the mixture was used for the GLC analysis as described in method 1.

Four consecutive injections of the nonvolatile acid standard mixtures were injected into the column and the resulting responses were used for calibration (Table 2).

Calibration was checked for accuracy by injecting known concentrations of acid, both singly and in combination. The reproducibility for the nonvolatile acids was seen to range from ± 4% for fumaric acid to ± 15% for pyruvic acid. The chromatogram obtained from the acid standard injection is shown in Fig. 1b.

Comparison of methods 1 and 2

In the first method, which has the minimum of sample pretreatment, the VFA may be detected and quantified with a reasonable degree of accuracy. This method, however, does lack sensitivity in the detection of the NVFA. In order to improve the detection of these substances, it is necessary to subject the culture supernatant to a slightly more complex extraction procedure. Using the method described, a single gas chromatograph may be utilised to analyse samples obtained by two different extraction techniques without changing columns or altering the analytical conditions. It should be pointed out that the detection of lactic acid in methylated samples may be complicated if even low levels of VFA are present in the sample. This is due to the overlap of the retention times of the pyruvic, lactic and some VFA.

Circumstances where it may be necessary to consider examining methylated samples include: (a) when no acids at all are detected in the aqueous extracts, (b) when only acetic acid is detected, and (c) in the case of Gram-negative anaerobic rods, when no butyric acid peak is present.

High-Performance Liquid Chromatography

High-performance liquid chromatography is, in principle, very similar to GLC, but it employs a liquid rather than the gaseous mobile phase. In HPLC (unlike GLC) both stationary and, to a lesser extent, the mobile phase have some affinity for the sample. Both systems are run in columns, but the higher viscosity of liquids greatly increases the force required to move the mobile phase through a packed column bed and partly accounts for the high pressure (up to 6000 psi) demanded in HPLC. Furthermore, the diffusion coefficients of sample components are smaller in liquids than in gases so that mass transfer of the mobile phase is restricted. As a result, very small particles are used for the stationary phase (5–10 μm in diameter) to reduce the diffusion distance and improve mixing of both phases. The use of such small particles also contributes to the high pressure requirements in HPLC. Both the stationary and mobile phases are critical factors in the separation of sample components by HPLC and must be chosen carefully. The best system currently available for the separation of acidic fermentation end products involves the use of cation-exchange resin columns in which the anions are separated using isocratic elution with mild acid (Turkelson & Richards 1978; Ehrlich *et al.* 1981; Bio-Radiations publications, No. 40E, 1982). The procedure described below is based on these publications. Because organic acids yield good ultraviolet (UV) absorption spectra and because the refractive index (RI) compared to the eluent is high, both RI and UV detectors can be used.

High-performance liquid chromatography was performed on a Waters Associates chromatograph (model ALC/GPC 244) fitted with a solvent delivery system (model 6000A), an injection system (model U6K), an RI detector (model R401) and a fixed wavelength detector (254 nm). The cation exchange resin column consisted of a 9.5-mm stainless-steel tube (7.8 i.d. × 300 mm) fitted with a microguard column cation trap kindly loaned by Bio-Rad Laboratories (Watford, Hertfordshire, UK). The column was operated at room temperature and the eluent was O.012 N H_2SO_4 with a flow rate of 0.7 ml min^{-1}. The operating pressure was between 800 and 1000 psi with a maximum setting of 1500 psi. The RI detector was set at × 16 and 100-μl samples were injected. Since this instrument was equipped with only a 254-nm UV detector and these organic acids absorb maximally at 210 nm, samples were rerun on an LKB (2150 pump and 2154 Valve Injector) HPLC system in order to obtain more accurate UV measurements. The same column and operating conditions were maintained for the LKB system.

Evaluation of method

The application of HPLC for the analysis of acid end products of metabolism is relatively new and no systematic analysis of the supernatants from anaerobic Gram-negative, nonsporing rods has been attempted. Preliminary studies in these laboratories indicate that under the present experimental conditions (using BM medium, Shah *et al.* 1979) RI detection is of greater value

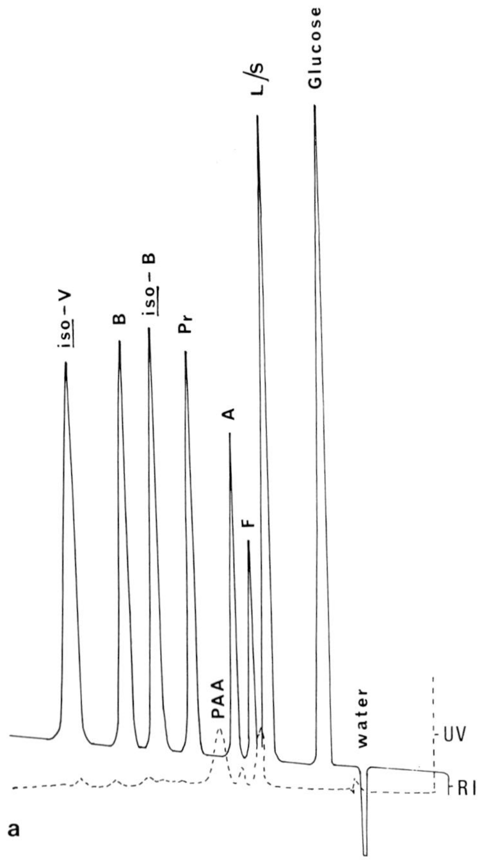

FIG. 2. (a) Separation of acid standards using RI and UV detectors (254); (b) Separation of acid standards using UV (206 nm); (c) Separation of acids from BM medium using RI and UV (254 nm); (d) Separation of acids from *Bacteroides putredinis* grown in BM medium using RI and UV (254 nm). Abbreviations as for Fig. 1. (*Continued on pp. 325 and 326.*)

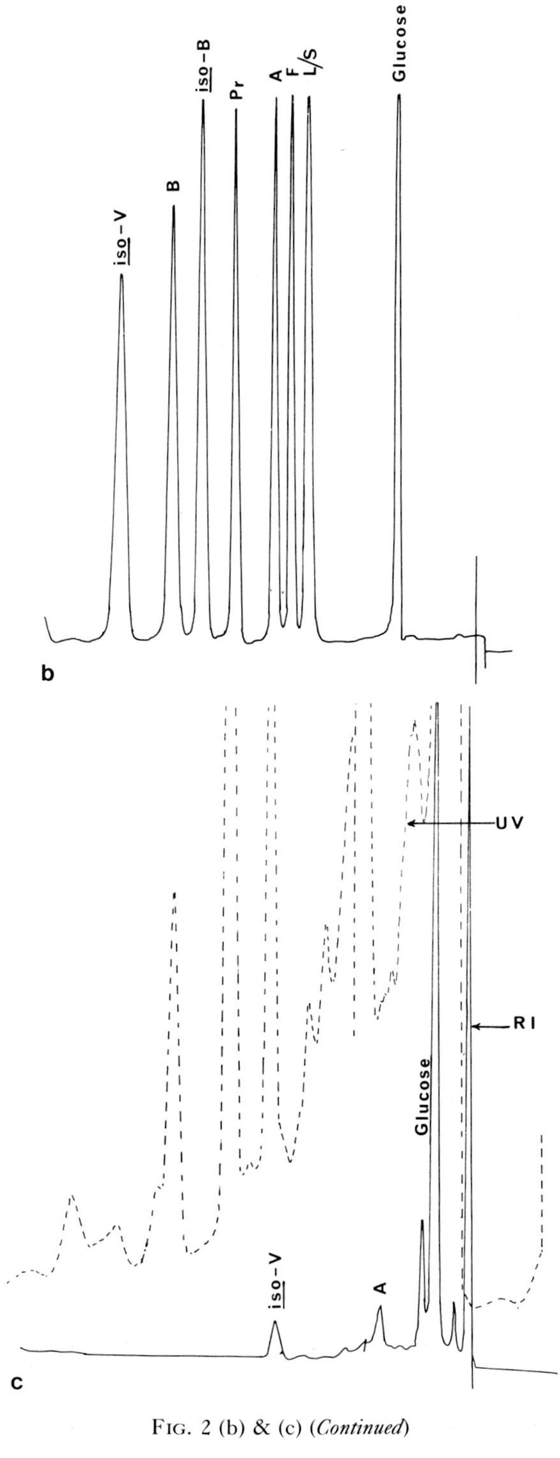

FIG. 2 (b) & (c) (*Continued*)

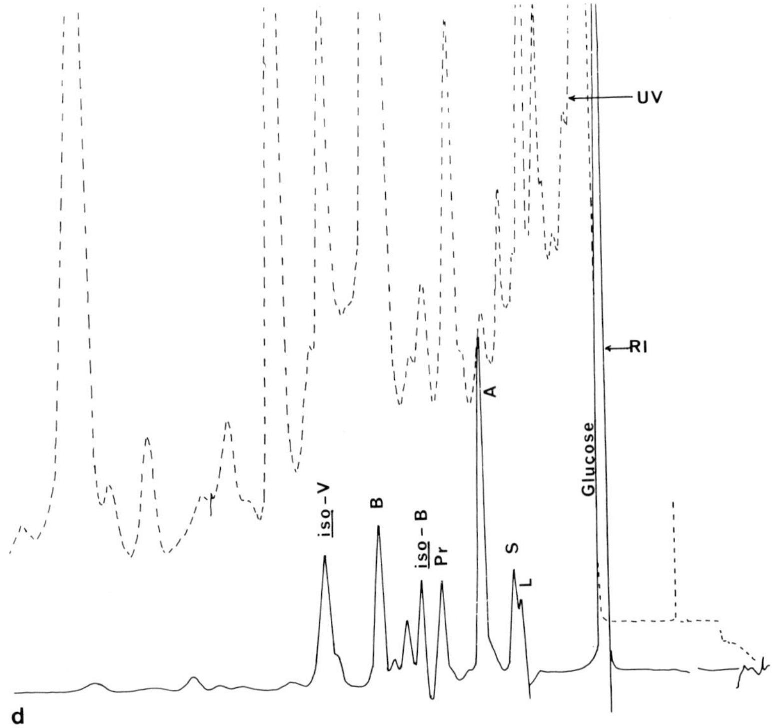

FIG. 2 (d) (*Continued*)

than UV. The separation of mixtures of standards (10 m*M*) is shown using RI and 254-nm UV detection (Fig. 2a) (Waters Associates) and UV (210 nm) detection (Fig. 2b) (LKB) using the same cation-exchange resin column (Bio-Rad Laboratories). Caproate, not shown in either figure, was detected about 53 min after injection. With acidic standards, UV (210 nm) appeared to be more sensitive, but with culture supernatants (BM medium), UV detection either at 210 or 254 nm yielded traces that were far too complex to resolve into their relevant constituent metabolic end products (Figs. 2c and 2d), and RI detection appeared to offer a better alternative.

Under the experimental conditions used, neither lactate nor succinate were resolved at room temperature. However, clear separation of these acids can apparently be achieved by maintaining a column temperature of 45°C during separation (Bio-Rad Laboratories, No. 40E, 1982). The level of detection of acids ranges between 0.1 and 0.5 m*M* and is therefore more sensitive than GLC. Indeed, HPLC may eventually become as common as GLC for the analysis of metabolic end products as more variations on the technique are tested and the high UV background of the medium is resolved.

Isotachophoresis

Isotachophoresis is a relatively new electrophoretic separation technique that can be utilized for qualitative and quantitative analysis of ionic species. In this technique, the ions being separated have the same migration velocity, hence the name (Greek *iso* = equal, *tacho* = velocity). In contrast, other electrophoretic techniques are based on ions moving at different velocities, the actual migration velocity being a function of the net charge of the ion.

Isotachophoresis uses the principle of electrophoresis in a discontinuous electrolyte system (i.e. two different electrolytes) and once equilibrium has been reached, the sample ions are separated by differences in their effective mobilities. The discontinuous electrolyte system has a leading electrolyte, having an ion with a high set mobility, and a terminating electrolyte, having an ion with a low set mobility. The two electrolytes should be chosen so that the mobilities of the sample ions under observation are lower than that of the leading electrolyte ion but higher than that of the terminating electrolyte ion. A counterion is chosen to give a good buffering capacity on the pH interval in which the separation takes place. The technique has been employed in a wide variety of analytical procedures including analysis of bacterial enterotoxins (Yotis & Catsimpoolas 1975), enzymes (Wadstrom *et al.* 1977), bacterial fermentation products (van der Hoeven *et al.* 1978), acids produced by *Candida albicans* (Samaranayake *et al.* 1982), and in dental plaque (a soft tenacious deposit, incorporating several metabolically active microorganisms that adhere to the tooth surface) (Edgar 1981; Geddes & Weetman 1981).

Instrumentation

The apparatus used was a LKB 2127 Tachophor (LKB Instruments, Bromma, Sweden), fitted with a thermal and a UV-light detection system. Separation was achieved in a 610-mm two-turn capillary (i.d. 0.5 mm) maintained at a constant temperature of 12°C.

Electrolytes and analytical conditions

All solutions were prepared with double-distilled deionised water. The leading electrolyte was 5 m*M* hydrochloric acid (AR Grade, BDH Chemicals Ltd., Poole, UK) adjusted to pH 4.0 by the addition of β-alanine (Biochemical Grade, BDH), which acted as the counterion. Hydroxypropylmethylcellulose (Sigma, Poole, UK) (0.1%, w/v) was added to the leading electrolyte to sharpen zone boundaries by reducing electroendosmosis. The terminating electrolyte was saturated (≃ 5 m*M*) *n*-octanoic acid (BDH Laboratory Grade). The detection current was 50 μA, although an initial current of 100 μA was employed for the first 15 min to shorten the analysis time. After each separation, the capillary and injection port were cleaned by rinsing with the leading and terminating electrolytes, respectively. Under these conditions acetate, butyrate, formate, lactate, propionate, pyruvate and succinate were identified and quantified.

Standardisation

A 5 m*M* standard solution containing acetic, butyric, formic, lactic, phosphoric, pyruvic, propionic and succinic acids was prepared. Standard calibration curves were then constructed by determining zone lengths obtained from injections of 0–5 μl volumes of the standard solution against the amount of sample injected.

Qualitative and quantitative analysis of acids

Qualitative analysis of sample components was performed by estimating the relative step heights of each component in the thermal detector tracing (i.e. the height of each zone above the leading electrolyte). The relative step heights of each component for a given operational system are constant, hence the components of a mixture can be identified by referring to the step heights found for the standard solution. Furthermore, "priming" the sample with a known anion can be used as an adjunct to identification of unknown ionic species when the thermal and UV-light tracings would indicate a lengthening of the particular zone in the isotachopherogram. Both of these methods were used for qualitative analysis of acid end products.

The calibration curves demonstrated highly significant relationships between zone lengths and sample size (correlation coefficient, $r = 0.99$ and $p = 0.001$). The standard calibration curves for acetate and succinate are shown in Fig. 3a, and a standard trace is shown in Fig. 3b. The time taken for the separation of a 1-μl sample of culture supernatant was ~35 min. The minimum quantifiable amount was 1 nmol, corresponding to a concentration of 1 m*M*. Since the BM medium contained detectable amounts of acetic, formic, lactic, phosphoric and propionic acids and succinic anions (see Fig. 3c), the samples were compared with uninoculated medium to determine which anions were released during fermentation. Figure 3d shows a slight increase in the formate zone and the larger increases in the succinate and acetate zones in the supernatant from *Bacteroides intermedius* T588. Similarly, an increase in the acetate and butyrate zones was observed in the supernatant from *Bacteroides gingivalis* W83 (Fig. 3e).

Isotachophoresis is also sensitive enough to detect small amounts of anions present in supernatants from nonfermentative species such as *Bacteroides coagulans* (ATCC 29798) (Fig. 3f). Indeed, using this technique the anions produced when the organism was grown without glucose can readily be detected (Figs. 3g and 3h). Traces of these anions cannot be measured reliably by GLC.

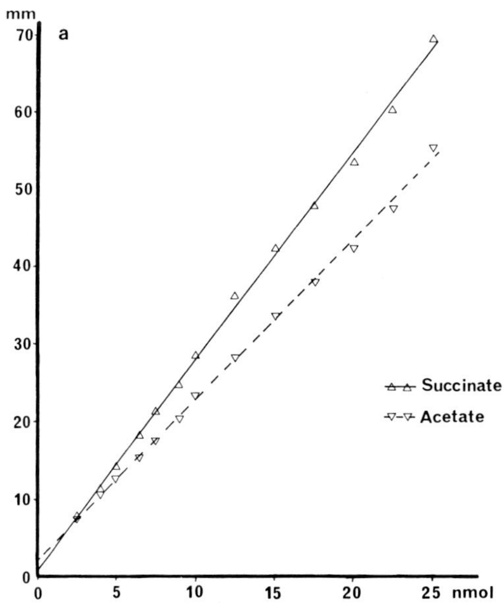

FIG. 3. (a) Standard calibration curves for succinate and acetate. Similar lines were constructed for the other anions in the standard solution. (b) Isotachopherogram of a standard solution (1 μl) containing 5 nmol each of formic (F), pyruvic (Py), phosphoric (Ph), lactic (L), succinic (S), acetic (A), propionic (Pr), and butyric (B) acids. LE, leading electrolyte (5 n*M* chloride at pH 4). TE, terminating electrolyte (5 n*M* *n*-octanoic acid). (c) Isotachopherogram of BM medium with glucose (1 μl). Abbreviations as for b. (d) Isotachopherogram of supernatant from fermentation of *Bacteroides intermedius* T588 in BM medium with glucose (1 μl). Abbreviations as for b. (e) Isotachopherogram of supernatant from fermentation of *Bacteroides gingivalis* W83 in BM medium with glucose (1 μl). Abbreviations as for b. (f) Isotachopherogram of supernatant from fermentations of *Bacteroides coagulans* ATCC 29798 in BM medium with glucose (1 μl). Abbreviations as for b. (g) Isotachopherogram of supernatant from fermentation of *Bacteroides intermedius* T588 in BM medium without glucose (1 μl). Abbreviations as for b. (h) Isotachopherogram of supernatant from fermentation of *Bacteroides gingivalis* W83 in medium without glucose (1 μl). Abbreviations as for b. (*Continued on pp. 330–333.*)

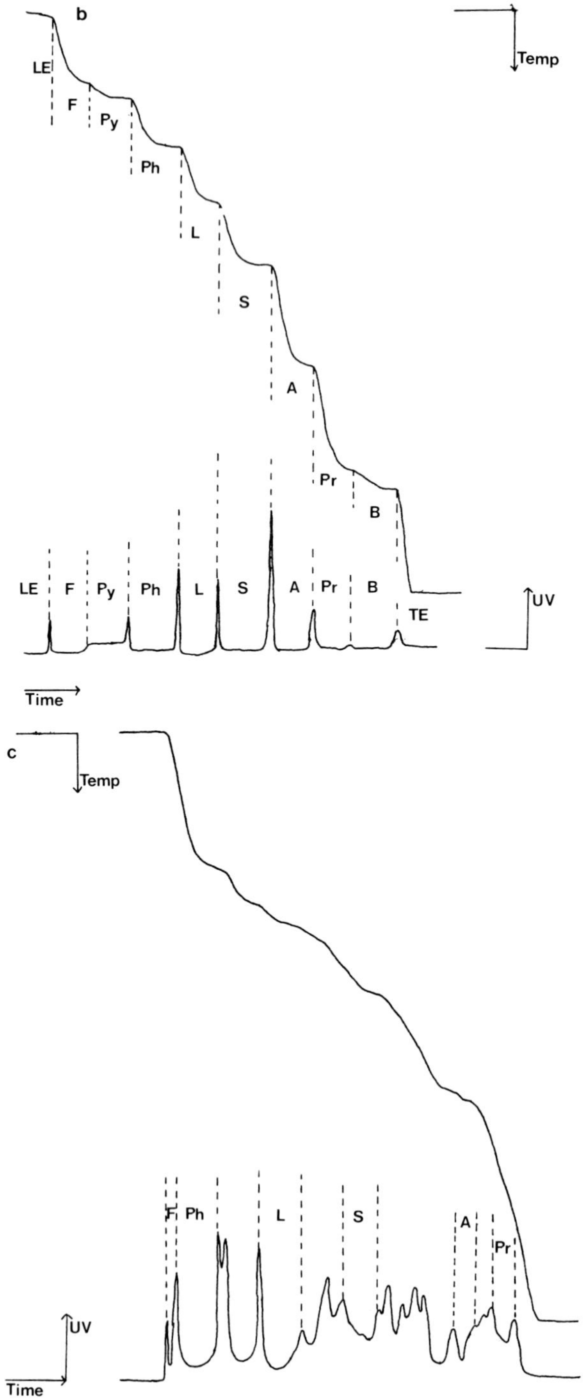

FIG. 3 (b) & (c) (*Continued*)

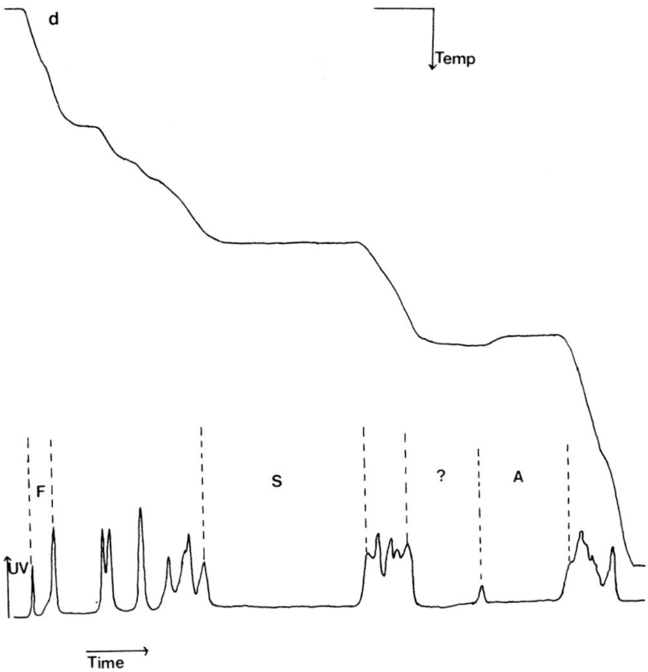

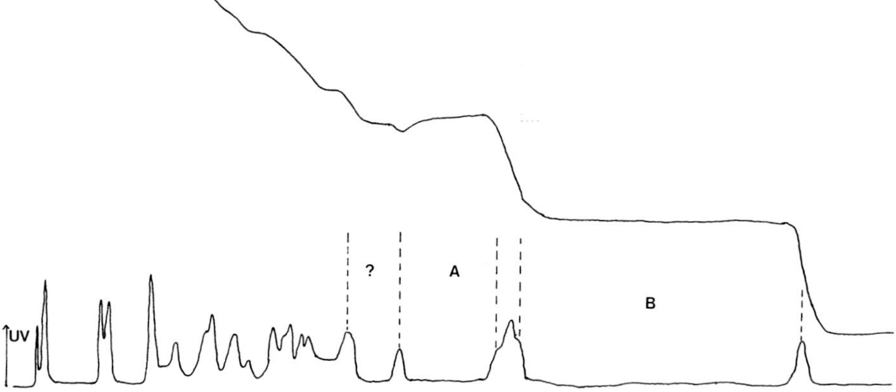

FIG. 3 (d) & (e) (*Continued*)

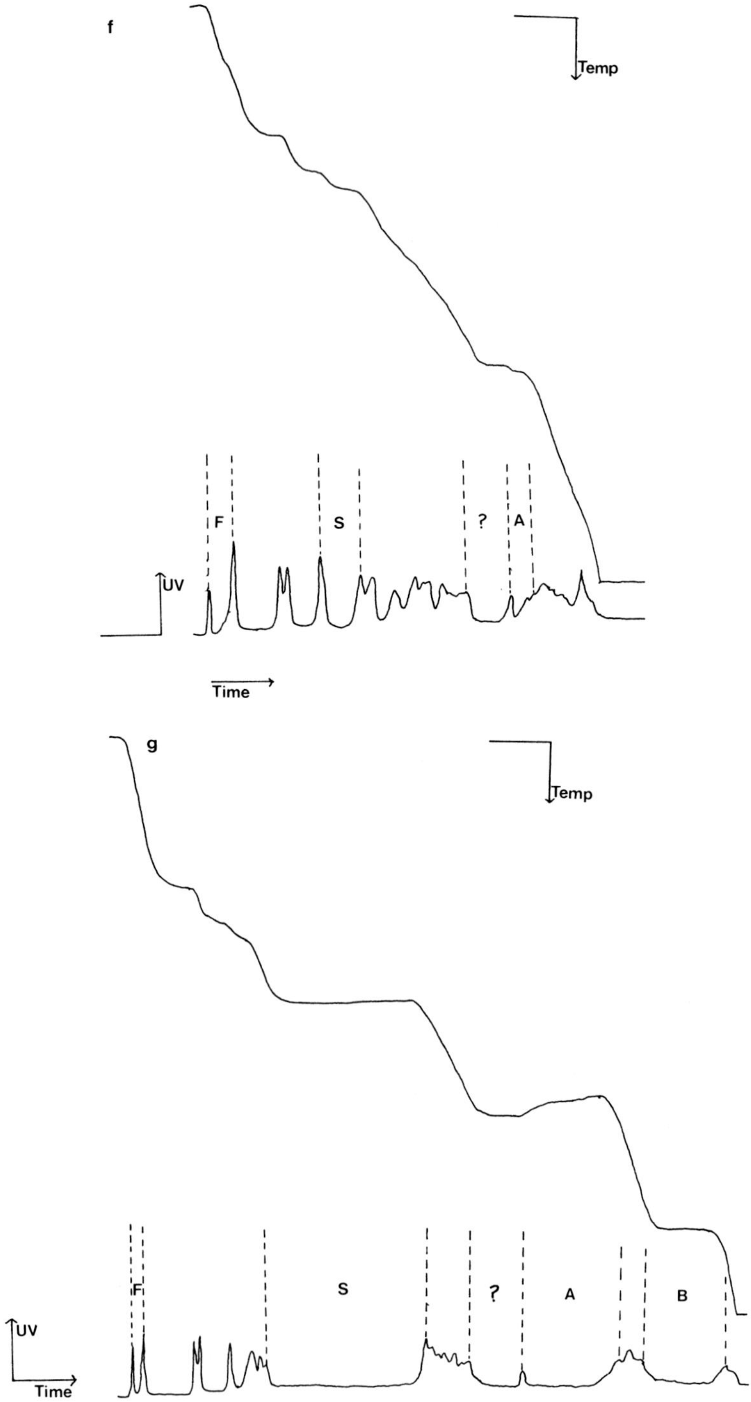

FIG. 3 (f) & (g) (*Continued*)

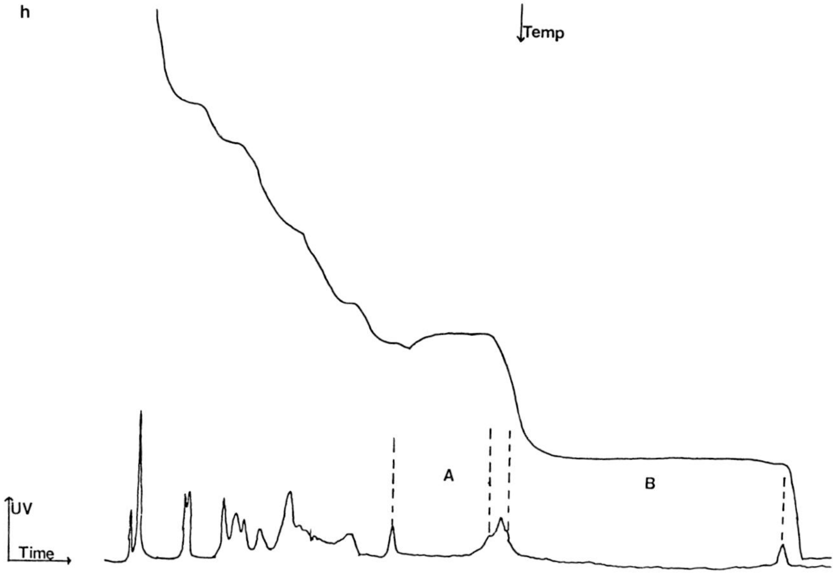

FIG. 3 (h) (*Continued*)

Conclusions

The present results indicate that isotachophoresis can be used successfully in the analysis of the acid end products of metabolism of anaerobic Gram-negative bacteria. The method has several advantages over GLC, and the separation of anions is highly reproducible. In particular, isotachophoresis is suitable for the determination of short-chain fatty acids (C_1 to C_4) in preference to GLC because the former technique does not require pretreatment or derivatisation of the sample. A major difficulty in GLC analysis of short-chain fatty acids has been the interaction of the acids with the column, causing loss of sample and "ghosting" (Geddes & Gilmour 1970). In addition, formic acid estimations are difficult by GLC owing to the acid's volatile nature. In isotachophoresis, the low-molecular-weight fatty acids (including formic acid), which are very soluble in water, can be determined without such difficulties.

Enzymatic Determination of L(+)- and D(−)-Lactic Acid in Bacterial Culture Supernatants

The various techniques described cannot be used to distinguish between the optical isomers of lactic acid. Both L(+)- and D(−)-lactic acid levels can conveniently be measured by a spectrophotometric method that is reliably sensitive to concentrations as low as 2 μg of acid in the assay.

In the case of L(+)-lactic acid, the spectrophotometric method relies on the principle that in the presence of the enzyme L(+)-lactate dehydrogenase (L-LDH), L(+)-lactic acid (L-lactate) is oxidised by nicotinamide adenine dinucleotide (NAD) to pyruvate (Eq. 1):

$$\text{L-lactate} + \text{NAD}^+ \xrightleftharpoons{\text{L-LDH}} \text{pyruvate} + \text{NADH} + \text{H}^+ \qquad (1)$$

Since the equilibrium of this reaction is in favour of L(+)-lactate, the pyruvate formed (Eq. 1) is trapped by a subsequent reaction (Eq. 2) catalysed by the enzyme glutamate–pyruvate transaminase (GPT) in the presence of glutamate as follows:

$$\text{pyruvate} + \text{L-glutamate} \xrightarrow{\text{GPT}} \text{L-alanine} + \text{2-ketoglutarate} \qquad (2)$$

The equilibrium reaction (Eq. 1) is therefore displaced in favour of pyruvate and NADH formation. The amount of NADH formed is stoichiometric with the concentration of L(+)-lactic acid, hence the increased NADH level can be monitored spectrophotometrically at its absorption maximum of 340 nm (or 334 or 365 nm). A similar set of reactions can be set up using D-lactate as the substrate in Eq. 1.

In the present study, a test combination kit (for 20 determinations) and the individual reagents listed below were obtained from Boehringer Mannheim. The method was based on the procedure described in a Chemotaxonomic Workshop Manual (Shah *et al.* 1983).

Solution for L(+)-lactic acid determination: (a) glycylglycine buffer, pH 10.0, containing L-glutamic acid (440 mg), 30 ml; (b) β-nicotinamide adenine dinucleotide (210 mg), requires 6 ml of water to rehydrate; (c) glutamate–pyruvate transaminase (1100 units), 0.7 ml; (d) L-lactate dehydrogenase (3800 units), 0.7 ml; (e) standard L(+)-lactic acid solution (20–400 μg/0.1 ml). (This is not included in the test combination kit.)

Solutions for D(−)-lactic acid determination: Solutions 1, 2, and 3 as above; (f) D-lactate dehydrogenase (0.5 mg/ml); (g) standard D(−)-lactic acid solution (2 to 40 μg/ml). (Neither the D-lactate dehydrogenase nor the D(−)-lactic acid were included in the test combination kit.)

The following reagents were added into two separate cuvettes:

	Cuvette 1 (blank) (ml)	Cuvette 2 (sample) (ml)
Solution 1	1.0	1.0
Solution 2	0.2	0.2
Water	1.0	0.9
Suspension 3	0.02	0.02
Sample solution (5 or sample)	—	0.1

The contents of the cuvettes were mixed and the absorbances read (A_1) after 5 min at 340 nm. The reaction was then initiated by the addition of 0.02 ml of solution 4. For D(−)-lactic acid determinations, 0.05 ml of D-lactate dehydrogenase was used instead of L-lactate dehydrogenase. The increased absorbance at 340 nm was monitored until the curve levelled off (~10 min). With D(−)-lactate dehydrogenase, the curve levelled off after ~30 min at 20°C or 20 min at 37°C. The increased absorbance (A_2) was determined, and the absorbance difference for both blank and sample ($A_2 - A_1$) was calculated. Finally the absorbance difference between the blank and the sample (ΔA) was determined.

The concentration of L(+)- and D(−)-lactic acid samples may be determined directly by reading the absorbance values of the test solutions from the appropriate calibration curve (Fig. 4).

Alternatively, the acid concentration of the samples may be calculated from the general formula:

$$C = \frac{V \times MW}{E \times d \times v \times 1000} \times \Delta A \ (\text{g/l})$$

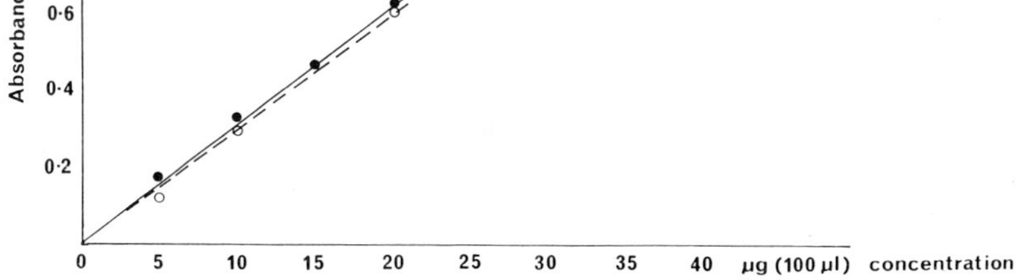

FIG. 4. Calibration curve for L(+)- and D(−)-lactic acids.

Where V = final volume (ml), v = sample volume (ml), MW = molecular weight of test compound, d = light path (cm) and E = extinction coefficient of NADH, which at

$$\text{Hg } 340 \text{ nm} = 6.3 \text{ mmol}^{-1} \text{ cm}^{-1}, \text{ at}$$
$$\text{Hg } 365 \text{ nm} = 3.4 \text{ mmol}^{-1} \text{ cm}^{-1}, \text{ and at}$$
$$\text{Hg } 334 \text{ nm} = 6.18 \text{ mmol}^{-1} \text{ cm}^{-1}.$$

For L(+)-lactic acid, therefore:

$$c = \frac{2.24 \times 90.1}{6.3 \times 1 \times 0.1 \times 1000} \times \Delta A$$

$$= 0.320 \times A[\text{g L(+)-lactic acid/L}]$$

For D(−)-lactic acid:

$$c = \frac{2.27 \times 90.1}{6.3 \times 1 \times 0.1 \times 1000} \times \Delta A$$

$$= 0.325 \times \Delta A[\text{g D(−)-lactic acid/L}]$$

The results in Table 3 show that L(+) and D(−)-lactic acids were produced by all of the Gram-negative anaerobic rods tested.

TABLE 3

Concentrations of D(−)- and L(+)-lactic acids (mM) in strains of representative species of Gram-negative anaerobic bacteria

	BM (without glucose)		BM (with glucose)	
	D(−)-lactate	L(+)-lactate	D(−)-lactate	L(+)-lactate
Leptotrichia buccalis NCTC 10249	2.1	2.3	3.3	2.2
L. buccalis T63	7.6	2.1	10.1	2.1
Fusobacterium nucleatum NCTC 10953	0	2.9	2.5	1.3
Mitsuokella multiacidus	0.77	2.3	42.2	20.3
Megamonas hypermegas	0.21	2.5	0.32	11.5
Bacteroides asaccharolyticus	1.36	0.06	0.21	1.56
B. coagulans	NT*	NT*	0.71	0.7
B. furcosus	0.05	2.9	0.14	9.2
B. gingivalis	0.15	0.46	0.48	5.2
B. intermedius	1.35	0.96	0.25	1.35
B. pentosaceus	NT*	NT*	0.16	1.2
B. praeacutus	0.18	3.66	0.32	2.9
"*B. variabilis*"	0.58	0.69	0.69	6.97

*NT, not tested.

Values of L(+)- *and* D(−)-*lactic acid determinations*

The analyses performed on culture supernatants prepared for GLC, HPLC, or isotachophoresis gave important additional information to the acid end product profile data. Both the L(+)- and D(−)-lactate dehydrogenases were found to be pure, since no contaminating complementary isomeric forms were detected. Data on D(−)-lactic acid levels as metabolic end products of Gram-negative anaerobes are rarely encountered in the literature. Our preliminary systematic studies indicate that the presence of D(−)-lactic acid, in particular, and the ratio of L(+)- to D(−)-lactic acid will prove to be useful chemotaxonomic markers for the study of Gram-negative anaerobic bacteria. It is known, for example, that *Bacteroides multiacidus (Mitsuokella multiacidus* (Shah & Collins 1982) produces high levels of lactic acid (Mitsuoka *et al.* 1974), but in the present study it was clearly shown that the major isomer produced is D(−)-lactic acid (Table 3). All of the Gram-negative anaerobic bacteria examined produced D(−)-lactic acid though generally lower in concentration than that of the L(+) isomer (Table 3).

General Discussion

Reliable and reproducible acid end product data depend on the cultural conditions under which microorganisms are grown. BM medium can be recommended (formulated by the second author, R.A.N.) for bulk growth and for the determination of acid end products of metabolism. This medium allows luxuriant growth of the most fastidious anaerobic strains but does contain low levels of several fatty acids (Fig. 3c). Comparison of test analyses with an uninoculated control is, therefore, necessary for quantitative studies. However, results obtained with BM medium compare favourably with media, such as peptone–yeast extract–glucose (PYG), employed in other laboratories (Sutter *et al.* 1972; Dowell & Hawkins 1974; Holdeman *et al.* 1977).

Broadly comparable results were obtained by each of the methods used in the present studies, but each technique has certain advantages and shortcomings. For example, the quantitative determination of non-VFA by GLC required prior methylation of samples, although semiquantitative methods have been developed by Carlsson (1973); the latter permit the detection of both VFA and non-VFA in a single isothermal run. High-pressure liquid chromatography offers a very simple and sensitive (1 m*M* acid) method of determining acid end products, but in the system employed, refractive index detectors had to be used because of the large number of UV-absorbing compounds in the culture medium. Another drawback of the system was due to difficulty in separating lactic and succinic acids at room temperature, although this may be achieved at 45°C. Isotachophoresis is the most rapid and sensitive (1 n*M*) method of analysis and in common with HPLC it does not require sample pretreatment. However, under the present running conditions *n*-butyric and isobutyric acids could not be separated. The detection of mixtures of L(+)- and D(−)-lactic acids could only be achieved by enzymatic methods. Although these assays are somewhat tedious and expensive to perform, they provide useful additional data that cannot be obtained by the other techniques reviewed in this chapter.

Although GLC has become the most widely used method for detection of acid end products of metabolism of Gram-negative anaerobic bacteria, it is clear from our preliminary studies that other analytical techniques can be adapted for this purpose. Further studies on the application of HPLC and isotachophoresis in microbiology should prove to be particularly rewarding.

References

CARLSSON, J. 1973 Simplified gas chromatographic procedure for identification of bacterial metabolic products. *Applied Microbiology* **25,** 287–289.

DEACON, A. G., DUERDEN, B. I. & HOLBROOK, W. P. 1978 Gas-liquid chromatographic analysis of metabolic products in the identification of *Bacteroidaceae* of clinical interest. *Journal of Medical Microbiology* **11,** 81–99.

DOWELL, V. R., & HAWKINS, T. M. 1974 Laboratory methods in anaerobic bacteriology. *CDC Laboratory Manual,* CDC Publication No. 77-8272. Georgia: Center for Disease Control, Atlanta.

DRUCKER, D. B. 1981 *Microbiological Applications of Gas Chromatography*. Cambridge: University Press.

EDGAR, W. M. 1981 Isotachophoretic analysis of major acids in human plaque before and after sugar consumption. *Zahn-Mund-und Keiferheilkunde* **69,** 744.

EHRLICH, G. G., GOERLITZ, D. F., BOURELL, J. H., EISEN, G. V. & GODSY, E. M. 1981 Liquid chromatographic procedure for fermentation product analysis in the identification of anaerobic bacteria. *Applied and Environmental Microbiology* **42,** 878–885.

GARIBALDI, J. A. & BAYNE, H. G. 1970 Production of acetoin and diacetyl by the genus *Salmonella*. *Applied Microbiology* **20,** 855–856.

GEDDES, D. A. M. & GILMOUR, M. N. 1970 The control of ghosting, a major source of error in gas liquid chromatographic determinations of C_2–C_5 acids. *Journal of Chromatographic Science* **8,** 394–397.

GEDDES, D. A. M. & WEETMAN, D. A. 1981 Organic acids in human dental plaque estimated by isotachophoresis. *Journal of Dental Research* **60,** 1192.

HAMMANN, R. & WERNER, H. 1980 Fermentation products (using g.l.c.) in the differentiation of non-sporing anaerobic bacteria. In *Microbiological Classification and Identification* eds. Goodfellow, M. & Board, R. G., pp. 257–271. London, New York, Toronto, Sydney & San Francisco: Academic Press.

HOLDEMAN, L. V., CATO, E. P. & MOORE, W. E. C. 1977 *Anaerobic Laboratory Manual,* 4th edn. Blacksburg, Virginia: Virginia Polytechnic Institute and State University Anaerobe Laboratory.

ILES, R. A., STEVENS, A. N. & GRIFFITHS, J. R. 1982 NMR studies of metabolites in living tissue. *Progress in NMR Spectroscopy* **15,** 49–200.

ILES, R. A., BUCKINGHAM, M. J. & HAWKES, G. E. 1983 H-Spin-echo n.m.r. detection of normal and abnormal metabolites in plasma and urine. *Biochemical Society Transactions* **11**, 374–375.

KROGER, A. 1977 Phosphorylative electron transport with fumarate and nitrate as terminal hydrogen acceptors. In *Microbial Energetics* eds. Haddock, B. A. & Hamilton, W. A., pp. 61–93. Cambridge: University Press.

MACY, J., PROBST, I. & GOTTSCHALK, G. 1975 Evidence for cytochrome involvement in fumarate reduction and adenosine—5[1] triphosphate synthesis by *Bacteroides fragilis* grown in the presence of haemin. *Journal of Bacteriology* **123**, 436–442.

MAYHEW, J. W. & GORBACH, S. L. 1977 Internal standards for gas chromatographic analysis of metabolic end products from anaerobic bacteria. *Applied and Environmental Microbiology* **33**, 1002–1003.

MITSUOKA, T., TERADA, A., WATANABE, K. & UCHIDA, K. 1974 *Bacteroides multiacidus*, a new species from the feces of humans and pigs. *International Journal of Systematic Bacteriology* **24**, 35–41.

MORRIS, J. G. 1975 The physiology of obligate anaerobiosis. *Advances in Microbial Physiology* **12**, 169–246.

MOUNTFORT, D. O. & ROBINSON, A. M. 1977 The role of menaquinone and b-type cytochrome in anaerobic reduction of fumarate by NADH in membrane preparations from *Bacteroides ruminicola* strain B4. *Journal of General Microbiology* **100**, 309–317.

REDDY, C. A. & BRYANT, M. P. 1977 Deoxyribonucleic acid base composition of certain species of the genus *Bacteroides*. *Canadian Journal of Microbiology* **23**, 1252–1256.

REDDY, C. A. & PECK, H. D. 1978 Electron transport phosphorylation coupled to fumarate reduction by H_2 and Mg^{2+} dependent adenosine triphosphatase activity in extracts of the rumen anaerobe *Vibrio succinogenes*. *Journal of Bacteriology* **134**, 982–991.

RIZZA, V., SINCLAIR, P. R., WHITE, D. C. & CUORANT, P. R. 1968 Electron transport system of the protoheme-requiring anaerobe *Bacteroides melaninogenicus*. *Journal of Bacteriology* **96**, 665–671.

SAMARANAYAKE, L. P., WEETMAN, D. A., GEDDES, D. A. M. & MACFARLANE, T. W. 1982 Isotachophoretic analysis of short-chain carboxylic acids produced by *Candida albicans*. *Microbios* **35**, 91–98.

SHAH, H. N. & COLLINS, M. D. 1980 Fatty acid and isoprenoid quinone composition in the classification of *Bacteroides melaninogenicus* and related taxa. *Journal of Applied Bacteriology* **48**, 75–87.

SHAH, H. N. & COLLINS, M. D. 1982 Reclassification of *Bacteroides multiacidus* (Mitsuoka, Terada, Watanabe & Uchida) in a new genus *Mitsuokella*, as *Mitsuokella multiacidus* comb. nov. *Zentralblatt für Bakteriologie. I. Abteilung Originale* **C3**, 491–494.

SHAH, H. N. & COLLINS, M. D. 1983 Genus *Bacteroides:* A chemotaxonomical perspective. *Journal of Applied Bacteriology* **55**, 403–416.

SHAH, H. N. & WILLIAMS, R. A. D. 1981 Metabolic studies on *B. asaccharolyticus* and *B. melaninogenicus subsp. intermedius*. *Journal of Dental Research* **60**, 1161.

SHAH, H. N., BONNETT, R., MATEEN, B. & WILLIAMS, R. A. D. 1979 The porphyrin pigmentation of subspecies of *Bacteroides melaninogenicus*. *Biochemical Journal* **180**, 45–50.

SHAH, H. N., NASH, R. A., WHILEY, R. A., HARDIE, J. M. & COLLINS, M. D. 1983 *Practical Workshop on Chemical Methods for Bacterial Classification and Identification*. London Hospital Medical College, Department of Oral Microbiology: University of London.

SHARPE, M. E. 1979 Identification of the lactic acid bacteria. In *Identification Methods for Microbiologists* eds. Skinner, F. A. & Lovelock, D. W., pp. 233–259. London, New York, Toronto, Sydney & San Francisco: Academic Press.

STOUTHAMER, A. H. 1980 Electron transport linked phosphorylation in anaerobes. In *Anaerobes and Anaerobic Infections* eds. Gottschalk, G., Pennig, N. & Werner, H., pp. 17–29. Stuttgart, New York: Gustav Fischer Verlag.

SUTTER, V. L., ATTEBERY, H. R., ROSENBLATT, J. E., BRICKNELL, K. S. & FINEGOLD, S. M. 1972 *Clinical Microbiology Techniques. Anaerobic Bacteriology and Workshop on Normal Oral and Bowel Flora.* Berkeley: University of California.

THAUER, R. K., JUNGERMANN, K. & DECKER, K. 1977 Energy conservation in chemotrophic anaerobic bacteria. *Bacteriological Reviews* **39**, 147–156.

TURKELSON, V. T. & RICHARDS, M. 1978 Separation of the citric acid cycle acids by liquid chromatography. *Analytical Chemistry* **50**, 1420–1423.

VAN DER HOEVEN, J. S., FRANKEN, H. C. M., CAMP, P. J. M. & DELLEBARRE, G. W. 1978 Analysis of bacterial fermentation products by isotachophoresis. *Applied and Environmental Microbiology* **35**, 17–23.

WADSTROM, Y., MOLLBY, R., OLSSON, B., SODERHOLM, J. & SMYTH, C. J. 1977 Preparative scale purification of bacterial enzymes and toxins by isoelectric focusing and isotachophoresis. In *Electrofocusing and Isotachophoresis* eds. Radola, B. J. & Graesslin, D. pp. 443–453. New York & Berlin: Walter de Gruyter.

WOLF, C. G. & TELFER, S. V. 1917 Contributions to the biochemistry of pathogenic anaerobes. *Biochemistry Journal* **11**, 197–213.

YAMADA, T. & CARLSSON, J. 1976 The role of pyruvate formate-lyase in glucose metabolism of *Streptococcus mutans*. In *Microbial Aspects of Dental Caries* eds. Stiles, H. M., Leosche, W. J. & O'Brien, T. C., pp. 809–819. Washington, D.C. & London: Information Retrieval Inc.

YOTIS, W. W. & CATSIMPOOLAS, N. 1975 Scanning isoelectric focusing and isotachophoresis of *Clostridium perfringens* type A enterotoxin. *Journal of Applied Bacteriology* **39**, 147–156.

A Combined Biochemical Approach to the Taxonomy of Gram-Positive Rods

V. E. DENT

Sub-Department Microbiology, Department of Botany, The Queen's University of Belfast, Belfast, Northern Ireland

R. A. D. WILLIAMS

Department of Biochemistry, The London Hospital Medical College, London, UK

Chemotaxonomic techniques have been used increasingly in recent years for the classification and identification of bacteria. Each method has its strengths and weaknesses, and may be applied usefully within particular taxa. A combination of such methods, therefore, is of greater value and strength than any single technique used in isolation. A collection of isolates obtained from the dental plaque of animals (Dent 1979; Dent & Marsh 1981) provided many bacteria for identification. The predominant plaque flora of animals, as in humans, consists of Gram-positive rods and filaments. Identification of such bacteria from dental plaque is difficult, and in this study a wide variety of techniques have been used to achieve this objective. The tests applied were also done on strains from various culture collections and personal sources in order that the wild-type strains from animals could be compared adequately. The results demonstrated here show the value of such a combined approach, especially in the *Actinomyces* and *Lactobacillus* genera.

Materials and Methods

Maintenance of cultures

Lactobacillus cultures were routinely maintained on plates containing (per litre) 60 g of defibrinated horse blood (Gibco), and 40 g of Blood Agar Base

0-12-289675-0

No. 2 (Oxoid), subcultured at 3-day intervals and incubated at 37°C in Gas Pak jars (BBL) with hydrogen–carbon dioxide generation sachets (Oxoid). *Actinomyces* strains were cultured in the same manner at 5-day intervals.

Physiological tests

For growth of lactobacilli, the MRS medium of de Man *et al.* (1960) and modifications thereof were used as follows: inocula for fermentation tests, MRS without beef extract but with glucose (5 g/litre); fermentation tests, MRS without glucose and beef extract at pH 6.5 with substrate (15 g/litre) sterilised by filtration; deamination of arginine, MRS without beef extract and ammonium citrate at pH 7.4, but with glucose (20 g/litre) and arginine hydrochloride (3 g/litre); hydrolysis of aesculin, MRS medium without beef extract and glucose, but with aesculin (5 g/litre) and ferric ammonium citrate (0.5 g/litre) adjusted to pH 7.4; gas from gluconate, MRS medium without beef extract and glucose, but with sodium gluconate (10 g/litre) adjusted to pH 5.5. Basal broth for actinomyces contained (per litre) 20 g of proteose peptone (Oxoid Ltd., London, UK), 5 g of yeast extract (Difco Laboratories, Detroit, Mich., USA), 5 g of sodium chloride, and 1 g of disodium hydrogen phosphate at pH 7.6. For the fermentation broths the basal medium was supplemented with filter-sterilised [Millex HA 0.45 μm; Millipore (UK) Ltd., Harrow] carbohydrates to give a final concentration of 10 g/litre. Reduction of nitrate and nitrite was tested in brain–heart infusion broth (Gibco Europe Ltd., Paisley, UK) supplemented with 0.5 g/litre of potassium nitrate or nitrite. Reduction of nitrate and nitrite was tested as described by Cowan & Steel (1974). The aesculin hydrolysis medium contained peptone (Oxoid) 10 g/litre, yeast extract (Difco) 5 g/litre, sodium acetate 10 g/litre, aesculin 5 g/litre, Tween 80 (B.D.H., Poole, UK) 1 g/litre, ferric ammonium citrate 0.5 g/litre, dipotassium hydrogen phosphate 0.5 g/litre, magnesium sulphate·$7H_2O$ 0.68 g/litre, manganous sulphate·H_2O 0.11 g/litre and ferrous sulphate·$7H_2O$ 0.034 g/litre. All broths were inoculated with four drops of starter culture containing 5 g/litre of glucose and incubated for 7 days at 37°C. Final pH values were measured with a Pye PW 9418 pH meter and aesculin hydrolysis was indicated by blackening of the medium.

Acid end-product analysis

Samples of the supernatants of growth media were analysed by the modification of Bowden *et al.* (1975) of the technique of Carlsson (1973). The analyses were carried out by Mr. R. A. Nash.

Purification and thermal denaturation of DNA

The extraction and purification of DNA from logarithmic phase cells was modified slightly from the method of Garvie (1976). Purified DNA was analysed by thermal denaturation in standard saline citrate (SSC) or dilute saline citrate (DSC) and the hyperchromicity at 260 nm was followed with a Gilford model 240 spectrophotometer and a model 2527 thermoprogramer.

For actinomyces strains, cells were grown on the medium described for SDS–PAGE analysis to which 40 units of benzylpenicillin/ml was added 1 h before harvesting. Cells were washed by resuspension in 0.15 *M* sodium chloride, 10 mM EDTA, pH 7.0 and then disrupted by suspending 5 g in 10 ml of *M* sodium chloride, adding 1 mg of proteinase (Protease XIV, Sigma, Poole, UK) and 1 ml of 25% sodium dodecyl sulphate and pressing at 8000 psi with a French pressure cell (Aminco, Silver Springs, Md., USA) into 100 ml of 1 *M* sodium chloride solution at 60°C. To this hot suspension 9 ml of 25% sodium dodecyl sulphate was added, and the whole was kept at 60°C for 10 min. The suspension was shaken with an equal volume of chloroform–isopentanol (24:1) for 15 min and centrifuged at 5000 rpm for 30 min. DNA was spooled from the aqueous layer with 2 volumes of cold ethanol, re-dissolved in DSC, and centrifuged at 10,000 rpm for 20 min. Thereafter the procedure of Garvie (1976) was followed. The molar ratio of bases was calculated according to the relationships $\% \text{ G} + \text{C} = 2.44\ (T_m - 69.3)$ (Mandel & Marmur 1968) for standard saline citrate or $\% \text{ G} + \text{C} = 2.44\ (T_m - 53.9)$ for dilute saline citrate. The equations $\% \text{ G} + \text{C} = 2.08\ (T_m = 106.4)$ (Owen & Lapage 1976) and $\% \text{ G} + \text{C} = 50.9 + 2.08\ (T_m - \text{Tm}$ *E. coli* strain B) (Owen & Hill 1979) were also used for comparative purposes.

DNA–DNA hybridization

DNA to be used in hybridization was purified by centrifugation in a caesium chloride gradient. DNA solution (8 ml) in SSC was mixed with 8 g of caesium chloride and 0.8 ml of ethidium bromide (5 mg/ml in water). The gradient was established by centrifugation at 150,000 *g* for 40 h. DNA was detected as an ethidium bromide-binding band. The DNA was dissolved in 0.1 × SSC, and ethidium bromide was removed by extraction with *n*-butanol. Pure DNA was spooled out with 2 volumes of ethanol and dialysed against 3 × SSC. The concentration of DNA was determined by absorption at 260 nm in a scanning spectrophotometer (Perkin Elmer Model 402). DNA sufficiently pure for hybridization had an absorbance ratio (280/260 nm) approaching 2.

Pure DNA was labelled with ^{3}H dCTP by nick translation (Amersham

International, nick translation kit TRK 700). The specific activity obtained varied between 8×10^5 and 1.2×10^6 cpm/μg DNA.

DNA was loaded onto nitrocellulose filters by the method described in Back & Stackebrandt (1978) and the hybridization procedures according to Back & Stackebrandt (1978) and Stackebrandt & Fiedler (1979) were followed. Hybridization in the *Lactobacillus animalis* system used 0.05 μg of ^{3}H-DNA in 200 μl of 3 × SSC–formamide (13%) against filter-bound DNA that contained between 9 and 15 μg of DNA per filter for 20 h at 60°C. In the *Actinomyces* systems, the same quantities of labelled DNA were used, but the concentration of formamide was 25% and 25.5% for *Actinomyces denticolens* and *Actinomyces viscosus*, respectively. The radioactivity that remained on the filter after washing and drying was counted in 3 ml of cocktail O (BDH) with an Intertechnique liquid scintillation counter (model SL3000).

SDS–polyacrylamide gel electrophoresis (SDS–PAGE)

Strains of lactobacilli were cultured in 500 ml of MRS medium for 48 h at 37°C in static conditions. Strains of actinomyces were grown for 3–4 days in a medium containing 19 g of brain heart infusion (Gibco), 7.5 g of proteose peptone (Oxoid), 4.5 g of trypticase (BBL, Becton Dickinson, Wembley, UK), 9 g of glucose and 3 g of yeast extract (Difco) dissolved in 950 ml of water at pH 6.5. Before inoculation, sterile sodium bicarbonate solution in 50 ml of water was added to give a final concentration of 5 g/litre. Cells were harvested by centrifugation at 6000 *g* and washed by recentrifugation three times in distilled water. Cells were resuspended in water and disintegrated in a Mickle shaker (Mickle Engineering Ltd., Gomshall, Surrey, UK) with glass beads (Grade 12, Jencons Ltd.) for 1.5 h. The supernatant was centrifuged at 17,000 *g* for 15 min and then at 150,000 *g* for 1 h to remove all traces of membranes. The cell-free extracts were electrophoresed in 10% polyacrylamide–SDS gels as described previously (Swindlehurst *et al.* 1977).

Purification and analysis of peptidoglycan

Cells were grown and harvested as above. Peptidoglycan was prepared from unbroken frozen cells by the rapid "screening" procedure of Schleifer & Kandler (1972). Hydrolysis prior to analysis of amino compounds was as previously described (Pask-Hughes & Williams 1978). For qualitative analysis, hydrolysates were examined by two-dimensional paper chromatography (Hardy *et al.* 1955) and high-voltage electrophoresis in pyridine–acetic acid buffer at pH 5.2 (Atfield & Morris 1961).

For quantitative analysis of amino acids and amino sugars, the single-column amino-acid analyser was used as described by Pask-Hughes & Williams (1978).

Analysis of cell wall carbohydrates

For the detection of wall carbohydrates, cell walls were prepared by shaking a suspension of cells in water with Grade 12 ballotini beads in a tissue disintegrator (Mickle Engineering, Ltd.) for 6 h at 4°C. Cell walls were washed from the ballotini, unbroken cells were removed by centrifugation at 3000 *g* and the walls sedimented at 17,000 *g* and suspended with 10 mg/100 ml of Protease XIV in a phosphate buffer (pH 8.0) at 37°C for 16 h. The residue was washed three times with water by recentrifugation and then freeze-dried. Hydrolysis was with 2 *N* sulphuric acid (0.5 ml per 10 mg) for 1 h at 100°C; the acid was neutralised with Dowex 2-X8 resin in the carbonate form. Qualitative analysis of carbohydrates involved paper chromatography with *n*-butanol–pyridine–water (60:40:30, v/v) and thin-layer chromatography (TLC) on silica gel impregnated with 0.03% (v/v) boric acid, using solvent systems 13 and 42 of Lato *et al.* (1968). Papers were dipped in aniline–phthalate (Partridge 1949), and TLC plates were sprayed with 4-aminobenzoic acid reagent (Menzies & Mount 1975). Sensitivity of detection was increased by examining the chromatograms with UV light (366 nm).

Results

Lactobacillus animalis

A numerically small proportion of the dental plaque isolates examined were lactobacilli, which were heterogeneous with respect to physiological reactions. A collection of isolates phenotypically resembling *Lactobacillus acidophilus* was found amongst these strains. They had an unusual pattern of polypeptides typified by relatively few densely staining bands after SDS–PAGE (Fig. 1). This group, with DNA containing 42–44% G + C, was clearly different from strains of *L. acidophilus* (32–37% G + C), a species that has been shown to be genotypically heterogeneous (Johnson *et al.* 1980; Lauer *et al.* 1980; Sarra *et al.* 1980). "*Lactobacillus acidophilus*-like" strains examined in the above studies having DNA containing above 40% G + C were also examined. The result was the description of *Lactobacillus animalis* (Dent & Williams 1982) comprising strains from the dental plaque of baboons and a patas monkey, together with isolates from mouse and dog intestines supplied by J. L. Johnson. DNA–DNA hybridization results to date

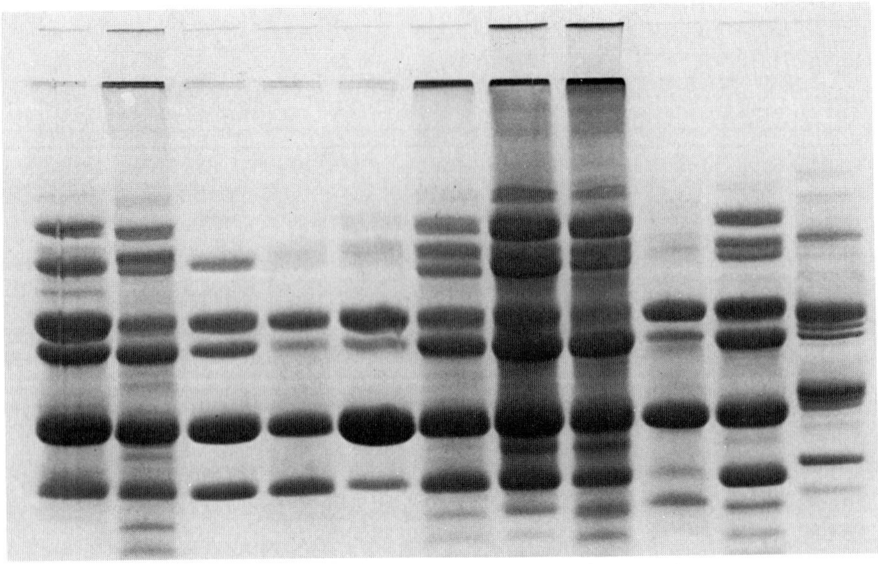

FIG. 1. Sodium dodecyl sulphate–polyacrylamide gel electrophoresis of soluble polypeptides of *Lactobacillus animalis*. From left to right, *Lactobacillus animalis* strains NCDO 2425 (1535), 1540, 1558, 2262, 1524, 1537, 1605, 1539, 8409, 1395 and *Lactobacillus acidophilus*-like strain PF2.

(Table 1) indicate that *L. animalis* comprises a single homology group and provide no evidence for dividing the species, despite a difference amongst the strains in the aesculin, amygdalin and arabinose tests. It also appears that *L. animalis* strains are quite distinct from the pig faecal isolates examined by Sarra *et al.* (1980) and the *L. acidophilus*-like isolates tested by Johnson *et al.* (1980). Current DNA hybridization studies will clarify further the relationships amongst these animal isolates.

Actinomyces

The strains of *Actinomyces* isolated from the dental plaque of animals were heterogeneous with respect to their physiological reactions and polypeptide patterns. All strains tested (> 100) had a peptidoglycan composition comprising 1 muramic acid, 1 glucosamine, 2 alanine, 2 glutamate, 1 lysine and 1 ornithine. Such a composition is consistent with the peptidoglycan structure of *Actinomyces naeslundii*, *Actinomyces viscosus*, *Actinomyces israelii* or *Actinomyces odontolyticus* but not with *Actinomyces bovis*, which has a unique peptidoglycan structure within the genus (Slack & Gerencser 1975). These molar ratios are

Table 1

*Properties of homofermentative lactobacilli isolated from the dental plaque of primates and the intestines of animals and designated Lactobacillus animalis**

Strain	Acid production from					Aesculin hydrolysis	DNA % G + C	% L isomer of lactate	Peptido-glycan type	% Homology with ^{3}H-DNA from strains		
	Amygdalin	Arabinose	Inulin	Ribose	Salicin					1535	1395	21S
NCDO 2425 (1535)	+	—	—	—	+	+	42.4	78.6	L-lys D-asp	100	79	7
1540	+	—	—	—	+	+	43.2	88.2	L-lys D-asp	101	89	6
1558	+	—	—	—	+	+	43.9	90.9	L-lys D-asp	89	85	7
1395	w[d]	—	—	w	+	+	43.7	81.5	L-lys D-asp	94	100	6
2262	—	+	w	w	w	—	42.4	93.3	L-lys D-asp	80	79	7
1524	—	+	—	w	+	—	42.9	90.6	L-lys D-asp	76	79	6
1537	—	+	w	w	+	—	44.4	83.9	L-lys D-asp	84	70	6
L. acidophilus-like strains												
21S	Pig caecum (P. G. Sarra)						42.6[a]	n.d.[c]	n.d.	6	7	100
N4	Pig caecum (M. Briggs)						41.8[a]	n.d.	n.d.	6	7	98
1406	Dental caries						45.1	44[b]	n.d.	6	4	4
1407	Dental caries						45.9	46[b]	n.d.	6	3	4
0776	Pig caecum						47.0[b]	n.d.	n.d.	5	5	4
1796	Chicken crop						44.4	43[b]	n.d.	10	10	6

*All the *Lactobacillus animalis* strains ferment glucose, fructose, maltose, cellobiose, lactose, melibiose and raffinose; none ferment xylose, rhamnose, melezitose, trehalose, sorbitol or mannitol; none produce ammonia from arginine nor CO_2 from hexoses or gluconate.

[a]Sarra *et al.* (1980).

[b]Johnson *et al.* (1980).

[c]Not determined.

[d]Weak reaction.

sufficiently unusual to establish a strain as a member of the major group of *Actinomyces* species.

Physiological tests precluded any of the isolates from being identified as *A. israelii* or *A. odontolyticus*. Strains were, therefore, designated putative *A. naeslundii* if catalase-negative, or *A. viscosus* if catalase-positive. No other physiological test will distinguish between these two species, and within each species physiological reactions are heterogeneous. In fact, it has been considered on the basis of a numerical taxonomic study that they are not separate species at all (Holmberg & Hallander 1973). A more recent examination concluded that the two species should remain separate and also indicated that *A. viscosus* contained four subgroups that corresponded to serological clusters (Schofield & Schaal 1981). The only DNA–DNA homology study available to date (Coykendall & Munzenmaier 1979) is unfortunately inconclusive. No supporting chemotaxonomic data were reported that might explain or clarify the unresolved taxonomic problems. It may well be that once the genospecies are defined, physiological reactions will become useful as diagnostic tests.

SDS–PAGE patterns of soluble polypeptides were used to survey the isolates. This permitted clusters of similar strains to be identified. Three such clusters were detected, whose polypeptide patterns were at once similar within the group and quite distinct from those of other strains. One cluster within the *A. naeslundii*-like subdivision has been described (Dent & Williams 1984) as *Actinomyces denticolens*. Two further clusters within the *A. viscosus*-like subdivision have also been investigated and are described in a preliminary form below.

Actinomyces denticolens

The SDS–PAGE patterns of *A. denticolens* (Fig. 2, Gel A) first drew our attention to the unusual properties of a group of strains isolated from the dental plaque of dairy cattle. This new taxon shows a wide spread of polypeptides throughout the molecular-weight range with no deeply staining bands. Other strains of the *A. naeslundii* subdivision of oral isolates were not all uniform in electrophoretic patterns. The most common type is illustrated in Fig. 2 (Gel A, tracks 7–10) for wild-type strains from animal dental plaque. Patterns of typical strains of *A. naeslundii* are depicted in Fig. 2, Gel B.

The cell-wall carbohydrates of *A. denticolens* include the 6-deoxyhexose sugar rhamnose, but not 6-deoxytalose (Table 2). This feature, which is not shared by *A. viscosus* or *A. naeslundii* (Slack & Gerencser 1975), serves to distinguish *A. denticolens* from *A. naeslundii*. However, the new taxa identi-

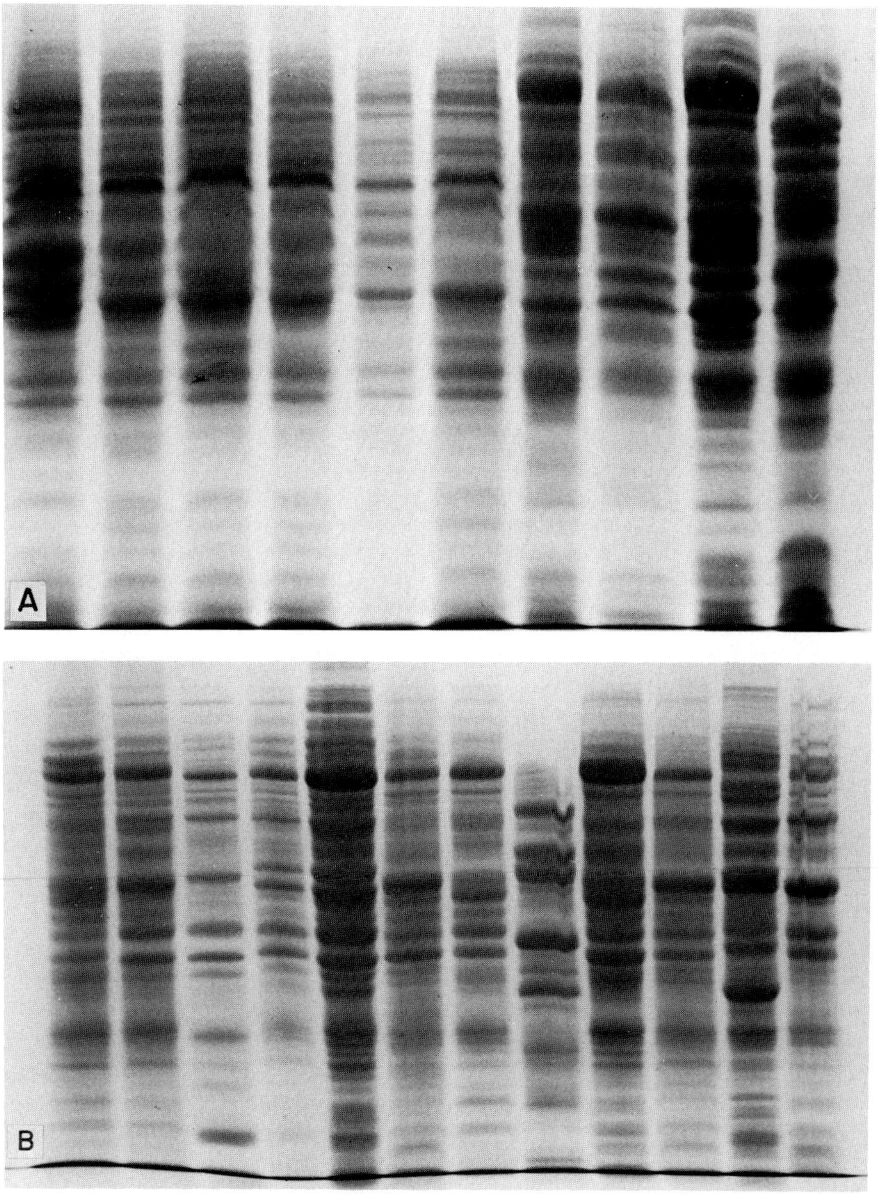

FIG. 2. Sodium dodecyl sulphate–polyacrylamide gel electrophoresis of soluble polypeptides from strains of *Actinomyces denticolens* and *Actinomyces naeslundii*. Gel A. From left to right, strains 4313, 4040, 4277, 4250, 4271, 4303, 12104, X600, WVU45 and T6. Gel B. From left to right, *A. naeslundii* strains 12104, X600, A336 (WVU820), B236, B25, TF11, WVU1048, 10301, WVU45, 4A05, A61 (WVU963) and 1958.

TABLE 2

*Properties of 6-deoxytalose-negative strains resembling Actinomyces naeslundii isolated from the dental plaque of cattle and designated Actinomyces denticolens**

Strains	Acid production from										DNA % G + C	Cell wall carbohydrates					% Homology with ^{3}H-DNA from strains			
	Amygdalin	Cellobiose	Glycerol	Inositol	Inulin	Mannitol	Mannose	Melezitose	Rhamnose	Ribose		Glucose	Galactose	Mannose	Rhamnose	6-deoxy-talose	11490[a]	4277	1958	T6
NCTC 11490 (4303)	—	—	—	w[b]	+	w	w	—	—	+	56.7	—	—	—	+	—	100	77	19	21
4313	—	w	w	—	+	—	w	w	w	+	66.7	—	—	—	+	—	106	73	19	21
4040	—	w	w	+	+	—	w	—	—	+	67.7	—	—	—	+	—	111	88	20	25
4277	w	—	—	w	+	+	w	w	—	+	67.5	—	—	—	+	—	109	100	25	33
4250	w	—	—	w	+	—	w	—	—	w	65.9	—	—	—	+	—	116	97	21	26
4271	—	—	—	+	—	+	—	—	—	—	67.7	—	—	—	+	—	n.d.	n.d.	n.d.	n.d.[c]
Actinomyces naeslundii 1958					n.d.[c]						65.9	—	t[d]	—	+	+	31	21	100	46
Actinomyces viscosus T6					n.d.						66.1	—	t	—	+	+	37	26	46	100

*All strains of *Actinomyces denticolens* ferment glucose, lactose, maltose, sucrose and salicin; none ferment arabinose, sorbitol, trehalose or xylose; all strains are catalase-negative, aesculin-positive and reduce nitrate but not nitrite.

[a]Apparent high homology with strains 11490 is due to unusually low binding to filters.

[b]Weak reaction.

[c]Not determined.

[d]Trace.

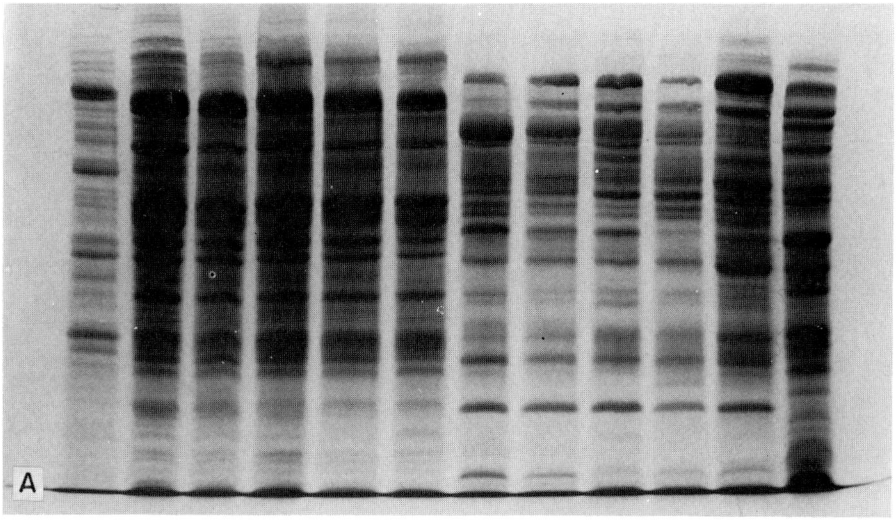

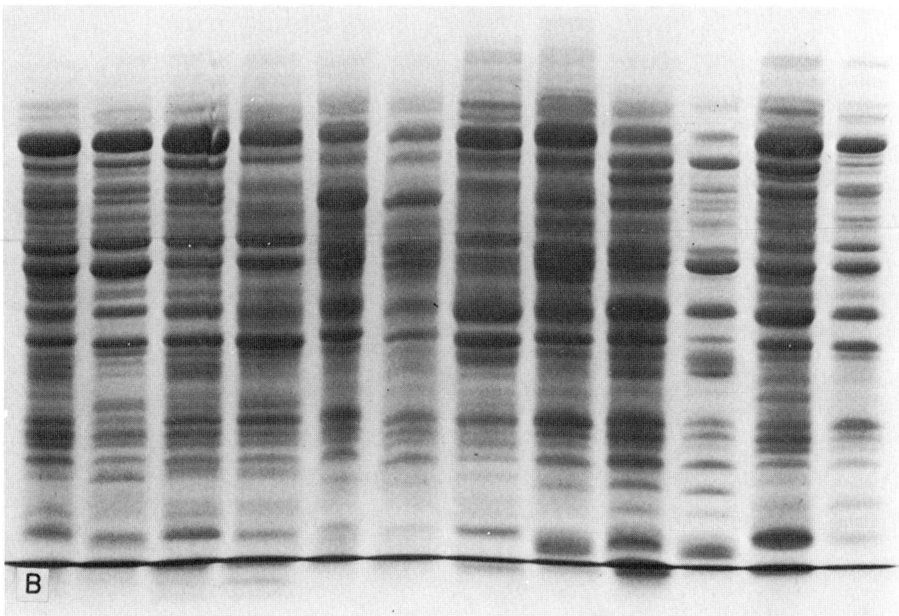

FIG. 3. Sodium dodecyl sulphate–polyacrylamide gel electrophoresis of soluble polypeptides from taxa resembling *Actinomyces viscosus*. Gel A. From left to right, 4036, electrophoretic group B (4503, 4528, 4444, 4563, 4476), electrophoretic group A (3960, 4587, 4159, 4276, 4422), and T6. Gel B. From left to right, *A. viscosus* strains W752, Be32, WA2, W1053, A328 (WVU440), WVU440, A79 (ATCC19246), A78 (ATCC15987), T6, NY1, WVU371, and 13426.

Table 3

*Properties of 6-deoxytalose-negative strains resembling Actinomyces viscosus isolated from the dental plaque of cattle**

Strain	Acid production from										DNA % G + C	Cell wall carbohydrates					% Homology with ^{3}H-DNA from strains				
	Arabinose	Maltose	Lactose	Mannose	Melibiose	Raffinose	Salicin	Sucrose	Trehalose	Xylose		Galactose	Glucose	Mannose	Rhamnose	6-deoxy-talose	4276	4444	1958	T6	
3960	+	+	+	+	w	+	—	+	w	+	66.9			n.d.			106	27	20	19	Electrophoretic group A (3960–4422)
4587	—	w	—	+	w	w	—	+	—	w	65.9	—	+	—	+	—	111	28	26	31	
4159	+	+	w	—	+	+	—	+	—	—	67.3	—	+	—	+	—	96	27	22	21	
4276	+	w	w	w	—	w	—	w	w	w	66.7	—	t	—	+	—	100	24	17	19	
4422	+	+	w	+	+	+	—	+	+	+	67.0	—	+	—	+	—	111	31	25	17	
4036	—	+	+	+	—	+	w	+	—	+	66.9	—	—	—	+	—	31	27	22	24	
4528	—	w	+	w	+	+	—	+	+	+	66.7	t	t	—	+	—	36	26	21	23	Electrophoretic group B (4528–4563)
4503	w	w	w	—	+	+	—	+	+	+	64.4	—	+	—	+	—	29	69	22	24	
4444	w	w	w	—	w	+	w	—	w	—	65.7	—	t	t	+	—	40	100	31	33	
4476	w	w	+	—	w	+	—	w	+	+	66.1	—	+	—	+	—	n.d.	n.d.	n.d.	n.d.	
4563	w	w	+	w	w	+	—	w	w	w	66.7	t	t	—	+	—	28	29	45	47	
Actinomyces naeslundii 1958					n.d.						65.9	—	t	—	+	+	19	25	100	30	
Actinomyces viscosus T6					n.d.						66.1	—	t	—	+	+	29	31	38	100	

*All strains fermented glucose; none fermented amygdalin, cellobiose, ribose, glycerol, inositol, mannitol, rhamnose or inulin.
Abbreviations: n.d. not determined; t, trace; w, weak positive.

fied amongst putative *A. viscosus* strains, which were, like *A. denticolens*, isolated from dental plaque of cattle, are also 6 deoxytalose-negative.

It is clear that *A. denticolens* strains may be distinguished from *A. naeslundii* by their weak fermentation of mannose and their failure to ferment trehalose. This indicates that physiological reactions may have a place in identification of species once the homology groups at present within *A. naeslundii* (and *A. viscosus*) have been characterised. DNA–DNA hybridization shows that *A. denticolens* is a distinct taxon, which shares only 20–30% homology with apparently typical strains of *A. naeslundii* and *A. viscosus* (Table 2).

Actinomyces *viscosus-like isolates*

The 51 catalase-positive *Actinomyces* from animal dental plaque were surveyed by SDS–PAGE. It was immediately obvious that two unusual types of polypeptide patterns could be identified (Fig. 3, Gel A) as different from the several patterns of *A. viscosus* strains. These strains could not be distinguished from typical *A. viscosus* strains by their physiological reactions, but were all isolated from cattle dental plaque and did not contain 6-deoxytalose as a constituent of the cell-wall carbohydrates. The two types of these cattle isolates comprise 5 strains of electrophoretic type A (Fig. 3, Gel A) and five of electrophoretic type B. Strain 4036 is a 6-deoxytalose-negative cattle isolate that does not conform to either of these patterns. Typical *A. viscosus* SDS–PAGE patterns are illustrated in Fig. 3, Gel B.

DNA–DNA homology indicates that electrophoretic group A, which we propose as a new taxon (Dent & Williams, in preparation), is homogeneous and distinct from *A. viscosus*, *A. naeslundii*, electrophoretic group B and strain 4036 (Table 3). Further DNA–DNA homology studies are in progress to reveal the relationships within electrophoretic group B and particularly within the much larger collection of isolates at present still designated as *A. viscosus* and as *A. naeslundii*.

Assessment of the Merits of Chemotaxonomic Techniques

Peptidoglycan composition

The components of peptidoglycan and their linkages are a highly stable feature of bacterial cell walls, and are not known to be altered by mutation within the time span of bacteriological studies. Peptidoglycan can be investigated at three levels: qualitative analysis, quantitative analysis and the complete determination of the linkage of the various components. Qualitative detection is often helpful in the classification and identification of bacteria. An example of this is provided by the presence of aspartate in the peptidoglycan of many lactobacilli. This amino acid is not present in the composition of many peptidoglycans and is easily detected as a prominent turquoise spot in the paper chromatographic system of Hardy *et al.* (1955). The peptidoglycan of most *Actinomyces* spp. can be tentatively identified by the unusual occurrence of both ornithine and lysine, although these same amino acids can also be found in another peptidoglycan linkage pattern shown by bifidobacteria (Schleifer & Kandler 1972). Quantitative analysis, however, provides more definitive information. The oral *Actinomyces* spp. contain equimolar ratios of ornithine and lysine but also have 2 mol of glutamate instead of 1. This provides further evidence of the unique nature of their peptidoglycan. The determination of the linkage pattern, however, is the ultimate study of peptidoglycan. This is done in few laboratories and is not used in surveys of large collections of bacteria. Compositional analysis cannot be regarded as unambiguous unless the possibility of differences occurring in the linkage pattern are considered. Peptidoglycan analysis is used at the species level since it must be expected that a species will be homogeneous with respect to peptidoglycan. Genera, however, can and often do contain species with different peptidoglycan structures.

Cell wall sugars

Cell-wall carbohydrates and serological studies, which often depend on differences in sugar composition, have a long history in classification and identification. The use of sugar qualitative analysis has been hampered by the poor resolution and variation of R_f obtained in many chromatographic systems. Thin layers containing borate have been used (Lato *et al.* 1968), and provide improved resolution and reproducibility. The observation of such plates under UV light enhances the sensitivity of the detection reagents. Classically, sugar analysis has been of value at the species or subspecies level and has proved helpful in this respect amongst the actinomyces studied (Tables 2 & 3).

Electrophoretic patterns of polypeptides

Electrophoresis of proteins involves an overall examination of the relative quantities of many gene products. Several techniques of extraction and electrophoresis are available (reviewed by Kersters & De Ley 1980). The method we have employed is the separation of polypeptides denatured by sodium dodecylsulphate in polyacrylamide gels. Two different uses have been made of such protein patterns. The first is the quantitative comparison of patterns, produced under highly standardised conditions, initially employing main-frame computers to assess similarities (Kersters & De Ley, 1980). Latterly, densitometers interfaced with microcomputers have made this technique more manageable, and the results of such quantitative polypeptide analysis have been shown to correlate well with DNA–DNA hybridization (Owen & Jackman 1982). The second use involves the visual recognition of polypeptide patterns as a surveying technique in studies also employing other taxonomic methods. We have found this method invaluable for the recognition of clusters of strains that have subsequently proved to deserve species status and the recognition of heterogeneity amongst supposed "species." Others have used the same strategy (Moore *et al.* 1980; Biavati *et al.* 1982) and found visually assessed polypeptide patterns to be more reliable than physiological reactions and almost identical in strains having high DNA–DNA homology.

DNA–DNA homology

Of the DNA hybridization techniques available, we have used the nitrocellulose disc method as described by Back & Stackebrandt (1978). Highly purified, high-molecular-weight DNA is required, and it is preferable that the labeled specimens should have a high specific activity. *Actinomyces* strains are difficult to lyse and the isolation of high-molecular-weight DNA sometimes fails and often gives low yields. Incorporation of ^{3}H-thymidine from the medium during growth is wasteful and produces DNA with low specific activities. The nick translation method of labelling purified DNA *in vitro* produced suitably high specific activities. This technique is relatively expensive, and therefore extensive homology studies based on hundreds of unsorted strains would be uneconomic as well as taxonomically unprofitable. The surveying of strains by SDS–PAGE has proved its value in screening collections of strains so that only appropriate clusters of bacteria need be tested by DNA homology. It is accepted that DNA–DNA hybridization is the ultimate technique for expressing similarities between bacteria at the genus level and below. However, it is difficult to apply this powerful method in isolation from other taxonomic techniques, and attempts to do so may produce quantitative data that are not helpful.

The authors wish to thank Cadbury Schweppes whose support for five years to V. E. Dent allowed the collection of animal oral bacteria to be made. The chemotaxonomic studies on these bacteria have been supported by the Medical Research Council (Grant numbers G 979295SB and G 8207598SB). We are grateful to Dr. E. Stackebrandt for his help with the DNA–DNA hybridization procedure.

References

ATFIELD, G. N. & MORRIS, C. J. O. R. 1961 Analytical separations by high voltage paper electrophoresis. Amino acids in protein hydrolysates. *Biochemical Journal* **81**, 606–614.

BACK, W. & STACKEBRANDT, E. 1978 DNS/DNS—Homologie—Studien innerhalb der Gattung *Pediococcus*. *Archives of Microbiology* **118**, 79–85.

BIAVATI, B., SCARDOVI, V. & MOORE, W. E. C. 1982 Electrophoretic patterns of proteins in the genus *Bifidobacterium* and proposal of four new species. *International Journal of Systematic Bacteriology* **32**, 358–373.

BOWDEN, G. H., HARDIE, J. M. & SLACK, G. L. 1975 Microbial variations in approximal dental plaque. *Caries Research* **9** 253–277.

CARLSSON, J. 1973 Simplified gas chromatographic procedure for identification of bacterial metabolic products. *Journal of Applied Microbiology* **25**, 287–289.

COWAN, S. T. 1974 *Cowan and Steel's Manual for the Identification of Medical Bacteria*, p. 152. Cambridge: University Press.

COYKENDALL, A. L. & MUNZENMAIER, A. J. 1979 Deoxyribonucleic acid hybridization among strains of *Actinomyces viscosus* and *Actinomyces naeslundii*. *International Journal of Systematic Bacteriology* **29**, 234–240.

DE MAN, J. C., ROGOSA, M. & SHARPE, M. E. 1960 A medium for the cultivation of lactobacilli. *Journal of Applied Bacteriology* **23**, 130–135.

DENT, V. E. 1979 The bacteriology of dental plaque from a variety of zoo-maintained mammalian species. *Archives of Oral Biology* **24**, 277–282.

DENT, V. E. & MARSH, P. D. 1981 Evidence for a basic plaque microbial community on the tooth surface in animals. *Archives of Oral Biology* **26**, 171–179.

DENT, V. E. & WILLIAMS, R. A. D. 1982 *Lactobacillus animalis* sp. nov., a new species of *Lactobacillus* from the alimentary canal of animals. *Zentralblatt für Bakteriologie, Parisitenkunde, Infektionskrankheiten und Hygiene. I. Abteilung Originale* **C3**, 377–386.

DENT, V. E. & WILLIAMS, R. A. D. 1984 *Actinomyces denticolens* Dent & Williams sp. nov.: a new species from the dental plaque of cattle. *Journal of Applied Bacteriology* **56**, 183–192.

GARVIE, E. I. 1976 Hybridization between the deoxyribonucleic acids of some strains of heterofermentative lactic acid bacteria. *International Journal of Systematic Bacteriology* **26**, 116–122.

HARDY, T. L., HOLLAND, D. O. & NAYLER, J. H. C. 1955 One-phase solvent mixtures for the separation of amino acids. *Analytical Chemistry* **27**, 971–973.

HOLMBERG, K. & HALLANDER, H. O. 1973 Numerical taxonomy and laboratory identification of *Bacterionema matruchotii*, *Rothia dentocariosa*, *Actinomyces naeslundii*, *Actinomyces viscosus*, and some related bacteria. *Journal of General Microbiology* **76**, 43–63.

JOHNSON, J. L., PHELPS, C. F., CUMMINS, C. S., LONDON, J. & GASSER, F. 1980 Taxonomy of the *Lactobacillus acidophilus* group. *International Journal of Systematic Bacteriology* **30**, 53–68.

KERSTERS, K. & DE LEY, J. 1980 Classification and identification of bacteria by electrophoresis of their proteins. In *Microbiological Classification and Identification* eds. Goodfellow, M. & Board, R. G., pp. 273–297. London: Academic Press.

LATO, M., BRUNELLI, B. & CIUFFINI, G. 1968 Bidimensional thin-layer chromatography of carbohydrates on silica gel impregnated with boric acid. *Journal of Chromatography* **34**, 26–34.

LAUER, E., HELMING, C. & KANDLER, O. 1980 Heterogeneity of the species *Lactobacillus acidophilus* (Moro) Hansen and Moquet as revealed by biochemical characteristics and DNA-DNA hybridisation. *Zentralblatt für Bakteriologie, Parisitenkunde, Infektionskrankheiten und Hygiene. I. Abteilung Originale* **C1**, 150–168.

MANDEL, M. & MARMUR, J. 1968 The use of ultraviolet absorbance-temperature profile for determining the guanine plus cytosine content of DNA. In *Methods in Enzymology* eds. Grossman, L. & Moldave, K., Vol. 12B, p. 195. New York: Academic Press.

MENZIES, S. & MOUNT, J. N. 1975 Advantages of silica gel as a medium for rapid thin layer chromatography of neutral sugars. *Medical Research Technology* **32**, 269–276.

MOORE, W. E. C., HASH, D. E., HOLDEMAN, L. V. & CATO, E. P. 1980 Polyacrylamide slab gel electrophoresis of soluble proteins for studies of bacterial flora's. *Applied and Environmental Microbiology* **39**, 900–907.

OWEN, R. J. & HILL, L. R. 1979 The estimation of base compositions, base pairing and genome sizes of bacterial deoxyribonucleic acids. In *Identification Methods for Microbiologists* eds. Skinner, F. A. & Lovelock, D. W., pp. 277–296. London: Academic Press.

OWEN, R. J. & JACKMAN, P. J. H. 1982 The similarities between *Pseudomonas paucimobilus* and allied bacteria derived from analysis of deoxyribonucleic acids and electrophoretic protein patterns. *Journal of General Microbiology* **128**, 2945–2954.

OWEN, R. J. & LAPAGE, S. P. 1976 The thermal denaturation of partly purified bacterial deoxyribonucleic acid and its taxonomic applications. *Journal of Applied Bacteriology* **41**, 335–340.

PARTRIDGE, S. M. 1949 Aniline hydrogen phthalate as a spraying reagent for chromatography of sugars. *Nature, London* **164**, 433.

PASK-HUGHES, R. A. & WILLIAMS, R. A. D. 1978 Cell envelope components of strains belonging to the genus *Thermus*. *Journal of General Microbiology* **107**, 65–72.

SARRA, P. G., MAGRI, M., BOTAZZI, V. & DELLAGLIO, F. 1980 Genetic heterogeneity among *Lactobacillus acidophilus* strains. *Antonie van Leeuwenhoek* **46**, 169–176.

SCHLEIFER, K. H. & KANDLER, O. 1972 Peptidoglycan types of bacterial cell walls and their taxonomic implications. *Bacteriological Reviews* **36**, 407–477.

SCHOFIELD, G. M. & SCHAAL, K. P. 1981 A numerical taxonomic study of members of the *Actinomycetaceae* and related taxa. *Journal of General Microbiology* **127**, 237–259.

SLACK, J. M. & GERENCSER, M. A. 1975 *Actinomyces, Filamentous Bacteria: Biology and Pathogenicity*. Minneapolis: Burgess Publishing Co.

STACKEBRANDT, E. & FIEDLER, F. 1979 DNA-DNA homology studies among strains of *Arthrobacter* and *Brevibacterium*. *Archives of Microbiology* **120**, 289–295.

SWINDLEHURST, C. A., SHAH, H. N., PARR, C. W. & WILLIAMS, R. A. D. 1977 Sodium dodecyl sulphate polyacrylamide gel electrophoresis of polypeptides from *Bacteroides melaninogenicus*. *Journal of Applied Bacteriology* **43**, 319–324.

Identification of Clinically Significant Actinomycetes and Related Bacteria Using Chemical Techniques

K. P. SCHAAL

Institute of Hygiene, University of Cologne, Cologne, Federal Republic of Germany

Diagnostic microbiological laboratories often encounter problems in identifying Gram-positive, non-spore-forming, filamentous or diphtheroidal bacteria from clinical sources. Difficulties are encountered mainly because the routine procedures commonly used for the identification of such clinical isolates are predominantly based on traditional physiological tests. These tests, although irreplaceable for the differentiation of bacteria at the species rank, usually provide too few characters for the clear separation of microorganisms at the generic level. This is especially true for the fermentative or oxidative actinomycetes, which usually need to be identified to the genus level before their species affiliations can be considered.

Chemotaxonomic techniques such as cell-wall, lipid or end-product analyses were primarily developed to improve the classification of certain bacterial groups but are equally useful for identification particularly to the generic level. Apart from acid end-product analyses, these techniques have rarely been employed in diagnostic laboratories because they were thought to be too expensive, elaborate or time-consuming for routine use. Objections such as the latter are no longer valid given the availability of comparatively simple, rapid and reliable modifications of the original analytical chemical procedures. Thus relatively simple methods exist for the determination of amino acids and sugars in whole-organism hydrolysates, for the demonstration and differentiation of mycolic acids in whole-cell methanolysates, and for the analysis of the end products of glucose fermentation.

In this chapter a set of chemical techniques routinely used at the Cologne Institute of Hygiene for the identification of actinomycetes and related bacteria are described. The application and practical value of the techniques are also considered together with their limitations.

0-12-289675-0

Preparation of Biomass

The comparatively large amounts of cell mass needed for chemical analyses must be free of contaminating material from culture media. For many microbes this requirement is readily fulfilled by using carefully washed biomass from fluid media. Broth cultures of actinomycetes, however, are especially prone to microbial contamination, which is more easily controlled when organisms are grown on agar media. On the other hand, the substrate mycelium of many actinomycetes can be difficult to remove from adhering agar. The procedure outlined below is recommended to overcome this problem.

Procurement of cell mass of aerobic and anaerobic actinomycetes

Brain–heart infusion agar (BHIA; Schaal 1977; Schaal & Pulverer 1981) is a good general-purpose medium for the growth of essentially all pathogenic actinomycetes and most morphologically related bacteria. *Actinomadura*, *Nocardia* and *Streptomyces* strains can also be cultivated on medium 79 (Prauser & Falta 1968) and fermentative actinomycetes on CC-medium (Heinrich & Korth 1967; Schaal & Pulverer 1981). These media are used in combination with a membrane culture technique. Thus autoclaved sheets of dialysis tubing placed on top of the agar surface are seeded with a heavy suspension of the test strain; the inoculated plates are incubated aerobically or anaerobically (GasPak jar) at 35°C until sufficient growth is obtained.

After incubation, the dialysis membranes plus growth are removed from agar surfaces with sterile forceps and transferred to Erlenmeyer flasks containing either distilled water or saline and a few glass beads (~5 mm diam). The organisms are killed either with formaldehyde or by heating flasks in a water bath at 75°C for 45 min. Growth is washed from the membranes by shaking in a rotary shaker at 200–250 rpm and collected by centrifugation. For hydrolysis or methanolysis, the sediment is washed in two changes of distilled water and the cells are dried by lyophilisation. Amino-acid and sugar analyses can also be carried out on cells dried overnight in absolute ethanol.

Detection of 2,6-Diaminopimelic Acid in Whole-Cell Hydrolysates*

Preparation of extracts

Dried cells (1–5 mg) are hydrolyzed with 1 ml of 6 *N* HCl in a sealed Pyrex tube at 100°C for 18 h. After cooling, the hydrolysate is filtered through paper and the residue is washed with 1 ml of distilled water. To remove most

*The method discussed here is modified from Becker *et al.* (1964) and Kutzner (1981).

of the HCl, the filtrate is then placed in a sand bath (cooler area) and the excessive fluid is evaporated without burning the residue. This procedure is repeated two or three times after further additions of distilled water. Alternatively, the filtrate can be dried on a rotary evaporator under reduced pressure at 40°C.

Thin-layer chromatography

The residue of the final washed extract is taken up in 0.3 ml of distilled water and 5 μl are spotted onto cellulose-coated, thin-layer chromatography (TLC) plastic sheets (Merck 5577). Separation of the amino acids is achieved by ascending, single-dimensional TLC in a solvent system containing methanol–pyridine–10 *N* HCl–water (80:10:2.5:17.5, v/v). Separation and identification of the diaminopimelic acid (DAP) isomers can be improved by a second development in the same direction and solvent mixture with intermediate air drying.

Identification of the diaminopimelic acid isomers

Amino-acid spots are visualized by spraying chromatograms with ninhydrin in acetone (0.1%, w/v) followed by air drying and heating at 100–110°C for about 2 min. The isomers of DAP produce slowly moving, olive-green spots that fade to yellow; hydroxy-DAP has the lowest R_f value and is followed by *meso*-DAP and LL-DAP, respectively. Most of the other amino acids present in whole-cell hydrolysates give purple spots. A standard solution containing *meso*-DAP (1%, w/v; usually contaminated with traces of LL-DAP) should always be included. D-Diaminopimelic acid is not separated from *meso*-DAP by this method.

Detection of Sugars in Whole-Cell Hydrolysates*

Preparation of extracts

Dried cells (50 mg) are hydrolyzed with 1 ml of 2 *N* H_2SO_4 in a sealed Pyrex tube held at 100°C for 2 h. The hydrolysate is neutralized (pH 5.0–5.5 with methyl red as an internal indicator) by adding barium hydroxide (0.5 g) and shaking the mixture several times; the resulting precipitate and the remaining $Ba(OH)_2$ are sedimented by centrifugation at 6000 rpm. The supernatant is then carefully removed and evaporated to dryness either by lyophilisation or on a rotary evaporator under reduced pressure at 40°C. The residue is

*The method discussed here is modified from Becker *et al.* (1965) and Kutzner (1981).

taken up in 0.4 ml of distilled water and pyridine (~1 ml) added. After thorough mixing, the pyridine phase is separated from the aqueous phase by centrifugation at 6000 rpm. The upper pyridine layer, which contains the purified sugar extract, is transferred to a small glass vial and the extraction procedure is repeated once. The combined pyridine extracts are concentrated in a stream of nitrogen at room temperature to give a final volume of 0.3 ml.

Thin-layer chromatography

The concentrated pyridine extract (5–10 μl) is spotted onto cellulose-coated TLC aluminium sheets (Merck 5552) and the sugars are separated by ascending, single-dimensional TLC using two solvent mixtures applied consecutively. Chromatograms are run in the first solvent, *n*-butanol–water–acetic acid (60:20:20, v/v), for 6 cm and dried in the air. This is followed by three developments, with intermediate air drying, in ethyl acetate–pyridine–water–acetic acid (100:35:25:5, v/v) to an optimal distance of about 12 cm.

Identification of sugars

The sugar spots are visualized by spraying chromatograms with an aniline–phthalate reagent (aniline, 2 ml; phthalic acid, 3.3 g; water-saturated *n*-butanol, 100 ml), air drying, and heating at 100°C for 2–5 min whereupon hexoses give brown, pentoses red, and 6-deoxy sugars grey-green spots. The R_f values increase in the order: galactose, glucose, mannose, arabinose, xylose, fucose, ribose, rhamnose, and 6-deoxytalose. A mixture of these sugars (1%, w/v) in pyridine should be used as a standard. When both solvent systems are applied consecutively, madurose (3-O-methyl-D-galactose; Lechevalier & Gerber 1970), which gives a brownish spot, moves to about the same position as fucose. The two sugars can be distinguished by running the extract in the second solvent mixture whereupon the madurose spot appears between arabinose and xylose and the fucose spot moves slightly farther than xylose.

Detection of Mycolic Acids in Whole-Cell Methanolysates*

Preparation of methanolysates

Dried cells (~50 mg) are placed in a suitable glass reaction vessel, which is sealed with a screw cap lined with inert polytetrafluoroethylene. Three ml of

*The method discussed here is taken from Minnikin *et al.* (1980).

a mixture containing methanol–toluene–sulphuric acid (30:15:1, v/v) are added and methanolysis is allowed to proceed in tightly closed vessels held at 75°C in an oven or heating block overnight (16–18 h). After cooling to room temperature, 2 ml of light petroleum (bp 60–80°C) are added and the mixture is thoroughly shaken prior to centrifugation at 2000 rpm for 10 min. Meanwhile, a small column (~1 cm) of ammonium hydrogen carbonate is prepared in a cotton wool–plugged Pasteur pipette and prewashed with 2 ml of diethyl ether. The upper light petroleum layer from the reaction vessel is pipetted onto the column and the eluent is collected in a small vial. The reaction mixture is extracted with a further 1-ml portion of light petroleum and the upper layer is added to the column. The pipette used for transfer and the column are washed with 1 ml of diethyl ether. The combined eluents are evaporated in a stream of nitrogen at about 40°C.

Thin-layer chromatography

Methanolysate residues are taken up in a few drops of light petroleum and 3 μl of each extract are spotted onto silica gel-coated aluminium sheets (Merck 5554). Separation of the methyl esters of the lipid components is achieved by single-dimensional ascending TLC, using light petroleum (bp 60–80°C)–acetone (95:5, v/v) as the solvent system. Better separation can be obtained by a triple development, with intermediate air drying, in the same solvent mixture. For more rapid analyses, smaller sheets (10 × 10 cm), cut from the original ones (20 × 20 cm), can be used.

Identification of mycolic acid methyl esters

Lipid spots are visualized by spraying chromatograms with a 10% ethanolic solution of molybdophosphoric acid followed by heating at 120°C for 15 min. Lipid components give blue-green spots. Fatty acid methyl esters appear slightly below the solvent front and the mycolic acid esters occupy various positions between the fatty acids and the origin. Typical mycolic acid patterns are identified by comparison with those from reference strains.

Determination of Fermentation End Products*

Culture conditions

Peptone–yeast-extract–glucose medium (PYG) or chopped meat–carbohydrate medium (CMC) (Holdeman *et al.* 1977) are heavily inoculated from an

*The method discussed here is taken from Holdeman *et al.* (1977).

actively growing broth culture using a Pasteur pipette. The inoculated tubes are incubated at 36°C for 3–7 days, or until visible growth is obtained, under anaerobic conditions (e.g. in GasPak jars).

Preparation of extracts for analysis of volatile fatty acids and alcohols

Cultures (1 ml) are pipetted into glass tubes and after the addition of 0.2 ml of 50% (v/v) H_2SO_4, 0.4 g of NaCl and 1 ml of diethyl ether, each tube is tightly closed with a screw cap lined with an inert polytetrafluoroethylene seal. The ether and culture medium are mixed by inverting tubes gently about 20 times and the ether is separated from the aqueous phase by centrifugation. The upper ether layers, which contain the end products, are carefully pipetted off the aqueous layers and transferred to other glass tubes. Anhydrous 4–20-mesh $CaCl_2$ is added to equal about one-fourth of the volume of ether in the tubes, which are closed and left to stand for about 5 min.

Preparation of extracts for analysis of nonvolatile acids

One-ml samples of the original cultures are pipetted into glass tubes as before, 2 ml of methanol and 0.4 ml of 50% H_2SO_4 are added and the tubes are sealed; then the mixture is heated at 60°C in a heating block or water bath for 30 min so that methyl esters can form. Next, water (1 ml) and chloroform (0.5 ml) are added, the tubes are closed and the contents are mixed by gentle inversion of each tube about 20 times. Any emulsions that form can be broken by brief centrifugation. The lower chloroform layer contains the methylated derivatives.

Gas–liquid chromatography

Comparatively simple and inexpensive instruments with either thermal conductivity or flame ionisation detectors can be used for end-product analyses. A Packard Model 427 gas chromatograph equipped with a thermal conductivity detector, a temperature programing unit, and a 2-m glass column packed with LAC-1-R 296 (WGA Düsseldorf, No. 6035) on Chromosorb-G as stationary phase has been successfully employed in Cologne. The operating conditions routinely followed are: injection block, 140°C; detector block, 150°C; column oven programmed to rise from 80 to 140°C at a rate of 2°C/min; 2× attenuation; helium carrier-gas flow rate, 120 ml/min; and recorder speed, 0.5 or 1.0 cm/min. Volatile and methylated products are run on the same column using the same operating conditions.

Ether or chloroform extracts (10–14 μl) are drawn into an appropriate injection syringe and, after wiping off the outside of the injection needle, injected onto the column. The elution times of the various compounds are compared with those of a standard mixture (Holdeman *et al.* 1977) treated in the same manner. Usually, qualitative interpretation of the results, e.g. determination of the presence or absence of characteristic end products, is sufficient for the identification of cultures but semiquantitative analyses using a standard curve of peak heights can also be performed. After testing approximately 20 methylated samples, columns should be reconditioned by injecting 14 μl of methanol.

Evaluation of Chemical Data

The simplified chemical techniques employed in Cologne are undoubtedly more expensive than most of the conventional biochemical and physiological tests commonly used in diagnostic microbiological laboratories, especially if the capital cost of the equipment is taken into account. Although chemical procedures may also be more elaborate than conventional tests, they usually yield results at least as rapidly as the latter. Thus, for example, DAP, sugars and mycolic acids can be detected in whole-cell extracts of aerobic actinomycetes within 4–5 days, including the time required to prepare cell mass, hydrolysates or methanolysates, and to run the chromatograms. Carbon-source utilization tests performed on the same organisms yield satisfactory results only within 7–14 days of incubation.

In principle, it should not be too difficult to introduce chemical methods like those described above into the routine programme of a diagnostic laboratory, although co-workers not used to this type of work may not be overenthusiastic at first. However, inasmuch as simplified chemical techniques give good results and are comparatively easy to learn, their introduction for routine purposes is by no means insurmountable.

The diagnostic potential of whole-organism and end-product analyses is not restricted to the identification of actinomycetes and related microbes. However, the pathogenic, Gram-positive, filamentous bacteria may be considered a litmus test for assessing the practical value of simple chemical techniques as actinomycetes inevitably cause diagnostic problems when isolated from clinical specimens.

Amino Acids and Sugars

The analysis of whole-cell hydrolysates for the presence or absence of DAP isomers provides especially useful information. The test involved is simple, comparatively rapid, easy to perform and is of considerable differential power. The whole procedure, including hydrolysis and TLC, yields reliable results within 24 h given sufficient cell mass. Interpretation of chromatograms usually does not present any problems since the spots produced by DAP isomers can readily be identified by low R_f values and green to yellow colour (Fig. 1).

The DAP content of whole-organism extracts can be directly related to cell-wall composition since this diamino acid is found only in the wall peptidoglycan. Gram-positive, filamentous and diphtheroidal bacteria can be assigned to three major groups on the basis of the DAP content.

Members of the first group, which lack DAP, include the fermentative actinomycete genera *Actinomyces* (Fig. 1, line 7), *Oerskovia, Rothia* and *Arcanobacterium*, *Bifidobacterium*, *Erysipelothrix* and *Lactobacillus* species (Tables 1 & 2). The second group is composed of both oxidative and fermentative microorganisms that contain the LL isomer of DAP. *Streptomyces* strains are the only obligately aerobic actinomycetes of clinical or diagnostic interest

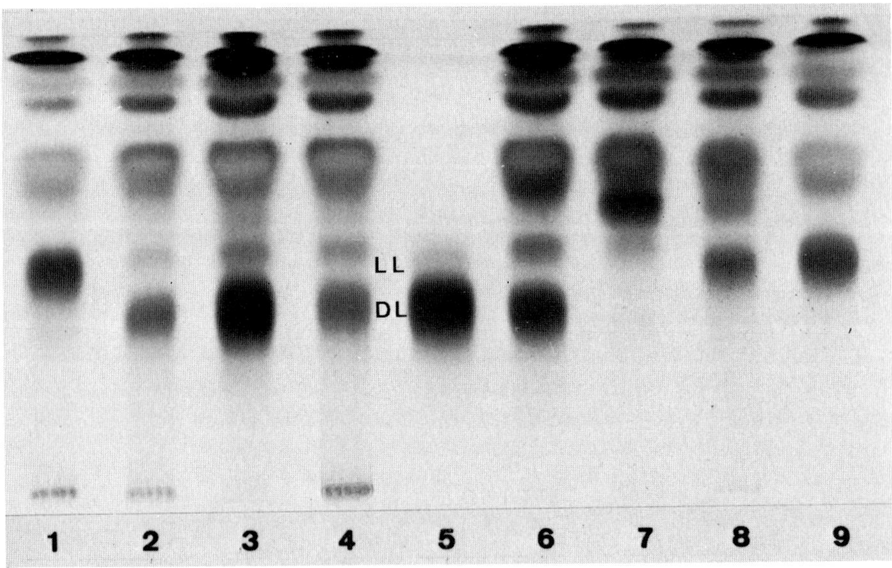

FIG. 1. TLC amino-acid analysis of whole-cell hydrolysates of selected actinomycetes and related bacteria. (1) *Streptomyces* sp. HIK N32 (LL-DAP); (2) *Nocardia asteroides* type A, HIK N3 (DL-DAP); (3) *Corynebacterium diphtheriae* Gf PW8 (DL-DAP); (4) *Corynebacterium xerosis* Gf C137 (DL-DAP); (5) DAP standard containing DL-DAP and traces of LL-DAP; (6) *Corynebacterium matruchotii* HIK A121 (DL-DAP); (7) *Actinomyces naeslundii* HIK A86 (no DAP); (8) *Propionibacterium avidum* HIK A161 (LL-DAP); (9) *Arachnia propionica* HIK A111 (LL-DAP). (HIK: Institute of Hygiene, University of Cologne, FRG; Gf: Dr. M. Goodfellow, University of Newcastle upon Tyne, UK).

that show this cell-wall component (Fig. 1, line 1; Table 1). Fermentative actinomycetes that contain LL-DAP exclusively belong to the genus *Arachnia* (Fig. 1, line 9; Table 2); the latter can be easily and reliably differentiated from morphologically and physiologically similar strains of *Actinomyces israelii* on this basis. Most *Propionibacterium* species (Fig. 1, line 8; Table 2) also have LL-DAP in their walls. Bacteria with *meso*-DAP in their peptidoglycan constitute the final group, which comprises the genera *Nocardia*, *Mycobacterium* and *Corynebacterium* (Fig. 1, lines 2, 3, 4, and 6; Table 1) together with *Actinomadura*, *Dermatophilus*, *Listeria*, *Micropolyspora*, *Nocardiopsis*, *Rhodococcus*, *Saccharopolyspora* and *Thermoactinomyces*. This group also contains certain *Eubacterium* and *Propionibacterium* strains and "*Bacterionema*" *matruchotii*, which has been transferred to the genus *Corynebacterium* as *Corynebacterium matruchotii* (Collins 1982; Goodfellow *et al.* 1982).

The sugar patterns derived from TLC analyses of whole-cell hydrolysates are more difficult to evaluate than the simple DAP patterns. The various diagnostic sugars may be derived from the cell wall, wall-associated structures, the cytoplasm or even from the culture medium. Contamination from the growth medium is usually not detected given the low amount involved when carefully washed cells are used. However, trace amounts of sugars from the medium may become apparent when highly concentrated extracts are employed.

The aerobic actinomycetes have been assigned to four whole-cell sugar patterns (Lechevalier 1968). Pattern A is characterized by the presence of arabinose and galactose and the absence of xylose and madurose. Organisms belonging to pattern B have madurose but neither arabinose, galactose nor xylose. Actinomycetes forming pattern C do not have any of the characteristic sugars mentioned, and those of pattern D have arabinose and xylose but lack galactose and madurose.

Pattern A is characteristic of the mycolic acid-containing taxa, *Corynebacterium*, *Mycobacterium*, *Nocardia* and *Rhodococcus* (Fig. 2, lines 2, 5 and 8; Table 1), as well as of *Micropolyspora* and *Saccharopolyspora*. *Actinomadura* and *Dermatophilus* strains exhibit sugar pattern B (Fig. 2, lines 4 and 6; Table 1), and *Nocardiopsis*, *Streptomyces* and *Thermoactinomyces* either lack all of the diagnostic sugars (pattern C) or do not have any characteristic sugars at all ("pattern NC": no characteristic sugars) (Fig. 2, line 3). *Micromonospora* is a good example of an actinomycete taxon belonging to sugar pattern D.

The sugar patterns of the fermentative actinomycetes are less well defined and are more heterogeneous than those of their oxidative counterparts. The classical, facultatively anaerobic actinomycetes such as the genera *Actinomyces* and *Rothia* do not contain detectable amounts of arabinose, xylose or madurose, and the same is true for *Arachnia*, *Bifidobacterium*, *Eubacterium*, *Lactobacillus*, *Listeria*, *Oerskovia* and *Propionibacterium*. The only filamentous,

TABLE 1

Differentiation of clinically significant aerobic actinomycetes and related bacteria using morphological, physiological and chemical characters[a]

Taxon	*Oerskovia*	*Corynebacterium*	*Mycobacterium*	*Nocardia*	*Nocardia brevicatena*	*Rhodococcus*	*Micropolyspora faeni*	*Saccharopolyspora*	*Streptomyces*	*Actinomadura*	*Dermatophilus*	*Nocardiopsis*	*Thermoactinomyces*
Physiological and morphological characters													
O/F Test[b]	F	F	O	O	O	O	O	O	O	O	O	O	O
Substrate mycelium	+	—	—	+	+	d	+	+	+	+	+	+	+
Fragmentation of substrate mycelium	+	—	—	+	d	+	d	—	—	—	+	+	—
Aerial mycelium	—	—	—	+	+	—	+	+	+	d	—	+	+
Motile spores	+	—	—	—	—	—	—	—	—	—	+	—	—
Endospores	—	—	—	—	—	—	—	—	—	—	—	—	+

Chemical characters (whole-cell hydrolysis)													
meso-DAP[c]	—	+	+	+	+	+	+	+	—	+	+	+	+
LL-DAP[c]	—	—	—	—	—	—	—	—	+	—	—	—	—
Arabinose	—	+	+	+	+	+	+	+	—	—	—	—	—
Galactose	+	+	+	+	+	+	+	+	—	—	—	—	—
Madurose	—	—	—	—	—	—	—	—	—	+	+	—	—
Xylose	—	—	—	—	—	—	—	—	—	—	—	—	—
Mycolic acids[d]	—	S	M	S	S	S	—	—	—	—	—	—	—
Wall chemotype[e]	VI	IV	IV	IV	IV	IV	IV	IV	I	III	III	III	III
Whole-cell sugar pattern[f]		A	A	A	A	A	A	A	NC	B	B	C	C

[a]Data from Cross (1981); Goodfellow & Schaal (1979); Kutzner (1981); Lechevalier & Lechevalier (1981a,b); Williams & Wellington (1981). +, present; —, absent; d, strain variation.
[b]O, oxidative; F, fermentative.
[c]DAP, 2,6-diaminopimelic acid.
[d]Whole-cell methanolysates contain S, a single spot pattern and M, a multispot pattern.
[e]Lechevalier & Lechevalier (1976).
[f]Lechevalier (1968); A, arabinose and galactose; B, madurose; C and NC, no characteristic sugars.

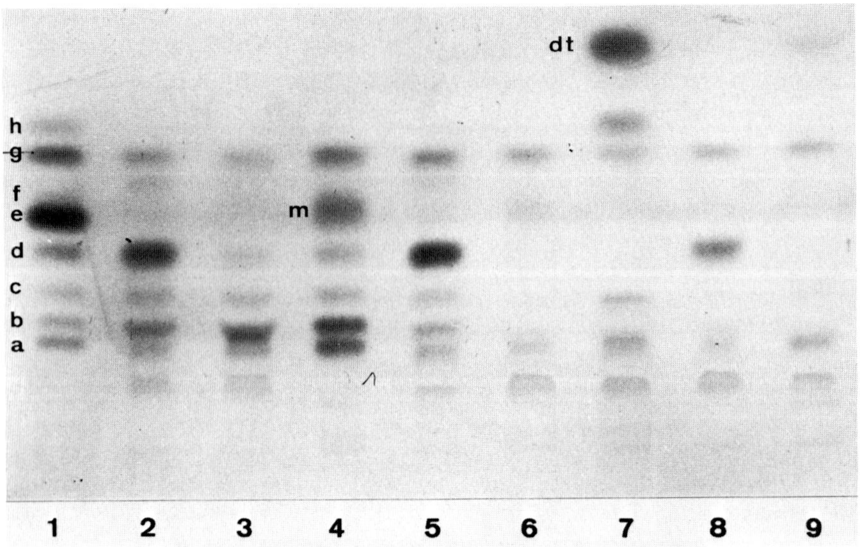

FIG. 2. TLC sugar analysis of whole-cell hydrolysates of selected actinomycetes and related bacteria. (1) Standard sugar solution (a, galactose; b, glucose; c, mannose; d, arabinose; e, xylose; f, fucose; g, ribose; h, rhamnose); (2) *Nocardia asteroides* HIK N3 (*galactose*, glucose, mannose, *arabinose*, ribose); (3) *Streptomyces* sp. HIK N32 (glucose and traces of other contaminating sugars); (4) *Actinomadura madurae* HIK N44 [*glucose*, *madurose* (m), ribose with small amounts of other contaminating sugars]; (5) *Mycobacterium avium* Gf D4 (*galactose*, glucose, mannose, *arabinose*, ribose); (6) *Actinomadura pelletieri* HIK N25 (glucose, *madurose*, ribose, and traces of contaminating galactose); (7) *Actinomyces viscosus* HIK A82 [*galactose*, *glucose*, *mannose*, ribose, *rhamnose*, *6-deoxytalose* (dt)]; (8) *Corynebacterium matruchotii* HIK A121 (*galactose*, *arabinose*, ribose); (9) *Rothia dentocariosa* HIK A142 (*galactose*, glucose, ribose). (For other abbreviations, see caption to Fig. 1.)

fermentative species characterized by the presence of arabinose in whole-cell or cell-wall hydrolysates is *Corynebacterium matruchotii*. A variety of other sugars, such as 6-deoxytalose, fucose, galactose, glucose or rhamnose can help to separate these genera and in some cases the species they contain. Thus, for example, hydrolysates of *Actinomyces viscosus* strains contain 6-deoxytalose, galactose, glucose, mannose and rhamnose but lack detectable amounts of fucose (Fig. 2, line 7). This combination reliably separates *A. viscosus* from all other species of *Actinomyces* as well as from *Arachnia*, *Rothia* (Fig. 2, line 9) and *Lactobacillus* (Table 2). Characteristic sugar patterns have been found for *Actinomyces bovis*, *A. israelii*, *A. naeslundii*, *A. odontolyticus*, "*A. suis*"/*A. denticolens* (Dent & Williams 1984) and *A. pyogenes* (Reddy *et al.* 1982; Slack & Gerencser 1975) (Table 2).

Analysis of whole-cell hydrolysates for DAP and sugars often allows unknown isolates to be classified to the correct wall chemotype (*sensu* Lechevalier & Lechevalier 1976) without the necessity of preparing purified cell-wall samples. Thus the presence of LL-DAP and the absence of characteristic sugars is typical of wall chemotype I, and *meso*-DAP, arabinose and xylose of cell-wall type II. A combination of *meso*-DAP and madurose, and probably *meso*-DAP and no characteristic sugar, is indicative of a wall chemotype III. Actinomycetes with a cell-wall type IV are defined by *meso*-DAP, arabinose and galactose; the latter can be detected in both cell-wall and whole-cell hydrolysates.

The procedure described for analyzing the sugar content of whole-organism hydrolysates is not only slightly more complicated than that for DAP determination but it is also more susceptible to technical error. In any event, however, only 2 h are required to carry out the hydrolysis and results can be obtained within one working day even though chromatographic development cannot be completed within 4 h.

The groups of filamentous actinomycetes and coryneform bacteria defined by amino-acid and sugar analysis of walls or whole cells show considerable heterogeneity. This is especially true for organisms with a wall chemotype IV *sensu* Lechevalier and Lechevalier (1976). The latter encompasses a large number of genera and species that show varying degrees of clinical significance, consequently their reliable identification in the diagnostic laboratory is of the utmost practical importance. Well-known pathogens such as *Corynebacterium diphtheriae* or *Mycobacterium tuberculosis* do not create too many problems, but less common organisms such as "atypical" mycobacteria, *Nocardia* and *Rhodococcus* strains are often much more difficult to identify.

Mycolic Acids

Many of the problems in assigning wall chemotype IV actinomycetes to genera can be overcome if mycolic acid analyses are routinely performed for this purpose. Minnikin and co-workers (1974, 1975, 1980, 1982a,b) have impressively shown the diagnostic potential of mycolic acids. For routine application, simple, single-dimensional TLC procedures (Minnikin *et al.* 1980) are usually satisfactory, although two-dimensional TLC, GLC, or mass spectrometry provide more detailed results and can be used to determine the chemical structures of the constituent mycolic acids (Minnikin *et al.* 1974, 1980, 1982a,b).

The simple, single-dimensional TLC procedure described has been used successfully in Cologne for several years. The technique, which yields results within 20 h, can be used to differentiate Gram-positive, filamentous or rod-shaped bacteria that contain mycolic acids (Fig. 3, lines 1–15) from those that lack these compounds (Fig. 3, lines 17 and 18). The latter include the

TABLE 2

Differentiation of fermentative actinomycetes and related bacteria, using physiological and chemical characters[a]

Taxon	*Actinomyces*	*Arachnia*	*Arcanobacterium*	*Corynebacterium*	*Corynebacterium matruchotii*	*Rothia*	*Bifidobacterium*	*Erysipelothrix*	*Eubacterium*	*Lactobacillus*	*Listeria*	*Propionibacterium*
Physiological characters												
Catalase	—[e]	—	—	+	+	+	—	—	—	—	+[f]	D
Nitrate reduction	D	+	d	D	+	+	—	—	D	—	—[g]	D
Chemical characters (whole-cell hydrolysis)												
meso-DAP[b]	—	—	—	+	+	—	—	—	D	—	+	d
LL-DAP[b]	—	+	—	—	—	—	—	—	—	—	—	+[h]
Arabinose	—	—	[i]	+	+	—	—	[i]	[i]	—	—	—
Galactose	D	+		+	+	+	[i]			+	—	[i]
Rhamnose	D	—		—	—	—				+	[i]	
6-Deoxytalose	D	—		—	—	—				—		
Fucose	D	—		—	—	—				—		
Mycolic acids[c]	—	—	—	+	+	—	—	—	—	—	—	—

End products from glucose fermentation[d]												
Acetic acid	+	+	+	+	+	+	+	+	+	(+)	[i]	+
Propionic acid	—	+	—	d	+	—	—	—	(+)	—		+
Isobutyric acid	—	—	—	—	—	—	—	—	—	—		—
n-Butyric acid	—	—	—	—	—	—	—	—	+	—		—
Isovaleric acid	—	—	—	—	—	—	—	—	—	—		d
n-Valeric acid	—	—	—	—	—	—	—	—	—	—		—
Isocaproic acid	—	—	—	—	—	—	—	—	—	—		—
n-Caproic acid	—	—	—	—	—	—	—	—	D	—		—
Pyruvic acid	—	—	—		d	(+)	—	—	—	—		—
Lactic acid	+	(+)	+	+	d	+	+	+	d	+		(+)
Succinic acid	+	(+)	d	[i]	—	(+)	(+)	+	(+)	(+)		(+)

[a]Data from Holdeman *et al.* (1977); Schaal & Pulverer (1981); Schleifer & Kandler (1972); Schofield & Schaal (1981); Slack & Gerencser (1975). +, present; —, absent; D, species differences; d, strain variation; (+), present in small amounts or absent.
[b]DAP, 2,6-diaminopimelic acid.
[c]From whole-cell methanolysates.
[d]In PYG medium (Holdeman *et al.* 1977).
[e]*Actinomyces viscosus* is positive.
[f]*Listeria grayi* is negative.
[g]*Listeria grayi* biovar *murrayi* is positive.
[h]*Propionibacterium lymphophilum* does not contain DAP.
[i]Incomplete or no data available.

genera *Actinomadura*, *Actinomyces*, *Arachnia*, *Arcanobacterium*, *Bifidobacterium*, *Dermatophilus*, *Erysipelothrix*, *Eubacterium*, *Lactobacillus*, *Listeria*, *Nocardiopsis*, *Oerskovia*, *Propionibacterium*, *Rothia*, *Saccharopolyspora*, *Streptomyces* and *Thermoactinomyces* (Tables 1 & 2). *Micropolyspora brevicatena*, which has been transferred to the genus *Nocardia* as *Nocardia brevicatena* (Goodfellow & Pirouz 1982), can readily be distinguished from *Micropolyspora faeni* on the basis of mycolic acid composition (Table 1).

Mycolic acid-containing bacteria can be classified into two major groups. The first group encompasses *Mycobacterium* species, which contain complex mixtures of mycolic acids (Fig. 3, lines 1–5, 11; Table 1). Indeed, mycobacteria may be classified into four groups on the basis of multispot patterns produced by their mycolic acid methyl esters on thin-layer plates (Minnikin *et al.* 1980, 1982a). The first pattern, which is characterized by methanolysates of *M. tuberculosis* (Fig. 3, line 1) and *M. bovis* (Fig. 3, line 4), can be identified by the presence of three fast-moving spots, which are not always well resolved by the recommended solvent mixture. The nature of the two additional spots seen in the *M. bovis* extract (Fig. 3, line 4) has not been determined, but they could represent an artifact. The second pattern is composed of three fast-moving and two more slowly moving components (Fig. 3, line 2). This general pattern is found in *M. avium*, *M. intracellulare*, *M. johnei* and *M. phlei*. *Mycobacterium fortuitum* (Fig. 3, line 3) and *M. smegmatis* are examples of a third pattern, which can be identified by three to four groups of lipid spots distributed over the whole distance through which mycolates move in the solvent system used. The fourth and simplest pattern, which consists of two fast-moving mycolic acid methyl esters, is characteristic of *M. chelonei* (Minnikin *et al.* 1982a). The reference strain of *M. kansasii* (HIK 1201) used in Fig. 3, line 5, gave a multispot pattern that could not be assigned to any of the four groups outlined above. In contrast, other *M. kansasii* strains examined yielded a multispot pattern similar to that of the *M. tuberculosis* group. *Mycobacterium* species HIK, a recent clinical isolate, has still to be identified with certainty but it has a mycolic acid pattern similar to that of *M. fortuitum*.

Corynebacterium, *Nocardia* and *Rhodococcus* species (Fig. 3, lines 6–10 and 12–15) generally produce a single spot that corresponds to a mycolic acid methyl ester. Mycolates from these taxa have different R_f values, which reflect differences in the molecular size, and they can be differentiated given sufficient standardisation of the chromatographic procedure. Mycolates from nocardiae (Fig. 3, lines 6 and 7) give spots with the highest mobility albeit lower than those produced by the fast-moving mycobacterial mycolates. The methyl mycolates of most rhodococci have a slightly lower R_f than those from nocardiae, but this difference is often difficult to demonstrate (Fig. 3, lines 8–10). Corynebacteria are usually easy to identify since they contain rela-

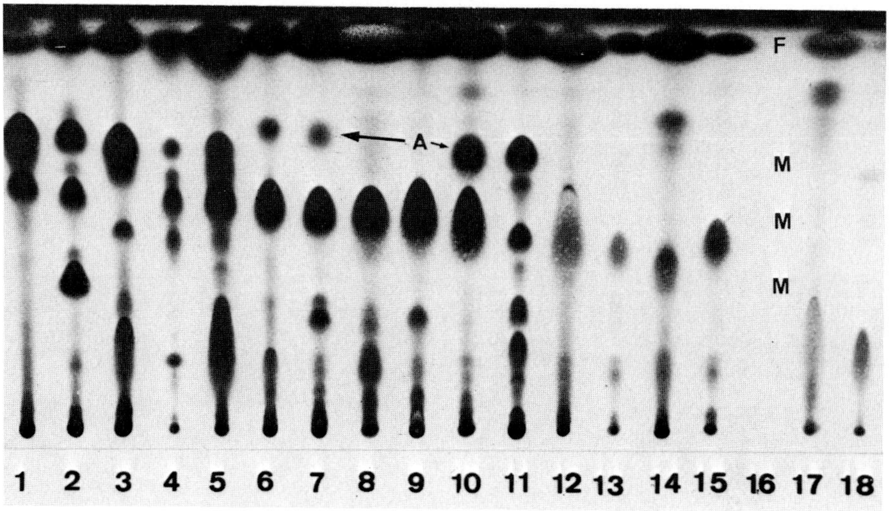

FIG. 3. TLC mycolic acid analysis of whole-cell hydrolysates of selected actinomycetes including corynebacteria and mycobacteria. (1) *Mycobacterium tuberculosis* Gf strain C; (2) *Mycobacterium avium* Gf D4; (3) *Mycobacterium fortuitum* Gf M62; (4) *Mycobacterium bovis* HIK 1011; (5) *Mycobacterium kansasii* HIK 1201; (6) *Nocardia asteroides* HIK N3; (7) *Nocardia amarae* Gf N782; (8) *Rhodococcus ruber* HIK N27; (9) *Rhodococcus equi* HIK D61; (10) *Rhodococcus erythropolis* Gf N11; (11) *Mycobacterium* sp. HIK recent clinical isolate; (12) *Corynebacterium diphtheriae* Gf PW8; (13) *Corynebacterium xerosis* Gf C137; (14) *Corynebacterium bovis* Gf C95; (15) *Corynebacterium matruchotii* HIK A121; (16) no sample; (17) *Streptomyces* sp. HIK N32 (no mycolic acids); (18) *Actinomyces naeslundii HIK A86* (no mycolic acids); (A) long-chain alcohols; (M) mycolic acid methyl esters; (F) fatty acid methyl esters. (For other abbreviation, see caption to Fig. 1.)

tively small amounts of mycolic acids, which give low R_f values (Fig. 3, lines 12–15).

End-Product Analysis

The analysis of the acid and alcohol end products of the fermentative metabolism of carbohydrates and/or amino acids has been widely used in clinical microbiological laboratories. Indeed, end-product determinations are indispensable for the identification of strictly anaerobic pathogens. The gas–liquid chromatography procedure described in the *V.P.I. Anaerobe Laboratory Manual* (Holdeman *et al.* 1977) is simple, reliable, and can be carried out in less than 2 h.

All of the Gram-positive, anaerobic or facultatively anaerobic bacteria considered in this chapter (Table 2) produce comparatively simple mixtures of end products. Nevertheless, most of the taxa concerned can be identified by qualitative or quantitative differences in their end-product profiles. *Arachnia*, *Corynebacterium* and *Propionibacterium* strains (Fig. 4, curves 1–3; Fig. 5, curves 3–5) are characterized by the formation of propionic and acetic acids and usually also produce lactic and succinic acids. This pattern differentiates them from *Actinomyces* strains, which yield acetic and lactic acids, and usually greater amounts of succinic acid (Fig. 4, curves 4 and 5; Fig. 5, curves 6 and 7). Bifidobacteria have a similar profile (Fig. 4, curve 6; Fig. 5, curve 2) but do not usually produce major amounts of succinic acid (Table 2). Homofermentative *Lactobacillus* strains are readily identified by the production of lactic acid as the sole end product. Heterofermentative lactobacilli form acetate and traces of succinate, but lactate remains the predominant organic

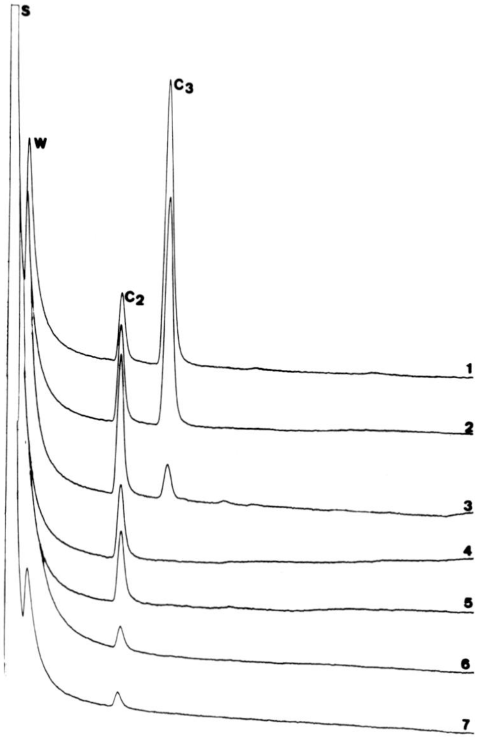

FIG. 4. GLC end-product profiles (volatile acids) of selected fermentative actinomycetes and related bacteria. (1) *Propionibacterium acnes* (clinical isolate); (2) *Arachnia propionica* HIK A111; (3) *Corynebacterium matruchotii* HIK A121; (4) *Actinomyces viscosus* HIK A82; (5) *Actinomyces pyogenes* NCTC 5224; (6) *Bifidobacterium bifidum* (clinical isolate); (7) *Lactobacillus casei* var. *rhamnosus* NCDO 243; NCTC: National Collection of Type Cultures, London, UK; NCDO: National Collection of Dairy Organisms; Reading, UK; HIK: Institute of Hygiene, University of Cologne, FRG; S: solvent; W: water; C_2: acetic acid; C_3: propionic acid.

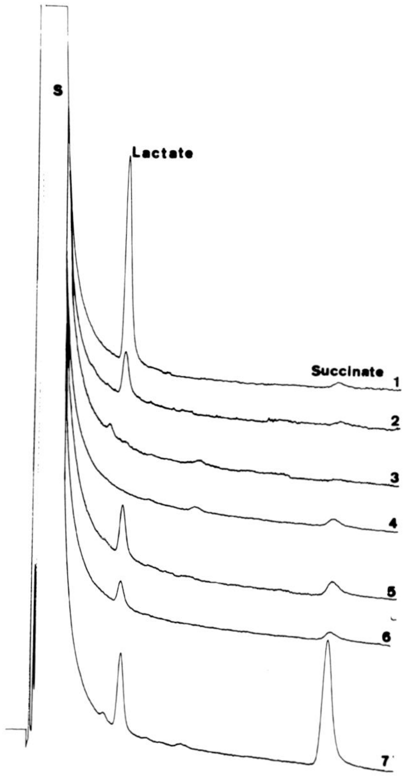

FIG. 5. GLC end-product profiles (methyl derivatives) of selected fermentative actinomycetes and related bacteria. (1) *Lactobacillus casei* var. *rhamnosus* NCDO243; (2) *Bifidobacterium bifidum* (clinical isolate); (3) *Propionibacterium acnes* (clinical isolate); (4) *Arachnia propionica* HIK A111. (5) *Corynebacterium matruchotii* HIK A121; (6) *Actinomyces pyogenes* NCTC 5224; (7) *Actinomyces viscosus* HIK A82; (S) solvent. (For other abbreviations see caption to Fig. 4.)

acid (Fig. 4, curve 7; Fig. 5, curve 1). Characteristic end products of *Eubacterium* species include *n*-butyric acid together with acetic, lactic and succinic acids; certain species may also form propionic or caproic acids and alcohols.

Conclusion

The diagnostic information that can be obtained from the chemical techniques described is summarized in Tables 1 and 2. These chemical data together with a few morphological and physiological characters form a reliable basis for differentiating various Gram-positive bacteria at the generic level. Aerobic actinomycetes and related organisms can be identified to the rank of genus by their type of metabolism, growth characteristics, ability to form spores, and DAP, sugar and mycolic acid patterns (Table 1). In turn, fermentative actinomycetes can be equally well characterized on the basis of catalase production, nitrate reduction, DAP, sugar and mycolic acid content, and fermentation end-product profiles (Table 2).

The chemical characteristics discussed are extremely stable compared with morphological or physiological features and are not usually subject to mutation or strain variation. It does, therefore, seem both admissible and advantageous to construct flow charts for the chemical identification of clinically significant actinomycetes and related organisms. The relevant aerobic actinomycetes including mycobacteria can be reliably differentiated by DAP, sugar and mycolic acid analyses (Fig. 6). When aerial mycelium formation and spore characteristics are included, unknown isolates can usually be identified to the genus level.

A similar flow chart for the fermentative filamentous and diphtheroidal bacteria (Fig. 7) is based upon amino-acid, sugar and lipid data, the presence or absence of catalase and end-product profiles. Indeed, some organisms can be identified to the species level solely by their sugar patterns.

It can be concluded that chemotaxonomic techniques should no longer be confined to research projects or reference laboratories. After appropriate simplification and modification, chemical techniques can be used for diagnostic purposes in microbiological laboratories and, because of their prac-

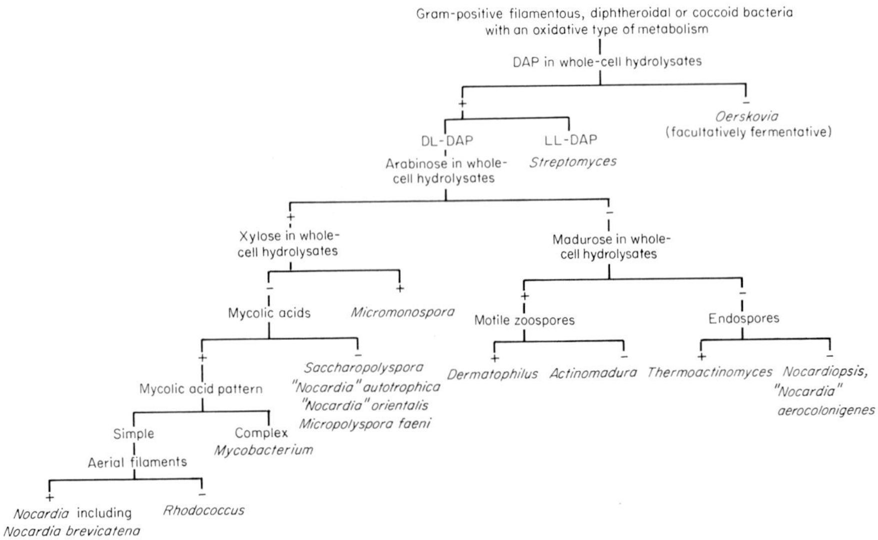

FIG. 6. Flow chart for the identification of clinically significant aerobic actinomycetes and related bacteria by chemical techniques.

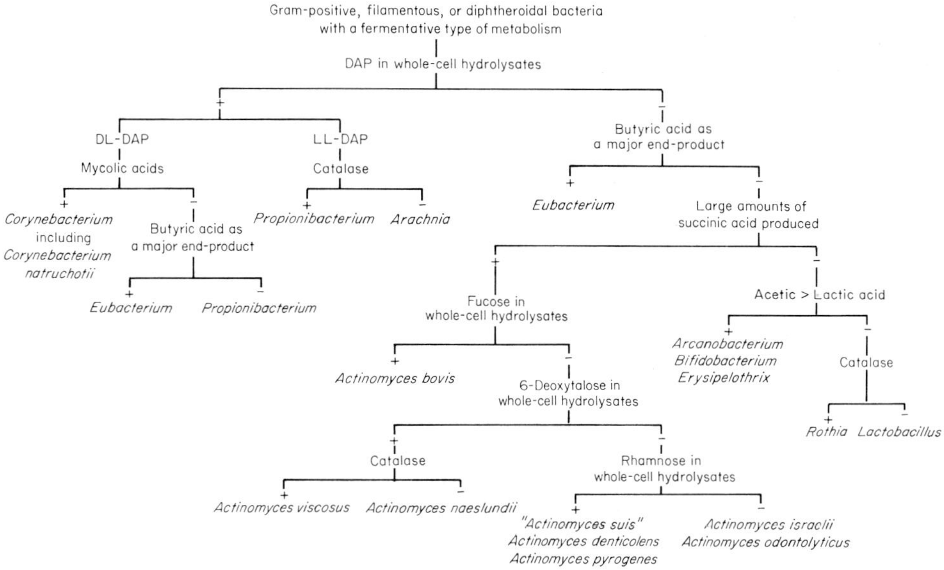

FIG. 7. Flow chart for the identification of clinically significant fermentative actinomycetes and related bacteria by chemical techniques.

ticability and considerable differential power, they can lead only to improved identification of unusual pathogens. The introduction of chemotaxonomic techniques into the clinical field may also be an important step toward the modernization of diagnostic procedures for infectious diseases in general.

The valuable and expert technical assistance of Miss E. Hobusch, Miss M. Pinkwart and Miss L. Fassbender is gratefully acknowledged. The author also wishes to thank Mrs. S. Glanschneider for photographic services and Miss E. Heidermann for typing the manuscript.

References

BECKER, B., LECHEVALIER, M. P., GORDON, R. E. & LECHEVALIER, H. A. 1964 Rapid differentiation between *Nocardia* and *Streptomyces* by paper chromatography of whole-cell hydrolysates. *Applied Microbiology* **12,** 421–423.

BECKER, B., LECHEVALIER, M. P. & LECHEVALIER, H. A. 1965 Chemical composition of cell-wall preparations from strains of various form-genera of aerobic actinomycetes. *Applied Microbiology* **13,** 236–243.

COLLINS, M. D. 1982 Reclassification of *Bacterionema matruchotii* (Mendel) in the genus *Corynebacterium,* as *Corynebacterium matruchotii* comb. nov. *Zentralblatt für Bakteriologie, Parasitenkunde, Infektionskrankheiten und Hygiene.* I. *Abteilung Originale* **C3,** 364–367.

CROSS, T. 1981 The monosporic actinomycetes. In *The Prokaryotes. A Handbook on Habitats, Isolation and Identification of Bacteria* eds. Starr, M. P., Stolp, H., Trüper, H. G., Balows, A. & Schlegel, H. G., pp. 2091–2102. Berlin & New York: Springer-Verlag.

DENT, V. E. & WILLIAMS, R. A. D. 1984 *Actinomyces denticolens* Dent & Williams sp. nov.: a new species from the dental plaque of cattle. *Journal of Applied Bacteriology* **56,** 183–192.

GOODFELLOW, M. & PIROUZ, T. 1982 Numerical classification of sporoactinomycetes containing *meso*-diaminopimelic acid in the cell wall. *Journal of General Microbiology* **128,** 503–527.

GOODFELLOW, M. & SCHAAL, K. P. 1979 Identification methods for *Nocardia, Actinomadura* and *Rhodococcus.* In *Identification Methods for Microbiologists* eds. Skinner, F. A. & Lovelock, D. W., 2nd edn. pp. 261–276. London: Academic Press.

GOODFELLOW, M., WEAVER, C. R. & MINNIKIN, D. E. 1982 Numerical classification of some rhodococci, corynebacteria and related organisms. *Journal of General Microbiology* **128,** 731–745.

HEINRICH, S. & KORTH, H. 1967 Zur Nährbodenfrage in der Routinediagnostik der Aktinomykose: Ersatz unsicherer biologischer Substrate durch ein standardisiertes Medium. In *Krankheiten durch Aktinomyceten und verwandte Erreger* ed. Heite, H. J., pp. 16–20, Berlin: Springer-Verlag.

HOLDEMAN, L. V., CATO, E. P. & MOORE, W. E. C. 1977 *V.P.I. Anaerobe Laboratory Manual,* 4th edn. Blacksburg, Virginia: Southern Printing Co.

KUTZNER, H. J. 1981 The family *Streptomycetaceae.* In *The Prokaryotes. A Handbook on Habitats, Isolation, and Identification of Bacteria* eds. Starr, M. P., Stolp, H., Trüper, H. G., Balows, A. & Schlegel, H. G., pp. 2028–2090. Berlin & New York: Springer-Verlag.

LECHEVALIER, H. A. & LECHEVALIER, M. P. 1981a Introduction to the order *Actinomycetales.* In *The Prokaryotes. A Handbook on Habitats, Isolation and Identification of Bacteria* eds. Starr, M. P., Stolp, H., Trüper, H. G., Balows, A. & Schlegel, H. G., pp. 1915–1922. Berlin & New York: Springer-Verlag.

LECHEVALIER, H. A. & LECHEVALIER, M. P. 1981b Actinomycete genera "in search of a family". In *The Prokaryotes. A Handbook on Habitats, Isolation and Identification of Bacteria* eds. Starr, M. P., Stolp, H., Trüper, H. G., Balows, A. & Schlegel, H. G., pp. 2118–2123. Berlin & New York: Springer-Verlag.

LECHEVALIER, M. P. 1968 Identification of aerobic actinomycetes of clinical importance. *Journal of Laboratory and Clinical Medicine* **71,** 934–944.

LECHEVALIER, M. P. & GERBER, N. N. 1970 The identity of madurose with 3-O-methyl-D-galactose. *Carbohydrate Research* **13,** 451–454.

LECHEVALIER, M. P. & LECHEVALIER, H. A. 1976 Chemical methods as criteria for the separation of nocardiae from other actinomycetes. *The Biology of the Actinomycetes* **11,** 78–92.

MINNIKIN, D. E., PATEL, V. & GOODFELLOW, M. 1974 Mycolic acids of representative strains of *Nocardia* and the "*rhodochrous*" complex. *FEBS Letters* **39,** 322–324.

MINNIKIN, D. E., ALSHAMAONY, L. & GOODFELLOW, M. 1975 Differentiation of *Mycobacterium, Nocardia* and related taxa by thin-layer chromatographic analysis of whole-organism methanolysates. *Journal of General Microbiology* **88,** 200–204.

MINNIKIN, D. E., HUTCHINSON, I. G., CALDICOTT, A. B. & GOODFELLOW, M. 1980 Thin-layer chromatography of methanolysates of mycolic acid-containing bacteria. *Journal of Chromatography* **188,** 221–233.

MINNIKIN, D. E., MINNIKIN, S. M., GOODFELLOW, M. & STANFORD, J. L. 1982a The mycolic acids of *Mycobacterium chelonei. Journal of General Microbiology* **128,** 817–822.

MINNIKIN, D. E., MINNIKIN, S. M. & GOODFELLOW, M. 1982b The oxygenated mycolic acids of *Mycobacterium fortuitum, M. farcinogenes* and *M. senegalense. Biochimica et Biophysica Acta* **712,** 616–620.

PRAUSER, H. & FALTA, R. 1968 Phagensensibilität, Zellwandzusammensetzung und Taxonomie von Actinomyceten. *Zeitschrift für allgemeine Mikrobiologie* **8**, 39–46.

REDDY, C. A., CORNELL, C. P. & FRAGA, A. M. 1982 Transfer of *Corynebacterium pyogenes* (Glage) Eberson to the genus *Actinomyces* as *Actinomyces pyogenes* (Glage) comb. nov. *International Journal of Systematic Bacteriology* **32**, 419–429.

SCHAAL, K. P. 1977 *Nocardia*, *Actinomadura*, and *Streptomyces*. In *CRC Handbook Series in Clinical Laboratory Science, Section E: Clinical Microbiology* eds. Seligson, D. & von Graevenitz, A, Vol. I, pp. 131–144. Cleveland, Ohio: CRC Press.

SCHAAL, K. P. & PULVERER, G. 1981 The genera *Actinomyces*, *Agromyces*, *Arachnia*, *Bacterionema*, and *Rothia*. In *The Prokaryotes. A Handbook on Habitats, Isolation and Identification of Bacteria* eds. Starr, M. P., Stolp, H., Trüper, H. G., Balows, A. & Schlegel, H. G., pp. 1923–1950. Berlin & New York: Springer-Verlag.

SCHLEIFER, K. H. & KANDLER, O. 1972 Peptidoglycan types of bacterial cell walls and their taxonomic implications. *Bacteriological Reviews* **36**, 407–477.

SCHOFIELD, G. M. & SCHAAL, K. P. 1981 A numerical taxonomic study of members of the *Actinomycetaceae* and related taxa. *Journal of General Microbiology* **127**, 237–259.

SLACK, J. M. & GERENCSER, M. A. 1975 *Actinomyces, Filamentous Bacteria*. Minneapolis, Minnesota: Burgess Publishing Co.

WILLIAMS, S. T. & WELLINGTON, E. M. H. 1981 The genera *Actinomadura*, *Actinopolyspora*, *Excellospora*, *Microbispora*, *Micropolyspora*, *Microtetraspora*, *Nocardiopsis*, *Saccharopolyspora*, and *Pseudonocardia*. In *The Prokaryotes. A Handbook on Habitats, Isolation and Identification of Bacteria* eds. Starr, M. P., Stolp, H., Trüper, H. G., Balows, A. & Schlegel, H. G., pp. 2103–2117. Berlin & New York: Springer-Verlag.

The Detection of Cytochrome Patterns in Bacteria

M. A. Carver and C. W. Jones

Department of Biochemistry, University of Leicester, Leicester, UK

The cytochromes were discovered almost 100 years ago during spectroscopic studies of mammalian muscle (MacMunn 1886). They were effectively rediscovered almost 40 years later by Keilin, who showed that the flight muscles of certain insects exhibited a characteristic absorption spectrum, which could be altered by the addition of an oxidising or reducing agent (Keilin 1925). In the following 10 years Keilin, Warburg, Tamiya and their colleagues carried out pioneering work on the cytochrome spectra of many species of bacteria, and reported many of the interspecies spectral differences, which have subsequently been examined in much more detail via more sophisticated spectrophotometric analyses (see, for example, Smith 1978) and which allow cytochrome patterns to be used as an increasingly powerful tool in bacterial taxonomy.

Properties and Functions of Cytochromes

Cytochromes are specialised haemoproteins, i.e. they are composed of a haem prosthetic group tightly bound to a protein. Haem consists of four pyrrole rings joined by methenyl bridges to form a planar ring molecule, porphyrin, at the centre of which is held a single iron atom. The individual cytochromes differ principally in the nature of their substituent groups on four of the carbon atoms that comprise the periphery of the haem molecule, and it is these substituents that are largely responsible for the different absorbance characteristics and redox potentials (E_0' or E_m) of the four types of cytochromes, namely, cytochromes *a* (haem *a*), *b* (haem *b* or protohaem), *c* (haem *c* or mesohaem) and *d* (haem *d* or chlorin). The central iron atom is usually in the low-spin configuration, which leads to the formation of an

CHEMICAL METHODS IN
BACTERIAL SYSTEMATICS

0-12-289675-0

octohedral coordination complex with a minimum of six ligands; four of these are the pyrrole nitrogens, whereas the fifth and sixth ligands are often nitrogen or sulphur atoms from histidyl and methionyl residues in the protein. In a few specialised cytochromes, such as cytochrome oxidases aa_3 or *d*, the sixth position is available to water, molecular oxygen and structurally related inhibitors such as carbon monoxide, sulphide or nitric oxide. In contrast, other specialised cytochromes including cytochrome c^1, cytochrome oxidase *o* (a *b*-type cytochrome) and cytochrome P450 (another *b*-type cytochrome, which acts as an oxygenase) exhibit rather different spin configurations and are often more akin to the oxygen transport haemoprotein, myoglobin. The haem–protein linkage is further stabilised in the *c*-type cytochromes by covalent, thioether bonds between specific vinyl groups on the periphery of the porphyrin ring and cysteinyl residues in the protein. In all of the cytochromes the iron undergoes one electron oxidation–reduction, oscillating between the reduced (Fe^{2+}) and oxidised (Fe^{3+}) states with their characteristic absorption spectra.

Cytochromes are involved in two major energy conservation processes in bacteria, namely, respiration and bacteriochlorophyll-dependent photosynthetic electron transfer. In both processes, reducing equivalents are transferred from a low redox potential donor to a higher redox potential acceptor with the concomitant release of an amount of free energy, which is proportional to the change in redox potential of the overall oxidation–reduction reaction. This free energy is initially conserved as a trans- or intramembrane electrochemical potential difference of H^+ (the protonmotive force, Δp or $\Delta\bar{\mu}H^+$; Mitchell 1966) and is subsequently used to drive various energy-dependent membrane functions including the synthesis of ATP (photosynthetic or respiratory chain phosphorylation) or pyrophosphate, the accumulative transport of selected solutes, cell motility and reversed electron transfer (see Garland 1977; Jones 1982).

Aerobic respiration in chemoheterotrophic and facultatively phototrophic bacteria is characterised by the transfer of reducing equivalents from an organic donor (principally NADH, but also other reductants such as succinate, lactate, or methanol) to molecular oxygen. In contrast, many facultatively anaerobic and obligately anaerobic chemoheterotrophs replace oxygen with alternative electron acceptors such as various oxyanions of nitrogen and sulphur, ferric iron, fumarate, trimethylamine *N*-oxide or carbon dioxide, and hence carry out anaerobic respiration. Finally, chemolithotrophic bacteria oxidise inorganic electron donors such as hydrogen, various nitrogen and sulphur compounds or ferrous iron using oxygen, or occasionally nitrate, on terminal electron acceptor. These various donors and acceptors, which are used as respiratory chain substrates by different types of organisms, or in some cases by the same organism growing under different conditions, afford a very wide range of redox potentials (e.g. $E'_0\,2H^+/H_2 = -420$ mV, cf. $\frac{1}{2}O_2/H_2O = 820$ mV). As a result of this, there can be immense energetic variations between different types of respiratory chains; compare, for example, the oxidation of NADH by fumarate ($\Delta E'_0 = 350$ mV) with the aerobic oxidation of hydrogen ($\Delta E'_0 = 1240$ mV). However, the wide range of redox potentials exhibited by bacterial cytochromes (Table 1) ensures that each of these various respiratory systems involves at least one, and usually several, cytochromes. Each of the latter acts either as an intermediate electron carrier (to transfer electrons from one redox component to another, e.g. cytochromes *b,c*) or as a terminal enzyme (to transfer electrons to an exogenous acceptor such as oxygen or nitrite, and hence to act as a cytochrome oxidase or nitrite reductase, respectively, e.g. cytochromes aa_3, *o*, a_1, *d* or cd_1).

When multiple cytochromes are present in a respiratory chain, redox potential considerations dictate that they act in the general order:

low E_m *c*-type —— *b*-type — high E_m *c*-type — *a*-type

(although many exceptions to this are known since only a very few species of bacteria synthesise low E_m *c*-type cytochrome). A large number of species do not synthesise significant concentrations of high E_m *c*-type cytochromes and some species are unable to synthesise *a*-type cytochromes (in which case cytochromes *o* or *d* act as the terminal oxidase during aerobic respiration). The synthesis of particular cytochromes also varies with the exact growth conditions, in particular, the availability of certain electron donors or acceptors.

Cellular location studies have indicated that, as expected from their known involvement in electron-transfer-dependent energy conservation, the cytochromes are associated with the bacterial cytoplasmic membrane. Most cytochromes can be classed as intrinsic membrane proteins, being firmly embedded within the membrane and requiring detergents for their removal. In contrast, some *c*-type cytochromes, including cytochrome cd_1, are only loosely attached to the periplasmic surface of the membrane and are easily removed by exposing whole cells to high ionic-strength conditions.

Analysis of Cytochrome Spectra

Cytochromes exhibit complex absorption spectra composed of α, β, γ (Soret), and δ bands. Of these, the α band is experimentally the most important since, although relatively small, it has a wavelength of maximum absorbance (λ_{max}) in the visible region of the spectrum, which is characteristic of the individual cytochrome (Table 1). It is principally via analysis of

TABLE 1

The cytochromes of bacterial respiratory chains

Haem type	Cytochrome	E_m (mV)	λ_{max} Reduced minus oxidised (nm)	λ_{max} Reduced plus CO minus oxidised (nm)	Function	Notes
a	aa_3	200–375	600–607(α)	428–438 (Soret)	Oxidase	*a* Nonautoxidisable
	a_a	160–500	593–598(α)	428–432 (Soret)	Oxidase?	
b	*b*	−104–110	556–565(α)	—	Intermediate electron carrier	
	o	−122–417	556–562(α)	416–418 (Soret)	Oxidase	
c	*c*	50–342	548–554(α)	—	Intermediate electron carrier	
	c^1	—	548(α)	560(α)	Intermediate electron carrier	Binds to CO but poor oxidase activity
	c^3	−324 to −284	550(α)	—	Intermediate electron carrier	Multihaem, low E_m
d	*d*	280	625–630(α)	635–640(α)	Oxidase	Isolated oxidase contains *b*
	cd_1	—	625–630(α)	630(α)	Nitrite reductase	Modified *d*, some oxidase activity

spectra in the α-band region that bacterial cytochrome patterns are determined. Like most respiratory chain components, the cytochromes exhibit their highest absorbance in the reduced rather than in the oxidised form, and hence purified cytochromes can easily be examined using a conventional split-beam spectrophotometer (in which a single light beam of desired wavelength is split into two identical beams that pass, respectively, through two cuvettes containing the test and reference samples). Using such an instrument, it is possible to measure very easily the absolute absorption spectrum of a particular cytochrome (i.e. the spectrum of a solution of the reduced cytochrome versus water).

The measurement of absolute absorption spectra using a split-beam spectrophotometer has certain limitations, however, in particular during the examination of cytochromes in highly turbid suspensions of whole cells or respiratory membranes. One way in which this turbidity problem can be overcome is to place identical suspensions in the test and reference cuvettes; the test suspension is then exposed to a reducing agent (e.g. sodium dithionite or a physiological reductant) and the reference suspension to an oxidising agent (e.g. potassium ferricyanide or a physiological oxidant). The resultant reduced-minus-oxidised difference spectrum exhibits the absorption bands that are characteristic of the cytochrome components present in the whole-cell or membrane suspension, albeit with slightly different wavelength maxima from those observed in absolute spectra.

The concept of measuring difference spectra rather than absolute spectra can be further extended to the detection of potential cytochrome oxidases. The latter catalyse the reduction of molecular oxygen to water, a reaction that is inhibited in a competitive manner by carbon monoxide, i.e. oxygen and carbon monoxide compete for the binding site on the ferrous iron. The binding of carbon monoxide thus perturbs the spectrum of the cytochrome oxidase(s), but not that of the other cytochromes present, and hence the measurement of a reduced + CO-minus-reduced difference spectrum allows the detection of any cytochrome oxidases present. It is important to note, however, that the mere ability to bind carbon monoxide does not necessarily mean that the cytochromes can reduce molecular oxygen at a rate that is sufficient for it to be regarded as a physiologically significant oxidase. In order to prove the latter, it is necessary to supplement evidence obtained from reduced + CO-minus-reduced difference spectra with those from photochemical action spectra (in which the photoalleviation of carbon monoxide-inhibited respiration is measured as a function of the wavelength of the alleviating light) (Castor & Chance 1959); unfortunately, further discussion of the latter is outside the scope of this chapter. It should also be noted that the measurements of difference spectra using turbid suspensions is only feasible provided that certain problems can be overcome. These problems include the necessity to use light of relatively narrow bandwidth, and hence of relatively low total energy content, in order to closely resolve adjacent or overlapping spectral bands. Thus since turbid suspensions tend to cause considerable scattering of the incident light beam, it is necessary to place the cuvettes containing the test and reference samples close to the photomultiplier and/or to use a particularly powerful light source in order to prevent an unacceptable loss of energy and hence of resolution.

An alternative approach to determining cytochrome spectra of turbid suspensions is to use the sophisticated technique of dual-wavelength spectrophotometry (Shibata 1976). This method, in contrast to split-beam spectrophotometry, entails the use only of a single sample through which one alternately passes two beams of light; in the scanning mode, one of these beams is maintained at a constant wavelength (usually an isosbestic point, λ_1) whereas the other scans over the desired wavelength range (λ_2). In this way the sample acts effectively as its own reference and so eliminates many of the problems that are normally associated with the analysis of turbid suspensions. The resultant spectra of oxidised, reduced and reduced + CO samples are essentially hybrids of absolute and difference spectra since they represent the absorbance of the sample at different wavelengths relative to that at a fixed wavelength. This technique can also be used to determine derivative spectra (see below).

The problems of resolving overlapping spectral bands and of exposing minor cytochrome components are solved principally through the use of low temperature and derivative facilities. In addition, a computer-linked spectral deconvolution method, based on best-fit analyses using simulated composite spectra has recently been introduced and has proved particularly successful, especially when carried out in tandem with a similar analysis of potentiometric data on the E_m values of the various cytochrome components (Van Wielink *et al.* 1982). The measurement of low-temperature difference spectra via split-beam spectrophotometry involves the insertion into each of the light beams of a specially adapted Dewar flask containing liquid nitrogen (77 K). The test and reference cuvettes are placed in the liquid nitrogen in such a way that the light beams are able to pass through the samples via unsilvered windows in the sides of the Dewar flasks. The effect of freezing the samples in this way is to increase significantly the effective pathlength of the light beams through the samples, and hence to increase the sharpness, absorbance and resolution of the resultant spectral bands (see, for example, Reid & Ingledew 1979; Poole *et al.* 1979). A similar technique can be used with dual-wavelength spectrophotometry.

The first derivative of an absorption spectrum is defined as $A/\delta\lambda$, the second derivative as $^2A/\delta^2\lambda$ and so on for higher orders (where A is the absorbance and λ is the wavelength). A derivative spectrum is usually produced from the original absolute spectrum or difference spectrum by processing the output signal from the spectrophotometer by electronic or numerical differentiation. Alternatively, a dual-wavelength scanning spectrophotometer may be used in which one light beam scans the sample at a constant wavelength interval ahead of the other (e.g. an interval of 2 nm); this method is restricted, however, to the direct generation only of first-order derivative spectra, and higher-order spectra can only be obtained by subsequent electronic or numerical differentiation. Derivative analysis is an extremely powerful tool for resolving overlapping bands with narrow bandwidths and hence for determining exactly the wavelength maxima of constituent peaks in complex spectra. Second-order analysis, and to an even greater extent fourth-order analysis, has proved particularly successful when applied to low-temperature reduced-minus-oxidised or reduced + CO-minus-reduced difference spectra of whole cells or respiratory membranes (see, for example, Shipp 1972; Scott & Poole 1982).

Analysis of Cytochromes by Staining or Fluorescence Procedures

The analysis of cytochromes using staining or fluorescence methods has so far been limited to the detection of *c*-type cytochromes. The staining procedure, the so-called oxidase test (Kovacs 1956), is an easy and rapid method for measuring the ability of bacterial colonies on agar plates to catalyse the oxidative synthesis of indophenol blue from *N,N*-dimethyl-*p*-phenylenediamine and α-naphthol with the concomitant reduction of oxygen to water (Fig. 1). The appearance of blue colonies within about a half minute of adding the substrates reflects the presence of a membrane-bound, high E_m *c*-type cytochrome linked to an active cytochrome oxidase, usually cytochromes aa_3 or *o* (Jurtshuk *et al.* 1975). Although the effective absence of either cytochrome *c* or cytochrome oxidase is enough to ensure a negative result, in practice the latter is associated in aerobic organisms with inability to synthesise cytochrome *c* and hence failure to catalyse the initial oxidation

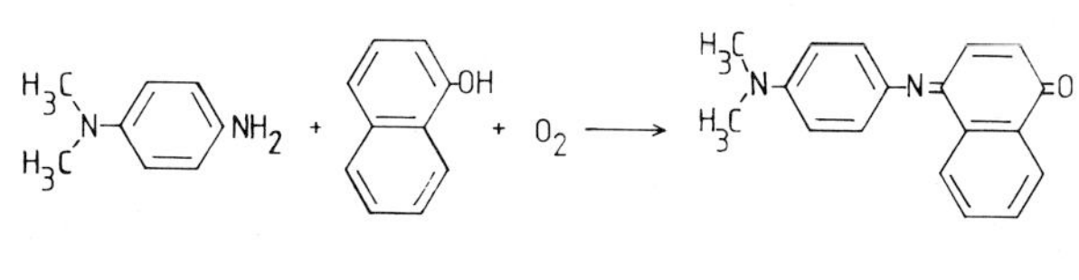

(N,N-dimethyl-p-phenylenediamine) (α-naphthol) (Indophenol blue)

FIG. 1. The oxidase test. Note that *N,N,N',N'*-tetramethyl-*p*-phenylenediame is often used in place of *N,N*-dimethyl-*p*-phenylenediamine.

step of the coupled reaction. In those organisms that synthesise an appropriate cytochrome *c*, and which are therefore oxidase positive, the speed at which the colonies become blue reflects the catalytic ability of the terminal respiratory chain.

The detection of *c*-type cytochromes by fluorescence procedures requires the initial separation of proteins in periplasmic or detergent-solubilised membrane fractions by polyacrylamide-gel electrophoresis in the presence of sodium dodecylsulphate (SDS–PAGE; see Swank & Munkres 1971; also Chapter 6, this volume). The *c*-type cytochromes have a characteristic red fluorescence when subjected to SDS–PAGE and excited by near-UV light. The gels can be photographed under near-UV light, and the negative obtained can be scanned in a gel scanner for MW determination (Wood 1980).

Methods

Methylophilus methylotrophus was grown and membranes were prepared as described in Dawson & Jones (1981) and Carver & Jones (1983). Cells were normally suspended in 20 m*M* glycylglycine, pH 7.0. Cytochrome c_L from *M. methylotrophus* was prepared as described by Crofts & Anthony (1980). Detergent extracts were prepared by incubating cell membranes at 10 mg of protein/ml with 1% Triton X100 and 1% Empigen for 1 h at 4°C. Non-solubilised material was removed by centrifugation at 140,000 *g* for 1 h.

Oxidised-minus-reduced difference spectra of whole cells were recorded on a Pye Unicam SP1800 dual-beam spectrophotometer with the cuvettes in the "close up" position near the photomultiplier. Oxidised-minus-reduced difference spectra, reduced-minus-reduced + carbon monoxide spectra, reduced-minus-reduced + nitric oxide spectra and first- to fourth-order derivative spectra of detergent extracts were recorded on a Perkin Elmer λ5 recording spectrophotometer. Samples were reduced with a few grains of sodium dithionite and oxidised by the addition of 0.5 m*M* potassium ferricyanide or 10 μl of 5-volume hydrogen peroxide.

Empigen was a kind gift from Dr. S. Collins, I. C. I. Billingham, but can be bought from Albright and Wilson. Triton X100, sodium dithionite and glycylglycine were purchased from B.D.H. Ltd. Other chemicals were purchased from Sigma.

Results and Discussion

Examination of the oxidised-minus-reduced difference spectrum of a bacterial cell suspension is the most simple and rapid way of identifying its cytochrome content. The α absorption bands of all of the commonly occurring cytochromes are easily identifiable and lie in the 500–650-nm region of the visible spectrum. Examples of these are shown in Fig. 2ab. *Methylophilus methylotrophus*, an obligately aerobic methylotroph grown under methanol limitation contains *c* type cytochromes ($\alpha\lambda_{max}$ = 550 nm), *b*-type cytochromes ($\alpha\lambda_{max}$ = 558 nm) and a cytochrome aa_3-type terminal oxidase ($\alpha\lambda_{max} \simeq$ 600 nm) (Fig. 2a). *Escherichia coli* grown anaerobically on a glycerol–fumarate medium contains *c*-type cytochromes($\alpha\lambda_{max}$ = 550 nm), *b*-type cytochromes ($\alpha\lambda_{max}$ = 560 nm), an a_1-type cytochrome ($\alpha\lambda_{max}$ = 585 nm) and a *d*-type cytochrome ($\alpha\lambda_{max}$ = 632 nm, minimum at 650 nm) (Fig. 2b). The features shown in these spectra are typical of those found in most bacteria.

The particulate nature of bacterial cell suspensions causes a large amount

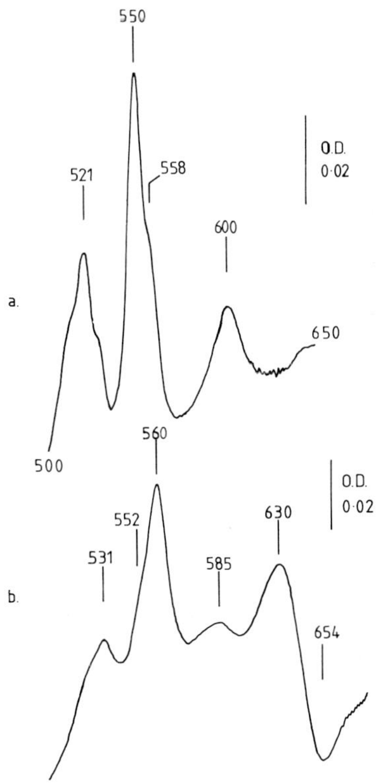

FIG. 2. (a) Oxidised-minus-reduced difference spectrum of *Methylophilus methylotrophus*. Cells were suspended at 5 mg dry wt/ml and spectra were recorded on a Pye Unicam SP1800 spectrophotometer. (b) Oxidised-minus-reduced difference spectrum of respiratory membranes from anaerobically grown *Escherichia coli*. Membranes were suspended at 5 mg protein/ml and spectra recorded as in 2a. Samples were reduced with a few grains of dithionite and oxidised by the addition of 0.5 m*M* potassium ferricyanide.

of light scattering and this can degrade the quality of spectroscopic information obtained from them. Several strategies are available for minimising this problem.

The simplest is to move the position of the reference and test cuvettes in the spectrophotometer close to the face of the photomultiplier. This serves to minimise the amount of light scattering and hence to maximise the amount of signal passing into the photomultiplier. The problem can also be relieved by careful choice of the cell density of the suspension under test. Too high a density causes too much scattering whereas too low gives too small a signal. Cell densities of ~5 mg dry wt/ml give an optimal balance between these two effects. Cell suspensions can be prevented from settling by the presence of 0.5% (w/v) of methylcellulose in the suspension medium.

An alternative strategy is to use cell membrane preparations, prepared by "French cell" pressure disruption or by sonication, rather than whole cells. These have a smaller particle size and, because of loss of cell contents, an effectively higher cytochrome content. Periplasmic or loosely bound cytochromes, notably cytochrome *c*, may be lost during these procedures, but they can be found and assayed in the supernatants to these membrane fractions.

Cells and cell membranes can be solubilised by treatment with cocktails of nonionic detergents such as Triton X100 and zwitterionic detergents like Empigen BB, yielding optically clear preparations that contain solubilised membrane cytochromes in essentially unaltered form (see Figs. 5–8).

Many organisms contain multiple *c*- and *b*-type cytochromes with overlapping α bands. Several methods are available for separating these, notably the use of liquid-nitrogen-temperature (77 K) spectra and derivative spectroscopy.

The effective pathlength of a beam of light passing through a sample frozen to 77 K is greatly increased by internal refraction and reflection by ice crystals. This results in enhanced detail being revealed in spectra obtained at this temperature. Multiple *b*- and *c*-type cytochromes are often separated as a series of shoulders on peaks (see Figs. 2ab and 3ab).

Derivative spectroscopy is an extremely powerful technique for resolving overlapping spectral bands. First- to fourth-order derivative spectra are the ones most commonly used. Figure 4a shows the oxidised-minus-reduced difference spectrum of the purified cytochrome c_L from *M. methylotrophus* and Figs. 4b–e show the first- to fourth-derivative spectra of the same. It can be seen from these that wavelength maxima in the oxidised-minus-reduced difference spectrum (521 and 550 nm) correspond to intercepts through zero in the first- and third-order spectra, to minima in second-order spectra, and to maxima in fourth-order spectra. Second- and fourth-order spectra thus provide the most useful means of analysing cytochrome spectra. They also

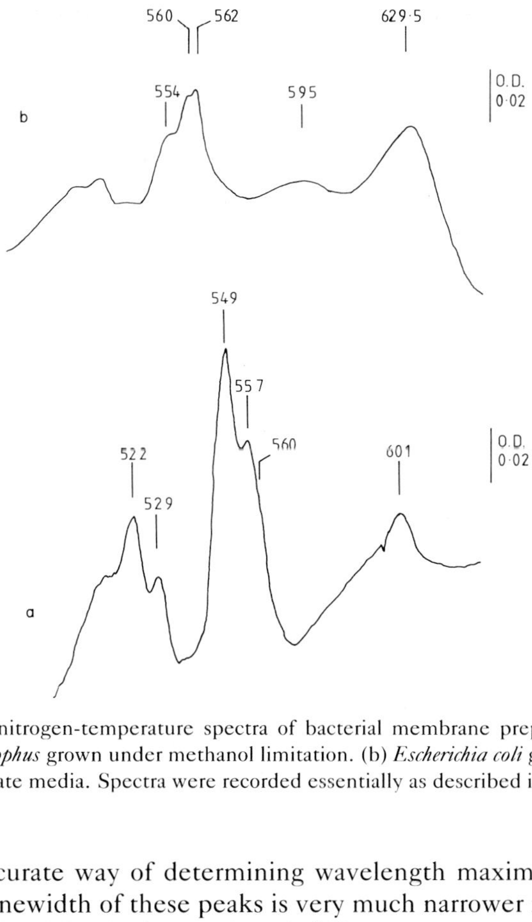

FIG. 3. Liquid-nitrogen-temperature spectra of bacterial membrane preparations. (a) *Methylophilus methylotrophus* grown under methanol limitation. (b) *Escherichia coli* grown anaerobically on glycerol fumarate media. Spectra were recorded essentially as described in Reid & Ingledew (1979).

provide an accurate way of determining wavelength maxima of absorption peaks as the linewidth of these peaks is very much narrower in these spectra than in normal oxidised-minus-reduced difference spectra.

The optimal separation of overlapping cytochrome peaks by derivative spectroscopy requires careful manipulation of the scanning parameters of the spectrophotometer, e.g. scan speed, slit width, delta wavelength setting ($\Delta\lambda$), response time and chart speed. This can clearly be seen in Fig. 5, which shows the oxidised-minus-reduced difference spectrum of detergent-solubilised *M. methylotrophus* membranes together with corresponding second-derivative spectra. At a $\Delta\lambda$ setting of 1 nm, the α bands of the *c*- and *b*-type cytochromes are completely separated (Fig. 5b, $\alpha\lambda_{min}$ 550 $\pm$ 558 nm, respectively), whereas the optimal separation of the broad α band of the *a*-type cytochromes is achieved using a $\Delta\lambda$ setting of 4 nm (Fig. 5c). This reveals the presence of at least two *a*-type cytochromes, probably a_1 and aa_3, a feature not obvious from inspection of the oxidised-minus-reduced dif-

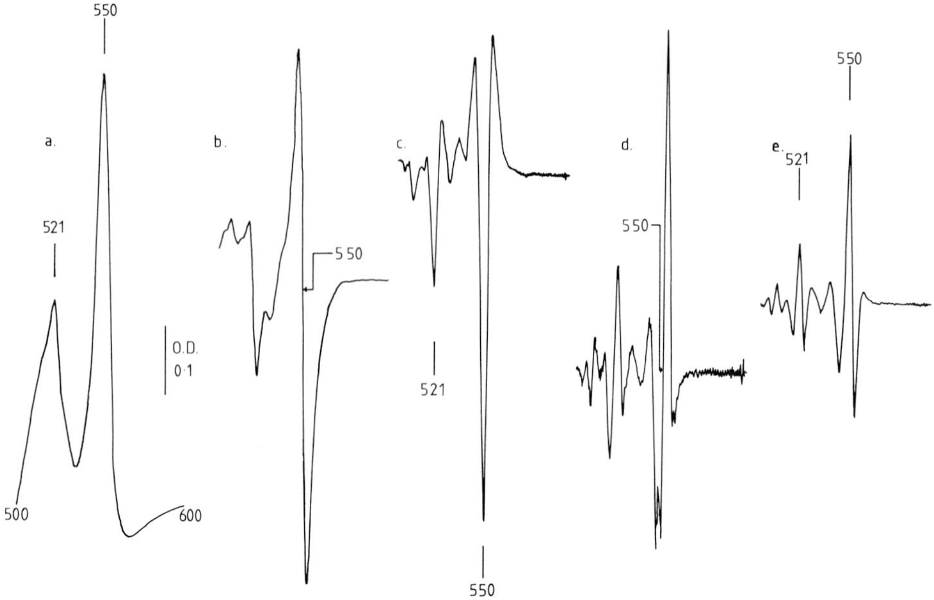

FIG. 4. (a) Oxidised-minus-reduced difference spectrum of cytochrome c_L from *Methylophilus methylotrophus;* (b) first-order derivative spectrum; (c) second-order derivative spectrum; (d) third-order derivative spectrum; (e) fourth-order derivative spectrum. Spectra were recorded on a Perkin–Elmer λ5 spectrophotometer. Slit width 2 nm, scan speed 60 nm/min, response time 0.5 sec, delta wavelength 1 nm, abscissa format 20 nm/cm. Samples were reduced with a few grains of sodium dithionite and oxidised with H_2O_2.

ference spectrum (Fig. 5a). Second-order analysis has also resolved the broad overlapping γ or soret bands, and several individual components are now visible, notably those at 415 nm (cytochrome *c*), 419 and 422 nm (cytochrome *b*), and 440 and 443 nm (probably cytochromes a_1 and aa_3).

Fourth-order spectra potentially reveal even greater spectral detail, but it is important when running them (and to a lesser extent second-order spectra) to minimise spectroscopic "noise" since this makes the identification of multiple and minor components difficult. This can be done by using optically clear samples and by minimising light scattering, as described previously. The positive identification of minor components is best achieved by running multiple spectra on multiple samples to minimise spurious peaks produced by light scattering. The fourth-order analysis of the α *b* and *c* peaks of detergent-solubilised *M. methylotrophus* is shown in Fig. 6ab. This reveals the power of the technique since the *b*-type cytochrome shoulder ($\sim\lambda_{max}$ 558 nm) in the oxidised-minus-reduced difference spectrum (Fig. 6a) is resolved into several components with wavelength maxima at 555, 557, 558.5 (cytochrome *o*) and 560.5 nm.

The size of peaks obtained in second- and fourth-order derivative spectra does not directly reflect the concentration of cytochrome in the sample since they depend not only on this but on the linewidth of the original absorption peak and the scanning parameters of the spectrophotometer. They are usually measured in arbitrary units and consequently cannot easily be used to quantitate or compare cytochrome levels in a sample.

Carbon monoxide and, to a lesser extent, nitric oxide have proved valuable in the identification of ligand-binding cytochromes and hence potential

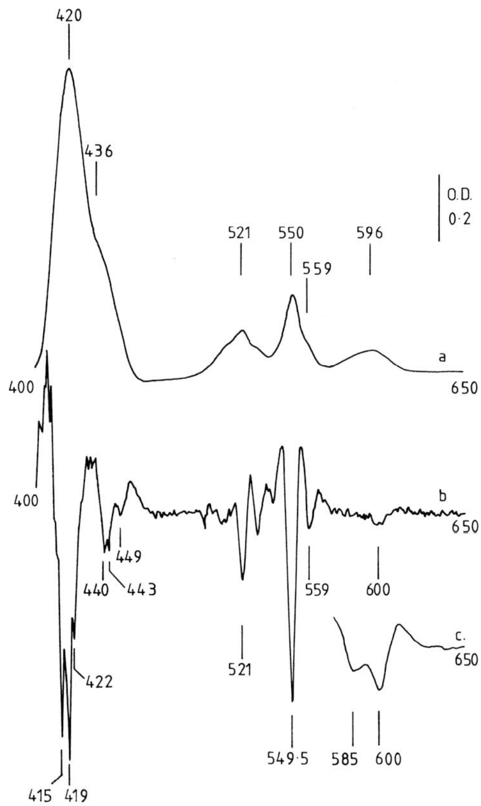

FIG. 5. (a) Oxidised-minus-reduced difference spectrum of detergent-solubilised *Methylophilus methylotrophus* membranes; (b, c) second derivative spectra. Spectra were recorded on a Perkin–Elmer λ5 spectrophotometer. (a) Slit width 1 nm, scan speed 120 nm/min, response time 0.5 sec; (b) slit width 1 nm, scan speed 60 nm/sec, response time 1 sec, delta wavelength 1 nm; (c) as (b) but delta wavelength set to 4 nm.

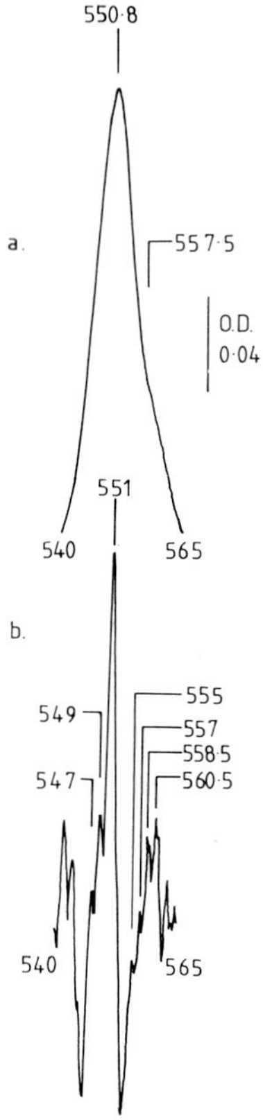

FIG. 6. (a) Oxidised-minus-reduced difference spectrum of the *c b* α bands of detergent-solubilised *Methylophilus methylotrophus* membranes; (b) fourth-derivative spectrum. Spectra were recorded on a Perkin–Elmer λ5 spectrophotometer. (a) Slit width 1 nm, scan speed 60 nm/sec, response time 0.5 sec; (b) slit width 1 nm, scan speed 60 nm/sec, response time 1 sec, delta wavelength 2 nm.

terminal cytochrome oxidases in bacterial systems. Both bind to certain cytochromes in the reduced state.

The simplest way to obtain CO or NO spectra is to bubble the gas through a reduced sample for 0.5–2 min before recording spectra using reduced sample as a reference. Samples may be reduced with a few grains of sodium dithionite or with respiratory substrates in excess.

The identification of individual ligand-binding cytochromes in whole cells, membranes, or solubilised membranes can be difficult since the major features of their spectra overlap and the various cytochromes react at markedly different rates with CO ($\frac{1}{2}$ reaction times vary from a few seconds to over an hour) (see Fig. 8). However, some of major features are outlined in Table 1.

Reduced-minus-reduced + CO spectra of *M. methylotrophus* reveal the presence of several CO binding cytochromes, notably one or two *a*-type cytochromes (troughs at 441, 585–7, and 601 nm), probably a_1 and a_3, a *b*-type cytochrome oxidase, C_LO oxidase (Carver & Jones 1983) (troughs at 420 and 558 nm) and a slower reacting *c*-type cytochrome (peak at 405 and a trough at 550 nm) (see Fig. 7). Similar though not identical features can be seen in reduced-minus-reduced + NO spectra (Fig. 8).

Spectrophotometric analysis can be rapid and simple and is consequently

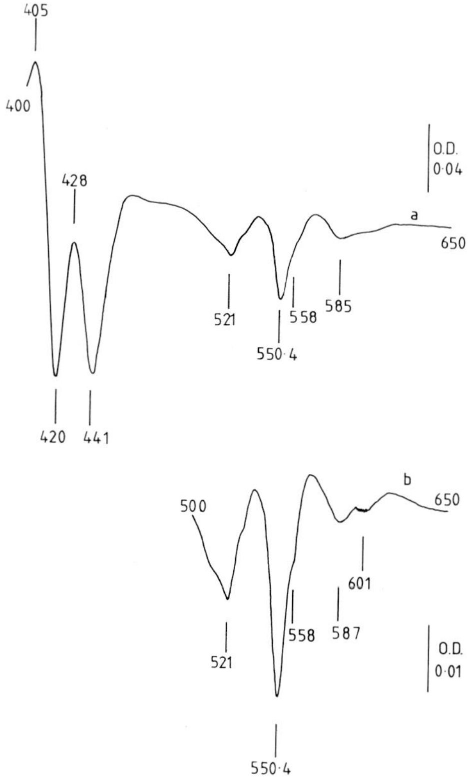

FIG. 7. Carbon monoxide reduced-minus-reduced difference spectra of detergent-solubilised *Methylophilus methylotrophus* membranes. Samples were reduced with a few grains of sodium dithionite prior to bubbling the sample cuvette with carbon monoxide for 30 sec. Liquid paraffin was layered over the surface of both sample and reference cuvettes to minimise reoxidation by air. Spectra were recorded after 5 min on a Perkin–Elmer λ5 spectrophotometer. Slit width 1 nm, scan speed 120 nm/sec, response time 0.5 sec.

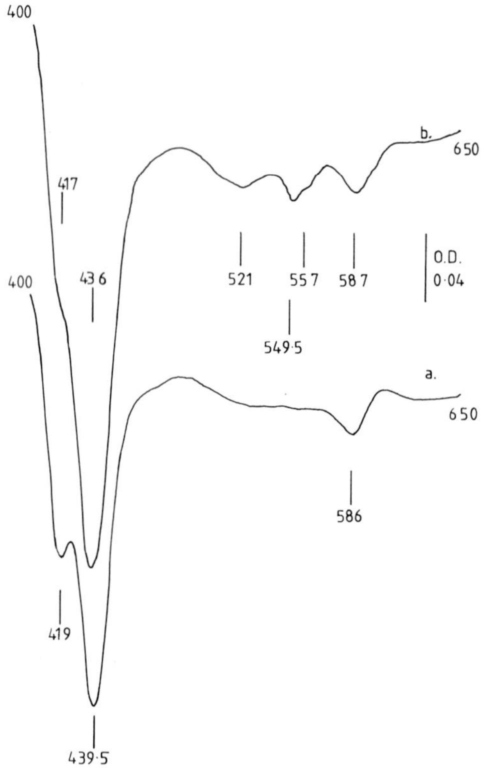

FIG. 8. Nitric oxide reduced-minus-reduced difference spectra of detergent-solubilised *Methylophilus methylotrophus* membranes. Samples were prepared as in the legend to Fig. 7 except that carbon monoxide was replaced by a few grains of sodium nitrite. Spectra were recorded as in Fig. 7. (a) 0.5 min after addition of sodium nitrite; (b) 10 min after addition of sodium nitrite.

ideally suited to the determination of the cytochrome composition of bacteria. It has proved crucial in the analysis and understanding of their respiratory chains and energy conservation pathways.

Work carried out in the authors' laboratory was supported by the Science and Engineering Research Council.

References

CARVER, M. A. & JONES, C. W. 1983 The terminal respiratory chain of the methylotrophic bacterium *Methylophilus methylotrophus*. *FEBS Letters* **155**, 187–191.

CASTOR, L. N. & CHANCE, B. 1959 Photochemical determinations of the oxidases of bacteria. *Journal of Biological Chemistry* **234**, 1587–1592.

CROFTS, A. R. & ANTHONY, C. 1980 The purification and properties of the soluble cytochromes *c* of the obligate methylotroph *Methylophilus methylotrophus*. *Biochemical Journal* **192**, 421–427.

DAWSON, M. J. & JONES, C. W. 1981 Respiration linked proton translocation in the obligate methylotroph *Methylophilus methylotrophus*. *Biochem. J.* **194**, 915–924.

GARLAND, P. B. 1977 Energy transduction in microbial systems. In *Microbial Energetics* eds. Haddock, B. A. & Hamilton, W. A., pp. 1–21. Cambridge: University Press.

JONES, C. W. 1982 *Bacterial Respiration and Photosynthesis* Walton-on-Thames, U.K.: Nelson.

JURTSHUK, P., MUELLER, T. J. & ACORD, W. C. 1975 Bacterial terminal oxidases. *Cleveland Rubber Company, Critical Reviews in Microbiology* **3**, 399–468.

KEILIN, D. 1925 On cytochrome, a respiratory pigment common to animals, yeast and higher plants. *Proceedings of the Royal Society of London, Series B* **98**, 312–319.

KOVACS, N. 1956 Identification of *Pseudomonas pyocyanea* by the oxidase reaction. *Nature, London* **178**, 703.

MACMUNN, C. A. 1886 Researches on myohaematin and the histohaematins. *Philosophical Transactions of the Royal Society of London* **177**, 167–298.

MITCHELL, P. 1966 Chemiosmotic coupling in oxidative and photosynthetic phosphorylation. *Biological Reviews of the Cambridge Philosophical Society* **41**, 445–502.

POOLE, R. K., WEARING, A. J. & CHANCE, B. 1979 The reaction of cytochrome *o* in *Escherichia coli* with oxygen. Low temperature kinetic and spectral studies. *Biochemical Journal* **184**, 379–389.

REID, G. A. & INGLEDEW, W. J. 1979 Characterisation and phenotypic control of the cytochrome content of *Escherichia coli*. *Biochemical Journal* **182**, 465–472.

SCOTT, R. I. & POOLE, R. K. 1982 A re-examination of the cytochromes of *Escherichia coli* using fourth-order finite difference analysis; their characterization under different growth conditions and accumulation during cell cycle. *Journal of General Microbiology* **128**, 1685–1696.

SHIBATA, S. 1976 Dual wavelength spectrophotometry. *Angewandte Chemie, International Edition in English* **15**, 673–679.

SHIPP, W. S. 1972 Cytochromes of *Escherichia coli*. *Archives of Biochemistry and Biophysics* **150**, 459–472.

SMITH, L. 1978 Bacterial cytochromes and their spectral characterization. In *Methods in Enzymology* eds. Fleischer, S. & Packer, L., Vol. 53, pp. 202–212. New York: Academic Press.

SWANK, R. T. & MUNKRES, K. D. 1971 Molecular weight analysis of oligopeptides by electrophoresis in polyacrylamide gel with sodium dodecylsulphate. *Analytical Biochemistry* **39**, 462–477.

VAN WIELINK, J. E., OLTMANN, L. F., LEEUWERIK, F. J., DE HOLLANDER, J. A. & STOUTHAMER, A. H. 1982 A method for *in situ* characterization of *b*- and *c*-type cytochromes in *Escherichia coli* and in complex III from beef heart mitochondria by combined spectrum deconvolution and potentiometric analysis. *Biochimica et Biophysica Acta* **681**, 177–190.

WOOD, P. M. 1980 Do photosynthetic bacteria contain cytochrome C_1? *Biochemical Journal* **189**, 385–391.

Subject Index

THE SOCIETY FOR APPLIED BACTERIOLOGY TECHNICAL SERIES

General Editor: F. A. Skinner

1 Identification Methods for Microbiologists Part A *1966*
Edited by B. M. Gibbs and F. A. Skinner
Out of print

2 Identification Methods for Microbiologists Part B *1968*
Edited by B. M. Gibbs and D. A. Shapton
Out of print

3 Isolation Methods for Microbiologists *1969*
Edited by D. A. Shapton and G. W. Gould

4 Automation, Mechanization and Data Handling in Microbiology *1970*
Edited by Ann Baillie and R. J. Gilbert

5 Isolation of Anaerobes *1971*
Edited by D. A. Shapton and R. G. Board

6 Safety in Microbiology *1972*
Edited by D. A. Shapton and R. G. Board

7 Sampling—Microbiological Monitoring of Environments *1973*
Edited by R. G. Board and D. W. Lovelock

8 Some Methods for Microbiological Assay *1975*
Edited by R. G. Board and D. W. Lovelock

9 Microbial Aspects of the Deterioration of Materials *1975*
Edited by R. J. Gilbert and D. W. Lovelock

10 Microbial Ultrastructure: The Use of the Electron Microscope *1976*
Edited by R. Fuller and D. W. Lovelock

11 Techniques for the Study of Mixed Populations *1978*
Edited by D. W. Lovelock and R. Davies

12 Plant Pathogens *1979*
Edited by D. W. Lovelock

13 Cold Tolerant Microbes in Spoilage and the Environment *1979*
Edited by A. D. Russell and R. Fuller

14 Identification Methods for Microbiologists (2nd Edn) *1979*
Edited by F. A. Skinner and D. W. Lovelock

15 Microbial Growth and Survival in Extremes of Environment *1980*
Edited by G. W. Gould and Janet E. L. Corry

16 Disinfectants: Their Use and Evaluation of Effectiveness *1981*
Edited by C. H. Collins, M. C. Allwood, Sally F. Bloomfield and A. Fox

17 Isolation and Identification Methods for Food Poisoning Organisms *1982*
Edited by Janet E. L. Corry, Diane Roberts and F. A. Skinner

18 Antibiotics: Assessment of Antimicrobial Activity and Resistance *1983*
Edited by A. D. Russell and L. B. Quesnel

19 Microbiological Methods for Environmental Biotechnology *1984*
Edited by J. J. Grainger and J. M. Lynch

20 Chemical Methods in Bacterial Systematics *1985*
Edited by M. Goodfellow and D. E. Minnikin